L. Henrikson Y.W. Brodin (Eds.)

Liming of Acidified Surface Waters

A Swedish Synthesis

AF181561

With 118 Figures

Springer

Dr. LENNART HENRIKSON

Municipality of Mark
S-51180 Kinna, Sweden

Professor Dr. Y. W. BRODIN

Swedish Environmental Protection Agency
S-17185 Solna, Sweden

ISBN-13:978-3-642-79311-0 e-ISBN-13:978-3-642-79309-7
DOI: 10.1007/978-3-642-79309-7

Library of Congress Cataloging-in-Publication Data. Liming of acidified surface waters : a Swedish synthesis / Y: W. Brodin, L. Henrikson, ed. p. cm. Includes index. ISBN-13:978-3-642-79311-0
1. Aquatic liming–Sweden. 2. Aquatic liming. I. Brodin, Y. W. 1954- . II Henrikson. L. (Lennart), 1948-. TD427.A27L55 1994 628.1'68–dc20 94-25161

This work is subject to copyright. All rights are reserved, whether the whole or part of the material is concerned, specifically the rights of translation, reprinting reuse of illustrations, recitation, broadcasting, reproduction on microfilm or in any other way, and storage in data banks. Duplication of this publication or parts thereof is permitted only under the provisions of the German Copyright Law of September 9, 1965, in its current version, and permissions for use must always be obtained from Springer-Verlag. Violations are liable for prosecution under the German Copyright Law.

© Springer-Verlag Berlin Heidelberg 1995
Softcover reprint of the hardcover 1st edition 1995

The use of general descriptive names, registered names, trademarks, etc. in this publication does not imply, even in the absence of a specific statement, that such names are exempt from the relevant protective laws and regulations and therefore free for general use.

Cover design: E. Kirchner, Heidelberg
Typesetting: Camera ready by authors

31/3130-5 4 3 2 1 0 - Printed on acid-free paper

L. Henrikson Y.W. Brodin (Eds.) Liming of Acidified Surface Waters

Foreword

Acidification is one of the most serious environmental problems facing Europe and North America. A development towards large-scale acidification is sadly enough also apparent in Asia and South America. Sweden belongs to the countries which suffer most from acidification. Vast areas of forest soils and around 20,000 lakes have been seriously affected.

In Sweden, emissions of sulphur have been reduced by 80% between 1980 and 1993. The effects on acid deposition are limited, however, as some 85% of the acid load on Sweden can be traced to emissions from sources outside Sweden, mainly from other European countries. The Swedish Government has therefore devoted major concern to international negotiations aimed at reducing acid emissions to levels which the environment can tolerate.

Besides international negotiations the Swedish Government has employed large-scale operative liming activities to counteract acidification of surface waters. Acidification cannot be solved by means of liming, but liming can prevent further environmental damage while awaiting substantial reductions of acid emissions. It has also to be realized that it will take several decades until recovery occurs in thousands of surface waters even if acid emissions would decrease considerably in the near future.

Governmental subsidies for aquatic liming have gradually increased and amount to SEK 200 million (c. US$ 25 million) in 1994. Sweden has allocated considerable more financial support to aquatic liming than all other countries together. The scientific results presented in this book show that the efforts have been worthwhile. The large-scale liming acitivities have resulted in substantially improved conditions in around 8,000 Swedish lakes and watercourses. Biodiversity has been considerably improved and elimination of endangered species has been avoided, which is in compliance with the RIO declaration on the protection of biological diversity. Liming has moreover been essential for recreational and household fishing to be maintained as the most popular leisure for the Swedish people.

Sweden is determined to continue the work to counteract acidification; straining harder to reduce emissions even more and to maintain large-scale liming activity for surface waters as long as necessary. The successively reduced acid emissions in Europe and successful liming of Swedish freshwaters show us that battles to combat environmental problems can be won if willpower and understanding are strong.

Rolf Annerberg
head of the Swedish Environmental Protection Agency

Preface

The Project

In the early 1960s, the inexplicable death of fish was observed in several lakes in south-western Sweden. Curious scientists managed to clarify that these events and fish deaths occurring in the same region as early as in the 1920s were correlated to acid rain mainly originating from long-range transport of acid emissions in other countries. Comprehensive research and experimental studies during the 1970s and 1980s made it clear that acidification of water and soils was a problem with global dimensions, particularly in Europe and North America.

The need to reduce acid emissions became evident, and in the past years growing attention has also been given to the need to develop less acidifying forestry practices. The efforts to reduce acid emissions have had some success during the past decade but direct environmental measures, among which liming is certainly the most important, will most likely be needed for several decades ahead.

Liming acidified surface waters should not be considered a solution to the acidification problem. It is, however, necessary to avoid decreased biodiversity and species elimination and to maintain recreational fishing while awaiting considerable reduction in acidifying emissions and more environmentally friendly forestry practices.

Liming of surface waters in Sweden commenced in 1976 and has developed into a large-scale operative activity financed by governmental subsidies. Sweden is so far the only country applying large-scale liming to mitigate the effects of acidification.

In conjunction with liming practices in Sweden considerable efforts have been assigned to research, development of liming strategies and monitoring of the effects of liming. Results have been published in international journals, but the main body of information occurs in "grey literature" i.e. reports/manuscripts written only in Swedish and generally not easily accessible to scientists. Taking into account the need to summarize the experience acquired over the years, including the grey literature, the Swedish Environmental Protection Agency decided to launch a comprehensive evaluation project which in popular parlance became known as "The Liming Bible". An incentive for this large-scale project, involving about 60 people in Sweden and a number of scientists from other countries, was also the increased interest in aquatic and forest soil liming arising in other countries.

A main aim of the project was to generalize the knowledge of physical-chemical and biological effects of liming separately as well as at an integrated ecosystem level, giving attention to both short- and long-term perspectives. Further goals included the effects of liming on mercury levels in fish, analyses of pertinent complementary measures and the socio-economic consequences of aquatic liming. The total fulfillment of the aims was not possible, but the important conclusion could be drawn that liming is an accurate, and probably an often cost-effective, way to mitigate the effects of acidification in surface waters. Well-performed liming operations may result in an ecosystem similar to that existing prior to acidification, but liming alone does not seem to result in a complete restoration of the ecosystem.

The style of the articles of the present book is a compromise between explicit scientific and popular scientific writing. This makes the information accessible not only to trained scientists but also to non-scientists possessing profound knowledge of environmental issues.

Acknowledgements

We wish to express our gratitude to all the co-authors of this book, and to those colleagues in the Nordic countries and abroad who reviewed the manuscripts. They represent the crème de la crème of those having knowledge and experience in surface water liming. We also thank those who contributed articles to this book and/or made it possible for the authors to find and analyse unpublished and grey literature. We further wish to acknowledge Tove Sandberg for editorial assistance, Michael Johns for English translations, and Johan Wihlke for most of the drawings and ideas for the cover of the book.

Finally, we are very grateful to our colleagues at the Muncipality of Mark and at the Environmental Protection Agency and, last, but not least, our wives and children. Their indulgence and encouraging smiles when faced with our, at times, desperate attempt to create order among reprints, reports, innummerable fax messages, illegible telephone notes, heaps of manuscripts and floppy discs were certainly appreciated.

Lennart Henrikson Y.-W. Brodin

Contents

Chapter 1
Liming of surface waters in Sweden - a synthesis
L. Henrikson and Y.W. Brodin

Summary ... 1
1. Introduction ... 3
1.1. From trial programme to large-scale liming ... 3
1.2. Previous documentation of liming activities in Sweden ... 3
1.3. The aims of the synthesis and the target group ... 4
1.4. Liming practices in other countries ... 4
2. Acidification loss of biological diversity and fish catch ... 6
3. Liming - a part of the Swedish strategy to counteract acidificatioation ... 8
3.1. Strategy ... 8
3.2. For how long is liming necessary? ... 10
3.3. Alternatives and complementary strategies ... 11
4. Aims and accomplishment ... 12
4.1. Aims ... 12
4.2. Extent and accomplishment ... 12
4.3. Strategies, methods and liming agents ... 15
5. The effects of liming ... 18
5.1. Desirable and undesirable effects ... 18
5.2. Water quality ... 20
5.3. Biology ... 23
5.4. Mercury in fish ... 25
5.5. Socio-economic consequences ... 25
6. Will liming restore ecosystems? ... 26
7. Complementary measures ... 30
8. Lack of knowledge and research needed ... 31
8.1. Ecological and socio-economic effects ... 32
8.2. Strategy and methodology ... 34
9. Conclusions ... 35
9.1. Overall conclusions ... 35
9.2. Recommendations for liming of lakes and water-courses ... 36

9.3. Concluding remarks 39
10. Acknowledgements 39
11. References 39

Chapter 2
Acidification of lakes and watercourses in a global perspective
Y.-W. Brodin

Summary 45
1. Acidification: a complex problem 46
2. Acidification in the aquatic environment 48
3. Causes of acidification 52
4. How much can the environment withstand? 54
5. Outlook and future action 56
6. References 59

Chapter 3
Acidification of Swedish freshwaters
Y.-W. Brodin

Summary 53 63
1. Introduction 64
2. Acidifying load 66
3. The extent of acidification 68
4. Biological impact 70
4.1. Biodiversity 70
4.2. Causes of changes in flora and fauna 73
4.3. Decomposition 74
4.4. Human health 75
5. Necessary action and research 75
6. References 76

Chapter 4
Strategies and methods for freshwater liming
W. Dickson and Y.-W. Brodin

Summary 81
1. Introduction 83
2. History 83
3. The accumulated liming requirements for soil and water 85
4. Purpose and aims 86
5. Criteria for state grants 87
6. Liming organization 88

6.1. Liming plans and application 88
6.2. Monitoring of effects 89
6.3. Staff 90
6.4. Changes in the liming grant 91
6.5. Extent of liming 91
7. Liming strategies and methods 92
7.1. Strategies 92
7.2. Methods 93
7.2.1. Lake liming 93
7.2.2. Wetland liming 96
7.2.3. Solid ground liming 97
7.2.4. Doser liming 99
7.2.5. Stream liming 100
7.2.6. Ditch liming 101
7.2.7. Road liming 101
7.2.8. Lime precipitation in wastewater treatment plants 101
8. Liming agents and fractions 102
8.1. Agents 102
8.2. Fractions of limestone and dolomite 106
8.3. Quality control of lake lime 108
9. Doses of lime and other neutralizing agents 108
9.1. Lime dissolution rates 108
9.2. Dosage recommendations 109
10. Cost 113
11. Complementary methods 114
12. References 116

Chapter 5

The effects of liming on water chemistry

A. Wilander, P. Andersson, H. Borg and O. Broberg

Summar 125
1. Introductio 126
1.1. Materia 127
2. Primary effects of liming 128
2.1. Primary water chemistry reactions 128
2.2. Acidic episodes 128
2.3. Lake liming 129
2.3.1. Whole-lake liming 129
2.3.2. Littoral liming 136
2.3.3. Soda 138
2.4. Liming of watercourses 139

2.5. Catchment liming 141
3. Secondary physical-chemical effects 146
3.1. Physical effects 146
3.2. Effects on water chemistry 147
3.2.1. Organic substances and optical characteristics 147
3.2.2. Phosphorus compounds 149
3.2.3. Nitrogen compounds 153
3.2.4. Trace metals 155
3.2.5. Aluminium 155
3.2.6. Iron and manganese 159
3.2.7. Other trace metals 159
3.3. Changes in sediment chemistr 160
4. Effects of liming in entire aquatic systems 162
4.1. River Fylleån 164
4.2. River Enningdalsälven 166
5. Effects of liming activities on water chemistry 167
5.1. Comparisons between limed and non-limed waters 167
5.2. Comparison with results from model calculations 169
6. Acknowledgements 170
7. References 170

Chapter 6
The effects of liming on microbial activity and the decomposition of organic material
G. Gahnström

Summary 179
1. Introduction 179
2. Bacterial density and microbial biomass 180
3. Microbial activity 182
3.1. Microbial activity in the water 182
3.2. Microbial activity in the sediment 182
4. Decomposition of coarse detritus 187
5. References 189

Chapter 7
The effects of liming on aquatic flora
S. Larsson

Summary 193
1. The conditions for flora in acidified and limed lakes 194
1.1. Acidification 194

1.2. Liming 196
2. Phytoplankton 197
2.1. Impact of acidification 197
2.2. Number of species and composition of species after liming 200
2.2.1. Clear-water lakes 200
2.2.2. Humic lakes 202
2.3. Biomass and primary production after liming 203
3. Epiphytic algae 205
3.1. Impact of acidification 205
3.2. Number and composition of species 206
3.3. Biomass and abundance 207
4. Macrophytes 208
4.1. Impact of acidification 208
4.2. Number and composition of species 209
4.3. Abundance and primary production 210
5. Wetland vegetation 213
5.1. Impact of acidification 213
5.2. Number and composition of species 213
5.3. Biomass and density 215
5.4. Potential long-term effects 215
6. References 216

Chapter 8
The effects of liming on aquatic fauna
E. Degerman, L. Henrikson, J. Herrmann and P. Nyberg
Summary 221
1. Introduction 223
2. Lakes 225
2.1. Fish and crayfish 225
2.1.1. Background - the impact of acidification 225
2.1.2. Number of species 226
2.1.3. Recruitment and number of individuals 226
2.1.4. The structure of fish communities 228
2.2. Zooplankton 229
2.2.1. Background - the impact of acidification 229
2.2.2. Species composition and diversity 231
2.2.3. The structure of the zooplankton community 234
2.3. Benthic fauna and pelagic insects 235
2.3.1. Background - the impact of acidification 235
2.3.2. Composition of species and number of species 236
2.3.3. Abundance 238

2.4. Birds 244
2.4.1. Background - the impact of acidification 244
2.4.2. The effects of liming 245
2.5. Frogs 246
2.5.1. Background - the impact of acidification 246
2.5.2. The effects of liming 246
3. Running waters 247
3.1. Fish 247
3.1.1. Background - the impact of acidification 247
3.1.2. Number of species 247
3.1.3. Recruitment and number of individuals 248
3.1.4. Growth and production of individuals 248
3.1.5. The structure of fish communities 249
3.2. Benthic fauna 249
3.2.1. Background - the impact of acidification 249
3.2.2. Number of species and density 250
3.2.3. Functional groups 254
4. Can the original fauna be restored after liming? 255
5. Factors affecting the post-liming development of fauna 258
5.1. Water chemistry 258
5.2. Liming strategies 260
5.3. Recolonization 263
6. References 266

Chapter 9
The impact of liming on aquatic communities
M. Appelberg
Summary 283
1. Introduction 284
2. Communities of organisms 285
3. Short-term and long-term effects of liming 286
4. Post-liming abiotic factors 288
5. Acidification and community development 289
6. Productivity and community structure 290
7. Biotic factors: competition as a controlling factor 292
8. "Bottom-up" effects 293
9. "Top-down" effects 294
10. Predation and changeable environments 297
11. Parasites and disease: an unknown factor? 298
12. Stability of post-liming communities 299
13. Trophic levels and food chains 301
14. Can acidified communities be restored? 303

15. Acknowledgements 304
16. References 304

Chapter 10
Liming strategies and effects: the Lake Gårdsjön case study

J.E. Svensson, L. Henrikson, S. Larsson and A. Wilander

1. Documentation 309
2. Geographical orientation 310
3. Acidification status 311
 3.1. Interactions in the pelagic zone 312
4. Liming strategy and re-introduction of fish 315
5. The effects of liming 315
 5.1. Water chemistry 315
 5.2. Pelagic organisms 318
  5.2.1. Post-liming interactions in the pelagic zone 321
  5.2.2. Interactions in the pelagic zone after re-introduction of fish 322
6. Conclusions 322
7. References 323

Chapter 11
Liming strategies and effects: the Lake Västra Skälsjön case study

P. Nyberg

1. Documentation 327
2. Geographical orientation 328
3. Acidification status 328
4. Liming strategy 329
5. Effects of liming 329
 5.1. Water chemistry 329
 5.2. Flora 330
 5.3. Zooplankton and benthic fauna 331
 5.4. Fish 333
6. Conclusions 337
7. References 337

Chapter 12
Liming strategies and effects: the Lake Stora Härsjön case study

M. Appelberg

1. Documentation 339
2. Geographical orientation 339
3. Acidification status 340
4. Liming strategy 343
5. The effects of liming 343
5.1. Phytoplankton 344
5.2. Zooplankton 345
5.3. Benthic fauna 346
5.4. Fish 346
5.5. Comparison of different trophic levels 348
6. Conclusions 349
7. References 350

Chapter 13
Liming strategies and effects: the Lake Gyslättasjön case study

M. Appelberg

1. Documentation 353
2. Geographical orientation 353
3. Acidification status 354
4. Liming strategy 356
5. Effects of liming 356
5.1. Water chemistry 356
5.2. Phytoplankton 357
5.3. Zooplankton 358
5.4. The littoral benthic fauna 358
5.5. Fish 358
6. Conclusions 360
7. References 361

Chapter 14
Liming strategies and effects: the River Högvadsån case study

I. Alenäs, E. Degerman and L. Henrikson

1. Documentation 363
2. Geographical orientation 364
3. Acidification status 365
4. Liming strategy 365

5. Effects of liming 366
5.1. Water chemistry 366
5.2. Salmon stock 368
6. Conclusions 370
7. References 371

Chapter 15
Liming strategies and effects: the Prästvallsbäcken Stream case study

P. Andersson

1. Documentation 375
2. Geographical orientation 375
3. Acidification status 376
4. Liming strategy 377
5. Effects of liming 378
6. Conclusions 381
7. References 381

Chapter 16
Liming effects on mercury concentrations in fish

M. Meili

Summary 383
1. Introduction 384
2. Methods 384
2.1. Study lakes 384
2.2. Treatments 385
2.3. Sampling and analyses 386
2.4. Calculations 386
3. Results and discussion 387
3.1. Pre-treatment situation 387
3.2. Liming 388
3.3. Development of Hg concentrations in pike after liming 389
3.4. Potential mechanisms 393
3.5. Cost-benefit analysis 395
4. Conclusions 395
5. Acknowledgements 396
6. References 396

Chapter 17
Supplementary measures to aquatic liming
B. Bergquist

Summary ... 399
1. Introduction ... 400
2. Measures to promote colonization and re-establish locally extinct species ... 402
2.1. Habitat improvement ... 403
2.2. Removal of migration obstacles and building of fish-ways ... 404
2.3. Restocking existing populations ... 405
2.4. Decimation fishing ... 408
2.5. Other measures ... 410
3. Reintroduction of locally extinct species ... 411
3.1. Salmonid fish (Atlantic salmon, trout, grayling and Arctic char) ... 411
3.2. Other species of fish ... 414
3.3. Crayfish and other invertebrates ... 416
4. Conclusions ... 417
5. References ... 418

Chapter 18
Socio-economic consequences of aquatic liming
B. Bengtsson and A. Bogelius

Summary ... 423
1. Introduction ... 424
2. Methods ... 428
3. Examples of the value of recreational salmon fishing, the impact of acid rain and the benefit of liming ... 429
3.1. Canada ... 429
3.2. Scotland ... 430
3.3. England ... 430
3.4. Iceland 430
3.5. Norway ... 430
3.6. Sweden ... 431
4. Cost-benefit studies of liming ... 431
4.1. The pre-liming recreational value of salmon and trout fishing in a Norwegian river ... 432
4.2. The socio-economic value of liming in Audna ... 432
5. Liming, commercial fishing, fish farming and aquaculture ... 433

5.1. Commercial fishing 433
5.2. Fish farming and aquaculture 434
6. The effects of liming on recreational fishing, leisure and tourism 435
6.1. Recreational fishing and liming in Sweden 435
6.1.1. Forest Service fishing projects 436
6.1.2. The Älvkarleby study 437
6.1.3. The Blekinge County Council study 437
6.1.4. The River Mörrumsån study 438
6.1.5. The West Coast study 438
7. Examples of the value of liming to recreational fishing 438
7.1. Introduction 438
7.2. Results and discussion 442
7.2.1. River Högvadsån 442
7.2.2. River Mörrumsån 444
7.2.3. Lake Lilla Le 445
7.2.4. Lake Unden 446
7.2.5. Fulufjället 447
7.2.6. Lake Bolmen 447
7.2.7. Hökensås 449
7.2.8. The Delsjö area 449
7.2.9. Lake Rottnen 450
8. Liming and mercury 451
9. Liming and caesium 451
10. Liming and environmental commitment 453
11. Acknowledgements 454
12. References 455

Chapter 1

Liming of surface waters in Sweden - a synthesis

Lennart Henrikson[1] and Y.-W. Brodin[2]

[1] Municipality of Mark, S-511 80 Kinna, Sweden
[2] Swedish Environmental Protection Agency, S-171 85 Solna, Sweden

Summary

A trial programme on liming of acidified lakes and watercourses in Sweden was conducted during the period 1976 to 1981. The positive experiences acquired resulted in the initiation of a large-scale operative activity from 1982 onwards, which is mainly financed by governmental subsidies. Approximately 6,300 lakes and 6,000 km of running water have been limed in Sweden, and the activities are among the greatest which have been applied in order to restore or preserve environments at risk in Sweden. Considerable resources have been assigned to research and monitoring of effects.

In order to summarize the experiences acquired, the Swedish Environmental Protection Agency initiated a comprehensive evaluation project. The results are described here, providing a synthesis mainly based on data from other chapters of this book.

Acquired experience shows that liming results in a considerably improved water quality and biological diversity. Liming is also generally favourable to fisheries. The water quality and organism communities following liming are usually rather similar to that of non-acidified water bodies, but liming will not rehabilitate the ecosystems completely. Substantial reductions in acidifying emissions and healthy forest practices are the only long-term strategies applicable to attain full restoration of aquatic ecosystems suffering from acidification.

Complementary measures are often needed to obtain the desired effects, e.g. elimination of obstacles for passage of fish and other organisms, reintroduction of species, and addition of nutrients, particularly phosphorus, in some surface waters. Accurate liming practices often result in a decreased mercury content in fish. Certain undesirable effects may occur after liming,

Liming of Acidified Surface Waters
L. Henrikson and Y.W. Brodin (Eds.)
© Springer-Verlag Berlin Heidelberg 1995

of which deterioration of limed wetlands is the most controversial.

The knowledge of the effects of liming on human society and national economy is sparse in Sweden, but there are indications from some studies that liming may be profitable with respect to economy.

Liming in freshwater to retain biological diversity and fisheries will probably be needed for at least the next 50 years, even if substantial reduction in acid emissions of sulphur and nitrogen in Europe were to occur in the coming years.

Limestone ($CaCO_3$) is the best and most frequently used neutralizing agent for treatment of acidified surface waters all over the world. To achieve desirable results, appropriate planning is essential, making use of a strategy counteracting episodic acidification and resulting in minimal variability in water quality. A combination of different strategies, i.e. direct lake liming, wetland liming and liming by means of dosers will produce the most adequate ecological results. Due to the potentially severe effect on vegetation, liming on wetlands should only be conducted after well-balanced evaluations regarding desirable and undesirable consequences taking into account the conservation value of different environmental categories.

Lime dosages which cause a considerably higher pH than under natural conditions, should generally be avoided. Reacidification of limed water bodies is also to be avoided.

Knowledge of the effects of freshwater liming is considerable, but far from complete. Long-term effects, i.e. more than 20 years, are unknown for natural reasons. It is not yet possible to generalize the effects of liming on a number of aquatic organisms or groups associated with water, e.g. micro-organisms, birds, amphibians and mammals. The long-term development of nutrient level and availability is ambiguous, and further studies are needed to answer the question whether nutrients should be added to limed lakes.

Further studies are also needed particularly on;

- production and decomposition processes,
- interaction between species (especially in running water),
- natural recolonization of species or reintroduction by man,
- effects of liming on micro-organisms,
- the significance of parasites and diseases,
- mechanisms governing mercury content in fish,
- socio-economic effects.

The national strategy is not to lime naturally acid surface waters. In northern Sweden, however, it is not sufficiently well known to what degree surface waters are naturally acid or acidified because of man's impact. This constitutes a major problem since several waters of the region are already being limed or planned to be limed.

Strategies must be developed regarding liming practices in lakes having a short water retention time and to counteract episodic acidification in watercourses experiencing considerable water-flow variations. Measurable biological goals, which can be quantified, are needed to adequately evaluate the effects of liming.

In conclusion, liming practices applied to Swedish surface waters have mainly produced desirable results. Similar conclusions have been drawn from less extensive aquatic liming activities in other countries. Current knowledge is sufficient to recommend continued large-scale operative liming activities in Sweden, to be run parallel with scientific and applied studies to increase knowledge and to optimize liming operations.

1. Introduction

1.1. From trial programme to large-scale liming

The possibility that fish mortality, occurring in Swedish surface waters as early as in the 1920s, could be attributable to acid rain was not understood until the 1960s. In the 1970s, the relationship between acid deposition and acidified lakes and watercourses was consolidated. Not only fish but also other animals were adversely affected. Mapping showed that the adverse effects of acidification were widespread. During the mid-1970s, the possibilities to use neutralizing agents in acidified freshwaters with the objective to prevent further acidification and to re-establish appropriate conditions for fish were discussed. This resulted in a trial programme on liming.

The trial programme commenced in 1976 and was managed by the National Swedish Board of Fisheries until 1981. The aims were (1) to restore acidified surface waters and to prevent further acidification in threatened waters of particular importance to fisheries and nature conservation; (2) to gain experience and obtain results on methods, liming agents and costs; and (3) to pay particular attention to positive as well as adverse effects on plants and animals. The results served as a basis for a decision on the direction of future liming activities. Positive experience resulted in large-scale operative activities from 1982 onwards.

A trial programme on liming acidified drinking-water wells has been accomplished in Sweden (Sundlöf 1994). A trial programme on forest-soil liming is in progress, with the aim of setting up large-scale activities within a few years time (Lindström *et al* 1993).

1.2. Previous documentation of liming activities in Sweden

The experience and results acquired from the trial period on aquatic liming have been summarized by the National Swedish Board of Fisheries and SEPA - the Swedish Environmental Protection Agency - (1981). Based on this experience, guidelines for liming of Swedish lakes and watercourses have been published by the National Swedish Board of Fisheries (1982a,b).

New, still valid, guidelines for liming Swedish lakes and watercourses were published by SEPA in 1988. SEPA has the responsibility for know-how and information on liming, which includes conferences, directed courses, a periodic publication, and research planning and administration. Surveys on extent, results etc. have been published by Bernes (1991) and Thörnelöf and Degerman (1991). These surveys cover only a limited number of results published in Swedish and in international journals. Therefore, SEPA decided to launch a comprehensive evaluation project covering Swedish activities on liming lakes and watercourses. The results are presented in this book. This first chapter is a synthesis based mainly on the information gained from the other contributions of the book, where the original references can be found.

A synthesis of Swedish research on acidification of aquatic ecosystems, which to a certain extent also deals with liming, has been published by Fleischer *et al* (1993). International surveys on liming and effects of liming have been published by Brocksen and Wisniewski (1988), Weatherley (1988), Olem (1991) and Olem *et al* (1991).

1.3. The aims of the synthesis and the target group

The aims of the present synthesis are:

1. To summarize the experience acquired from Swedish liming activities concerning strategies and methods, effects on water chemistry and biology, physical-chemical and biological mechanisms, complementary measures, prospects to reduce mercury concentrations in fish, and socio-economic effects.
2. To form a basis for continued research and development, as well as a revision of current guidelines and directives for planning and practices concerning liming operations.

The synthesis also addresses international results and experience from liming and acidification.

The target group for this synthesis, and the book as a whole, is primarily scientists and those working professionally on environmental issues in Sweden and other countries. Other people interested in the addressed issues may also find this book useful.

1.4. Liming practices in other countries

Liming of lakes to increase the yield of fish has been carried out for a long time in Sweden (Dickson and Brodin, Chap. 4) and several other countries (cf. Hasler *et al* 1951, Waters 1956). At the present time, several countries practice limestone treatments of acidified lakes and watercourses (Fig. 1).

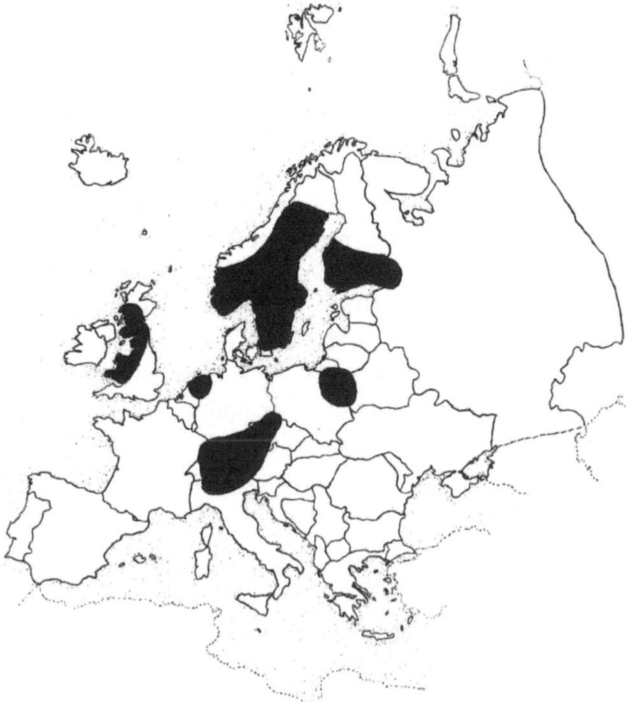

Figure 1. Areas in Europe where liming of acid water bodies has been conducted (based on literature and personal communications).

Only Sweden and Norway, however, have chosen large-scale liming as a national strategy to preserve species threatened by acidification.

Norway has gradually built up an extensive liming programme, which in 1994 will entail a cost of about a quarter of the Swedish activities. Hundreds of Norwegian lakes and watercourses have been limed for scientific purposes, but also in order to preserve valuable stocks of fish. Scientific results have been published on several occasions and summarized also by Baalsrud *et al* (1985), Rosseland and Hindar (1988), Hindar and Rosseland (1991), and Hegna (1992). A manual on liming has also been produced (Directorate for Nature Management 1990).

Research on the effects of liming and liming methods is in progress in a number of countries (e.g. Olem *et al* 1991). Finland has limed more than 100 lakes and watercourses. Several of these have been scientifically studied (e.g. Alasaarela *et al* 1990, Niinioja *et al* 1990, Rask 1990).

In Great Britain, projects dealing with liming of lakes and watercourses have been carried out in Scotland (e.g. Howells and Brown 1992) and

Wales (e.g. Ormerod *et al* 1990, National River Authority *et al* 1992). A comprehensive documention is available for Loch Fleet in Scotland (Howells and Dalziel 1992a).

About 150 lakes and 10 watercourses have been limed in the USA until 1991 (Olem 1991). A number of water bodies are part of major projects such as "The Acid Precipitation Mitigation Program (APMP)" (Schreiber and Villella 1991) and "Living Lake Aquatic Liming Program" (Brocksen 1991). In Canada, comprehensive studies have been conducted on limed lakes in Ontario (Keller *et al* 1990, Keller 1992) and some watercourses containing salmon in Nova Scotia (Watt 1986). Published scientific results are available from the mentioned projects in the USA and Canada. Manuals have also been published (e.g. Brocksen *et al* 1992, Weigmann *et al* 1992).

In other countries, such as Italy, Poland, Germany, Austria and the Netherlands, liming treatments to counteract acidic conditions have only been conducted to a limited extent. Lake Orta in Italy was limed to neutralize acidification caused by an industrial effluent containing ammonium (Calderoni *et al* 1991).

2. Acidification - loss of biological diversity and fish catch

Acidification is to a major extent caused by emissions of oxides of sulphur and nitrogen. Land use, in particular forestry, may also play a significant role. Acidification is a large-scale environmental problem with a series of different consequences, i.e. extensive changes in water chemistry and biology in thousands of lakes and watercourses in Europe and North America (e.g. Fleischer *et al* 1993, Brodin, Chap. 2 and 3). Sweden is among the countries most significantly harmed by acidification. The quality of soils and surface waters has changed over entire catchment areas, which has influenced the prerequisites for living creatures.

Time will change every ecosystem, but acidification brings about changes so fast that species having limited possibilities to adapt to new conditions will be wiped out or experience appreciably decreased numbers. The process of acidification initially means stress where sensitive species are eliminated due to the toxic effect of low pH and different species of aluminium. Elimination of so-called key organisms, fish in particular, brings about changed conditions with respect to predation and competition. This will, in turn, lead to a series of effects which will change the ecosystem even more. The final phase will be an ecosystem with an altered structure, i.e. a lower number of species and a considerably changed species composition. Certain processes, particularly decomposition but possibly also primary production, can be affected. All this will result in a loss and alteration of biodiversity as well as a reduction in biological production which means decreased potential for man's utilization of the aquatic ecosystem.

Figure 2. Most acidified lakes are oligotrophic, and liming treatment has mainly focused on clearwater forest lakes (drawing: Eva Engblom).

Biodiversity can be studied at the ecosystem, population, species and genetic level. All levels are affected by acidification. At the ecosystem level, oligotrophic freshwater - the most common aquatic ecosystem in the Nordic countries - is particularly affected (Fig. 2). The effects occur over large areas.

At the species level, acidification is a - sometimes major - threat for several species in Europe and North America. There are even species which are endangered in a global perspective. This is true for the spring-spawning cisco *(Coregonus trybomi)* classified as endangered by IUCN (International Union for the Conservation of Nature). The species is known from 6-7 lakes in the world, 4 of these are in Sweden. Continued acidification will eliminate the species in Sweden, and the threat of extinction would increase considerably as there is only one lake - which is limed - where the species has a viable population. At the national level, freshwater pearl mussel *(Margaritifera margaritifera,* Bivalvia) is classified as vulnerable on the national Red List of threatened and rare species in Sweden. Acidification exerts a significant threat within a complex of problems, which includes eutrophication and water regulation. Also for the genetically distinct relict populations of Arctic char *(Salvelinus alpinus)* in Sweden, acidification is a major problem, and the species is classified as "care-demanding" on the national Red List. Among fish-eating birds, red-throated diver *(Gavia stellata)*, black-throated diver *(Gavia arctica)* and osprey *(Pandion haliaetus)* are classified as "care-demanding" on the national Red List, i.e. the species is not threatened in a short-term perspective but requires specific consideration in order to secure its long-term survival on a national basis. Decreased abundance of fish or increased risk of exposure to toxic metals through the intake of fish from acid lakes are among the problems to be considered for these species. In regions seriously affected by acidification, local elimination is a threat also for populations of other species which do not appear on any national Red List. This is problably the case for the stonefly *Dinocras cephalotes* (Plecoptera) in south-western Sweden. Even common Nordic species such as roach *(Rutilus rutilus)* and minnow *(Phoxinus phoxinus)* can be wiped out locally.

Loss of genetic variability has been observed particularly for different species of fish, e.g. the threat against natural populations of Atlantic salmon *(Salmo salar)* and brown trout *(Salmo trutta)* in Europe. This is probably also the case for other animal species with small and/or isolated populations, e.g. crustaceans defined as "glacial relicts".

In thousands of lakes in Europe and North America, fish populations are affected by acidification. As a consequence of declining fish production, fishing opportunities for man have been lost to a major extent. This is naturally disquieting as fishing is one of the most popular recreational pastimes in Europe, and not unimportant to the national economy of Sweden and other countries (Bengtsson and Bogelius, Chap. 18). Acidification thus produces negative effects on society. In the future, fish production in surface waters might regain significant interest as a national food supply.

Acidification usually means increased accumulation of mercury in fish. A drop in the lake water pH by one unit may double the mercury concentration in fish (Meili, Chap. 16). High concentrations of mercury (>1 mg/kg wet weight) are estimated in northern pike *(Esox lucius)* - a popular sport fish species in Sweden - in approximately 10,000 lakes over vast areas of Sweden. Acidification is a potential menace to human health as consumption of fish with a high mercury content may have detrimental effects, especially on children. Enhanced concentrations of metals, particularly mercury, also exert a threat to birds such as divers and osprey and other top predators having fish as a food resource.

3. Liming - a part of the Swedish strategy to counteract acidification

3.1. Strategy

As described above, acidification is one of the most serious and extensive environmental problems facing Europe and North America. The only solution in the long run is naturally to reduce acid emissions of compounds containing sulphur and nitrogen, bringing about decreased acid deposition.

In Sweden only, the accumulated acidification is comparable to 90 million tonnes of limestone, entailing a cost for liming agents and measures of approximately 25 billion Swedish kronor (SEK)(Dickson and Brodin, Chap. 4). Acid deposition amounts to 60 million tonnes of limestone, and forestry, particularly the removal of biomass, amounts to 30 million tonnes. This accumulation and the natural sluggish reaction of the soils to changed conditions mean that the deleterious effects of acidification on the environment and economics will be long-lasting even if acidic emission were cut immediately to zero.

The Swedish strategy to counteract acidification is based on: (1) a major reduction in acidic emissions in Sweden, (2) international lobbying and agreements aimed at reductions in emissions in Europe, and (3) liming of particularly valuable aquatic ecosystems to preserve conditions while awaiting decreased acidic deposition. The acidification effect caused by forestry has also received attention over the last years.

Different arguments against aquatic liming have been put forward in debates in Sweden and other countries. The most common arguments are that undesired secondary effects may occur and long-term effects are unknown. Moreover, as liming often brings about adequate ecological results, this could be used as an argument for not developing costly techniques to reduce acid emissions. It could be argued that it would be less expensive to treat surface water with limestone than to reduce acid emissions. These arguments have gained less support in Sweden and other countries concurrently with the increased understanding of the harmful effects of acidification also on forest soils, materials and objects of art. It is important both to reduce acid emissions and to preserve plants and animals by other means such as liming (e.g. Dickson 1982, Wright 1982).

Liming is the most large-scale restoration activity applied to Swedish lakes and watercourses, and at the same time the largest deliberate manipulation of aquatic ecosystems in Sweden. People have in their criticism stressed the daring venture to initiate such a large-scale operative activity before the effects have been scientifically scrutinized. Some knowledge, obtained mostly from activities favouring fisheries, on liming in practice was, however, available prior to initiation of the trial programme in 1976. It was furthermore assessed that such major environmental and fishery values were at risk and that there was no time left to await scientific lucidity. This criticism should be looked upon as legitimate, but the experience provided by the trial programme 1976-1981 and the knowledge acquired thereafter have shown that liming mainly leads to the desired effects.

3.2. For how long is liming necessary?

An appreciable reduction in acid emission in Europe will probably result in a gradual improvement of the water quality and ecological conditions in acidified surface waters. The recovery may, however, be a very sluggish process as it is governed by the chemical weathering rates of the soils (Skeffington 1992). Lakes located on thin soils in Scotland have experienced a comparatively fast rise in pH and alkalinity within a few years after a decrease in acidic deposition (Battarbee *et al* 1990). Forsberg *et al* (1985) observed an increase in pH by 0.3-0.4 units in lakes of south-western Sweden as a result of a 20% drop in wet deposition of sulphate. In acidified regions with thick soils, the normalization process may not take place until some decades after a drop in acidic deposition.

That decreased acidic deposition results in improved forest soil and freshwater conditions has also been shown experimentally in the Norwegian RAIN project (Reversing Acidification In Norway) (Wright *et al* 1993). Considerably improved lake water quality was observed a few years after a major reduction in acid emissions from a smeltery in Ontario, south-eastern Canada (Keller *et al* 1992b). This quick recovery might have been attributable to an increased weathering rate of the surrounding soils (Skeffington 1992). There has also been a comparatively fast recovery of the aquatic flora and fauna, but recolonization of species has not been completed (Keller *et al* 1992a, Griffiths and Keller 1992), similar to the response of flora and fauna of limed lakes in Sweden (Degerman *et al*, Chap. 8).

Brodin (1994) analyzed future effects of acidification in Sweden based on alternative scenarios on emission abatement in Europe. In order to prevent Swedish surface waters from becoming even more acidified in the next 10-20 years, acid emissions in Europe need to be cut by approximately 70% in comparison with 1990. Acid deposition in Sweden will drop by around 30% between 1990 and 2010 if international agreements and national abatement plans are followed. However, even if a reduction by 70% could be realized by the year 2010, several acidified soils and water bodies will probably be found in southern Sweden even after a period of 50 years from now (Bernes 1991, Brodin 1994). Consequently, in order to preserve biodiversity and propitious fishing conditions, continued liming of Swedish freshwater systems is necessary for at least another 50 years.

The future situation is estimated to be even more serious for Norway. Even if countries conform to international agreements, surface water in 60,000 km^2 of Norway will suffer from the harmful effects of acidification on fish (Henriksen 1992, Henriksen and Hindar 1993). At present, 10,000 km^2 are affected by acidification. The critical loads for harmful environmental effects would be exceeded over vast regions in Norway even if sulphur deposition could be decreased by 90%. In some areas, liming is probably a permanent measure for the survival of fish.

3.3. Alternatives and complementary strategies

Experiments aimed at finding acid-tolerant strains of brown trout in Norway have been conducted within the project REFISH (Restoring Endangered Fish In Stressed Habitats). Differences between different strains have been noted; certain strains can withstand the conditions in surface waters affected by acidification (Kroglund *et al* 1992, Rosseland and Kroglund 1992). Variable results were obtained from the introduction of brook trout *(Salvelinus fontinalis)* into lakes in the Adirondack region, north-eastern USA (Olem 1991).

Whether introduction of acid-tolerant species should be regarded as an alternative to liming is naturally related to the purposes of liming. If the aims are limited to retain fish production for recreational purposes, introduction of acid-tolerant strains of fish may be looked upon as an alternative. It should, however, be noted that a high accumulation of metals, such as mercury, may still be a problem, and the fish caught may be unacceptable as food for human beings. If the aims are to restore or preserve original populations of plant and animal species, introduction of acid-tolerant strains provides no alternative to liming.

Groundwater with a pH $>$ 7 was pumped into a watercourse in the USA, and this resulted in an increase in pH 4.9 to 6.0 (Gagen *et al* 1989). Favourable effects were also gained when the same method was applied to a lake (Garrison and Rose 1990).

Current forestry practices contribute to the acidification of soils and water. Changes in land use, such as increased numbers of deciduous trees such as birch *(Betula* spp.) and burning of clearings, improve the conditions (Dickson and Brodin, Chap. 4).

A number of European countries have added lime to forest soils in order to prevent decreased forest productivity. Large-scale activities occur in Germany (Matzner *et al* 1985), however, not with the purpose of mitigating conditions in acid surface waters. Experiments using lime to counteract acidifiction of forest soil have been carried out in Sweden, and there are plans to turn this into a large-scale activity in southern Sweden (Lindström *et al* 1993). Such large-scale activities might provide an important complement to liming of lakes or adjacent wetlands within a number of decades (Dickson and Brodin, Chap. 4). Trial programmes on forest-soil liming are also occurring in Finland and Norway.

Based on present knowledge, the conclusion to be drawn is that dispersal of limestone or other neutralizing agents is the only realistic possibility to counteract the harmful effects of freshwater acidification, while awaiting substantial emission abatements.

4. Aims and accomplishment

4.1. Aims

The aims of the Swedish liming practices according to SEPA (1988) can be summarized as follows:

- The biological aims are to detoxicate the water for the continued existence or recolonization of natural flora and fauna.
- The chemical aims are to raise the pH above 6.0 and alkalinity above 0.1 meq/l.

It is furthermore stressed that "overdosage" is to be avoided, and that pH and alkalinity should not be allowed to fall below 6.0 and 0.05 meq/l, respectively, at any occasion throughout the year.

Thus, the overall aims of freshwater liming are to maintain or restore biodiversity and fisheries. A revision of the aims is planned in the near future, taking into account the results of the present book and the improved knowledge of acidification in naturally acid water bodies.

4.2. Extent and accomplishment

Liming treatment of surface waters has developed into a large-scale environmental protection activity in Sweden. Until 1993, around 8,400 water bodies (mainly lakes) have been limed (Fig. 3). From 1982 onwards, SEPA, in co-operation with the National Swedish Board of Fisheries, is responsible for general planning, allocation of governmental subsidies to counties, and research and development. The county administrative boards have the task of regional planning, decisions on subsidies for liming projects and monitoring of effects. Municipalities are usually responsible for liming operations.

During the period 1986 to 1989, comprehensive experimental projects were carried out in Sweden with the aim to elucidate whether liming could result in decreased mercury concentrations in fish (Håkanson *et al* 1990, Meili, Chap. 16). The project is continuing on a limited scale.

The total costs for governmental subsidies for liming until 1993 were approximately SEK 1 billion. This includes administration, planning, biological restoration (cf. Bergquist, Chap. 17), and local monitoring of effects (Fig. 4; cf. Dickson and Brodin, Chap. 4). Comparatively large amounts have also been allocated for research, national and regional monitoring of effects, and evaluation projects (Fig. 5).

Around 100 lakes and 50 watercourses throughout Sweden have been designated as national reference waters. These water bodies are chosen primarily with the objective to ascertain possible trends attributable to the deposition of air pollutants or climate change, but they also serve as control objects for limed lakes and watercourses.

***Figure 3**. Lakes and watercourses in Sweden which have been limed at least once during the period 1985-1993. The total number of waters is approximately 8,400; most of them are lakes.*

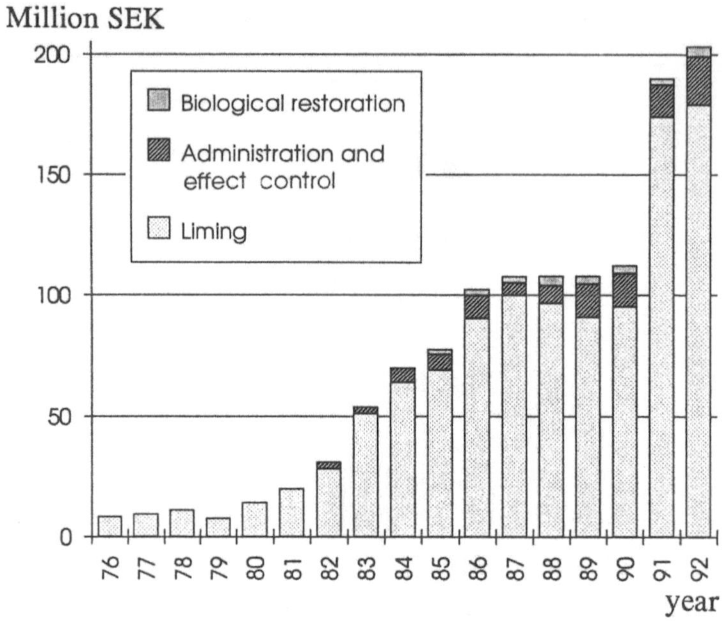

***Figure 4.** Governmental subsidies for surface water liming, administration, follow-up of effects and biological restoration after liming in Sweden from 1976/1977 to 1992/1993. 1 SEK is about 0.13 US dollars (July 1994).*

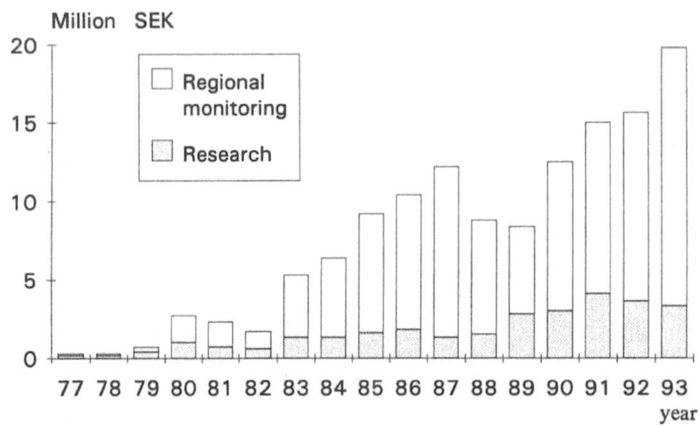

***Figure 5.** Grants for research and regional monitoring of surface water liming in Sweden, from 1976/1977 to 1992/1993. 1 SEK is about 0.13 US dollars. (July 1994).*

4.3. Strategies, methods and liming agents

A number of different strategies, methods and neutralizing agents have been used for freshwater liming during the 15 years of large-scale liming application in Sweden (Dickson and Brodin, Chap. 4). The strategies have developed from liming of individual lakes to liming treatment of entire or large parts of a watershed. Dickson and Brodin (Chap. 4) have analysed the efficiency of different strategies (cf. also Wilander *et al*, Chap. 5). The knowledge acquired is summarized below.

Full-scale liming, i.e. liming of all acidified soils and waters within a watershed, is the most appropriate long-term strategy. To achieve positive effects in running waters takes time, however, due to the sluggish process of soil neutralization in the watershed. This strategy initially entails high costs, and might be considered too expensive if the aims are to protect the aquatic flora and fauna only. In total, probably 10-20% of the alkaline materials added to the soils will benefit the aquatic ecosystem.

Liming on wetlands is often the second-best strategy to alleviate acidification of lakes and running waters. This means fixation of toxic metals in the wetlands and moderation of acid surges. If suitable wetlands are available, liming of 1-2% of the catchment area may be enough to achieve satisfactory results in the aquatic ecosystem. Under favourable conditions, 50-70% of the applied lime will have a positive effect on the acidified lake or watercourse, but the figure is generally lower.

Lake liming is comparatively inexpensive, and may be the only strategy necessary if the lake area of the catchment constitutes more than 10% and the hydraulic residence time of the lake is less than 0.2 years. The strategy, however, is often not efficacious enough to neutralize the supply of acidic water from the surroundings during periods of snowmelt or heavy rainfall. According to Wilander *et al* (Chap. 5), the mean utilization rate for direct lake liming in Swedish lakes is about 63%; for a quarter of the lakes the rate is below 38%, and for another quarter the rate exceeds 90%.

Liming by means of dosers may be an effective and inexpensive strategy to mitigate acidication of watercourses and tributaries to lakes with a short retention time. A basic demand is naturally an adequate location and accurate handling of the dosing device. Lime utilization is generally as high as 70-90%.

Treatment of all acidified lentic and lotic systems within a watershed is optimized by a combination of different strategies. It has to be observed, however, that the strategy chosen should take into consideration liming efficiency as well as potential undesirable effects such as damaged flora of limed wetlands. Figure 6 provides a survey of the applicability of different liming strategies.

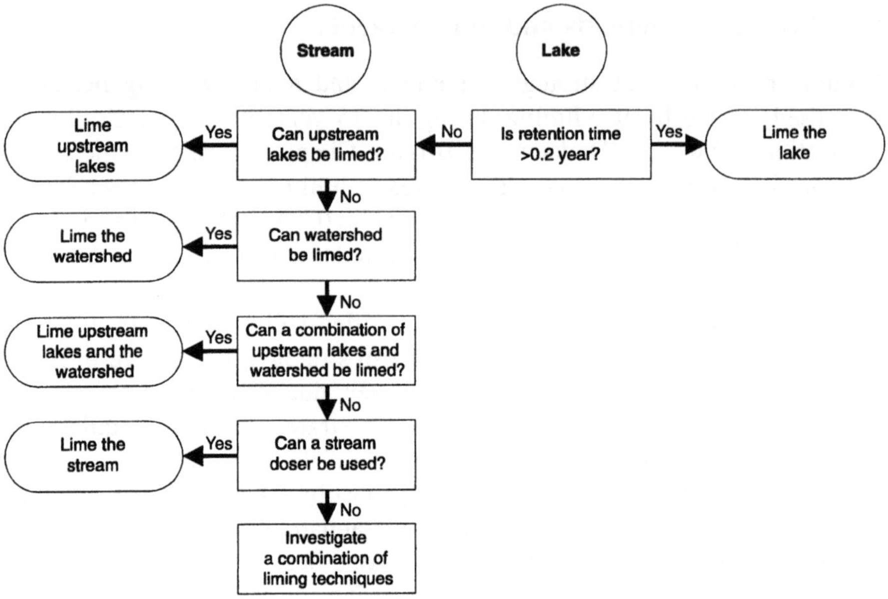

***Figure 6.** Recommended decision strategy for liming of surface waters in Sweden (SEPA 1988).*

Methods for lime spreading in lakes make use of specialized boats where lime is mixed with lake water to a slurry before spreading. Spreading on the ice of the lake makes it easier to place the lime where it is most needed. Helicopters are in most cases used for wetland liming. A number of different dosers have been developed for direct dosing of lime into running water.

The strategy or treatment methods have substantial importance for biological development if the chemical goals are not achieved, i.e. if the liming operation is not sufficient to detoxicate the water (Degerman *et al*, Chap. 8). This is most evident when lime is added to watercourses. Liming by means of dosers only is generally not enough to assure a desired biological response. Wetland liming, on the other hand, is probably a safer strategy. The best results for running water are achieved by a combination of different strategies, i.e. liming of tributary lakes, wetland liming and liming dosers. Acid surges due to high-water flow and non-functioning dosers can still cause problems, however. Inflow of acid and aluminium-rich water from the surrounding soils and tributaries may be harmful to sensitive species and prevent colonization.

Limestone ($CaCO_3$) has proved to be the most appropriate neutralizing agent, taking into account effects, treatment, and cost. Limestone is the main agent used to counteract acidification in Sweden and other countries. Supplies of limestone in Sweden are sufficient for national demands within a foreseeable future. Only 10% of the quarried lime in Sweden is used for lake and watercourse treatments. There has been a development towards finer fractions (0-0.2 mm). For wetland liming, however, coarser fractions (0-1 or 0-2 mm) have been proposed in order to reduce wind dispersal, increase lime dissolution efficiency and to reduce the costs (Dickson and Brodin, Chap. 4). The limestone used for freshwater liming in Sweden is subjected to official quality control.

A number of neutralizing materials other than limestone can be used for freshwater liming. Sodium carbonate (Na_2CO_3) is, however, the only other product worth mentioning. Sodium carbonate was harvested earlier into the bottom sediments, but is nowadays added as pellets on the bottom (Lindmark 1984, 1990, 1994). Dickson and Brodin (Chap. 4) suggest that sufficiently pure waste products should be used to alleviate acidification instead of dumping.

In Sweden, the recommendations in the manual "Guidelines for Liming" (SEPA 1988) are generally used to calculate the required dosages of lime (Dickson and Brodin, Chap. 4). The dose calculations are adjusted to water pH and lake retention time, which in reality means 10-75 g/m^3 lake water. For watercourses, 10-30 g limestone per m^3 is proposed. Lake liming, adding doses of 10-30 g/m^3, should be repeated before the pH falls below 6. For wetland liming, 0.3-1.0 t/ha drainage area is recommended, which equals 3-50 t/ha limed wetland area. A number of mathematical models can also be applied to calculate lime dosages. A survey of available models used for these purposes in the Nordic countries and North America has been published by Sverdrup and Warfvinge (1988).

The percentage of lime dissolved can be used as a measure of efficiency. There is no relationship between dosage and dissolution rate (Wilander *et al*, Chap. 5). According to Sverdrup and Warfvinge (1988), smaller particle size and low pH result in an enhanced liming efficiency.

Acid water is continuously supplied to the limed freshwater body, which means that the lime treatment must be repeated. The lake's retention time is used to determine the interval between lime treatments. The average retention time for Swedish lakes is around 1.7 years (Wilander *et al*, Chap. 5). Wetlands are generally limed within intervals of 1 or 2 years. For lake liming, Wilander *et al* (Chap. 5) showed that coarser fractions, i.e. 0-0.5 mm particle size, than the commonly used particle size of 0-0.2 mm result in a longer duration, while the dosages seem to be of minor importance.

In order to follow up the effects of aquatic liming, all Swedish liming projects are equipped with a physical-chemical monitoring programme for strategically chosen sites. Monitoring of biological effects has become more common in recent years, and not only follows the development of fish populations but also includes analyses of several other parameters, e.g. studies of lotic benthos. Besides these monitoring programmes within individual liming projects, specially designed projects aimed at detailed analyses of the effects of aquatic liming on water chemistry and biota in order to throw light upon ecological mechanisms have been conducted (Appelberg *et al* 1993, Appelberg and Aldén 1992, Stenson *et al* 1993).

5. The effects of liming

5.1. Desirable and undesirable effects

Liming of acidified waters entails considerable alteration of the ecosystem. Addition of calcium carbonate results in an instantaneous strong change in the lake water pH, alkalinity and calcium concentrations. This will bring about a series of secondary changes in the water quality, which, in turn, is the primary reason for the biological changes.

The concept of "negative and positive effects" is commonly used to describe the changes due to liming. It is, however, more appropriate to use the terms desirable and undesirable effects. Desirable effects are e.g. changes resulting in detoxication of the water and the development of an ecological community similar to that of the original one. The elimination or decrease in species favoured by acidification such as water mosses (*Sphagnum* spp.) and water boatman (Corixidae) after liming and fish introduction is also desired.

The profound knowledge and experience regarding acidification acquired over the years make it generally possible to estimate the effects of liming with high accuracy. Unexpected effects may, however, still occur. In most cases where desirable effects not have been achieved (incomplete, but not undesired results), this is attributable to an inaccurate liming strategy, inappropriate liming treatment or factors not directly connected with liming such as difficulties in recolonization.

Desirable and undesirable effects are summarized in Table 1. Most effects of aquatic liming in Sweden are expected and desired, e.g. alleviated water toxicity, return of previously present species, increased fish productivity, improved fisheries, and reduced risk of human-health effects due to a decreased mercury content in fish. Limestone treatment has also not caused major undesired effects in Norway (Baalsrud *et al* 1985, Hindar 1992). The same conclusion is drawn for liming of lakes in the Adirondacks, USA (Porcella 1989). Howells and Dalziel (1992b) distinguished between "target effects" and "non-target" effects in a report on the effects of liming in Loch Fleet, Scotland; they concluded that non-target effects were minimal.

***Table 1.** Desirable and undesirable effects of aquatic liming on water chemistry, biology and socio-economics in Sweden (cf. text).*

Desirable effects	**Undesirable effects**
Water chemistry	*Water chemistry*
* increased and often restored pH and alkalinity are general features * increased calcium content is generally experienced * improved carbon level in most cases * decreased concentration of toxic metals such as aluminium generally occurs * increased accessibility of phosphorus to algae often takes place * fixation of metals in limed wetland generally occurs	* precipitation of metals in water mixing zones may sometimes take place, e.g. downstream dosers * increased conductivity and phosphorus concentration in extremely oligotrophic waters may occur, mainly in mountain areas
Biology	*Biology*
* recolonization of species generally takes place * species numbers and diversity increase in most cases * a normalization of decomposition processes is probable in most cases * re-establishment of natural food webs seems generally to occur * enhanced fish production in most cases * re-establishment of functional groups such as shredders is a common feature * elimination of threatened species can often be prevented * decrease or elimination of species favoured by acidification generally occurs	* vegetation of limed wetlands is often considerably changed/damaged * species new to the water may appear and occasionally influence natural food webs * precipitation of metals may sometimes exert a temporary stress to species * algal blooms and mass development of plants may occasionally occur * disturbance of bird breeding on a few occasions
Socio-economics	*Socio-economics*
* fisheries are generally improved or restored * economic benefit for fisheries (recreational as well as commercial fishing) is probable in several cases * decreased mercury concentration in fish on most occasions	* increased mercury content in fish in a few lakes * turbid bath water for some time

Realizing the huge influence caused by lake liming, some undesired effects can naturally be expected. Most unwanted effects are, however, transient such as turbid bath water in lakes, or limited such as metal precipitation downstream from dosers. Other undesired effects, such as disturbance to bird breeding and lime banks downstream from dosers, can be avoided if an accurate strategy is chosen, e.g. the season for liming treatment and localization of the doser. Algal blooms and mass development of alternate-flowered water-milfoil *(Myriophyllum alterniflorum)* may occasionally occur as an unexpected and unwanted effect in limed surface waters.

In conclusion, it seems probable that undesired effects occur more frequently, the more harmful the effects of acidification have been. Seriously acidified freshwater bodies tend to experience undesired effects, as the biological regulation systems are strongly disturbed.

Deterioration of wetland ecosystems is the most significant undesirable effect of liming. Wetland liming is an effective and ecologically auspicious strategy to mitigate acidification of freshwater, but sensitive wetland species disappear and competition between species is radically altered, which results in the development of "new" plant communities (Larsson, Chap. 7). It is likely that large doses of lime bring about more harmful effects than small doses, but this has not been thoroughly documented as yet. There is a risk of increased peat decomposition (Rafstedt 1991), resulting in completely changed hydraulic conditions, and finally the wetland may prove unsuitable for liming purposes.

5.2. Water quality

In addition to an increased pH and calcium content, liming results in an increased conductivity. This effect and a number of secondary water chemical effects of liming are discussed in Wilander *et al* (Chap. 5). They stress that the secondary effects are related to the water quality before liming and the amount of lime added.

Liming results in decreased aluminium concentrations in the water. The degree of decrease depends on pH conditions and the content of organic matter in the water. The total content of aluminium is hardly changed in organic-rich lentic systems, but there is a shift to less toxic species of aluminium. Under aluminium-rich conditions, toxic species of aluminium may cause damage to fish immediately after liming (Degerman *et al*, Chap. 8). Long-lasting problems due to metal precipitation can occur in water mixing zones downstream from lime dosers, or where acid tributaries reach a limed watercourse (Lessmark *et al* 1986, Rosseland and Hindar 1991, Weatherley *et al* 1991). Liming often results in decreased concentrations of iron and manganese in the water. The results may, however, differ considerably due to variable redox potentials and the content of humic substances, which both strongly influence the solubility of the metal in question. A consequence of liming is, moreover, the shift to less toxic compounds or species of iron and manganese. Concentrations of other trace metals in the water also generally drop after liming.

The content of organic substances (water colour) decreases in polyhumic water bodies after liming, whereas increases as well as decreases can occur in waters less rich in humus. Addition of lime results in increased turbidity and a transient decrease in transparency. Changes in plankton production and precipitation of humus substances may lead to long-term decreases in the transparency of some lakes, whereas others experience an increase.

A transient elevation of the phosphorus content of the water may occur in limed lakes (Olsson and Pettersson 1993). The phosphorus content of the lime is generally 0.02-0.04% (dry weight), but the increase may also be caused by a release of phosphorus bound in the sediments or in decaying plants such as *Sphagnum* and benthic algae. The phosphorus content may, however, sometimes drop as a consequence of precipitation of complexes containing aluminium, organic substances and limestone.

In the long run, the phosphorus content often remains unchanged or decreases in limed as well as unlimed water bodies (Olsson and Pettersson 1993). This is also expected as phosphorus is more strongly fixed in acidified soils. In spite of this, the algal biomass increases with time in limed lakes (Larsson, Chap. 7). This is a consequence of the increased accessibility of phosphorus to algae, which e.g. is related to less harmful effects of metals, particularly aluminium, in limed lakes.

Concentrations of nitrogen may increase or decrease in limed freshwater. Organic nitrogen seems to reach short-lived levels of elevated concentrations, which in part is due to increased biological activity in the sediments or mineralization. Primary production is generally not large enough to cause a decrease in nitrogen concentration, but drops may sometimes occur during periods of high algal production.

Deposits of limestone occur downstream from many liming dosers. The bottom structure is changed, but the ecological implications have not been elucidated (Degerman *et al*, Chap. 8). Deposits of limestone may, however, be important to buffer against acid conditions during periods when the lime-dosing device is out of order.

A number of models can be used to predict the physical-chemical development following liming treatment rather accurately, and to prognosticate the duration of the treatment (e.g. Sverdrup and Warfvinge 1988).

Thanks to liming, the chemical conditions have been substantially improved in acidified lakes in Sweden (Fig. 7), while this is less obvious for watercourses. Studies by Bernes (1991) and Wilander *et al* (Chap. 5)

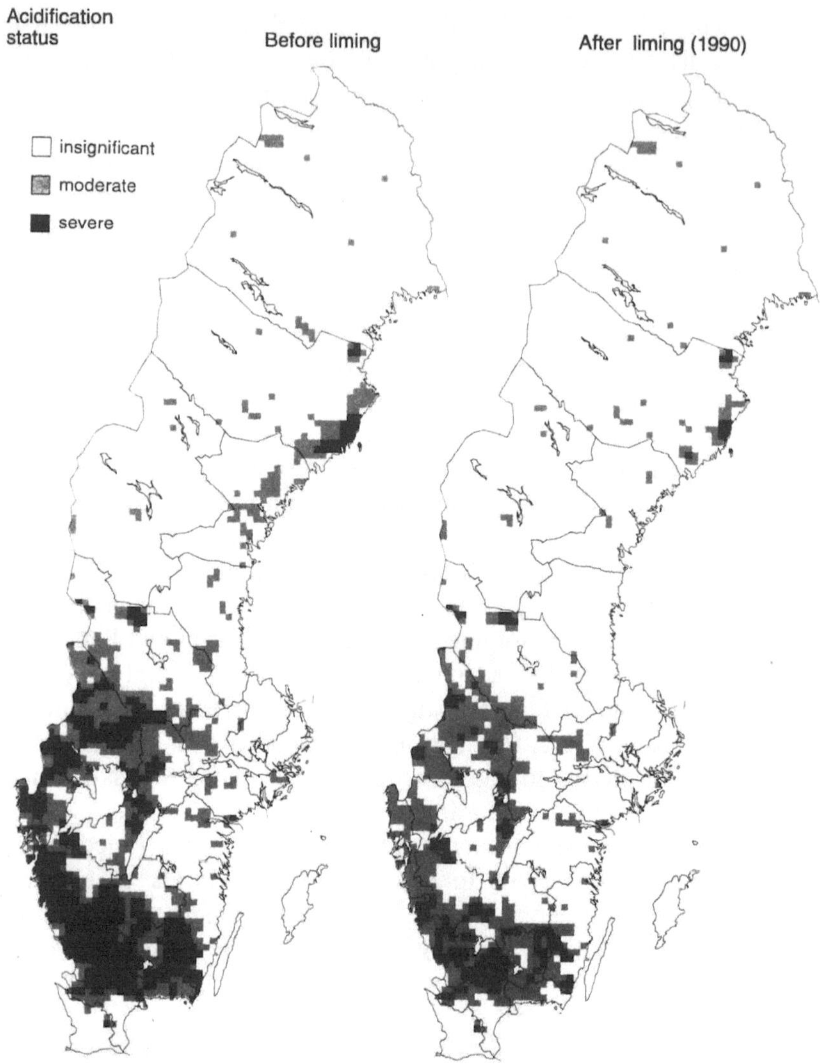

Figure 7. *Acidification status of Swedish lakes before and after liming, indicating that liming has improved ecological conditions in most regions and thousands of lakes (from Bernes 1991). County borders are indicated.*

imply, however, that further improvement can be achieved. Based on an analysis of information available in the Swedish data-base on liming, Wilander *et al* (Chap. 5) claimed that the aims with respect to water chemistry are better met at the present time than about 10 years ago, when large-scale activity began in Sweden. Studies on direct liming of lakes for the 10-year period showed that the chemical goals were reached for 90%

of the lakes 2 months after liming, but after 6 months this dropped to only 60%. The decrease with time may be attributable to reacidification of lakes with a short retention time. A limited number of the liming projects were complete failures (i.e. no change in water chemistry) or resulted in a pH and alkalinity far above the target (Bernes 1991). Harmful biological effects due to "overdosage" have not been observed, but overdosage is considered as inappropriate with respect to potential ecological effects and costs (SEPA 1988).

There are several explanations as to why the chemical goals are not reached or exceeded. Some waters are difficult to lime, such as lakes with a short retention time and watercourses experiencing large variations in flow. There are also examples of inaccurate planning, e.g. too low dosages when liming wetlands, failure to repeat lime treatment in due time, and too large doses with too long intervals. Technical problems and inadequate supervision particularly with respect to lime dosers occur as well as insufficient control of lime spreading.

5.3. Biology

As the water quality largely determines the conditions for life in aquatic habitats, liming will result in changes in the biota. This is a gradual process, which implies that short-term effects can differ substantially from long-term ones. To distinguish between short- and long-term effects is, however, not self-evident; 2 years means several generations of plankton species, whereas long-lived species such as fish need half a century to build up a similar number of generations. Appelberg and Degerman (1991) observed that changes within fish communities occurred during 16 years after liming. The prospects of reconstructing a community of the kind existing prior to acidification is to a considerable extent related to the detoxication of the water after liming.

Results presented in other chapters of the present book (Appelberg, Chap. 9, Degerman *et al*, Chap. 8, Larsson, Chap. 7) indicate a number of changes in the structure of the organism community and the complexity of the food web. Improved water quality is the primary reason for the increase in species diversity and recolonization of species. A surplus of nutrients, limited competition and predation can result in mass development of opportunistic species during the first years after liming. The long-term effects are, however, to a major extent governed by competition and predation. Given time, the community structure will stabilize in a greater or lesser degree within the new abiotic limits set by liming. Not only the structure of the ecosystem is changed, but also processes within the lake, such as increased rate of decomposition of coarse detritus as a consequence of an increased number of shredders. Microbial activity can also be favoured in the long run if the supply of organic matter increases (Gahnström, Chap. 6). Liming furthermore results in increased fish production and algal biomass. Functional groups of benthos in running waters, which have been harmfully affected by acidification, seem generally to be re-established after liming.

Transient biological "shock effects" following liming have been observed, e.g. decreased biomass of phytoplankton, decreased numbers of zooplankton and a retarded decomposition rate. Other unexpected and undesired biological effects have been described previously.

Liming generally produces the most dramatic biological changes in gravely acidified freshwater bodies (Degerman *et al*, Chap. 8). This is expected, as the conditions in these waters differ most from non-acidic ones. The considerable difference may also be attributable to pronounced disturbances of the relationship between species in heavily acidified lakes and watercourses. Changes in fish populations play a central role in the development of the organism community after liming. Liming of lakes devoid of fish will not bring about major changes in the invertebrate fauna until recolonization or reintroduction of fish has occurred. This effect on the invertebrate fauna is also true for strongly acidified lakes containing impoverished populations of fish prior to liming.

Lime treatment before serious effects on water quality and biology have occurred results in less dramatic ecological changes (Henrikson *et al* 1985, Degerman *et al*, Chap. 8). Protective liming in an initial acidification phase is thus the most appropriate strategy to preserve species and to retain an ecosystem as natural as possible (Henrikson *et al* 1990). This is in agreement with the findings of Howell *et al* (1991) and Raddum and Fjellheim (1994). Such lime-treatment operations should, however, not be allowed to result in unnaturally high pH, alkalinity or calcium concentration.

As it not is feasible to analyse the effects on flora and fauna in most limed Swedish surface waters from a financial and practical point of view, it is impossible to generally judge whether the national biological aims have been achieved. There is also a lack of appropriate criteria to determine whether liming has been biologically successful. Most of the studies that have been carried out do, however, show that species sensitive to acid conditions increase in numbers after liming and that natural relationships between species are restored gradually (Appelberg, Chap. 9, Degerman *et al*, Chap. 8, Larsson, Chap. 7). There are, furthermore, several studies indicating that liming prevents elimination of local populations of endangered species and thus loss of biodiversity.

The problems related to inadequate biological recovery and difficulties in recolonization are addressed in Sect. 7.

5.4. Mercury in fish

The mercury content in fish is usually high in acidified lakes (Mannio *et al* 1986, Håkanson 1991, Lathrop *et al* 1991). Liming mostly results in a decreased mercury content in fish (Andersson *et al* 1989, 1991, Håkanson *et al* 1990).

Based on a study on northern pike from 46 limed Swedish lakes, Meili (Chap. 16) observed that - after a period of 4-7 years - different liming strategies produced about the same effects on water chemistry, but significant differences in the mercury content of fish. Direct lake liming resulted in a drop in the mercury concentration by 20%, while wetland liming resulted in a drop of 10%. An important observation was also that a significant decline in the mercury content only occurred in lakes where lime treatment resulted in a pH increase of more than 0.5 units, and only if the pH exceeded 6. Liming to pH levels > 6.5 did not bring better results than liming to levels between 6 and 6.5.

Meili furthermore noted major differences in the response of individual lakes. Substantial increases in the mercury content in fish occurred in a lake where the lime treatment did not produce a pH > 6. This is in line with the findings of Andersson *et al* (1989), who showed that mercury concentrations in heavily acidified lakes may be lower than in moderately acidified lakes. Liming of such lakes may result in a raised mercury content in fish, and this may also be the case for lakes receiving water from limed lakes.

5.5. Socio-economic consequences

As for environmental protection activities in general, it is very difficult to give an adequate answer to the question whether liming is profitable in monetary terms. The economic value of preserving populations of species by means of liming cannot be assessed by means of currently available economic approaches, and therefore may not be compared with the cost for liming operations. The overall national goal for Swedish liming is to preserve biodiversity. In practice, however, liming is generally applied for fishery purposes, particularly recreational and household fishing. This is generally not in conflict with the overall national goals as liming for fisheries simultaneously results in desirable effects on biodiversity.

Successful chemical liming operations will favour fisheries, yet thorough studies on the effects on socio-economics have not been carried out in Sweden. Nine Swedish aquatic-liming projects have nevertheless been studied by Bengtsson and Bogelius (Chap. 18), who put the cost for liming, fishery conservation and measures promoting fisheries in relation to benefit expressed as recreational value. The difference between cost and recreational value was defined as net profit. Five of the projects were considered profitable, one broke even and three were looked upon as probably unprofitable. For one of the projects, containing several small and attractive lakes for fishing, the net profit amounted to SEK 8.5 million per year. Net profit proved to be negative, about -SEK 250,000 per year, for a large lake demanding high costs for liming and subjected to limited fishing. In the first case, the relation between profit and cost was c. 4.3 and in the second case c. 0.25. Also, commercial salmon fishing in the sea has gained economic benefit due to the positive effects on salmon spawning in limed rivers of south-western Sweden (Appelberg *et al* 1989).

A number of studies on national economy and liming of water bodies containing salmon and trout have been carried out in Norway (e.g. Navrud 1990, 1993a,b,c, Krokan 1992). It is generally established that liming is profitable for the national economy. For four selected liming projects, Navrud (1993a,b,c) assessed that each Norwegian krone on liming resulted in a net profit for national economy of 1.19, 1.69, 4-5 and 12-18 Norwegian kronen, respectively.

Liming has produced other appreciated effects, which are not easily evaluated, e.g. more concern and interest for fisheries in freshwater (Bengtsson and Bogelius, Chap. 18). Several members of fishing or nature conservation associations, and schools, have worked together in order to improve conditions in surface waters in the neighbourhood. This brings about a positive, however scarcely measurable, social and educational spin-off. It seems feasible that liming practices have contributed to enhance environmental awareness among the Swedish people. Research on problems addressing acidification and liming has also produced an increased insight into how oligotrophic aquatic ecosystems function.

The drop in mercury content in fish in limed freshwaters is favourable for fisheries. However, the question whether liming is a cost-effective measure to reduce mercury in fish is still not adequately answered.

6. Will liming restore ecosystems?

Whether liming in reality results in ecosystem restoration is frequently discussed. The question arises as to what conditions can be designated as "natural" or "pristine". Consideration of the ecosystem in a long-term historic perspective provides a different basis for an answer to the question than a perspective of 50-100 years. For clearwater lakes in southern Sweden, Renberg *et al* (1993; Fig. 8) distinguished four developmental periods with different pH's since the last glaciation in Europe: (1) gradual decrease from pH 7 to 5.5 (12,000 B.C. - 2,300 B.C.), (2) increase to pH > 6 attributable to human influence (2,300 B.C. - 1900 A.D.), (3) anthropogenic acidification to pH 4.5 (1900 - present time) and, (4) increase to pH 6-7 as a result of liming (1970 - present time). For liming purposes, the environmental conditions prevailing about 50-100 years ago - i.e. before the surface waters were significantly damaged by anthropogenic acidification - should be regarded as "natural".

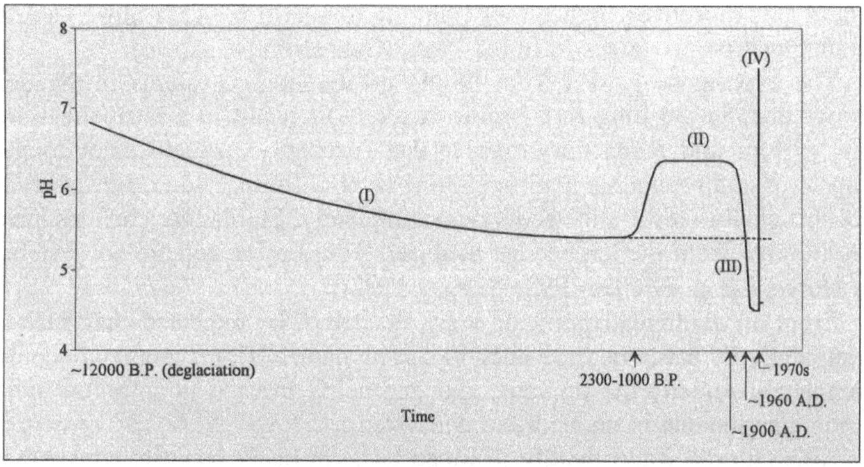

Figure 8. *Schematic diagram of pH development in acidic lakes in southern Sweden during the postglacial period (cf. text)(from Renberg et al 1993).*

The overall goal of Swedish liming activities is to preserve or restore natural environmental conditions. It is, however, impossible from a scientific point of view to determine whether liming can restore the environmental conditions prevailing 50-100 years ago. This is mainly related to the fact that nobody knows the exact ecosystem conditions before anthropogenic acidification deleteriously influenced the aquatic conditions. This lack of knowledge is also related to the fact that all ecosystems experience a developmental process even if external conditions remain more or less constant (cf. Brinck *et al* 1988). Accurate evaluations of the effects of an induced impact, such as liming, cannot be made until a certain stabilization within the ecosystem has occurred, which, in the case of aquatic liming, may take 10-15 years or even longer (Appelberg, Chap. 9). Biological documentation for such a long time has only been carried out for a few limed waters.

It is known that liming of water bodies mostly means a return of fish species found prior to the period of anthropogenic acidification (Degerman *et al*, Chap. 8). Information on other organisms is sparse, but unique well-documented material is available for non-biting midges (Chironomidae) in Lake Västra Skälsjön in southern Sweden. The species composition 5-7 years after liming was more similar to that of the phase prior to anthropogenic acidification (1943) than that of the phase of severe anthropogenic acidification (Degerman *et al*, Chap. 8, Nyberg, Chap. 11). Similar relationships have also been revealed by comparison of benthos and zooplankton in limed and non-acidified (unlimed) lakes (Degerman *et al*, Chap. 8). There are, moreover, indications pointing to a shift towards more natural decomposition processes in limed lakes (Gahnström, Chap. 6).

The experience gained from liming of aquatic ecosystems in Sweden shows that, in the long run, liming is likely to result in a normalization, viz. a flora and fauna very similar that (probably) present prior to the impact of anthropogenic acidification (Fig. 9). Liming alone, however, is not sufficient to restore the ecosystem completely. Similar conclusions have been drawn from the knowledge acquired from liming aquatic ecosystems in Norway (e.g. Hindar 1992, Sandøy 1992).

From an ecological point of view, it cannot be expected that such a comparatively plain measure such as adding neutralizing agents only could completely remedy the immense damage being exerted upon abiotic and biotic components in an acidified ecosystem.

Although the water quality of limed lakes in many respects equals that of non-acidified lakes more than acidified ones (Wilander *et al*, Chap. 5), gained knowledge shows difficulties in restoring natural environmental conditions (e.g. Weatherley 1988, Blomqvist *et al* 1993, Appelberg, Chap. 9). Liming has its limits with respect to e.g. accurate dosage and spreading in accordance with hydrological variability between years in lakes as well as watercourses. It is, furthermore, difficult to retain an acceptable water quality in littoral zones of lakes, owing to the continuous supply of acidified water via runoff from unlimed soils and tributaries. This is particularly evident during periods of heavy rainfalls or snowmelt. The littoral zone is of major importance for an appreciable number of species.

As previously described, more powerful fixation of nutrients, such as phosphorus, in the soils is a feature accompanying acidification. The consequently diminished supply of nutrients also to limed freshwater bodies may be of considerable importance to the aquatic community, as the development following liming to a very high extent is governed by lake productivity, i.e. accessibility of nutrients and their exploitation. Olsson and Pettersson (1993) claimed, however, that a decreased content of phosphorus was not evident in acidified lakes, but high aluminium concentrations could curtail the accessibility of phosphorus for algae.

The continuous process of reacidification and a sudden increase in pH and other chemical parameters when lime is added imply a more variable environment for the lake's organisms than experienced prior to the period of anthropogenic acidification. This will influence the development of the organism community. Such aquatic disturbances may abate the competitive

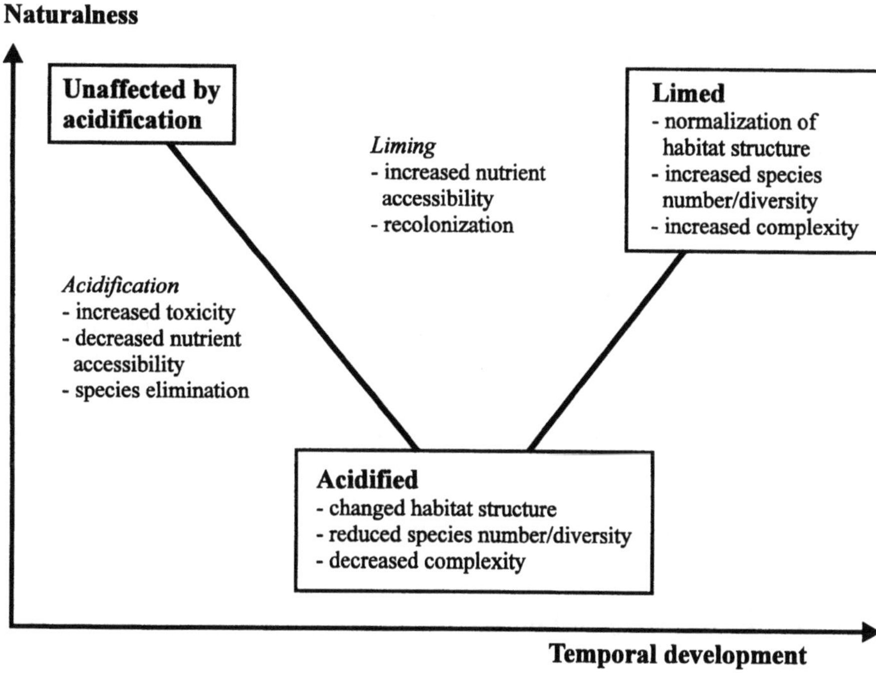

Figure 9. *Generalization of changes occurring in freshwater exosystems influenced by acidification and liming (based on an idea of Atle Hindar). Acidification results in decreased ecosystem complexity. Successful liming entails a "normalization", i.e. the complexity and function of the ecosystem will be largely similar to that of unaffected ecosystems, yet the ecosystems will not become identical and liming alone might not restore natural conditions completely. The neutralization process requires substantial time to be completed (sometimes more than 15 years is needed, cf. text), and will depend on e.g. recolonization and the restoration of food webs. The development following liming is initially largely governed by abiotic mechanisms, but in the long run biotic mechanisms such as predation and competition will form the ecosystem.*

influence of dominant species and, in this way, allow space for less competitive species, resulting in an increased species diversity (Appelberg, Chap. 9), but the opposite may also occur.

Another important explanation to the difficulties in regaining a natural ecosystem is that liming alone is not sufficient in regions where the damage caused by acidification is substantial. Recolonization of sensitive species is an extended process if the species in question has been wiped out from an entire watershed (Bergquist, Chap. 17). It is further unlikely that the original colonization sequence will be repeated, which means other prerequisites for e.g. competition and, eventually, the structure of the aquatic community. As fish are top predators and strongly influence the structure of the lower trophic levels, the recolonization potential of fish species is extremely important (Appelberg, Chap. 9).

Liming of freshwater will, thus, not completely restore original conditions, nor will other kinds of restoration of damaged environments, such as strongly eutrophic lakes. According to Brinck *et al* (1988), different final stages can be anticipated but probably none will be identical to that prior to man's impact. One reason for this is the complex of problems involved in recolonization, not only with respect to liming but also generally when damaged ecosystems are to be restored (e.g. Cairns 1988). It must be noted, that regaining a completely natural ecosystem in limed water bodies is impossible for other reasons as well. Most acidified waters in Sweden and other European countries are also influenced by human activity in other respects, particularly forestry and water regulation.

7. Complementary measures

As previously mentioned, biological recovery might be a long-term process. Slow recovery, or the lack of it, can be due to migration obstacles or a limited colonization capacity of the species (Bergquist, Chap. 17).

Complementary measures in conjunction with liming can speed up the recovery process. Bergquist (Chap. 17) defined these measures as "biological restoration", and distinguished two categories: (1) measures aimed at facilitating natural recolonization and re-establishment of species which have been eliminated, and (2) direct re-introduction of species by man. Examples are elimination of migration obstacles, biotope reconstruction, and replenishment of species occurring in restricted numbers. If eliminated species do not return after 5 - 10 years (for crayfish 8-10 years), complementary measures should be taken. It is important to use strains with genetic characteristics concordant or at least highly similar to those of species resident in the water body. Introduction of species which never have occurred in the water body in question is not considered biological restoration.

In limed watercourses, passages for fish have been constructed, dams have been removed and biotope-conservation measures have been carried out. Trout, salmon and grayling *(Thymallus thymallus)* have been re-introduced and, to a very limited degree, invertebrates such as freshwater pearl mussel.

Crayfish *(Astacus astacus)*, arctic char and roach are the most common species re-introduced to restore ecological conditions in limed lakes. Intensive fishing in order to reduce populations of certain fish species, e.g.

European perch *(Perca fluviatilis)*, has been applied to some lakes.

At present, too little time has passed to evaluate the success of these complementary measures for lakes and watercourses, but experience so far indicates that measures favouring natural recolonization and re-establishment are to be preferred over direct introduction of strains by man (Bergquist, Chap. 17). There is an obvious lack of practical experience concerning complementary measures, but experience will increase within the framework of the current work on biological restoration in Sweden. In order to take target-oriented measures, there is a need for increased knowledge and intensified research.

Supply of nutrients, especially phosphorus, is another complementary measure which can be conducted to restore the productivity of limed water bodies. Such measures, however, are used by only a few research projects (e.g. Blomqvist *et al* 1993). This is mainly related to the fact that careful analyses and calculations are required for nutrient addition into lakes, and this is not generally feasible within the large-scale liming practices in Sweden.

Adding limestone powder to the different stages of wastewater treatment is a complementary method used for liming treatment of freshwater bodies in Sweden (Dickson and Brodin, Chap. 4). Treated water from wastewater plants could also be used to supply adjacent limed lakes with nutrients in a controlled way. To make use of nutrients from waste treatment plants is naturally only conceivable for a limited number of limed lakes and watercourses.

8. Lack of knowledge and research needed

As apparent from this chapter and other chapters of this book on liming aquatic ecosystems in Sweden, the large-scale liming activities in Sweden mainly result in the ecological effects desired. For purposes of optimization of future activities, it is nevertheless interesting to address the lack of knowledge and shortcomings which naturally exist. In the analysis below, uncertainties and undesirable effects are therefore not commensurate with the overall picture of liming practices for aquatic ecosystems in Sweden.

Knowledge of the response of ecosystems to liming treatment is, as has been described, considerable, but there are, for natural reasons, still significant uncertainties with respect to long-term effects, and to some degree also short-term effects. Some people blame the lack of knowledge of long-term effects on insufficient monitoring, particularly with respect to biology, but also water chemistry to some degree. Research has not been sufficient nor co-ordinated well enough to compensate for the shortcomings with regard to effect monitoring following liming.

In the two following sections, lack of scientific knowledge and uncertainties in operational activities are described and discussed. Future liming activities can be optimized by adopting a more target-oriented scientific and developmental approach.

8.1. Ecological and socio-economic effects

It can be concluded that it is still not feasible to scientifically describe and explain the mechanisms behind different processes and causality between chemical, biochemical and biological changes in limed ecosystems in detail.

Table 2 provides an overview of topics where enhanced knowledge is important for optimization purposes. These topics can serve as a basis for discussions on continued scientific work on liming and subsequent effects (cf. also Fleischer *et al* 1990).

Table 2. *Topics related to liming of aquatic ecosystems in Sweden or other countries needing further scientific work.*

☐ **Long-term effects, i.e. more than 20 years**
As large-scale liming activities in Sweden have occurred since 1976 only, knowledge of long-term effects is lacking. This is a general uncertainty which concerns the development of populations of different species, communities and processes such as primary production and decomposition. A better understanding of these issues may e.g. serve as a basis for evaluations of the need for nutrient supply to limed ecosystems.

☐ **Concentrations and accessibility of nutrients, and primary production**
The available results do not provide a clear-cut view on development regarding nutrients such as phosphorus and nitrogen in limed water bodies. These nutrients are crucial for development, as particularly phosphorus governs primary production and the productivity of the ecosystem, and thus the structure of the plant and animal community.

☐ **Decomposition of organic matter**
Decomposition has been studied in a few liming projects only, and the results are not straightforward. Different decomposition processes are essential for the circulation of nutrients within the limed water body.

☐ **Interactions between groups of organisms or species, particularly in watercourses**
Several studies indicate that biological interactions govern the final structure of the aquatic community. This field needs further studies, particularly with respect to running water.

☐ **The structure and function of limed wetlands**
The vegetation of a limed wetland may be subjected to devastating effects. There is a lack of knowledge of e.g. the response of the peat layer and, in this conjunction, the potential long-term effects on

hydrology. In 1993, SEPA initiated a research programme addressing this issue.

- ☐ **The development of groups of organisms**
  Knowledge is very sparse on the effects of liming on groups such as bacteria and other micro-organisms, periphyton, amphibians, birds and mammals.
- ☐ **The significance of parasites and diseases**
  This is an almost unexplored field in regard to limed or acidified surface waters.
- ☐ **Metals**
  The sublethal effects of metals on organism groups are insufficiently known. For the optimization of liming, mechanisms governing the uptake of mercury in fish need to be understood.
- ☐ **Recolonization and reintroduction of species**
  Recolonization capacity and accessory mechanisms need to be further studied in order to apply appropriate complementary measures. It should be clarified, especially with respect to invertebrates, which species need complementary measures. To evaluate suitable strains for reintroduction, genetic differences between populations need to be studied in more detail.
- ☐ **Effects of "overdosage" of lime**
  Deliberate or unintentional overdosages of lime occur under certain circumstances. It is unclear how this affects organisms and communities, e.g. in ion-poor water bodies.
- ☐ **Synergic effects of acidification, nitrogen loading and liming**
  This is a topic which has rarely been studied, but is of interest for the understanding of the connection to other environmental problems, such as eutrophication, ground-level ozone and the greenhouse effect.
- ☐ **Social effects and consequences for the national economy**
  Scientific cost-benefit studies addressing national economy and social aspects are urgently needed.
- ☐ **Reacidification**
  The consequences of potential reacidification, i.e if liming is ceased, are poorly known, e.g. what will happen to metals precipitated in the sediments?

Liming treatment of Swedish surface waters is to be continued in the foreseeable future, but it is nevertheless interesting and important to discuss what will happen if liming is ceased, viz. reacidification. The few studies which have addressed this theme have e.g. shown mass development of the algae *Mougeotia* and other signs of ecosystem instability (Alenäs *et al* 1991).

Future scientific projects on liming should focus more on the ecosystem as a whole, based on general ecological principles. Better knowledge of the response of entire ecosystems or communities can provide a sound basis for the development of models useful for assessments of the long-term development after liming. Such models may also be helpful in clarifying the reasons for unexpected changes, such as algal blooms and invasion of "new" species. To disclose and understand the mechanisms behind these phenomena, better knowledge at the population and species level is required.

Liming activities are to be regarded as long-termed, considering the importance of restoring or keeping ecosystems as natural as possible. The achievements should also be better co-ordinated and, as far as possible, focus on liming projects where accurate background information is available. In addition to these activities in conjunction with liming, results gained from other kinds of environmental supervision, e.g. the national and regional programmes of environmental monitoring in Sweden, should be used as well as supervision programmes designed for discharges from industrial plants or human settlements.

In order to manage such a comprehensive programme as described above, immediate measures are needed to deal with the scarcity of lakes and watercourses that can serve as reference objects. Over extensive regions of southern and central Sweden, it is difficult to find representative surface waters, especially running water, which have not been directly or indirectly influenced by liming. This is a shortcoming for scientific and applied studies on the effects of liming. There is thus a need to increase the number of unlimed, reference waters where conditions are in accordance with typically limed water bodies.

8.2. Strategy and methodology

As apparent from the various chapters of this book, insufficient liming results on water quality and biology are often related to strategies and methods applied for liming. Therefore, for liming in practice it is very important to intensify the development of strategies and methods; some important issues where applied reasearch and development are needed are listed in Table 3.

Table 3. Issues in conjunction with liming practices requiring further applied research and development.

☐ **Strategies and methods for liming of lakes with short retention time and watercourses experiencing substantial flow variations**
It is difficult to maintain a stable water quality in such surface waters, which may result in undesired biological effects.

☐ **Strategies and methods to eliminate harmful effects of acid episodes**
There is no appropriate strategy available to mitigate the biological

effects caused by acid episodes in all parts of a lake.

- ☐ **Operational safety and systems for supervision of liming dosers**
  With the aim to reduce the risk of biological damage when dosers are out of order, improved techniques and more accurate working systems for supervision and care are required.
- ☐ **Prospects to reduce harmful effects in limed wetlands**
  As harmful effects occur in limed wetlands, there is a need to develop better liming techniques, which might imply changed particle size, spreading of pellets, and adding mixtures of lime and water (slurry). Current experiments include such techniques as well as techniques reducing wind dispersal when helicopters are used for spreading.
- ☐ **Optimization of lime-treatment practices to achieve optimum lime utilization and minimize lime deposits downstream from liming dosers**
- ☐ **Optimization of lime dosages in relation to spreading, biological effects and cost**
- ☐ **Risk assessment for spreading of diseases when species are reintroduced**
- ☐ **Measurable biological goals which can be quantified**

In order to evaluate whether liming treatment has been successful from a biological point of view, measurable goals for the effects of liming on the plant and animal community must be elaborated.

9. Conclusions

9.1. Overall conclusions

Aquatic liming supported by governmental subsidies is one of the greatest large-scale restoration and preservation activities ever applied to damaged or threatened environments in Sweden. Table 4 provides a summary of the overall conclusions drawn from aquatic liming in Sweden. It is based on information and analyses previously mentioned and from other chapters of the present book.

Table 4. *Overall conclusions from the large-scale liming activities in Sweden during the period 1976 to 1993.*

---

1. Chemical conditions of water have been significantly improved in approximately 90% of all lake liming operations, which in most cases also involve favourable effects on flora and fauna.

2. Liming mostly results in preserved or restored fisheries, which probably in several cases have been profitable to the national economy.
3. Liming mostly results in preservation and restoration of biodiversity.
4. Liming has prevented population decrease or elimination of several endangered or rare freshwater species.
5. The water quality and biology of adequately limed ecosystems will develop towards that of non-acidified ones, but liming does not result in complete recovery of the ecosystem. The normalizisation process may take several decades to be completed.
6. In spite of the abundant scientific and applied studies available, it is not yet possible to present a generalized view on certain long-term ecological effects occurring in limed ecosystems. This is also true for mechanisms involved in changes.
7. Complementary measures are sometimes necessary in limed water bodies, e.g. reintroduction of species to quickly restore an at least fairly normal species composition, or adding nutrients to restore lake productivity.
8. Lake liming to $pH > 6$ may result in a decreased mercury content in fish, but the response of different lakes varies considerably. The mercury content in fish may even increase in some lakes.
9. An overwhelming majority of all liming operations result in desired and expected effects in the aquatic ecosystem. Undesirable effect may occur, however; the most serious damage occurs in limed wetlands.
10. Liming of aquatic ecosystems is most probably necessary for at least another 50 years. This being the case even if a major reduction in acid emissions were to take place over a few years time.

9.2. Recommendations for liming of lakes and watercourses

Experience gained from liming practices, particularly in Sweden, is summarized in Table 5.

Table 5. Recommendations for liming of lakes and watercourses based on the experience from the large-scale activities in Sweden covering the period 1976 to 1993.

---

1. Accurate planning, taking the entire or at least a considerable part of the watershed into account, is important for optimization and to eliminate conflicts with other conservation interests, e.g. wetland protection.
2. Liming treatment at an early stage of acidification results in moderate ecological changes and is therefore recommended for protection of biological diversity.

3. Limited variations in water quality should be the goal. This can be achieved by means of e.g:
   - using small doses of lime frequently rather than large doses and long intervals.
   - liming of littoral zones in lakes to alleviate acid surges.
   - repeated liming treatment before there is a risk of biological damage due to reacidification.
4. Overdosage should generally be avoided, except for rare cases under special circumstances (cf. text).
5. Complementary measures to liming are often required, such as biological restoration and elimination of obstacles for the passage of fish and other animals.
6. A combinatition of different strategies such as direct lake liming, wetland liming, and liming by means of dosers will generally bring about the best ecological results.
7. Wetland liming is only to be adopted when other strategies have turned out to be insufficient. Careful evaluations and comparison of the value of the wetland versus that of the surface water are needed, as limed wetland can be permanently damaged.
8. Otherwise accurate liming operations may fail to produce desirable results if the watershed is influenced by major land-use activities, such as clear-felling and forest-soil fertilization.
9. Limestone is definitely the most adequate and, consequently, the mainly used neutralizing agent for surface waters. Other neutralizing agents can nevertheless be used.
10. Reacidification should not be allowed in limed surface waters, as there is an obvious risk to lose what has been achieved.
11. Governmental subsidies should not be granted for liming of naturally acid waters which not have become more acid because of human impact.

The overall aim of the large-scale liming activities in Sweden is to regain or retain conditions as natural as possible. Owing to the results presented in this chapter and other chapters of this book, this indicates that liming should result in natural pH and alkalinity levels, or somewhat higher if evidence is available that this is needed to avoid deleterious effects of acid surges during periods of snowmelt or heavy rainfall. This also implies that a substantial overdosage - e.g. a rise in pH and alkalinity considerably above natural levels - is mostly inappropriate from both an economic and ecological point of view, bearing in mind also the lack of knowledge on the ecological effects of overdosage.

There are, however, certain circumstances under which substantial overdosage, resulting in pH levels considerably above natural conditions, may be appropriate. If the main aim is to protect an endangered species, a safety margin may be needed to precipitate toxic metals such as aluminium more effectively, or to protect the species of concern from shortage of nutrients and competition from species indirectly favoured by lower pH levels. Thorough studies and careful evaluations need to be conducted, taking into account the benefits of an overdosage in relation to adverse effects that could be caused. In this connection, it is important to note that a considerable overdosage of lime does not seem to result in a further drop in the mercury content in fish.

In this context it seems appropriate also to present a brief account of two of the most energetically debated issues in conjunction with the large-scale liming activities in Sweden, viz. liming of possibly naturally acid lakes and watercourses, and the damage caused to limed wetlands.

The national strategy is not to lime acid surface waters if they are not further acidified because of man's activity. Several scientists claim that this strategy is not in line with the liming of a substantial number of naturally acidic surface waters - not further acidified by human activity - in northern Sweden (Bishop *et al* 1990, Bishop 1991, Kullberg *et al* 1993). They further claim that (1) liming these ecosystems may result in dramatic changes in a well-functioning ecosystem, (2) the acidic conditions in these surface waters are attributable to the high content of organic acids in the soils, and not significantly caused by deposition of acid air pollutants, (3) there is no evidence to support the idea that the region in question is acidified because of man's activity, and finally, (4) a high content of sulphur in the soils is a natural feature for several of the areas along the coast, which naturally results in very acidic conditions in surface waters. On the other hand, Bernes (1991) and Brodin (1991, 1994) have carried out model calculations indicating that several naturally acid surface waters in northern Sweden may have become even more acid because of man's impact. Degerman *et al* (1992) take the same view. Palaeoecological studies of Scottish lakes also show that naturally acid lakes may become substantially more acid when exposed to acid deposition for some decades (Batterbee *et al* 1990). Those in favour of liming surface water in these regions of northern Sweden further claim that critics have not taken into account that (1) land use, e.g. forestry, may have significantly contributed to the acidic conditions in the surface waters, and (2) the decreased pH of precipitation over the last decades produces acid episodes under periods of snowmelt which may be considerably more acid than those of natural conditions (Brodin unpubl. data).

Liming wetland as a measure to combat acidification in lakes and watercourses has received strong criticism (e.g. Löfroth 1991, Rafstedt 1991). The conflict involved in wetland liming is also observed in other countries (e.g. Woodin and Skiba 1990, Howells and Dalziel 1992b). It is a conflict within nature conservation, as certain wetlands may be partly or completely destroyed by liming, when other ecosystems, i.e. acidified surface waters, are to be saved. It should, however, not be categorically claimed that liming is not allowed on wetlands, nor that wetland liming can be carried out without taking harmful effects on the vegetation into account. In liming practices it is important that liming is preceded by appropriate planning, possibly entailing complementary studies. This is needed in order to make well-balanced evaluations of the desirable and undesirable consequences in relation to the value of the different categories of environments. Liming of wetland areas classified as valuable can generally be avoided.

9.3. Concluding remarks

This chapter and the other chapters of the present book show that the large-scale liming activities in Sweden nearly always bring about desirable effects. The cost of liming is moderate, and the ecological results are sometimes even profitable in terms of national economy. Similar conclusions are drawn from the less extensive liming activities in other countries. It is therefore recommended to continue large-scale liming activities in Swedish surface waters, to be carried out hand in hand with appropriate scientific and applied studies to increase knowledge and to optimize liming operations.

10. Acknowledgements

The authors wish to express their gratitude to colleagues in Sweden and abroad who provided valuable comments and suggestions to the present paper, in particular, Paul Andersson, Magnus Appelberg, Hans Borg, Erik Degerman, William Dickson, Mats Eriksson, Siegfried Fleischer, Jörgen Hellberg, Arne Henriksen, Jan Herrmann, Atle Hindar, Frode Kroglund, Tove Sandberg, Torbjörn Svenson, Eva Thörnelöf and Anders Wilander. Most drawings were by Johan Wihlke.

11. References

Alasaarela, E., Havu, J., Heikkinen, K. and Weppling, K. 1990. Neutralization of acidified watercourses. - In: Kauppi, P., Anttila, P. and Kenttämies, K. (Eds.) Acidification in Finland. Springer, Berlin Heidelberg New York, pp. 1117-1125.

Alenäs, I., Andersson, I.B., Hultberg, H. and Rosmarin, A. 1991. Liming and reacidification reactions of a forest lake ecosystem, Lake Lysevatten, in SW Sweden. - Water Air Soil Poll. 59: 55-77.

Andersson, P., Grahn, P., Hörnström, E., Nyberg, P. and Dahlquist, K. 1989. Kalkning och gödsling i Rammsjön och Ämten, Örebro län 1980-1987. Ett försök att minska kvicksilverhalten i fisk. [Liming and fertilization of Lake Rammsjön and Lake Ämten, county of Örebro, 1980-1987. An attempt to decrease mercury levels in fish.] - Swedish Environm. Prot. Agency 3584.

Andersson, P., Borg, H. and Kärrhage, P. 1991. Kalkningseffekter på kvicksilverhalten i fisk. Läget efter genomförda åtgärder i 19 typsjöar. [Liming effects on the mercury concentration in fish. The situation after measures in 19 study lakes.]- Swedish Environm. Prot. Agency 3883.

Appelberg, M. and Aldén, U. 1992. Integrerad uppföljning av kalkningens effekter på sjöar och vattendrag - en treårsrapport. [English summary: Integrated monitoring in limed lakes and rivers - a three year study.] - Inform. Inst. Freshw. Res., Drottningholm (4): 1-60.

Appelberg, M. and Degerman, E. 1991. Development and stability of fish assemblages after lime treatment. - Can. J. Fish. Aquat. Sci. 48: 545-554.

Appelberg, M., Degerman, E., Johlander, A. and Karlsson, L. 1989. Liming increases the catches of Atlantic salmon on the west coast of Sweden. - Nordic J. Freshw. Res. 65: 44-53.

Appelberg, M., Henrikson, B.-I., Henrikson, L. and Svedäng, M. 1993. Biotic interactions within the littoral community of Swedish forest lakes during acidification. - Ambio 22: 290-297.

Baalsrud, K., Hindar, A., Johannessen, M. and Matsow, D. (Eds.) 1985. Liming of acid water. Liming project, final report. - Dep. Environm. Dir. Nat. Managem, Norway.

Battarbee, R.W., Mason, J., Renberg, I. and Tallings, J.F. (Eds.) 1990. Palaeolimnology and lake acidification. - Phil. Trans. R. Soc., London B327.

Bernes, C. 1991. Monitor 12. Acidification and liming of Swedish freshwaters. - Swedish Environm. Prot. Agency Informs.

Bishop, K.H. 1991. Is there more to acidity in organic-rich surface waters than air pollution? An example from northern Sweden. - Vatten 47: 342-347.

Bishop, K.H., Grip, H. and O'Neill, A. 1990. The origins of acid runoff in a hillslope during storm events. - J. Hydrol. 116: 35-61.

Blomqvist, P., Bell, R.T., Olofsson, H., Stensdotter, U. and Vrede, K. 1993. Pelagic ecosystem responses to nutrient additions in acidified and limed lakes in Sweden. - Ambio 22: 283-289.

Brinck, P., Nilsson, L.M. and Swedin, U. 1988. Ecosystem redevelopment. - Ambio 17: 84-89.

Brocksen, R.W. 1991. The Living Lakes program: Five years of liming. - In: Olem, H., Schreiber, R.K., Brocksen, R.W. and Porcella, D.B. (Eds.) International lake and watershed liming practices. - The Terrene Inst., Washington D.C., USA, pp. 15-23.

Brocksen, R.W. and Wisniewski, J. (Eds.) 1988. Restoration of aquatic and terrestrial systems. - Kluwer Acad. Publ., Dordrecht Boston London.

Brocksen, R.W., Marcus, M.D. and Olem, H. 1992. Practical guide to managing acidic surface waters and their fisheries. - Lewis Publ., Chelsea, Michigan, USA.

Brodin, Y.-W. 1991. Acidification of surface waters in northern Sweden. Current status - future - countermeasures. - Vatten 47: 334-337.

Brodin, Y.-W. 1994. Acidification - an everlasting problem or is there hope? - Swedish Environm. Prot. Agency 4132.

Cairns, J. 1988. Restoration and the alternative, a research strategy. - Rest. Manag. Notes 6: 65-67.

Calderoni, A., Mosello, R. and Quirci, A. 1991. Chemical response of Lake Orta (Northern Italy) to liming. - Arch. Hydrobiol. 122: 421-439.

Degerman, E., Engblom, E., Lingdell, P.-E., Melin, E. and Olofsson, E. 1992. Försurning i fjällen? [English summary: Acidification in the mountains?] - Inform. Inst. Freshw. Res., Drottningholm (1): 1-112.

Dickson, W. 1982. Kalkningens etik .... eller vad rätt har man att kalka? [The ethics of

liming ....or who has the right to lime?] - Vatten 38: 421.
Directorate for Nature Management 1990. Håndbok i kalking av surt vann. [Guidelines for liming of acid waters] - Direct. Nat. Managem., Norway, Handbook 1 (sec. ed.).
Fleischer, S., Andersson, G., Dickson, W., Muniz, I. and Herrmann, J. 1990. Surface water acidification - effects and remedial measures. Research programme for the period 1988/89 - 1992/93. - Swedish Environm. Prot. Agency 3685.
Fleischer, S., Andersson, G., Brodin, Y.-W., Dickson, W., Herrmann, J. and Muniz, I. 1993. Acid water research in Sweden - knowledge for tomorrow? - Ambio 22: 258-263.
Forsberg, C., Morling, G. and Wetzel, R.G. 1985. Indications of the capacity for rapid reversibility of lake acidification. - Ambio 14: 164-166.
Gagen, G.J., Sharpe, W.E., DeWalle, D.R. and Kimmel, W.G. 1989. Pumping alkaline groundwater to restore a put-and-take fishery in a stream acidified by atmospheric deposition. - N. Amer. J. Fish. Manag. 9: 92-100.
Garrison, P.J. and Rose, W.J. 1990. Increasing the pH of acidic lakes by augmenting the groundwater input: an alternative to liming. - Internat. Conf. Acid deposition: its nature and impacts, Glasgow, UK, R. Soc. Edinburgh, Abstract p. 605.
Griffiths, R.W. and Keller, W. 1992. Benthic macroinvertebrate changes in lakes near Sudbury, Ontario, following a reduction in acid emmissions. - Can. J. Fish. Aquat. Sci. 49 (Suppl. 1): 25-32.
Håkanson, L. 1991. Measures to reduce mercury in fish. - Water Air Soil Poll. 55: 193-216.
Håkanson, L., Andersson, P., Andersson, T., Bengtssson, Å., Grahn, P., Johansson, J.-Å., Jönsson, C.-P., Kvarnäs, H., Lindgren, G. and Nilsson, Å. 1990. Åtgärder mot höga kvicksilverhalter i insjöfisk. Slutrapport för kvicksilverdelen av projektet: Kalkning-Kvicksilver-Cesium. [Measures to reduce mercury in lake fish. Final report from the project Liming-Mercury-Caesium.]- Swedish Environm. Prot. Agency 3818.
Hasler, A.D., Brynildson, O.M. and Helm, W.T. 1951. Improving conditions for fish in brown-water bog lakes by alkalinization. - J. Wildl. Manag. 15: 347-352.
Hegna, K. (Ed.) 1992. Vassdragskalking - strategi och effekter. Referat fra FoU-seminar i 1992. [Liming of surface waters - strategies and effects. Report from scientific seminar in 1992.] - Direct. Nat. Managem., Norway, DN-notat 1992-5.
Henriksen, A. 1992. Er kalking kommet for å bli? [Has liming come to stay?] - In: Hegna K. (Ed.) Vassdragskalking - strategi och effekter. Referat fra FoU-seminar i 1992. [Liming of surface waters - strategies and effects. Report from scientific seminar in 1992.] - Direct. Nat. Managem., Norway, DN-notat 1992-5. pp. 95-98.
Henriksen, A. and Hindar, A. 1993. Miljøtiltak i vann: Kan vi beregne kalkbehovet for Norge? [Environmental measures in water: Is it possible to calculate the need of lime for Norway?] - Liming of lakes and streams, Seminar at Haugesund, October 1993.
Henrikson, L., Nyman, H.G., Oscarson, H.G. and Stenson, J.A.E. 1985. Changes in the zooplankton community after lime treatment of an acidified lakes. - Verh. Internat. Verein. Limnol. 22: 3008-3013.
Henrikson, L., Larsson, S., Nyman, H. and Oscarson, H.G. 1990. Are we liming the lakes for the benefit of an undisturbed ecosystem or for a high pH-level. - In: Internat. Conf. Acid deposition: its nature and impacts, Glasgow, R. Soc. Edinburgh, Abstract p. 607.
Hindar, A. 1992. Hvordan kalker man økonomisk og økologisk riktig? [How to lime appropriately from an economic and ecological points of view?] - In: Hegna K. (Ed.) Vassdragskalking - strategi och effekter. Referat fra FoU-seminar i 1992. [Liming of surface waters - strategies and effects. Report from scientific seminar in 1992.] - Direct. Nat. Managem., Norway, DN-notat 1992-5, pp. 43-53.
Hindar, A. and Rosseland, B.O. 1991. Liming strategies for Norwegian lakes. - In: Olem, H., Schreiber, R.K., Brocksen, R.W. and Porcella, D.B. (Eds.) International lake and

watershed liming practices. - The Terrene Inst., Washington D.C., USA, pp. 173-192.

Howell, E.T., Coker, G., Booth, G.M., Keller, W., Neary, B., Nicholls, K.B., Tomassini, F.D., Yan, N., Gunn, J.M. and Reitveld, H. 1991. Ecosystem responses of pH 5.9 lake trout to whole lake liming. - In: Olem, H., Schreiber, R.K., Brocksen, R.W. and Porcella, D.B. (Eds.) 1991. International lake and watershed liming practices. - The Terrene Inst., Washington D.C., USA.

Howells, G. and Brown, D.J.A. 1992. Acidification and liming in context. - In: Howells, G. and Dalziel, T.R.K. (Eds.) Restoring acid waters: Loch Fleet 1984-1990. - Elsevier Appl. Sci., London, pp. 3-22.

Howells, G. and Dalziel, T.R.K. (Eds.) 1992a. Restoring acid waters: Loch Fleet 1984-1990. - Elsevier Appl. Sci., London.

Howells, G. and Dalziel, T.R.K. 1992b. The Loch Fleet project and catchment liming in perspective. - In: Howells, G. and Dalziel, T.R.K. (Eds.) Restoring acid waters: Loch Fleet 1984-1990. - Elsevier Appl. Sci., London, pp. 393-411.

Keller, W. 1992. Introduction and overview to aquatic acidification studies in the Sudbury, Ontario, Canada, area. - Can. J. Fish. Aquat. Sci. 49 (Suppl. 1): 3-7.

Keller, W., Dodge, D.P. and Booth, G.M. 1990. Experimental Lake Neutralization Program: Overview of neutralization studies in Ontario. - Can. J. Fish. Aquat. Sci. 47: 410-411.

Keller, W., Gunn, J.M. and Yan, N.D. 1992a. Evidence of biological recovery in acid-stressed lakes near Sudbury, Canada. - Environ. Poll. 78: 79-85.

Keller, W., Pitbaldo, J.R. and Carbone, J. 1992b. Chemical responses of acidic lakes in the Sudbury, Ontario, area to reduced smelter emissions, 1981-89. - Can. J. Fish. Aquat. Sci. 49 (Suppl. 1): 25-32.

Kroglund, F., Dalziel, T., Rosseland, B.O., Lien, L., Lydersen, E. and Bulger, A. 1992. Restoring endangered fish in stressed habitats ("ReFish") Project, 1988 to 1991. Progress report. - Norwegian Inst. Water Res., Rep. 30.

Krokan, P.S. 1992. Sammfunnøkonomisk lønnsomhet av kalking av Nisser. [The socio-economic value of liming River Nisser.] - Direct. Nat. Managem., Norway, Rep.

Kullberg, A., Bishop, K.H., Hargeby, A., Jansson, M. and Petersen, R.C. 1993. The ecological significance of dissolved organic carbon in acidified lakes. - Ambio 22: 331-337.

Lathrop, R.C., Rasmussen, P.W. and Knauer, D.R. 1991. Mercury concentrations in wall-eyes from Wisconsin (USA) lakes. - Water Air Soil Poll. 56: 295-307.

Lessmark, O., Degerman, E., Johlander, A. and Sjölander, E. 1986. Effekter av kalkning på fisk omedelbart nedströms doserare. [English summary: Effects of liming on fish immediately downstream of lime dosers.] - Inform. Inst. Freshw. Res., Drottningholm (5).

Lindmark, G.K. 1984. Acidified lakes: ecosystem response following sediment treatment with sodium carbonate. - Verh. Internat. Verein. Limnol. 22: 772-779.

Lindmark, G.K. 1990. Soda ash briquettes in lake sediments - a new approach in reclamation of acidified waters. - Verh. Internat. Verein. Limnol. 24: 173-178.

Lindmark, G.K. 1994. Soda ash for mitigation in acidified waters. - Verh. Internat. Verein. Limnol. 25: (in press)

Lindström, I., Samuelsson, H. and Wijk, S. 1993. Skogsmarkskalkning. Resultat från en fyraårig försöksperiod samt förslag till åtgärdsprogram. [Liming of forest soils. Results from a four-year trial period and suggestion of measure programme.] - Nat. Board Forest. 1993:6.

Löfroth, M. 1991. Våtmarkerna och deras betydelse. [English summary: The wetlands and their importance.] - Swedish Environm. Prot. Agency 3824.

Mannio, J., Verta, M., Kortelainen, P. and Rekolainen, S. 1986. The effect of water quality on the mercury concentration of northern pike *(Esox lucius* L.) in Finnish forest lakes and reservoirs. - Nat. Board Waters, Helsinki, Water Res. Inst. 65: 32-43.

Matzner, E., Khanna, P.K., Meiwes, K.J. and Ulrich, B. 1985. Effects of fertilization and

liming on the chemical conditions and element distribution in forest soils. - Plant Soil 87: 405-415.
National Rivers Authority, Institute of Terrestrial Ecology, Institute of Hydrology, University of Hydrology, University of Wales College of Cardiff, University College of Swansea and Forestry Commission 1992. LlynBrianne acid waters project. - Nat. Rivers Auth. EAR/91/7.
National Swedish Board of Fisheries 1982a. Råd och riktlinjer för kalkning av sjöar och vattendrag. [Guidelines for liming of lakes and streams.] - Inform. Nat. Swedish Board Fish. (1).
National Swedish Board of Fisheries 1982b. Råd och riktlinjer för kalkning av sjöar och vattendrag. Förslag till anbudshandlingar, avtal mm. för kalkningsentrepenader. [Guidelines for liming of lakes and streams. Recommendations for tenders, agreements etc.] - Inform. Nat. Swedish Board Fish. (2).
National Swedish Board of Fisheries and Swedish Environmental Protection Agency 1981. Kalkning av sjöar och vattendrag 1977-1981. [English summary: Liming of lakes and streams 1977-1981.] - Inform. Inst. Freshw. Res., Drottningholm (4).
Navrud, S. 1990. Nytte-kostnadsanalyse av vassdragskalking. En studie i Audna. [Cost-benefit analysis of liming of streams. A study in River Audna.] - Direct. Nat. Managem., Norway, DN-notat 1990-5.
Navrud, S. 1993a. Sammfunnsøkonomisk lønnsomhet av å kalke mindre fiskevann. Nytte og kostnadsanalyse av kalking i innlandsfiskevassdrag. En studie av Lauvann og seks vann i Gjerstadskogene. [The socio-economic value of liming in small fishing-lakes. Cost-benefit analyses of liming in fishing-lakes. A study of lake Lauvann and six lakes in the area of Gjerstadskogene.] - Direct. Nat. Managem., Norway, 1993-3.
Navrud, S. 1993b. Sammfunnsøkonomisk lønnsomhet av å kalke Audna. [The socio-economic value of liming of River Audna.] - Direct. Nat. Managem., Norway, 1993-4.
Navrud, S. 1993c. Sammfunnsøkonomisk lønnsomhet av å kalke Vegår. [The socio-economic value of liming of River Vegår.] - Direct. Nat. Managem., Norway, 1993-5.
Niinioja, R., Ahtiainen, M. and Holopainen, A.-L. 1990. Liming of the acidic Lake Valkealampi in eastern Finland: effects on water chemistry and phytoplankton. - In: Kauppi, P., Anttila, P. and Kenttämies, K. (Eds.) Acidification in Finland. Springer, Berlin Heidelberg New York, pp. 1127-1143.
Olem, H. 1991. Liming acidic surface waters. - Lewis Publ., Chelsea, Michigan, USA.
Olem, H., Schreiber, R.K., Brocksen, R.W. and Porcella, D.B. (Eds.) 1991. International lake and watershed liming practices. - The Terrene Inst., Washington D.C., USA.
Olsson, H. and Pettersson, A. 1993. Oligotrophication of acidified lakes - a review of hypotheses. - Ambio 22: 312-317.
Ormerod, S.J., Weatherley, N.S. and Merrett, W.J. 1990. Restoring acidified streams in Upland Wales: a modelling comparison of the chemical and biological effects of liming and reduced sulphate deposition. - Environm. Poll. 64: 67-85.
Porcella, D.B. 1989. Lake Acidification Mitigation Project (LAMP): an overview of an ecosystem pertubation experiment. - Can. J. Fish. Aquat. Sci. 46: 246-248.
Raddum, G.G. and Fjellheim, A. 1994. Invertebrate community changes caused by reduced acidification. - In: Steinberg, C.E.W. and Wright, R.F. (Eds) Acidification of freshwater ecosystems: implication for the future. J. Wiley & Sons, pp. 345-354.
Rafstedt, T. 1991. Effekter av kalkning. [Effects of liming.] - Nat. Swedish Board Fish., Inform. (1): 28-33.
Rask, M. 1990. Iso Valkjärvi resarch: an introduction to a multidisciplinary lake liming study. - Finnish Fish. Res. 12: 25-34.
Renberg, I., Korsman, T. and Andersson, N.J. 1993. A temporal perspective of lake acidification in Sweden. - Ambio 22: 264-271.

Rosseland, B.O. and Hindar, A. 1988. Liming of lakes, rivers and catchments in Norway. - Water Air Soil Poll. 41: 165-188.

Rosseland, B.O. and Hindar, A. 1991. Mixing zones, a fishery management? - In: Olem, H., Schreiber, R.K., Brocksen, R.W. and Porcella, D.B. (Eds.) International lake and watershed liming practices. - The Terrene Inst., Washington D.C., USA, pp. 161-173.

Rosseland, B.O. and Kroglund, F. 1992. Variasjon i toleranse overfor surt vann hos ørret. [Variation in the tolerance of acid water in trout.] - In: Hegna, K. (Ed.) Vassdragskalking - strategi och effekter. Referat fra FoU-seminar i 1992. [Liming of surface waters - strategies and effects. Report from scientific seminar in 1992.] - Direct. Nat. Managem., Norway, DN-notat 1992-5, pp. 23-29.

Sandøy, S. 1992. Innledning. [Introduction.] - In: Hegna, K. (Ed.) Vassdragskalking - strategi och effekter. Referat fra FoU-seminar i 1992. [Liming of surface waters - strategies and effects. Report from scientific seminar in 1992.] - Direct. Nat. Managem., Norway, DN-notat 1992-5, pp. 15-16.

Schreiber, R.K. and Villella, R. 1991. An overview of the U.S. Fish and Wildlife Service Mitigation Research Program. - In: Olem, H., Schreiber, R.K., Brocksen, R.W. and Porcella, D.B. (Eds.) International lake and watershed liming practices. - The Terrene Inst., Washington D.C., USA, pp. 25-40.

SEPA (Swedish Environmental Protection Agency) 1988. Kalkning av sjöar och vattendrag. [Liming of lakes and watercourses.] - Swedish Environm Prot. Agency, Off. Guidel. 88:3.

Skeffington, R.A. 1992. Soil/water acidification and the potential for reversibility. - In: Howells, G. and Dalziel, T.R.E. (Eds.) Restoring acid waters: Loch Fleet 1984-1990. Elsevier Appl. Sci., London, pp. 23-38.

Stenson, J.A.E., Svensson, J.-E. and Cronberg, G. 1993. Changes and interactions in the pelagic community in acidified lakes in Sweden. - Ambio 22: 277-282.

Sundlöf, B. 1994. Kalkning som åtgärd mot surt brunnsvatten. [Liming acidified drinking water.] - Swedish Environm. Prot. Agency, Rep. (in press)

Sverdrup, H. and Warfvinge, P. 1988. What is left for researchers in liming? A critical review of state-of-the art acidification mitigation. - Lake Reserv. Manag. 4: 87-97.

Thörnelöf, E. and Degerman, E. 1991. Lake liming in Sweden, case studies and general experiences. - In: Olem, H., Schreiber, R.K., Brocksen, R.W. and Porcella, D.B. (Eds.) International lake and watershed liming practices. - The Terrene Inst., Washington D.C., USA, pp. 41-60.

Waters, T.F. 1956. The effects of lime application to acid bog lakes in northern Michigan. - Trans. Am. Fish. Soc. 86: 329-344.

Watt, W.D. 1986. The case for liming some Nova Scotia salmon rivers. - Water Air Soil Poll. 31: 775.

Weatherley, N.S. 1988. Liming to mitigate acidification in freshwater ecosystems: a review of the biological consequences. - Water Air Soil Poll. 39: 421-437.

Weatherley, N.S., Rutt, G.P., Thomas, S.P. and Ormerod, S. 1991. Aluminium toxicity to fish in mixing zones. - Water Air Soil Poll. 55: 345-354.

Weigmann, D.L., Helfrich, L.A., Josephson, D.C. and Speenburgh, R.M. 1992. Guidelines for liming acidified lakes and ponds. - Virginia Water Resources Res. Center, Virginia Polytech. Inst. State Univ., Blacksburg, USA.

Woodin, S. and Skiba, U. 1990. Liming fails the acid test. - New Scientist 10 March 1990, pp. 50-54.

Wright, R.F. 1982. The ethics of liming. - Vatten 38: 422.

Wright, R.F., Lotse, E. and Semb, A. 1993. RAIN project. Results after 8 years of experimentally reduced acid deposition to a whole catchment. - Can. J. Fish. Aquat. Sci. 50: 258-268.

Chapter 2

Acidification of lakes and watercourses in a global perspective

Y.-W. Brodin

Swedish Environmental Protection Agency, S-171 85 Solna, Sweden

Summary

In the approximately 10,000 years since the last ice age, human activities have caused far more extensive acidification in Europe and North America than natural processes. Only 5-10% of the present acidifying load in Europe comes from natural sources. The majority of the anthropogenic load consists of sulphur emissions from electricity and heat generation. In several areas, airborne nitrogen emissions and forestry also play a significant role; nitrogen acidification is on the increase, while the role of sulphur is declining.

Acidification is one of the foremost environmental problems in Europe and North America and has caused damage to forested land, surface waters, historical buildings, monuments and technical structures. In Europe, 50,000-100,000 lakes have been affected by acidification. Over 85% of all acidified lakes and watercourses are to be found in six countries: Sweden, Norway, Canada, the USA, Scotland and Finland.

During the 1980s, emissions of acidifying substances in Europe and North America declined by 20-30%. It is estimated that a similar reduction will be possible during the 1990s if international agreements are adhered to. In order to prevent further acidification in Europe during coming decades, however, emissions of both sulphur and nitrogen must be cut by 70-75% between 1990 and 2010. The cost of achieving such a reduction is estimated at approximately 0.3-0.4% of the total annual European Gross Domestic Product (GDP), i.e. about three times as much as today.

Even if long-term reductions in emissions were to be achieved, it would take several decades for many of the lakes and watercourses to recover. Supplementary action is therefore being taken in the form of liming, which is being performed extensively in Sweden and on a smaller scale in Norway, the USA, Canada, Scotland, Finland, Wales and Germany.

Liming of Acidified Surface Waters
L. Henrikson and Y.W. Brodin (Eds.)
© Springer-Verlag Berlin Heidelberg 1995

1. Acidification: a complex problem

Acidification is not simply a new man-made phenomenon. It is also a natural process. In northern Europe and North America, most areas of land and water have become increasingly acidic during the 10,000-15,000 years since the last ice age. The main evidence of this process is to be found in studies of the composition of species of diatoms in lake sediment (Davies 1987, Charles *et al* 1990), and in theoretical studies of chemical processes in the soil (van Breemen *et al* 1983, Rechcigl and Sparks 1985). Natural acidification is primarily caused by increasing humus formation and successive leaching of nutrients from the soil.

In certain places around the world, there are freshwaters and soils that are much more acidified than any that have suffered from human impact. What is probably the most naturally acidic lake in the world is to be found in a volcanic area in eastern Japan (Yoshimura 1934). This lake has a pH of about 1.4, i.e. it is much more acidic than any of the lakes and watercourses in Europe and North America which have become acidified as a result of human activity. Despite this extreme acidity, there are a few species of plant (algae) and animal (midges) life in this Japanese lake. Certain related species are to be found in acidified lakes in Europe and North America.

Natural processes are, however, incapable of achieving such drastic and extensive acidification as has been caused by human activity during the 1900s. The impact of acidification on forests, water, human health, the economy, man-made structures, historical buildings and monuments means that it can be regarded as one of the most serious environmental problems in Europe and North America (Fig. 1). Acidification is comparable to the other severe environmental problems in these parts of the world: the transformation of the cultural landscape, eutrophication of soil, water and coastal regions, high levels of ground-level ozone, increased ultraviolet radiation as a result of stratospheric ozone depletion, and the greenhouse effect.

Seen in a global perspective, eutrophication (particularly as a result of phosphorus deposition) and water regulation measures present a more serious environmental problem for the world's lakes and watercourses than acidification. In extensive areas of northern Europe, the eastern USA and Canada, however, it is acidification which is the most serious environmental problem for surface waters. The debit side of the acidification account includes depletion of biodiversity, hazardous levels of mercury in fish and the loss or depletion of fish stocks and the consequent impact on commercial and leisure fishing.

Studies from the Netherlands and elsewhere indicate that acidification in combination with nitrogen eutrophication is a major problem in wetlands such as bogs, mires and fens (Hutchinson and Meema 1987).

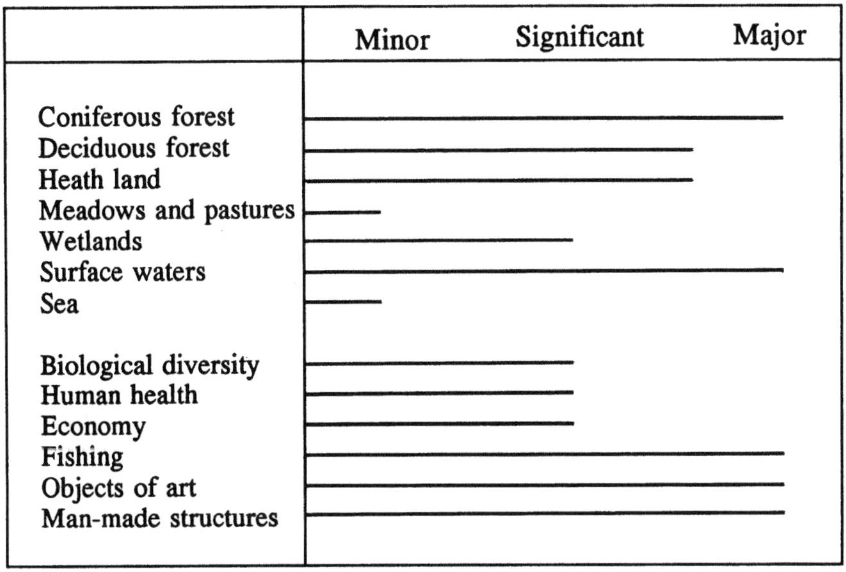

Figure 1. *Degree of severity (current status and outlook for the coming decade) for different acidification problems in Europe. Based mainly on Ellenberg (1985), McLaughlin (1985), Hutchinson and Meema (1987), WHO (1987), Merilehto et al (1988), Roberts et al (1989), Alcamo et al (1990), Chadwick and Hutton (1991) and Butlin (1991). No corresponding data are available for North America, but in most cases there are unlikely to be any major differences.*

The most spectacular consequences of acidification can be observed in coniferous forests in highland areas of Poland, Germany and the Czech Republic, where extensive areas of the forest have died (McLaughlin 1985, Nilsson and Duinker 1987). Other forested areas in Europe and North America have also suffered serious damage from acidification, although forest dieback is only to be found in certain limited areas in the vicinity of industrial installations (Hutchinson and Meema 1987, Roberts *et al* 1989). The main damage in this case is to spruce, but local populations of pine, oak and beech have also suffered severely.

Other vegetation in European forests has also been seriously affected by acidification. Far-reaching changes in the vegetation have been found in the Netherlands, England and the Kola Peninsula in Russia, as well as in the other countries mentioned above (Ellenberg 1985, Terengenza 1986, Bobbink *et al* 1992).

Soil acidification in combination with nitrogen eutrophication, high levels of sulphur dioxide and atmospheric ozone is probably the main cause of damage to trees and other vegetation in the forests (Ulrich and Pankrath 1983, Roberts *et al* 1989). Forestry methods also play an important part, and in certain cases local damage may also be related to high levels of chlorine pollution and ultraviolet radiation (Frank and Frank 1986, Ulrich and Meyer 1987).

It is not clear whether or not acidification represents a problem for naturally unforested land (raised bogs, moors, mountainous regions above the tree line, etc., which are relatively unaffected by human activity). Studies indicate, however, that acidification combined with eutrophication attributable to airborne nitrogen may have affected lime-deficient moor lands in highland areas (Heil and Diemont 1983, Degerman *et al* 1991, Lee *et al* 1992).

In terms of cultural heritage, acidifying atmospheric pollution represents a very serious problem. The damage is worst in built-up areas and affects both buildings and works of art such as statues, church windows, paintings, ceramics and textiles (Butlin 1991). Other man-made items such as iron structures, cables and water pipes are also being destroyed much faster as a result of acidifying atmospheric pollution (Kucera 1990).

Acidifying substances also affect human health. Hazardous levels of sulphur and nitrogen oxides are found in the main urban areas, primarily affecting people with asthmatic, cardiac and respiratory illness (WHO 1987, Watson *et al* 1988). Mexico City, Athens and Tokyo are among the worst hit of the world's cities in this respect. In terms of atmospheric pollution, however, carcinogenic substances are a greater health problem than acidifying substances. Acidification can also indirectly damage human health by contributing to the occurrence of high mercury levels in fish, high aluminium and copper levels in drinking water from wells, selenium deficiency and increased levels of cadmium in crops grown on acidified land (WHO 1986, Svensson *et al* 1987, Martyn *et al* 1989).

2. Acidification in the aquatic environment

Acidification of soil, lakes and watercourses is one of the most serious environmental problems in Europe and North America, as it has been for many decades. In other parts of the world, acidification damage to the natural environment is still only moderate, but there is a serious threat of extensive damage within the next few decades in many countries, such as China, Brazil and Venezuela (Rodhe and Herrera 1988, Zhao and Xiong 1988, Galloway 1989).

Southern Scotland was the first major region to be hit by extensive acidification as a result of long-distance pollutants, possibly in combination with the impact of forestry. At the time Charles Darwin was writing his books on the origin of species in the mid-19th century, the pH value of many Scottish freshwaters was already falling with resultant changes in the fauna and flora (Fig. 2). In the early years of the present century, acidifica-

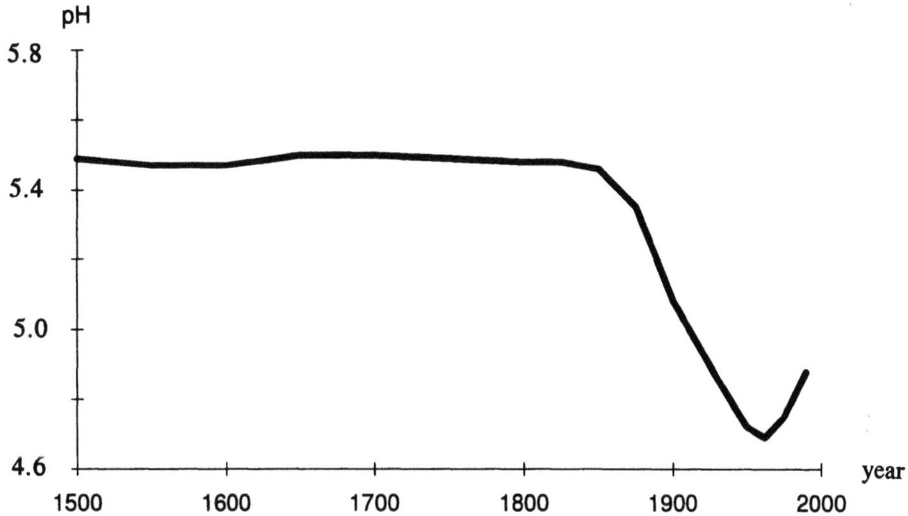

Figure 2. *Typical pH value trends for lakes in southern Scotland from 1500-1985, based on analysis of diatoms in sediment in Round Loch of Glenhead (Battarbee et al 1988).*

tion was also affecting lakes and watercourses in southern Norway and along the west coast of Sweden. The composition of the fauna and flora changed (Berge *et al* 1990, Renberg *et al* 1993) and after about 1910 fish production fell sharply in several freshwater systems (Dahl 1921, Muniz 1984).

Palaeoecological studies, particularly of diatoms in lake sediments, indicate that acidification spread across northern Europe and eastern North America between 1930 and 1960 (Charles *et al* 1990, Davies 1987). Acidification damage to lakes and watercourses did not appear on a large scale in Europe and North America until the 1960s, however. It then accelerated during the 1970s in pace with increasing emissions of acidifying sulphur dioxides.

In most cases, no major change of the situation in aquatic environments in Europe and North America took place during the 1980s, apart from lakes and watercourses in Scotland and parts of eastern Canada, where certain improvements took place - probably as a result of significant reductions of acidifying emissions from neighbouring areas (Hutchinson and Havas 1985, Battarbee *et al* 1988). Certain areas deteriorated, including smaller watercourses in certain southern areas of the Swedish-Norwegian mountains and in certain regions of eastern Finland (Huttunen *et al* 1990, Degerman *et al* 1991). The reason for this deterioration is not clear, but the delayed effects of the high sulphur load during the 1970s and climatic fluctuations may have played a significant part.

One basic reason for the extensive acidification of lakes and watercourses is to be found in the soil acidification which has occurred in the forests of Europe and North America during the twentieth century. Soil acidification is mainly caused by the successively increasing sulphur load from the air, although forestry and airborne nitrogen also play an important part in certain areas. It is not known when this process of acidification started, however. There is little direct proof of increasing soil acidification; assessments are mainly based on comparisons of soil samples from different periods during the twentieth century (von Butzke 1981, Glatzel *et al* 1984, Hallbäcken and Tamm 1986) and mathematical simulation models (Alcamo *et al* 1990, Sverdrup *et al* 1992). Most predictions based on mathematical models suggest that soil acidification is still on the increase in many parts of Europe and is affecting ever deeper layers (Ulrich and Meyer 1987, Alcamo *et al* 1990).

Ion balance analyses suggest that 100,000-200,000 of Europe's lakes and perhaps a corresponding number of watercourses have higher sulphur levels and lower alkaline levels than would have been the case had they been unaffected by atmospheric pollution (Henriksen *et al* 1988, Kortelainen and Mannio 1988, Merilehto *et al* 1988, Bernes and Thörnelöf 1990, Alcamo *et al* 1990). In 50,000-100,000 of these lakes, the change in the ion balance is so severe that their pH values have fallen, i.e. they have become acidified.

Apart from certain Mediterranean countries, Byelorussia and Iceland, freshwaters which have been acidified as a result of human activity can probably be found throughout Europe (Fig. 3). Over 80% of all the lakes which have been acidified as a result of human activity are to be found in five countries: Sweden, Norway, Scotland, Russia (mainly Kola peninsula) and Finland. In the southern areas of the first three, more than half of all the small freshwater bodies, lakes and watercourses would have been seriously damaged by acidification had they not been limed. On a local scale acidification is to be found elsewhere in Europe, in England, Germany, the Netherlands and Slovakia, for example. In North America, it mainly occurs along the east coast of the USA and Canada (Fig. 4).

The greatest number of acidified lakes and watercourses are to be found in coniferous regions with lime-deficient soils. This is the case in most areas of Sweden, Finland, Russia, the USA and Canada. Acidified freshwaters are also to be found on unforested land, such as lowland moors (England, Denmark and the Netherlands) and moors in mountainous or highland regions of Norway, Sweden, Slovakia and Scotland, for instance (Merilehto *et al* 1988). One typical feature of most acidified freshwaters is that they are deficient in nutrients. Many have probably become even more so as a result of acidification (Grahn *et al* 1974, Jansson *et al* 1986).

Figure 3. Areas of Europe where forested land and surface water have been - or suspected of being - damaged by acidification. Based mainly on data from Merilehto et al (1988), McLaughlin (1985), Alcamo et al (1990) and Hettelingh et al (1991).

Figure 4. Areas of North America where surface waters have been damaged by acidification (after Merilehto et al 1988).

Certain lakes and watercourses are only acidified for a short period of the year, during the few days or weeks when the snow melts during the spring and early summer. In certain cases, similar situations can arise during periods of heavy rain - during the autumn, for instance. These "acid surges" constitute a particular problem in smaller lakes and watercourses in snowy highland regions, particularly in Sweden, Norway and Canada (Bergmann and Welch 1985, Jacks *et al* 1986). Relatively brief acid surges can cause just as much biological damage as if the water were acidified all year round (Degerman *et al* 1991).

3. Causes of acidification

Acidification in Europe and North America is primarily caused by sulphur dioxide emissions to the air from electricity and heat generation using fossil fuels such as oil and coal. In coastal regions, shipping also makes a significant contribution to atmospheric deposition of acidifying substances. Estimates suggest that emissions of sulphur in most European countries cause 50-80% of soil and water acidification (Dickson 1986, Henriksen and Brakke 1988, Alcamo *et al* 1990, Hettelingh *et al* 1991). Annual emissions of sulphur in Europe amount to 20-25 million tonnes, of which only about 5% comes from natural sources (Finkel *et al* 1986, Alcamo *et al* 1990). World-wide sulphur emissions from natural sources are estimated to be approximately equal to those caused by human activity. The sea and active volcanoes are the main natural sources.

In an area stretching from central Europe to southern Sweden, emissions of nitrogen in the form of nitrogen oxides (mainly from motor vehicles) and ammonia (mainly from agriculture and cattle-farming) play a significant part in acidification. In the vicinity of built-up areas and agricultural areas, emissions of nitrogen may cause more than half of the acidification (Malanchuk and Nilsson 1989). There is also a clear increase in the significance of nitrogen (Henriksen and Brakke 1988).

The sulphur and nitrogen oxides which cause acidification can be dispersed widely. In total, about 5% of the acidifying load in central Finland originates in Hungary, England and France, for instance (Iversen *et al* 1991). Emissions in the USA and Canada constitute about 5% of the sulphur load in Norway and 3% of the sulphur load in Sweden (Tarrasón 1992). Further clear evidence of long-distance transport is to be found in the sulphur content of ice from Greenland. The content has increased considerably during recent decades as a result of increased emissions from North America and Europe (Finkel *et al* 1986). The third important acidifying compound, ammonia, is normally only transported half as far as sulphur and nitrogen oxides (Asman and Janssen 1987).

Germany, Russia and the UK top the list of the European countries which emit acidifying substances. Together with Poland and the Ukraine, these three countries account for more than 50% of emissions (Table 1). In most European countries, more than 40% of the airborne acidifying load comes from abroad. One important exception in this respect, however, is Great Britain, where 85% of the contribution is domestic in origin. In Iceland, Norway and Austria the domestic contribution is as low as 5-7%.

The fourth important factor in soil and water acidification is forestry. Acidification particularly attacks the upper layers of the soil and is primarily attributable to: 1) removal of base cations when trees are felled;

Table 1*. Export and import of acidifying atmospheric pollution in European countries, 1985-1991: A) the top ten sulphur exporters; B) the top ten nitrogen (nitrogen oxides and ammonia) exporters; C) countries with the highest domestic contributions to domestic sulphur load; D) countries with the lowest domestic contributions to domestic sulphur load. Based on data from Sandnes and Styve (1992).*

A)

Country	million tonnes	%
Germany	3.4	16
Russia	2.9	14
Poland	1.9	9
Great Britain	1.9	9
Ukraine	1.5	7
Italy	1.2	6
Spain	1.1	5
Romania	0.9	4
France	0.6	3
Hungary	0.6	3

B)

Country	million tonnes	%
Russia	4.6	15
Germany	4.2	13
Great Britain	3.1	10
France	2.5	8
Italy	2.1	7
Poland	1.9	6
Ukraine	1.9	6
Spain	1.3	4
Greece	0.8	3
Holland	0.8	3

C)

Country	%
Great Britain	85
Spain	73
Germany	72
Russia	60
Italy	60

D)

Country	%
Iceland	5
Norway	5
Austria	7
Switzerland	11
Albania	13

2) increased forest growth, i.e. increased uptake of base cations in the trees; 3) the use of nitrates in fertilizers and 4) spruce plantations in places where deciduous forest, meadows or agricultural land previously existed (Jones *et al* 1986, Ulrich and Meyer 1987, Berdén *et al* 1987). In northern Sweden, more than half of the acidification of the upper soil layers depends on forestry (Berdén *et al* 1987). The part played by forestry in the acidification of freshwaters is difficult to estimate, but palaeoecological data (Nilsson *et al* 1982, Burns *et al* 1984, Battarbee *et al* 1990) and mathematical models (Bernes 1991, Henriksen *et al* 1992, Kämäri *et al* 1992) suggest that in most cases it represents considerably less than half.

Temporary or lasting climate fluctuations may have an indirect impact on the acidification situation. Studies of lake sediment from northern Europe and Canada indicate that sudden climatic change can cause lasting reductions in the pH of a lake and also have a drastic impact on the composition of fauna (Walker *et al* 1990). The relatively wet climate of the 1980s may be one important explanation of the fact that the acidification situation in lakes and watercourses in the Nordic region has not improved despite reductions in acidifying substances in Europe by about 25% during the 1980s. The most important reason, however, may be that the reaction of the soil to the change in load is considerably delayed, and that it may thus take several decades before any change in the pH of lakes and watercourses takes place.

4. How much can the environment withstand?

The first detailed studies of critical loads for different ecosystems and geographical areas appeared during the 1970s. The term critical load states how much human activity an area can withstand before it can be considered to suffer damage (Gorham 1976, Almer *et al* 1978, Henriksen 1979). During the past decade, a number of studies have been performed in order to calculate the critical load for acidifying substances in forested land and lakes in Europe and North America. The results of these studies provide important material for international agreements on reductions in atmospheric pollution (Hettelingh *et al* 1991).

One shortcoming in the assessment of limit values for critical loads is the sparse knowledge concerning what different groups of organisms and individual species can survive in the long term without serious biological damage arising. Inadequate insight into the part played by forestry in the acidification of surface waters is another shortcoming. There is, however, a high degree of concordance between areas in Europe where acidification damage to plants and animals has been observed and areas where the critical load has been exceeded (Brodin, unpubl. data).

Sweden, Norway and Finland are among the countries which are considered to be least capable of withstanding an acid load (Fig. 5). About 2-5% of the lakes and watercourses in these countries are estimated to be able to withstand not more than 1-3 kg of sulphur per hectare and year (Henriksen *et al* 1992), which is 1-2 kg more than the natural load (adjusted according to data from Finkel *et al* 1986, Alcamo *et al* 1990, Iversen *et al* 1991).

In almost all countries in Europe, areas can be found where the critical load is being exceeded - which means that acidification damage has arisen or may arise within the next few decades (Fig. 6). Critical loads are particularly exceeded in areas of central Europe which have suffered the most serious soil acidification and the most extensive damage to forest and vegetation, i.e. Germany, Poland, the Czech Republic and the Netherlands (McLaughlin 1985, Ellenberg 1985).

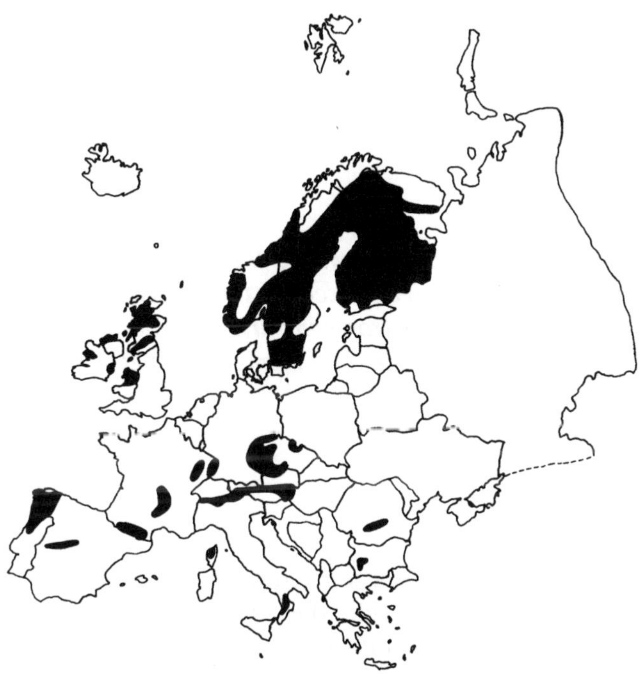

Figure 5*. Land and water areas in Europe which are particularly susceptible to acidifying load. This is based primarily on the chemical weathering capacity of soils. Based on data from Chadwick and Kuylenstierna (1991), Hettelingh et al (1991), Henriksen et al (1992) and Kämäri et al (1992).*

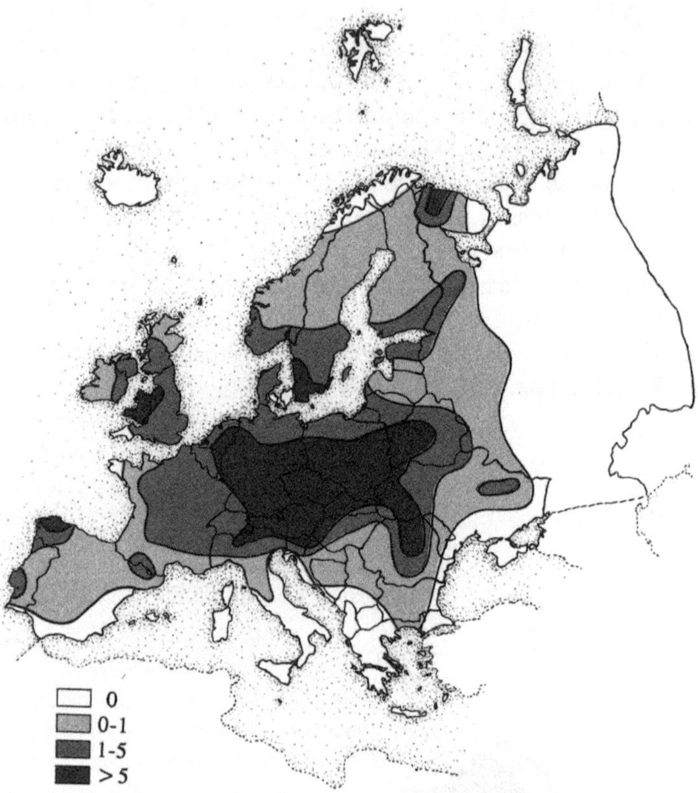

Figure 6. Land and water areas in Europe where the current acidifying deposition (oxides of sulphur and nitrogen) exceeds the critical load (the load at which serious damage to the environment arises). The acidifying impact of forestry is not included. Based on data from Alcamo et al (1990) and Hettelingh et al (1991).

5. Outlook and future action

International agreements and national decisions concerning reductions in emissions of acidifying substances provide the main source of hope concerning future improvements in the acidification situation in Europe and North America. Many countries in Europe and Canada have signed agreements to reduce sulphur emissions by at least 30% during the 1980-1993 period, with no increase thereafter, and not to increase nitrogen oxides emissions during the 1987-1994 period. Certain countries with high levels of sulphur emissions, including the USA, the UK, Spain and Poland, have not signed the agreement.

Sulphur emissions in Europe declined by about 28% from the mid-1970s, when they were at a peak, up to 1991 (Sandnes and Styve 1992). Emissions of nitrogen increased by about 5% during the same period, however. International agreements have probably played a major part in reducing sulphur emissions. Assuming that international agreements and national environmental decisions are implemented, acidifying emissions of both sulphur and nitrogen in Europe will decline by a further 20-30% during the 1990s (Amann *et al* 1992).

Predictions based on mathematical models show, however, that a reduction of 20-30% during the 1990s will not be sufficient to prevent further acidification in Europe during the next 20 years (Alcamo *et al* 1990, Kämäri *et al* 1992). Soil acidification is expected to increase throughout Europe and thousands of new freshwaters may be affected (Fig. 7).

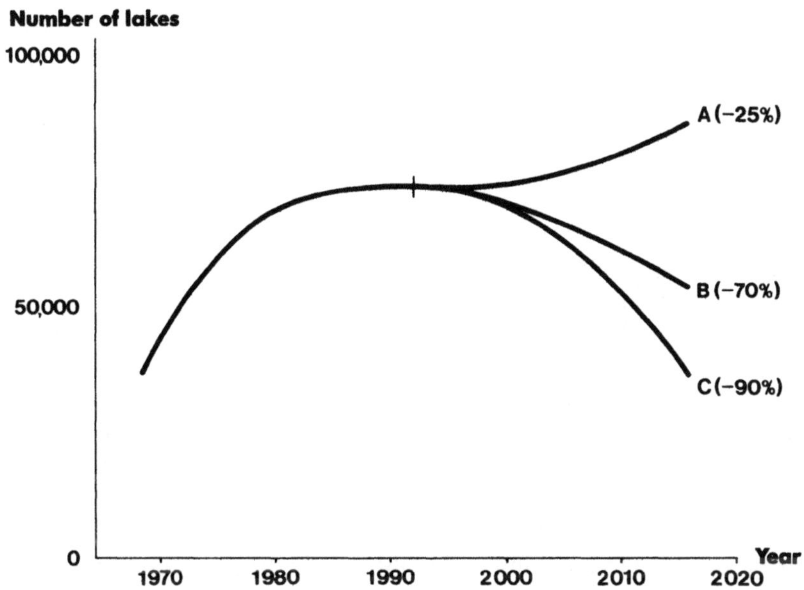

***Figure** 7. The number of acidified lakes in Europe in 1992 and subsequent decades in the following scenarios: A) reduction of acidifying emissions by 25% (1990-2005); B) reduction by 70% (1990-2005); C) reduction by 90% (1990-2005). Acidification is defined as a decline in alkalinity by at least 25%, down to below 0.05 meq/l, along with a decline in pH value by more than 0.2 units. Based on data from Kämäri (1986), Merilehto et al (1988), Alcamo et al (1990), Bernes (1991), Chadwick and Hutton (1991), Hettelingh et al (1991), Henriksen et al (1992), Kämäri et al (1992) and Brodin (unpubl. data).*

In order to prevent the situation deteriorating in any European country during the 1990-2010 period, acidifying emissions during the same period will have to decline by 65-75% (Alcamo et al. 1990, Chadwick and Hutton 1991). A reduction of this order could prevent increasing soil acidification and lead to local improvements of lakes and watercourses in Scotland, southern Norway and western Sweden, for instance, during the next 20 years. The pH in precipitation would probably increase significantly and acid surges would be much less of a problem for small watercourses and lakes in mountainous and highland regions of Norway, Sweden, Poland and Slovakia.

In order to achieve a reduction in emissions of 70% or more with undiminished energy consumption, energy must be used much more efficiently, electricity and heat generation and industrial processes must be subjected to better purification, and more than 80% of the cars in Europe must be fitted with catalytic converters. The cost of implementing these measures has been estimated at 0.3-0-4% of Europe's total annual GDP, which is three times today's cost of reducing acidifying emissions (Amann *et al* 1992).

In order to reduce the load in all areas of Europe to below the critical load values, i.e. loads at which severe damage is not expected to arise, European emissions of sulphur and nitrogen must be cut by around 85% compared with the 1990 levels. This is technically feasible, but the cost will be high, perhaps more than 1% of the European annual GDP (Chadwick and Hutton 1991). If emissions were to be reduced by 85% during the next 10 years, it would already result in major improvements during the first decades of the 21st century (cf. Kämäri *et al* 1992, Amann *et al* 1992). Predictions based on mathematical models suggest, however, that it may take more than 50 or even 100 years before several of the most severely affected areas return to their pre-1950 levels (Kämäri 1986, Alcamo *et al* 1990). There are also fears that certain ecosystems where the load is particularly severe have been damaged to such an extent that they can never be restored to anything resembling their original condition. One example is to be found in the area of dead forest in the "black triangle" of the northern Czech Republic, south-western Poland and north-western Germany (Ulrich and Pankrath 1983, McLaughlin 1985).

International agreements on reductions in emissions of acidifying substances will hopefully lead to such far-reaching reductions in sulphur and nitrogen emissions in northern Europe and North America that there will at least be no increase in acidification. A new agreement on sulphur was reached during 1994. An agreement on nitrogen may be completed by 1995/1996.

The knowledge that it will take many decades before Europe can recover appreciably from acidification highlights the need for other action. Even if sweeping reductions in emissions were to be achieved, recovery would take a long time. The best supplementary action is to lime the soil, lakes and watercourses. This will speed up the recovery of damaged freshwaters and also prevent damage from arising in freshwaters which are under threat but not yet seriously affected, pending such reductions (Henrikson and Brodin, Chap. 1). Large-scale liming of lakes and watercourses has at present only been undertaken in Sweden (e.g. Bernes 1991, Henrikson and Brodin, Chap. 1). Norway, the USA, Canada, Finland, Scotland, Wales, Poland and Germany are also liming their lakes and watercourses, but on a fairly limited scale (see for example Hindar and Rosseland 1988, Alasaarela *et al* 1990, Olem 1991). Acidified forested land is also being limed in parts of Germany and on a small scale in Sweden and Finland.

6. References

Alasaarela, E., Havu, J., Heikkinen, K. and Weppling, K. 1990. Neutralization of acidified watercourses. In: Kauppi, P., Anttila, P. and Kenttämies, K. (Eds.) Acidification in Finland. Springer, Berlin Heidelberg New York, pp. 1117-1125.

Alcamo, J., Shaw, R. and Hordijk, L. 1990. The RAINS model of acidification - science and strategies in Europe. Kluwer Acad. Publ., Dordrecht.

Almer, B., Dickson, W., Ekström, C. and Hörnström, E. 1978. Sulphur pollution and the aquatic ecosystem. In: Nriagu, I.O. (Ed.) Sulphur in the environment. - J. Wiley & Sons, New York, pp. 271-286.

Amann, M., Bertok, I., Cofala, J., Klaasen, G. and Schöpp, W. 1992. Strategies for reducing sulfur dioxide emissions in Europe. - Internat. Inst. Applied. Syst. Anal. Status Rep. May 14, 1992.

Asman, W.A.H. and Janssen, A.J. 1987. A long-range transport model for ammonia and ammonium in Europe. - Atmos. Environ. 21: 2099-2119.

Battarbee, R.W., Flower, R.J., Stevenson, A.C., Jones, V.J., Harriman, R. and Appleby, P.G. 1988. Diatom and chemical evidence for reversibility of acidification of Scottish lochs. - Nature 332: 530-532.

Battarbee, R.W., Mason, J., Renberg, I. and Tallings, J.F. (Eds.) 1990. Palaeolimnology and lake acidification. - Phil. Trans. R. Soc., London B327.

Berdén, M., Nilsson, S.I., Rosén, K. and Tyler, G. 1987. Soil acidification - extent, causes and consequences. - Swedish Environm. Prot. Agency 3292.

Berge, F., Brodin, Y.-W., Cronberg, G., El-Daoushy, F., Høeg, H.I., Nilssen, J.P., Renberg, I., Rippey, B., Sandøy, S., Timberlid, A. and Wik, M. 1990. Palaeolimnological changes related to acid deposition and land-use in the catchment of two Norwegian softwater lakes. - Phil. Trans. R. Soc., London B327: 385-389.

Bergmann, M.A. and Welch, H.E. 1985. Spring meltwater mixing in small arctic lakes. - Can. J. Fish. Aquat. Sci. 42: 1789-1798.

Bernes, C. 1991. Monitor 12 - Acidification and liming of Swedish freshwaters. - Swedish Environm. Prot. Agency Informs.

Bernes, C. and Thörnelöf, E. 1990. Present and future acidification of Swedish lakes: model calculation based on extensive survey. In: Kämäri, J. (Ed.) Environmental impact models to assess regional acidification. - Kluwer Acad. Publ., Dordrecht, pp. 83-107.

Bobbink, R., Boxman, D., Fremstad, E., Heil, G., Houdijk, A. and Roelofs, J. 1992. Nitrogen eutrophication and critical loads for nitrogen based upon changes in flora and

fauna in (semi-)natural terrestrial ecosystems. In: Grennfelt, P. and Thörnelöf, E. (Eds.) Critical loads for nitrogen. - Nordic Council Ministers Miljørep. 1992:41, pp. 111-159.
Burns, J.C., Coy, J.S., Tervet, D.J., Harriman, R., Morrison, B.R.S. and Quine, C.P. 1984. The Loch Dee project; a study of the ecological effects of acid precipitation and forest management on an upland catchment in southwest Scotland. - Fish. Managem. 15: 145-167.
Butlin, R.N. 1991. Effects of air pollutants on buildings and materials. - Proceed. R. Soc. Edinburgh, 97B: 255-272.
Chadwick, M.J. and Hutton, M. 1991. Acid depositions in Europe - environnmental effects, control strategies and policy options. Information Press, Oxford.
Chadwick, M.J. and Kuylenstierna, J.C.I. 1991. The relative sensitivity of ecosystems in Europe to acidic depositions. A preliminary assessment of the sensitivity of terrestrial and aquatic ecosystems. - Perspect. Ecol. 1: 71-93.
Charles, D.E., Battarbee, R.W., Renberg, I., van Dam, H. and Small, J.P. 1990. Paleo-ecological analysis of lake acidification trends in North America and Europe using diatoms and chrysophytes. In: Norton, S.A., Lindberg, S.E. and Page, A.L. (Eds.) Acid precipitation. Vol. 4. Soils, aquatic processes, and lake acidification. Springer, Berlin Heidelberg New York, pp. 207-276.
Dahl, K. 1921. Undersökelse över örretens utdöen i det syd-vestlig Norges fjellvande [Studies on the decline of trout in the mountain waters of south-western Norway.] - Norsk Jaeger Fiskeforen. Tidsskr. 49: 249-267.
Davies, R.B. 1987. Palaeolimnological diatom studies of acidification of lakes by acid rain: an application of quaternary science. - Quatern. Sci. Res. 6: 147-163.
Degerman, E., Engblom, E., Lingdell, P.-E., Melin, E. and Olofsson, E. 1991. Försurning i fjällen? [English summary: Acidification in the mountains?] - Inform. Inst. Freshw. Res., Drottningholm (2).
Dickson, W. 1986. Critical load of nitrogen on surface waters. In: Nilsson, J. (Ed.) Critical loads for sulphur and nitrogen. - Nordic Counc. Minst., Miljørap. 1986:11: 210-221.
Ellenberg, H. 1985. Veränderungen der Flora in Mitteleuropa unter dem Einfluss von Düngung und Immissionen. - Schweiz. Z. Forstwes. 136: 19-39.
Finkel, R.C., Langway, C.C. and Claussen, H.B. 1986. Changes in precipitation chemistry at Dye 3, Greenland. - J. Geophys. Res. 910: 9849-9855.
Frank, H. and Frank, W. 1986. Photochemical activation of chloroethenes leading to destruction of photosynthetic pigments. - Experientia 42: 1267-1269.
Galloway, J.N. 1989. Atmospheric acidification: projections for the future. - Ambio 18: 161-166.
Glatzel, G., Kilian, W., Sterba, H. and Stöhr, D. 1984. Waldbodenversauerung in Österreich - Veränderungen der pH-werte von Waldböden während der letzten Dezennien. - Österreichischer Forstverein. Forschungsber., Vienna, Rep. 1984.
Gorham, E. 1976. Acid precipitation and its influence upon aquatic ecosystems- an overview. - Water Air Soil Poll. 6: 457-481.
Grahn, O., Hultberg, H. and Landner, L. 1974. Oligotrophication - a self accelerating process in lakes subjected to excessive supply of acid substances. - Ambio 3: 93-94.
Hallbäcken, L. and Tamm, C.-O. 1986. Changes in soil acidity from 1927 to 1982-1984 in a forest area of south-west Sweden. - Scand. J. Forest Res. 1: 219-232.
Heil, G.W. and Diemont, W.M. 1983. Raised nutrient levels change heathland into grassland. - Vegetatio 53: 113-120.
Henriksen, A. 1979. A simple approach for identifying and measuring acidification of freshwater. - Nature 278: 542-545.
Henriksen, A. and Brakke, D.F. 1988. Increasing contribution of nitrogen to the acidity of surface waters in Norway. - Water Air Soil Poll. 42: 183-201.

Henriksen, A., Lien, L., Traaen, T.S., Sevaldrud, I.S. and Brakke, D.F. 1988. Lake acidification in Norway - present and predicted chemical status. - Ambio 17: 259-266.

Henriksen, A., Kämäri, J., Posch, M. and Wilander, A. 1992. Critical loads of acidity to Nordic surface waters. - Ambio 21: 356-363.

Hettelingh, J.-P., Downing, R.J. and de Smet, P.A.M. 1991. Mapping critical loads for Europe. - Coord. Cent. Effects, Bilthoven, The Netherlands, Techn., Rep. 1.

Hindar, A. and Rosseland, B.O. 1988. Liming acidic waters in Norway: national policy and research and development. - Water Air Soil Poll. 41: 17-24.

Hutchinson, T.C. and Havas, M. 1985. Recovery of previously acidified lakes near Coniston, Canada, following reductions in atmospheric sulphur and metal emissions. - Water Air Soil Poll. 28: 319-333.

Hutchinson, T.C. and Meema, K.M. (Eds.) 1987. Effects of atmospheric pollutants on forests, wetlands and agricultural ecosystems. - Springer, Berlin Heidelberg New York, pp. 267-276.

Huttunen, P., Kenttämies, K., Liehu, A., Liukkonen, M., Nuotio, T., Sandman, O. and Turkia, J. 1990. Palaeoecological evaluation of the recent acidification of susceptible lakes. In: Kauppi, P., Anttila, P. and Kenttämies, K. (Eds.) Acidification in Finland. Springer, Berlin Heidelberg New York, pp. 1071-1090.

Iversen, T., Halvorsen, N.E., Mylona, S. and Sandnes, H. 1991. Calculated budgets for airborne acidifying components in Europe. - EMEP MSC-W (Norway) 1/91.

Jacks, G., Olofsson, E. and Werme, G. 1986. An acid surge in a well-buffered stream. - Ambio 5: 282-285.

Jansson, M.G., Persson, G. and Broberg, O. 1986. Phosphorus in acidified lakes: the example of Lake Gårdsjön, Sweden. - Hydrobiologia 39: 81-96.

Jones, V.J., Stevenson, A.C. and Battarbee, R. 1986. Lake acidification and the land-use hypothesis; a mid-postglacial analogue. - Nature 322: 157-158.

Kämäri, J. 1986. Critical deposition limits for surface waters assessed by a process-oriented model. In: Nilsson, J. (Ed.) Critical loads for nitrogen and sulphur. - Nordic Counc. Minist., Miljørap. 1986:11: 121-142.

Kämäri, J., Amann, M., Brodin, Y.-W., Chadwick, M.J., Henriksen, A., Hettelingh, J.-P., Kuylenstierna, J.C.I., Posch, M. and Sverdrup, H. 1992. The use of critical loads for the assessment of future alternatives for acidification. - Ambio 21: 377-386.

Kortelainen, P.J. and Mannio, J. 1988. Natural and anthropogenic acidity sources for Finnish lakes. - Water Air Soil, Poll. 42: 341-352.

Kucera, V. (Ed.) 1990. Effects of water and soil acidification on corrosion. - Nordic Counc. Minist., Miljørap. 1990:9.

Lee, J.A., Caporn, S.J.M. and Read, D.J. 1992. Effects of increasing nitrogen deposition and acidification on heathlands. In: Schneider, T. (Ed.) Acidification research: evaluation and policy applications. Elsevier Sci. Publ. B.V., Amsterdam, pp. 97-106.

Malanchuck, J. and Nilsson, J. (Eds.) 1989. The role of nitrogen in the acidification of soils and surface waters. - Nordic Counc. Minist., Miljørap. 1989:10.

Martyn, C.N., Osmond, C., Edwardson, J.A., Barker, D.J.P., Harris, E.C. and Lacey, R.F. 1989. Geographical relation between Alzheimer's disease and drinking water. - Lancet (1989): 59-62.

McLaughlin, S.B. 1985. Effects of air pollution on forests: a critical review. - J. Air Poll. Cont. Assoc. 35: 512-534.

Merilehto, K., Kenttämies, K. and Kämäri, J. 1988. Surface water acidification in the ECE region. - Nordic Counc. Minist., Miljørap. 1988:14.

Muniz, I.P. 1984. The effects of acidification on Scandinavian freshwater fish fauna. - Phil. Trans. R. Soc. London B305: 517-528.

Nilsson, S. and Duinker, P. 1987. The extent of forest decline in Europe. A synthesis of

survey results - Environment 29: 4.
Nilsson, S.I., Miller, H.G. and Miller, J.D. 1982. Forest growth as a possible cause of soil and water acidification: an examination of the concepts. - Oikos 39: 40-49.
Olem, H. 1991. Liming acidic surface waters. Lewis Publ., Chelsea, Michigan, USA.
Rechcigl, J.E. and Sparks, D.L. 1985. Effect of acid rain on the soil environment: a review. - Comm. Soil Sci. Plant. Anal. 16: 653-680.
Renberg, I., Korsman, T. and Anderson, J. 1993. A temporal perspective of lake acidification in Sweden. - Ambio 22: 264-271.
Roberts, T.M., Skeffington, R.A. and Blank, L.W. 1989. Causes of type 1 spruce decline in Europe. - Forestry 62: 179-222.
Rodhe, H. and Herrera, R. (Eds.) 1988. Acidification in tropical countries. - Scope Rep. 36. J. Wiley & Sons, Chichester, England.
Sandnes, H. and Styve, H. 1992. Calculated budgets for airborne acidifying components in Europe, 1985, 1987, 1988, 1989, 1990 and 1991. - EMEP/MSC-W (Norway) 1/92.
Svensson, B.G., Björnham, Å., Schütz, A., Lettevall, U., Nilsson, A. and Skerfving, S. 1987. Acidic deposition and human exposure to toxic metals. - Sci. Tot. Environ. 67: 101-115.
Sverdrup, H., Warfvinge, P., Frogner, T., Håøya, A.O., Johansson, M. and Andersen, B. 1992. Critical loads for forest soils in the Nordic countries. - Ambio 21: 348-355.
Tarrasón, L. 1992. Contributions to sulphur background deposition over Europe: results for 1988. - EMEP/MSC-W (Norway) Note 5/92.
Terengenza, N.J.C. 1986. Acidification effects on the flora of Cornwall. - Cornwall Conserv. Rev. 1/86.
Ulrich, B. and Meyer, H. 1987. Chemischer Zustand der Waldböden Deutschlands zwischen 1920 und 1960, Ursachen und Tendenzen seiner Veränderung. - Ber. Forsch. Zent. Waldökosyst., Univ. Göttingen, Germany B6.
Ulrich, B. and Pankrath, S. (Eds.) 1983. Effects of accumulation of air pollutants in forest ecosystems. Reidel Publ. Comp., Dordrecht.
Van Breemen, N., Mulder, J. and Driscoll, C.T. 1983. Acidification and alkalinization of soils. - Plant Soil 75: 283-308.
Von Butzke, H. 1981. Versauern unsere Wälder? Erste Ergebnisse der Überprüfung 20 Jahren alter pH-Wert-Messungen in Waldböden Nordrhein-Westfalens. - Der Forst- & Holzwirt 21: 542-547.
Walker, I.R., Smol, J.P., Engstrom, D.J. and Birks, H.J.B. 1990. An assessment of Chironomidae as quantitative indicators of past climatic change. - Can. J. Fish. Aquat. Sci. 48: 975-987.
Watson, A.Y., Bates, R.R. and Kennedy, D. 1988. Air pollution, the automobile, and public health. National Acad. Press, Washington, USA.
WHO 1986. Health impact of acidic deposition. - Sci. Tot. Environm. 52: 157-187.
WHO 1987. Air quality guidelines for Europe. - WHO Reg. Publ., European Ser. 23.
Yoshimura, S. 1934. The most acid-water lake of the world; Katanuma by Narugo Hot Springs. - Science 4: 498-499.
Zhao, D. and Xiong, J. 1988. Acidification in south-western China. In: Rodhe, H. and Herrera, R. (Eds.) Acidification in tropical countries. - Scope Rep. 36. J. Wiley & Sons, Chichester, England, pp. 317-346.

Chapter 3

Acidification of Swedish freshwaters

Y.-W. Brodin

Swedish Environmental Protection Agency, S-171 85 Solna, Sweden

Summary

Estimates indicate that at least 20% of plant and animal species have died out in the 15,000-25,000 European lakes where the pH has fallen by more than 0.5 units as a result of anthropogenic acid deposition. Severe depletion of species is also to be found in watercourses in several areas of Europe, but there is a lack of data upon which to base quantitative assessments.

No species has yet been reported extinct as a result of acidification, but it is important to bear in mind that few data are available upon which to determine whether or not a species has died out or is under threat.

In Sweden, at least 20% of species has probably disappeared from 5,000-7,000 acidified lakes and thousands of acidified watercourses. The number of lakes would probably have been higher (7,000-9,000) without liming. Significantly acidified lakes and watercourses can be found almost throughout the country. The most severely affected waters are to be found in south-western Sweden where about 80% of waters would have been significantly damaged by acidification had they not been limed.

Acidified lakes and watercourses are often characterised by a sharp increase in filamentous algae and *Sphagnum* moss, and a significant decline in the number of species of phytoplankton. The number of animal species is declining in all groups, in lakes and watercourses alike, mainly among molluscs, bivalves, crustaceans and mayflies. At a pH of below 5, only a few species remain in these groups. In acidified lakes, organic material decomposes more slowly on the surface of the sediment, but it is unclear whether this also applies within the sediment.

Lakes and watercourses where the critical load for acid deposition has been exceeded can be found throughout Sweden. This indicates that there is a clear risk of acidification affecting more waters than it already has done. Estimates show that if the acidifying load remains unchanged, the

Liming of Acidified Surface Waters
L. Henrikson and Y.W. Brodin (Eds.)
© Springer-Verlag Berlin Heidelberg 1995

number of significantly acidified lakes and watercourses in Sweden may rise by about 50% within the next 20 years. In order to prevent Swedish surface waters from deteriorating, emissions in Europe must be reduced by 60-70%. This can be compared with the 30% reduction in emissions which will result from current international agreements and the action which has been planned at a national level.

In order to prevent increasing depletion of fauna and flora and rising mercury levels in fish, large-scale liming operations to prevent acidification of Swedish freshwaters must continue for at least the next 30-50 years.

1. Introduction

Acidification of Swedish lakes and watercourses already began about 10,000 years ago at the end of the last ice age. By the mid-19th century, the pH levels of more or less pristine lakes had fallen by up to one unit, although the process probably became slower each millennium (Fig. 1). This natural process of acidification is primarily caused by leaching of

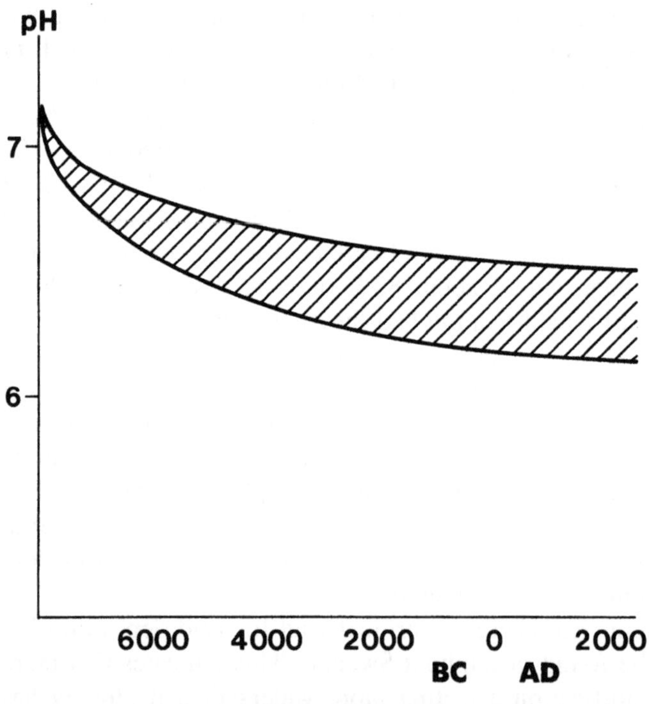

Figure 1. Natural pH trends in lakes in the Swedish forest (not including areas rich in calcium) during the 10,000 years since the last ice age, based mainly on studies of diatoms in lake sediment.

accumulated nutrients from the soil and an increasing buildup of humus in forest land (van Breemen and van Dijk 1988). In the text below "acidification" means acidification due to human impact unless otherwise stated.

Without anthropogenic impact, changes in the pH levels of Swedish lakes and watercourses during the present century would probably have been too small to be detected even by the most sensitive instruments. In reality, air pollution and, to some extent, forestry have caused the pH levels of many waters to fall by about one unit, particularly in southern Sweden - thousands of times more than that caused by natural processes. The pH decrease is generally considerably more pronounced than the pH increase due to land use occurring some 100/1000 years ago in several lakes, at least in southern Sweden (Renberg *et al* 1993).

Sweden is one of the countries which has been hardest hit by acidification. Together with Norway, Sweden has more acidified lakes than any other country in the world. Acidification has probably caused ecological changes in almost half of Sweden's 85,000 lakes and probably a corresponding proportion of watercourses (Brodin 1994).

One basic cause of acidification of lakes and watercourses is the extensive acidification of forest land in the southern half of Sweden (Hallbäcken and Tamm 1986, Falkengren-Grerup 1987). In south-western Sweden, the soil in both deciduous and coniferous forests has become more than ten times more acid in many cases - i.e. the pH has fallen by more than one unit. Soil acidification has given rise to forest damage and significant changes in the vegetation (Torstensson and Liljelund 1989). Because of new forestry methods and the high nitrogen load, however, the productivity of the forest is higher than ever before. If emissions of acidifying substances are not reduced, soil acidification and eutrophication by nitrogen could eventually lead to significant forest damage and possibly even to forest dieback in southern Sweden (Nihlgård 1990, Sverdrup *et al* 1992). Over and above the far-reaching ecological damage this would cause, it would also constitute a major loss to the national economy.

Extensive research on acidification has been under way in Sweden since the late 1960s. It was Swedish researchers who first clarified the relationship between transboundary air pollution and increasing acidification (Odén 1968). Research since then has resulted in several thousand scientific publications which have played a major part in securing comprehensive action to reduce acidifying emissions in Sweden. Swedish research has significantly influenced emission reductions in other countries in Europe and North America. Fleischer *et al* (1993) have presented comprehensive results of Swedish research into freshwater acidification.

2. Acidifying load

Together with Norway and Finland, Sweden is one of the countries which is least able to withstand the load from acidifying substances. Rocks and soil which weather very slowly and thus have a relatively poor ability to produce alkaline substances to provide a buffer against the anthropogenic load are found all over Sweden.

The critical load for long-term impact on the most susceptible lakes and watercourses in Sweden is thought to correspond to about 2-3 kg sulphur per hectare and year (Dickson 1986, Henriksen *et al* 1992, Kämäri *et al* 1992).

Lakes and watercourses where the current load of acidifying substances exceeds the critical load can be found throughout Sweden (Fig. 2). In south-western Sweden, more than every second lake and watercourse is

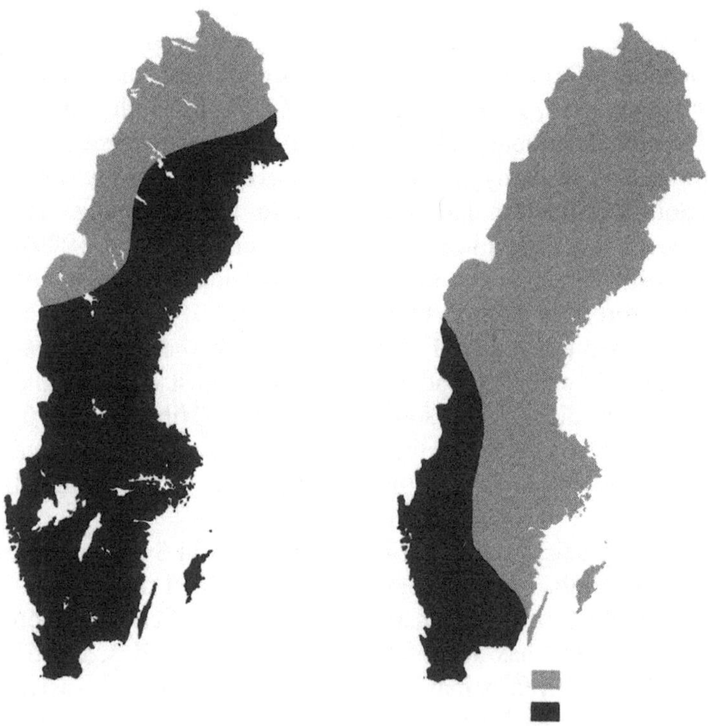

***Figure** 2. Areas in Sweden where 10-50% and > 50% of lakes and watercourses are exposed to an acid load above the critical load for long-term impact, showing the situation in 1990-1991 and taking the effects of airborne deposition and forestry into account (cf. Brodin 1994).*

subjected to more than the critical load - up to eight times as much in certain cases - although there is considerable local variation. At high altitudes in the northern Swedish mountains the load is often several times higher than in more low-lying regions (Degerman *et al* 1991). The load on spruce forests may be up to 70% higher than on other types of forest and several times higher than on unforested land (Ivens *et al* 1989, Lövblad *et al* 1992). This means that surface waters which receive large quantities of water from mountain or highland spruce forests are often far more exposed to acidifying loads than other waters in the same geographical area.

The load of acidifying substances and their potential environmental impact can also be related to the pH levels of precipitation. Prior to the impact of human activities, the pH of precipitation over northern Europe was probably about 5.5-6.0 (McNeely and Gummer 1984). Studies suggest that serious biological acidification damage may occur if the pH of precipitation falls below 5.0 (Dickson 1990, Degerman *et al* 1991). It is therefore worth noting that the pH levels in precipitation over Sweden are below 5.0 except in certain minor areas in the mountains along the Norwegian border.

In most cases, more than 70% of the acidification of lakes and watercourses in Sweden is probably caused by sulphur deposition (based on Dickson 1988, Henriksen and Brakke 1988, Alcamo *et al* 1990, Bernes 1991). About 65% of sulphur deposition in Sweden can be traced to emissions in countries elsewhere in Europe. Germany, Poland and the UK are the main contributors to sulphur deposition in Sweden (Table 1). The domestic Swedish share of the load is 10-15%. It is impossible to determine the country of origin of 10-15% of the deposition in southern Sweden and 30-40% in northern Sweden. Swedish emissions constitute 1-2% of the acidifying airborne load on the environment in Finland, Norway, Estonia and Denmark, but this figure falls below 0.3% in other countries.

Nitrogen deposition causes freshwater acidification, particularly in areas where it is greater than the amount that vegetation is able to consume. This is the case in areas of south-western Sweden, where considerable leaching of nitrogen into lakes and watercourses has been observed (Wiklander *et al* 1991). On the whole, however, nitrogen probably accounts for not more than 20% of the acidification in most lakes and watercourses in Sweden (Dickson 1988, Henriksen and Brakke 1988, Grennfelt and Thörnelöf 1992). During certain periods of heavy rain or thaw, for instance, nitrogen may play as important a part in acidification and its impact as sulphur.

Of the nitrogen deposition in Sweden, 60% comes from emissions of nitrogen oxides and 40% from ammonia (Sandnes 1993). Germany and the UK are the major countries of origin, apart from Sweden itself (Table 1).

Forestry is also a cause of soil and water acidification in Sweden. Little is known about the contribution of forestry to the acidification of water, but in most cases it is considered to be significantly less than that of airborne

Table 1. The countries which represent the main sources of acidfying (oxides of sulphur and nitrogen and ammonia) deposition in Sweden in percentages for 1992 (Sandnes 1993).

	Sulphur	Nitrogen
Germany	20	18
Sweden	12	19
UK	11	11
Poland	7	5
Denmark	6	9
France	2	3
Russia	3	1
Czech Republic	3	1
Other countries	14	19
Indeterminate + natural	22	14

deposition of acidifying substances (Bernes 1991, Nilsson 1993). Acidification takes place in several ways, but the main one is probably increased removal of base cations from the forest in conjunction with felling and planting of spruce on land which previously consisted of deciduous forest, agricultural land or meadows (Berdén *et al* 1987, Ulrich and Meyer 1987).

3. The extent of acidification

It has been estimated that about half of all the lakes in Sweden has been affected by acidification caused by human activity (Brodin 1994). About 14,000 or 15% of the lakes can be regarded as being significantly acidified (Bernes and Thörnelöf 1990, Bernes 1991). Without liming, the number, however, would have been above 17,000. Significant acidification means that pH and alkalinity have fallen so much that a significant change in the ecosystem has taken place, with the result that at least some species of plants and animals have been decimated or have completely died out. Such biological consequences are expected to occur if the alkalinity of the water falls by more than 25% and reaches a figure below 0.05 meq/l (Wiederholm 1989, Bernes 1991). With an alkalinity of 0.05, the water generally has a pH of between 5.5 and 6.3.

About one-fifth of all the watercourses in Sweden can be regarded as being significantly acidified. This estimate is based on county council surveys during the early 1980s and has been confirmed by studies of water chemistry, benthic fauna and fish during the late 1980s (Dickson 1988, Lingdell and Engblom 1990, Degerman *et al* 1991). Less data are available for watercourses than for lakes, however. In northernmost Sweden it is difficult to make such estimates since it is unclear whether the acidity level of many waters is natural or anthropogenic in origin (Bishop 1991).

The acidification of Swedish lakes and watercourses has not changed to any great extent in most areas since the late 1970s, with the exception of small watercourses in parts of the southern mountain regions, where the reduction in the number of species of benthic fauna and increased precipitation of iron and aluminium suggest that acidification damage is on the increase (Degerman *et al* 1987, Engblom and Lingdell 1991).

An overview of the situation for lakes and watercourses in Sweden is shown in Fig. 3. Areas containing acidified lakes and watercourses can be

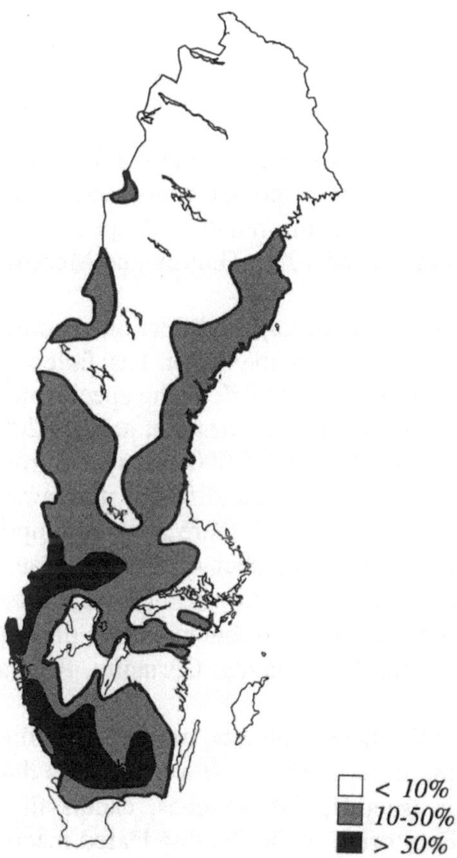

Figure 3. *Areas in Sweden where < 10%, 10-50% and > 50% of lakes and watercourses are significantly affected by acidification.*

found in almost all parts of the country. In many cases in the northern Swedish inland and the mountainous regions, acidification takes place only during a short period each year, when the snow melts in the late spring and early summer. Although these reductions in pH are short-lived, the biological impact can be very severe (Degerman *et al* 1991, Herrmann *et al* 1993). Such "acid surges" can also arise in other parts of Sweden, both in conjunction with winter and spring thaws and heavy rainfall in the autumn.

Acidification has primarily affected the smaller lakes and watercourses in Sweden. About 15% of lakes with a surface area of less than 1 km^2 are significantly acidified (Bernes 1991). Only 5-7% of lakes with a surface area in excess of 1 km^2 are significantly acidified, however, together with a few isolated lakes with a surface area of more than 10 km^2. The acidified lakes and watercourses are generally situated in areas of poor soil dominated by coniferous forest.

4. Biological impact

4.1. Biodiversity

Any change in the pH of the water involves a change in the conditions for plant and animal communities in the lake or watercourse concerned. At pH below 7, each reduction involves a risk that the number of species will decline (Eilers *et al* 1984, Økland and Økland 1986, Harvey and McArdle 1986, Herrmann *et al* 1993).

In many areas of Europe and North America, acidification resulting from human activity has meant that the pH in the water has fallen by 0.5-1.5 units. A fall of this sort can mean that 20-50% of the species may disappear (Eilers *et al* 1984, Brodin 1994). This implies that at least 20% of all the species have died out from the 15,000-25,000 lakes in Europe where the pH has fallen by more than 0.5 units. Acidified watercourses have probably suffered a similar depletion, although there is no data upon which to make quantitative assessments of the number of affected watercourses in Europe. The main depletion of species in acidified waters has probably occurred in Scotland, Norway and Sweden, but extensive damage has also occurred in areas of Finland, the Netherlands, Germany, Russia, Italy, Wales and Slovakia.

The main depletion affects planktonic algae, molluscs, mussels, mayflies and crustaceans. Few species in these groups survive at a pH of less than 5. Other susceptible groups include isoetids, fish, leeches, caddis flies, stoneflies, and wheel animalcules (Rotatoria), while floating leafed macrophytes, emergent macrophytes, midges, dragonflies, mites, water bugs and beetles do not suffer to the same extent.

Even in very acid lakes with original pH of below 5.5, dramatic changes in fauna and flora can be caused by acid deposition despite only a moderate change in pH (Kerekes *et al* 1985, Renberg *et al* 1990), and the fact that these lakes have had fauna and flora which have adapted to a relatively acid environment for thousands of years.

The impact of acidification on plant and animal life in Swedish lakes and watercourses has been relatively well investigated (Fleischer *et al* 1993, Appelberg *et al* 1993). Acidification has meant that 5,000-7,000 Swedish lakes have suffered a reduction in pH of more than 0.5 units (Bernes 1991). On the basis of field studies and literature on the tolerance of different species to low pH values and high aluminium levels, it can be assumed that these lakes have lost more than 20% of their plant and animal diversity as a result of acidification - more than 70% of them in the southern half of Sweden. The number of lakes which have lost more than 20% of their species would probably have risen to about 7,000-9,000 without liming.

This depletion has mainly affected planktonic algae, crustaceans, fish, mussels, bivalves, leeches and mayflies. In many lakes along the west coast of Sweden, the number of species of planktonic algae has declined by more than 50% (Eriksson *et al* 1983).

Mussels, bivalves and leeches have probably disappeared completely from most of the 3,000-5,000 lakes where the pH has fallen below 5 (Brodin 1990a). The roach *(Rutilus rutilus)*, which is susceptible to acidification, may have disappeared from almost 20% of the acidified Swedish waters, i.e. from more than 2,000 lakes (Table 2), while northern pike *(Esox lucius)* and perch *(Perca fluviatilis)* have fared somewhat better.

***Table** 2. Species of fish which have died out in acidified lakes and watercourses in Sweden. Based on samples and interviews during the 1980s (Brodin 1990a).*

Species	Number of surface waters studied	Waters where the species has disappeared no.	%
Perch	1614	130	8
Northern pike	1566	144	9
Roach	1350	248	18
Brown trout	387	57	15

A considerable body of material is available for assessing the impact of acidification on flora and fauna in Swedish watercourses (Engblom and Lingdell 1991). A fall in the pH of 0.5 units within the interval of 4.5-7.0 may mean that 10-50% (about 20% on average) of all species disappear.

The changes are greatest at pH of about 6.5-7.0 and below 5.5. They are more dramatic among animal species (a decline of about 40% for a pH reduction of 1 unit) than among plants (approximately 25% for a pH reduction of 1 unit), as shown in Fig. 4. The most susceptible groups are the same as in lakes. The brown trout *(Salmo trutta)*, which is important to Swedish leisure fishing, has probably disappeared from about 10-20% of the watercourses it would normally inhabit (Table 2).

Acidification therefore means that species have been lost in many lakes and watercourses, but it has not been proven to result in the extinction of any species. It is, however, important to point out that few data are available on extinct or threatened species in many groups such as planktonic algae and insects. It should also be mentioned that, in many cases, the threat of extinction has been lifted by liming (Olem 1991, Henrikson and Brodin, Chap. 1).

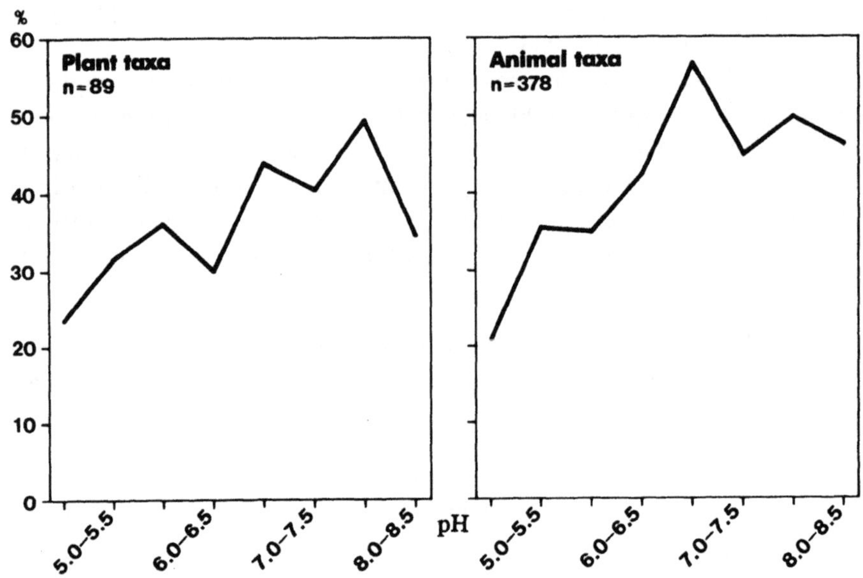

Figure 4. Number of taxa of A) plants and B) animals in Swedish watercourses at different pH value intervals. Most of the taxa correspond to species but in certain cases they indicate groups of species such as worms (Oligochaeta) and midges (Ceratopogonidae and Chironomidae). Each pH value interval covers 30 randomly selected watercourses with a breadth of 2-5 m (data from Lingdell and Engblom).

4.2. Causes of changes in flora and fauna

The reduction in the number of species of planktonic algae in acidified lakes is probably the result of a combination of low pH, diminishing nutrient availability (as a result of a reduced level of phosphorus), reduced carbonic acid levels, increased levels of aluminium, altered levels of humic substances and a change in grazing patterns, particularly for animal plankton (Eriksson *et al* 1983).

The production of benthic algae often shows a sharp increase in acidified freshwater environments. One typical feature of acidified watercourses in Europe, particularly in Norway and Sweden, is an enormous increase in the production of filamentous green algae and, in certain cases, diatoms (Hendrey 1976, Degerman *et al* 1991, Herrmann *et al* 1993). These changes may be the result of a combination of preference for acid environments, reduced grazing as a result of depletion of the benthic fauna and reduced microbial breakdown (Elwood and Mulholland 1989). It is also thought that the increased nitrogen level in these environments may be a significant factor in promoting growth (Herrmann *et al* 1993).

Another typical feature of acidified environments is the spread of bulbous rush *(Juncus bulbosus)* and mosses, particularly in lakes (Grahn 1977, Havas and Hutchinson 1983, Roelofs 1983). In northern Europe, *Sphagnum* moss may take over and form carpets in water with pH below 5. In Swedish lakes, for instance, such mass growth has suffocated isoetids, which have declined in number and distribution (Lazarek 1985, Larsson, Chap. 7). The spread of *Sphagnum* moss and bulbous rush is mainly considered to be attributable to the fact that, unlike most other water plants, they can use carbon dioxide as their source of carbon instead of carbonic acid (Elwood and Mulholland 1989). In waters where the pH has fallen below 5, the carbon dioxide level increases sharply, while that of carbonic acid falls.

The combination of low pH and high levels of certain forms of aluminium is probably the main reason why the number of animal species declines in acidified lakes and watercourses (Dickson 1978). The results of laboratory experiments and field studies have shown that this combination is probably one of the main explanations for the decline in populations of fish belonging to the salmon family such as trout, salmon *(Salmo salar)* and whitefish *(Coregonus* spp.) in acidified waters (Leivestad 1982, Stoner *et al* 1984, Skogheim *et al* 1984). It has also been found that aluminium and low pH have adverse effects on aquatic crustaceans and insects (Burton and Allen 1986, Ormerod *et al* 1987, Herrmann *et al* 1993).

Significantly higher levels of metals other than aluminium may also occur in acidified waters or in the organisms which live in or around them (Borg 1988). Precipitation of manganese and iron poses a serious threat to fauna, particularly in the mountainous regions of Sweden and Norway

(Andersson and Nyberg 1984). The impact of increased concentrations of other potentially toxic metals such as cadmium and zinc on fauna and flora is unclear.

It has not been clearly established why bivalves and mussels are not usually found in waters having a pH less than 5. Apart from the direct effects of low pH, poor calcium and carbonic acid availability may also play a significant part (Økland and Økland 1986).

Changes in fauna in acidified waters may often be attributable to changes in competitive and predatory patterns. A sharp increase in the dominance of certain large predatory insects such as dragonflies, water bugs and beetles is a typical feature of many severely acidified lakes (e.g. Henrikson and Oscarson 1981, Bendell and McNicol 1987). The lack of fish is considered to be one important explanation for this change in the fauna (Eriksson *et al* 1980).

Amphibians and birds which are associated with the aquatic environment may also be adversely affected by acidification. Several studies of amphibians carried out in the USA, Sweden and elsewhere indicate that certain species may be declining as a result of acidification (Pough 1976, Andrén *et al* 1988, Baker *et al* 1990). The declining populations of fish and small animals in acidified waters may be one important reason why certain species of bird such as osprey *(Pandion haliaetus)*, dipper *(Cinclus cinclus)* and diver *(Gavia* spp.) have declined locally in Canada, Sweden and Wales (Barr 1986, Ormerod *et al* 1986, Eriksson 1987).

Little is known about the impact of certain significant factors on fauna and flora in acidified waters. These include oligotrophication (depletion of the nutrient supply) and changes in the amount of humic substances in the water. Oligotrophication would appear to have occurred in Gårdsjön in south-western Sweden, which has been the subject of several studies (Grahn *et al* 1974, Jansson *et al* 1986), but the extent to which oligotrophication occurs as a general phenomenon in acidified waters is not clear (Olsson and Pettersson 1993).

4.3. Decomposition

Materials on the surface sediment decompose more slowly in acidified lakes. This has been reported from many areas of Europe and North America, particularly in waters where the pH has fallen below 6.0 (Otto and Svensson 1983, Allard and Moreau 1986). The reasons are not known in detail, but there is thought to be a link between low pH and the decline in the fauna (particularly insects and crustaceans) which processes coarse materials. It is uncertain whether changes in microbial activity are of any significance in this context. Some researchers have found a link between low pH and falling microbial activity (Mackey and Kersey 1985, Elwood and Mulholland 1989) but this is contradicted by the findings of others (Gahnström, Chap. 6).

Little research has been done on the impact of acidification on decomposition in sediment and these findings are also contradictory. In some cases, no change in bacterial decomposition has been reported (Kelly *et al* 1984, Gahnström and Fleischer 1985), while in others an increased amount of detritus as a result of decreased decomposition has been reported (Brezonik *et al* 1990). The number and composition of animals living in the sediment whose function is to decompose materials are also important. Certain studies claim that sediment fauna is probably not particularly affected by acidification (Ormerod *et al* 1987, Elwood and Mulholland 1989), but palaeoecological studies have shown that the number of such faunas falls sharply in lakes which have been acidified for more than 30 years (Brodin 1990b).

4.4. Human health

When soil and water become acidified, mercury accumulates much more in aquatic fauna. Mercury levels in pike in 12,000-14,000 acidified Swedish lakes are so high - above the limit of 0.5 mg/kg in pike weighing 1 kg - that such fish are considered unsuitable for human consumption (Andersson *et al* 1987). Perch, pike-perch *(Lucioperca lucioperrca)* and burbot *(Lota lota)* from the same waters also have excessive mercury levels, whereas levels in salmon and some other species are usually below the limit.

In Sweden, almost exclusively in acidified waters, the levels of radioactive caesium in fish are so high that people are advised against eating fish. Radioactive fallout of this kind originated from the nuclear accident in Chernobyl in the Ukraine in 1986.

5. Necessary action and research

International agreements and national decisions aimed at reducing emissions of acidifying substances - if they lead to active measures - will mean that acid deposition in Sweden will be reduced by about 35% during the 1990s (Brodin, unpubl.). This reduction is probably too small to significantly change the overall situation during the period from 1990 to 2010, however. Calculations show that the number of acidified lakes and watercourses may even increase in some areas in southern Sweden (Brodin 1994). Improved conditions may occur in some mountainous regions in northern Sweden where acid surges is a major problem.

The Swedish environmental objective for acid load is that atmospheric deposition should be reduced by approximately 70% as compared with 1990 deposition levels (Swedish Environmental Protection Agency 1994).

International cooperation in the field of cost-effective reductions in acidifying emissions offers some hope that acid deposition in Sweden may be reduced by 60-80% during the period from 1990 to 2010. Strenuous and often successful efforts to develop environmental technology can also be seen as a source of hope in this respect. Sweden has done well in terms of sulphur emissions, with a reduction of 80% during the 1980-92 period. Like most countries, however, Sweden has enjoyed little success in reducing nitrogen emissions.

Even if acid deposition were to be reduced by as much as 60-80% during the next 20 years, there will still be thousands of severely acidified lakes and watercourses in Sweden for at least another 50 years (Brodin 1994). This means that other action is needed, beyond that of simply reducing emissions. The most important supplementary action consists of soil and water liming.

In addition to cutting emissions and practical measures in the form of liming, further research is required in order to provide new insights and generate material upon which to base decisions regarding further action. Important topics for research include the role of forestry in the acidification of surface waters and the threat of further acidification, particularly in northern Sweden. Research is also urgently needed into the role played by nitrogen and humic substances, as well as into liming methods, techniques and strategies (Fleischer *et al* 1993).

6. References

Alcamo, J., Shaw, R. and Hordijk, L. 1990. The RAINS model of acidification - science and strategies in Europe. Kluwer Acad. Publ., Dordrecht, The Netherlands, 1991.

Allard, M. and Moreau, G. 1986. Leaf decomposition in an experimentally acidified stream channel. - Hydrobiologia 139: 109-117.

Andersson, P. and Nyberg, P. 1984. Experiments with brown trout *(Salmo trutta* L.) during spring in mountain streams at low pH and elevated levels of iron, manganese and aluminium. - Rep. Inst. Freshwat. Res., Drottningholm, 61: 34-47.

Andersson, T., Brydsten, L., Håkanson, L. and Nilsson, Å. 1987. Kvicksilver i svenska sjöar. [Mercury in Swedish lakes.] - Swedish Environm. Prot. Agency 3292.

Andrén, C., Henrikson, L., Olsson, M. and Nilsson, G. 1988. Effects of pH and aluminium on embryonic and early larval stages of Swedish frogs *Rana arvalis*, *R. temporaria* and *R. dalmatina*. - Holarct. Ecol. 11: 127-135.

Appelberg, M., Henrikson, B.-I., Henrikson, L. and Svedäng, H. 1993. Biotic interactions within the littoral community of Swedish forest lakes during acidification. - Ambio 22: 290-297.

Baker, J.P., Bernard, D.P., Christensen, S.W., Sale, M.J., Freda, J., Heltcher, K., Marmorek, D., Rowe, L., Scanlon, P., Suter, G., Warren-Hicks, W. and Welbourn, P. 1990. Biological effects of changes in surface water acid-base chemistry. In: Irving, P.M. (Ed.) Acidic deposition: state of science and technology. vol. II. Aquatic processes and effects. NAPAP Rep. 13, USA.

Barr, J.F. 1986. Population dynamics of the common loon *(Gavia immer)* associated with

mercury-contaminated waters in northwestern Ontario. - Can. Wildlife Serv. Occas. Paper 56: 1-25.
Bendell, B.E. and McNicol, D.K. 1987. Fish predation, lake acidity and the decomposition of aquatic insects assemblages. - Hydrobiologia 150: 193-202.
Berdén, M., Nilsson, S.I., Rosén, K. and Tyler. G. 1987. Soil acidification - extent, causes and consequences. - Swedish Environm. Prot. Agency 3292.
Bernes, C. 1991. Monitor 12 - Acidification and liming of Swedish freshwaters. - Swedish Environm. Prot. Agency Informs.
Bernes, C. and Thörnelöf, E. 1990. Present and future acidification of Swedish lakes: model calculation based on extensive survey. - In: Kämäri, J. (Ed.) Environmental impact models to assess regional acidification. IIASA, Reidel & Co, pp. 83-107.
Bishop, K. 1991. Is there more to acidity in organic-rich surface waters than air pollution? An example from northern Sweden. - Vatten 47: 342-347.
Borg, H. 1988. Trace metals in Swedish forest lakes - factors influencing the distribution and speciation in water: - Acta Univ. Upsaliensis 1988-145: 1-21.
Brezonik, P.L., Webster, K.E. and Perry, J.A. 1990. Effects of acidification on benthic community structure and benthic processes in the Little Rock Lake, Wisconsin. - Verh. Internat. Verein. Limnol. 24: 445-448.
Brodin, Y.-W. (Ed.) 1990a. Effekter av svavel-och kvävebelastning på skogsmark, yt- och grundvatten. [English summary: Effects of sulphur and nitrogen deposition on forest soil, surface water and groundwater.] - Swedish Environm. Prot. Agency 3762.
Brodin, Y.-W. 1990b. Midge fauna development in acidified lakes in northern Europe. - Phil. Trans. R. Soc., London B327: 295-298.
Brodin, Y.-W. 1994. Acidification - an everlasting problem or is there hope? - Swedish Environm. Prot. Agency 4242.
Burton, T.M. and Allen, J.W. 1986. Influence of pH, aluminium, and organic matter on stream invertebrates. - Can. J. Fish. Aquat. Sci. 43: 1285-1289.
Degerman, E., Lindgren, G., Lingdell, P.-E. and Nyberg, P. 1987. Kartering av strömfauna och fisk i mindre vattendrag i Norrlands inland och fjälltrakter i relation till försurning. [Survey of stream-fauna and fish in small watercourses in the central and mountainous region of Norrland in relation to acidification.] - Inform. Inst. Freshw. Res., Drottningholm (2).
Degerman, E., Engblom, E., Lingdell, P.-E., Melin, E. and Olofsson, E. 1991. Försurning i fjällen? [Acidification in the mountains?] - Inform. Inst. Freshw. Res., Drottningholm (2).
Dickson, W. 1978. Some effects of acidification of Swedish lakes. - Verh. Internat. Verein. Limnol. 20: 851-856.
Dickson, W. 1986. Some data on critical load for sulphur on surface waters. - In: Nilsson, J. (Ed.) Critical loads for sulphur and nitrogen. Nordic Counc. Minist. Miljørap. 1986:11, pp. 143-158.
Dickson, W. 1988. Försurningsläget i svenska vatten och kalkningsbehovet. [Acidification of Swedish surface waters and liming requirements.] - Vatten 44: 300-304.
Dickson, W. 1990. Acidification of lakes and streams in northern Sweden. - Nordic Counc. Minist. Miljørap. 1990:2: 94-99.
Eilers, J.M., Lien, G.J. and Berg, R.G. 1984. Aquatic organisms in acidic environments: a literature review. - Dep. Nat. Res., Wisconsin, Techn. Bull. 150.
Elwood, J.W. and Mulholland, P.J. 1989. Effects of acidic precipitation on stream ecosystems. - In: Adriano, D.C. and Johnson, A.H. (Eds.) Biological and ecological effects. Acidic precipitation, vol. 2. Springer, Berlin Heidelberg New York, pp. 85-135.
Engblom, E. and Lingdell, P.-E. 1984. The mapping of short-term acidification with the help of biological indicators. - Inst. Freshwat. Res., Drottningholm, 61: 60-68.

Engblom, E. and Lingdell, P.-E. 1991. Försurning och bottenfaunaförändringar i Sverige. [Acidification and changes in benthic fauna in Sweden.] - Vatten 47: 348-351.
Eriksson, F., Hörnström, E., Mossberg, P. and Nyberg, P. 1983. Ecological effects of lime treatment of acidified lakes and rivers in Sweden. - Hydrobiologia 101: 145-164.
Eriksson, M.O.G. 1987. Some effects of freshwater acidification on birds. - ICBP Techn. Publ. 6: 1-8.
Eriksson, M.O.G., Henrikson, L., Nilsson, B.-I., Nyman, H.G., Oscarson, H.G., Stenson, J.A.E. and Larsson, P.K. 1980. Predator-prey relations important for the biotic change in acidified lakes. - Ambio 9: 248-249.
Falkengren-Grerup, U. 1987. Long-term changes in pH of forest soils in southern Sweden. - Environ. Poll. 43: 79-90.
Fleischer, S., Andersson, G., Brodin, Y.-W., Dickson, W., Herrmann, J. and Muniz, I. 1993. Acid water research in Sweden - knowledge for tomorrow?. - Ambio 22: 258-263.
Gahnström, G. and Fleischer, S. 1985. Microbial glucose transformation in sediment from acid lakes. - Ecol. Bull. (Stockholm) 37: 287-292.
Grahn, O. 1977. Macrophyte succession in Swedish lakes caused by deposition of airborne acid substances. - Water Air Soil Poll. 7: 295-305.
Grahn, O., Hultberg, H. and Landner, L. 1974. Oligotrophication - a self-accelerating process in lakes subjected to excessive supply of acid substances. - Ambio 3: 93-94.
Grennfelt, P. and Thörnelöf, E. (Eds.) 1992. Critical loads for nitrogen. - Nordic Counc. Minist., Miljørep. 1992:41.
Hallbäcken, L. and Tamm, C.-O. 1986. Changes in soil acidity from 1927 to 1982-1984 in a forest area of south-west Sweden. - Scand. J. Forest Res. 1: 219-232.
Harvey, H.H. and McArdle, J.M. 1986. Composition of the benthos in relation to pH in the LaCloche lakes. - Water Air Soil Poll. 30: 529-536.
Havas, M. and Hutchinson, T.C. 1983. The Smoking Hills: natural acidification of an aquatic ecosystem. - Nature 301: 23-27.
Hendrey, G.R. 1976. Effects of pH on the growth of periphytic algae in artificial stream channels. - SNSF, Oslo-Ås, Norway, Internat. Rep. 25.
Henriksen, A. and Brakke, D.F. 1988. Increasing contribution of nitrogen to the acidity of surface waters in Norway. - Water Air Soil Poll. 42: 183-201.
Henriksen, A., Kämäri, J., Posch, M. and Wilander, A. 1992. Critical loads of acidity to Nordic surface waters. - Ambio 21: 356-363.
Henrikson, L. and Oscarson, H.G. 1981. Corixids (Hemiptera - Heteroptera), the new top predator in acidified lakes. - Verh. Internat. Verein. Limnol. 21: 1616-1620.
Herrmann, J., Degerman, E., Gerhardt, A., Johansson, C., Lingdell, P.-E. and Muniz, I. 1993. Acid-stress effects on stream biology. - Ambio 22: 298-307.
Ivens, W., Kauppi, P., Lövblad, G. and Westling, O. 1989. Throughfall monitoring as a means of monitoring deposition to forest ecosystems - evaluation of European data. - Utrecht Univ. Rep. AD-1989-01.
Jansson, M.G., Persson, G. and Broberg, O. 1986. Phosphorus in acidified lakes: the example of Lake Gårdsjön, Sweden. - Hydrobiologia 39: 81-96.
Kämäri, J., Amann, M., Brodin, Y.-W., Chadwick, M.J., Henriksen, A., Hettelingh, J.-P., Kuylenstierna, J.C.I., Posch, M. and Sverdrup, H. 1992. The use of critical loads for the assessment of future alternatives for acidification. - Ambio 21: 377-386.
Kelly, C.A., Rudd, J.W.M., Furutani, A. and Schindler, D.W. 1984. Effects of lake acidification on rates of organic matter decomposition in sediments. - Limnol. Oceanogr. 29: 687-694.
Kerekes, J., Tembley, C. and Pollock, P. 1985. Organic versus anthropogenic acidity in the tributaries of the Kejimujik watersheds in western Nova Scotia. - In: Internat. Symp., Acidic Precipitation, 1985, Muskoka, Ontario, Canada, Abstracts, p. 225.

Lazarek, S. 1985. Epiphytic algal production in the acidified Lake Gårdsjön, SW Sweden. - Ecol. Bull. (Stockholm) 37: 213-218.

Leivestad, H. 1982. Effects from acid rain on fish. - In: Schneider, T. and Grant, L. (Eds.) air pollution and nitrogen oxides. Elsevier Sci. Publ. Comp., Amsterdam, pp. 567-576.

Lingdell, P.-E. and Engblom, E. 1990. Rena och oförsurade vatten finns dom? [Can pristine and non-acidified waters still be found?] - Swedish Environm. Prot. Agency 3708.

Lövblad, G., Amann, M., Andersen, B., Hovmand, M., Joffre, S. and Pedersen, U. 1992. Deposition of sulphur and nitrogen in the Nordic countries: present and future. - Ambio 21: 339-347.

Mackey, R.J. and Kersey, K.E. 1985. A preliminary study of aquatic insect communities and leaf decomposition in acid streams near Dorset, Ontario. - Hydrobiologia 122: 3-13.

McNeely, R. and Gummer, W.D. 1984. A reconnaissance survey of the environmental chemistry in east-central Ellesmere Island, N.W.T. - Arctic 37: 210-223.

Nihlgård, B. 1990. Effekter på skog [Effects on forest]. - In: Brodin, Y.-W. (Ed.) Effekter av svavel-och kvävebelastning på skogsmark, yt- och grundvatten. [English summary: Effects of sulphur and nitrogen deposition on forest soil, surface water and groundwater.] - Swedish Environm. Prot. Agency 3762, pp. 65-83.

Nilsson, S.I. 1993. Acidification of Swedish oligotrophic lakes - interactions between deposition and forest growth and effects on lake-water quality. - Ambio 22: 272-276.

Odén, S. 1968. The acidification of air and precipitation and its consequences on the natural environment. - Swed. Nat. Sci. Res. Council, Ecol. Comm. Bull. 1: 1-68.

Økland, J. and Økland, K.A. 1986. The effects of acid deposition on benthic animals in lakes and streams. - Experientia 42: 471-486.

Olem, H. 1991. Liming acidic surface waters. Lewis Publ., Chelsea, Michigan, USA.

Olsson, H. and Pettersson, A. 1993. Oligotrophication of acidified lakes - a review of hypotheses. - Ambio 22: 312-317.

Ormerod, S.J., Allison, N., Hudson, D. and Tyler, S.J. 1986. The distribution of breeding dippers *(Cinclus cinclus* (L.); Aves) in relation to stream acidity in upland Wales. - Freshwat. Biol. 16: 501-507.

Ormerod, S.J., Boole, P., McCohan, C.P., Weatherley, N.S., Pascoe, D. and Edwards, R.W. 1987. Short-term experimental acidification of a Welch stream: comparing the biological effects of hydrogen ions and aluminium. - Freshwat. Biol. 17: 341-356.

Otto, C. and Svensson, B. 1983. Properties of acid brown water streams in south Sweden. - Arch. Hydrobiol. 99: 15-36.

Pough, F.H. 1976. Acid precipitation and embryonic mortality of spotted salamanders, *Ambystoma maculatum*. - Science 192: 68-70.

Renberg, I., Brodin, Y.-W., Cronberg, G., El-Daoushy, F., Oldfield, F., Rippey, B., Sandøy, S., Wallin, J.-E. and Wik, M. 1990. Recent acidification and biological changes in Lilla Öresjön southwest Sweden, and the relation to atmospheric pollution and land-use history. - Phil. Trans. R. Soc., London B327: 391-396.

Renberg, I., Korsman, T. and Anderson, N.J. 1993. A temporal perspective of lake acidification in Sweden. - Ambio 22: 264-271.

Roelofs, J.G.M. 1983. Impact of acidification and eutrophication on macrophyte communities in soft waters in the Netherlands - Aquat. Bot. 17: 139-155.

Sandnes, H. 1993. Calculated budgets for airborne acidifying components in Europe. - EMEP MSC-W Rep. 1/93 (Norway).

Skogheim, O.K., Rosseland, B.O. and Sevaldrud, I. 1984. Deaths of spawners of Atlantic salmon in River Ogna, SW Norway, caused by acidified aluminium-rich water. - Rep. Inst. Freshwat. Res., Drottningholm 61: 195-202.

Stoner, J.H., Gee, A.S. and Wrade, K.R. 1984. The effects of acidification on the ecology

of streams in the Upper Tywi catchment in West Wales. - Environ. Poll. 35: 125-157.
Sverdrup, H., Warfvinge, P., Frogner, T., Håøya, A.O., Johansson, M. and Andersen, B. 1992. Critical loads for forest soils in the Nordic countries. - Ambio 21: 348-355.
Swedish Environmental Protection Agency 1994. Strategy for sustainable development - enviro '93. - Swedish Environm. Prot. Agency 4234.
Torstensson, P. and Liljelund, L.-E. 1989. Flora- och faunaförändringar i terrestra miljöer - orsakade av luftföroreningar och försurning. [Changes in terrestrial flora and fauna attributable to air pollutants and acidification.] - Swedish Environm. Prot. Agency 3604.
Ulrich, B. and Meyer, H. 1987. Chemischer Zustand der Waldböden Deutschlands zwischen 1920 und 1960, Ursachen und Tendenzen seiner Veränderung. - Ber. Forsch. Zent. Waldökosyst., Univ. Göttingen, Germany B6.
Van Breemen, N. and van Dijk, H.F.G. 1988. Ecosystem effects of atmospheric deposition of nitrogen in The Netherlands. - Environ. Poll. 54: 249-274.
Wiederholm, T. 1989. Bedömningsgrunder för sjöar och vattendrag. Bakgrundsdokument 1. Näringsämnen, syre, ljus, försurning. [Assessment bases for lakes and watercourses. Background document 1. Nutrients, oxygen, light, acidification.] - Swedish Environm. Prot. Agency 3627.
Wiklander, G., Nordlander, G. and Anderson, R. 1991. Leaching of nitrogen from a forested catchment at Söderåsen in southern Sweden. - Water Air Soil Poll. 55: 263-282.

Chapter 4

Strategies and methods for freshwater liming

William Dickson and Y-W Brodin

Swedish Environmental Protection Agency, S-171 85 Solna, Sweden

Summary

Liming is an effective method to counteract the chemical and biological effects of acidification. Sweden and Norway are the only countries that maintain large-scale liming operations as a national strategy to restore ecological conditions in acidified lakes and watercourses. The number of limed surface waters in Sweden amounts to nearly 8,000. Several hundred surface waters are limed in Norway, while liming is conducted on a limited scale, i.e. less than 200 surface waters, in other countries.

To counteract the total acidification of Sweden during the 20th century around 90 million tonnes of limestone are required. The cost of purchasing, transporting and dispersing this amount of lime is approximately 25 trillion Swedish krona (SEK)(c. USD 3 trillion), which can be compared with the rougly SEK 200 million presently used each year to lime surface waters in Sweden.

The best form of liming is often a combination of methods. All liming operations require monitoring. Liming is a chemical form of treatment which achieves its optimal biological effects after a few years.

In order to comply with the overall goal of Swedish liming activities, viz. to restore ecosystems to their natural conditions as far possible, full-scale liming of the catchment soil is the best method in the long term. This strategy, however, is rarely applied, mainly because of the high initial costs involved and the long time needed to achieve the desired effects in the surface water. To achieve adequate results in the runoff water, doses in southern Sweden of at least 5 t/ha of runoff area are needed. Total limestone uptake to the benefit of the surface water will after full-scale liming be perhaps 20% before the treatment needs to be repeated.

Wetland liming is costly but seals metals such as iron, aluminium, and manganese before they reach the water. This method also suppresses the acid surges during high water periods which occur more frequently the

Liming of Acidified Surface Waters
L. Henrikson and Y.W. Brodin (Eds.)
© Springer-Verlag Berlin Heidelberg 1995

more acid the soils and the wetland become. However, if there are suitable discharge areas, liming of 1-2% of the runoff area could be effective for at least a couple of years. The first liming dose can be limited to approximately 300 kg/ha of runoff area and the annual doses can be subsequently reduced by approximately 100 kg/ha of runoff area. The doses on the limed wetland areas are nevertheless significant: 10-30 t/ha. The total lime exploitation can in favourable cases be 50-70%, but it is often significantly lower. Harmful effects on wetland vegetation can occur.

Lake liming is the most common and cheapest method for liming surface waters in Sweden and other countries. It is the only method used when the lake area percentage is more than 10%. However, the method cannot cope with acid surges from unlimed areas in catchment areas close to the lakeside. The duration of lake liming is mostly a few years. In lakes with a retention time of less than 1 year, the duration is often equivalent to 4-5 turnovers, while the duration in lakes with a retention time of several years is equivalent to 2-3 turnovers. The total assimilation of added lime or sodium carbonate is often 50-80%.

Doser liming can be an effective and cheap method of treating running waters and lakes with short turnover times. Around 300 lime dosers were in operation in Sweden in 1993. The dosers require regular checks and should be located several metres upstream from the primary liming object. They must be located where they can be refilled and the annual dosage must be so high that the price of the doser can be justified. The total uptake of lime can be regulated and should therefore normally be high, 70-90%.

A number of other liming methods or strategies to mitigate acid conditions in surface waters are available, i.e. road liming as an alternative to salting to reduce skidding, lime precipitation in wastewater treatment plants, ditch liming, and spreading of coarsely ground lime into watercourses. These methods can be used to complement the methods described above, but they presently contribute less than 1% of the lime added to acidified freshwater in Sweden.

A number of agents to increase pH and alkalinity of surface waters are available. In Sweden, limestone ($CaCO_3$) or dolomite ($CaMg(CO_3)_2$) are used in more than 95% of all liming operations. For treatment of surface waters, limestone powder in fractions of 0-0.5 mm for lakes, more finely ground for dosers and slightly coarser fractions for wetlands are probably optimal in the light of price, dispersion qualities, dissolution and duration. Sodium carbonate (Na_2CO_3) is used in some Swedish lakes.

Thousands of tonnes of basic products from food production, egg and mussel shells, and sludge from sulphate pulp production are currently being dumped. Many of the products meet the purity requirements and should be used to lime forest soils. Over a period of a few years, this would have an effect on the runoff water.

There are no satisfactory alternative methods to liming, but other measures can be complementary particularly in order to counteract nutrient depletion in the surface water, e.g. increased numbers of deciduous trees, spreading of ash formed when burning biofuels, and addition of treated wastewater.

1. Introduction

Substantial reductions in acidic emissions of sulphur and nitrogen and forestry practices not causing nutrient depletion and soil acidification are the only ways to abolish acidification as a serious environmental problem in Europe and North America.

Current national reduction plans and international agreements indicate that the acid deposition in Sweden may decrease by 25-35% during the period from 1990 to 2005. This will, however, not prevent increased acidification in Swedish lakes and watercourses (Bernes 1991, Brodin 1994). Even if considerably more far-reaching steps to cut emissions were to be taken and forest practices were to produce significantly less soil acidification, the acidification of surface water will still remain a substantial problem for many decades ahead. Other measures are thus also necessary of which liming is certainly the most important.

This chapter addresses strategies and methods used within large-scale liming practices for surface waters in Sweden, taking also into consideration liming conducted in other countries. The aim is to synthesize the knowledge acquired so far and to put forward ideas which can be useful for future liming practices in Sweden and in other countries. The effects of liming on water chemistry, biology, socio-economy and several other issues in conjunction with freshwater liming are addressed in other chapters of the present book.

2. History

Liming of small lakes and ponds in Sweden began as early as in the 19th century, with the aim of increasing fish production. At that time, freshwater fishing formed a vital part of the household economy. The limed waters were primarily used for farming carp *(Cyprinus carpio)* or hatching pike *(Esox lucius)* and pike-perch *(Lucioperca lucioperca)*, which were subsequently moved to the larger lakes. Unslaked lime, CaO, was normally used for liming and produced a significant increase in pH, precipitation of iron and humic substances and a mobilization of important nutrients from the beds of the ponds (Lundberg 1893, Haglund 1911, Tiberg 1912).

"It has been proven that liming of ponds really does bring about quite a considerable change in the quality of the water. The fact that following heavy liming the colour of the water undergoes a significant change - from brown to colourless - can be easily observed. There almost certainly appears to be the possibility that a tolerable water quality has been produced for carp as a result of liming, even if we are unable to understand how the fish organism is affected." (Ahlgren 1918).

When angling became popular in the 1950s, liming of forest lakes became more common, the aim being to provide a good living environment for Atlantic salmon *(Salmo salar)* and brown trout *(Salmo trutta)*. Acid deposition in south-west Sweden had become acute and several thousand lakes in the area were already affected, although the causal connection had yet to be discussed. Such "dead" clearwater lakes with a pH of less than 5 became demonstrably good, nutrient-rich trout waters when the pH was raised above 6.

The most common liming agents at that time were cheap residual industrial products. Lime sludge, a calcium carbonate sludge, was produced by the sulphate paper mills which were largely located near the forest rivers inland. Large dumps and depots were established below the mills, which had a very beneficial effect on the pH of the downstream water. In the production of acetylene, AGA lime was produced, a calcium hydroxide sludge with a high pH which was sometimes used for liming lakes. The steel industry also produced other residual products rich in lime and silicate. T-lime was a cheap and very effective silicate lime which was produced in considerable quantities in the manufacture of ferrochromium. Unfortunately, it also contained chrome and nickel. Sugar beet processing also produced lime sludge which could be used for liming lakes.

A standard requirement when liming lakes at this time was that the lime should be cheap, as the user was responsible for the goods, transport and dispersion. However, for smaller lakes, finely ground limestone powder was used, often called agricultural lime (Berzins 1962).

***Figure** 1. Several alternative methods have been applied for dispersion of lime in lakes. Drawing by J. Wihlke based on photo by L. Henrikson.*

In the 1960s and 1970s, more extensive liming was introduced to save and improve the fish stocks in south-west Sweden. Large lakes were also limed. Alternative liming agents were presented and tested, as were alternative methods of dispersion such as dispersion on solid ground, wetlands, in lakes and on ice (Fig. 1) (e.g. Stenberg 1965, Lundin and Lundh 1968, Schmuul 1970, Wilander and Ahl 1972, Andersson 1974a, 1974b, Bernhoff 1974, Larsson and Lindquist 1974, Sander and Eriksson 1974, Dickson 1974, Grahn and Hultberg 1975, Holmberg 1975, Levander 1975, Malmer 1975, Ministry of Agriculture 1976). The results were promising, which led to trials - with the NBF (the National Board of Fisheries) as principal - being conducted between 1977 and 1982 (Broberg 1978, Dickson 1979, Karlén 1979, NBF and SEPA 1979, 1981, Bengtsson *et al* 1980). After this SEPA (the Swedish Environmental Protection Agency) and the county councils assumed primary responsibility for large-scale, national liming activities in Sweden.

3. The accumulated liming requirements for soil and water

The majority of Sweden's bedrock and soil types are among those with the lowest levels of base elements and calcium in the world. Natural forests, afforestation and several centuries of forestry have depleted the soil and made the runoff water prone to acidification (Dickson 1985, Rosén 1988a, Nilsson 1993). Almost a century of acid deposition has further reduced soil resistance and many lakes and watercourses have been acutely damaged by acid deposition. In many areas the pH of the soil has fallen during the 20th century; by 0.5-1.0 pH units in southern and central Sweden and slightly less further north (Falkengren-Grerup 1986, Tamm and Hallbäcken 1988).

Taking into consideration the annual deposits of acidifying sulphur and nitrogen in Sweden, there is at present a theoretical liming requirement equivalent to approximately 1 million tonnes of limestone each year (EMEP 1992, Dickson 1992). Acidification resulting from forestry - purely through pulpwood extraction and sawn timber - is, at the same time, equivalent to a loss of base elements from woodlands corresponding to approximately 0.5 million tonnes of lime per year (Dickson 1990, 1991). Considering that acid deposition and biomass extraction increased throughout the 20th century up to the 1970s, with hardly any change since, the accumulated acidification during the 20th century is significant.

In agriculture, liming has been practised for over 100 years as a means of improving the cultivation conditions and compensating for acidification by crops, fertilizers and leaching. During the 1970s and 1980s, the annual Swedish consumption of limestone was 0.5-1 million tonnes, although this has fallen in recent years to 300,000 tonnes as a result of restructuring measures and poor profitability.

In forestry, liming is still modest. Only 5,000 tonnes per year are added in the form of dolomite in the commercial fertilizers used in the central and northern parts of the country. To compensate for the part played by forestry in acidification, the lime being added at present needs to be increased 100-fold. However, large-scale trials are being instituted, mainly to compensate for acid deposition in the southern part of the country. In recent years, large-scale trials with the aid of state grants have resulted in the dispersion of 15,000 tonnes of lime per year in southern Sweden.

Since the beginning of this century, 20 million tonnes of sulphur have been deposited through air pollution, representing an accumulated liming requirement of 60 million tonnes of limestone. Biomass extraction from the forests over the same period is equivalent to a removal of base elements equivalent to approximately 30 million tonnes of limestone. The accumulated liming requirement during the 20th century is thus around 90 million tonnes of limestone. The cost of purchasing, transporting and dispersing this amount of lime is approximately USD 3,000 million, which can be compared with the roughly USD 25 million used today to lime lakes and watercourses in Sweden. For the forested areas of southern Sweden, this means that the soil during the 20th century accumulated a liming requirement of approximately 5 t/ha, for central Sweden approximately 3 t/ha and for northern Sweden approximately 1.5 t/ha. The area of 120,000 km^2 - approximately a quarter of Sweden's land surface - from which harmful acidifying water is at present running, has a liming requirement of approximately 25 million tonnes of limestone (Dickson 1988).

4. Purpose and aims

Large-scale liming in Sweden was preceded by extensive trials from 1977 to 1981 (NBF and SEPA 1981). The aims were as follows:
* To use various means of restoring acidified water areas and prevent continued acidification of the endangered waters which are of particular significance to fish stocks and nature conservation.
* To gather experience and results regarding methods, liming agents and the cost of measures taken to establish a basis for decisions concerning the orientation and extent of future liming operations.
* To take particular note of positive and possible negative effects on flora and fauna during the course of the above work.

Many of the liming projects started during the trial period are still in progress and experience from these forms the basis for the present body of knowledge in Sweden concerning liming of lakes and watercourses. Many studies on effects and methods have been published by NBF and SEPA (1981) in other contexts and have also been used as a basis for this chapter.

During the 1982-1988 period, liming was carried out mainly according to the advice and guidelines issued by the NBF (1982), concentrating mainly on fish and crayfish.

The purposes of operations today, as formulated in SEPA's general recommendations (1988), are as follows:

* The biological aim of liming is to detoxify the water so that the natural flora and fauna can subsist or recolonize the limed waters.

* The chemical aim of liming is to raise the pH above 6.0 and the alkalinity above 0.1 meq/l, thus providing an acceptable buffer. In addition, there is a precipitation of metals during liming, thereby reducing their toxic effect. Excess dosing should be avoided and environments with a naturally low alkalinity should be retained. Alkalinity should therefore not exceed 0.2-0.3 meq/l after liming.

Waters which are acid owing to a naturally high humus content and low alkalinity (<0.05 meq/l) have become even more acid by precipitation in large areas of Sweden (Bernes 1991). Even these waters should be limed, at least up to their natural alkalinity, so that the original flora and fauna can recolonize. The water should at no time during the year have a pH of less than 6.0 or an alkalinity lower than 0.05 meq/l. Even brief acid surges can cause considerable damage (Degerman *et al* 1992).

The purpose and aims of the activities are thus to restore acidified areas worth protecting and to protect environments not yet destroyed, through environmental measures.

From the beginning, work was directed at combatting the effects of acid deposition on runoff water. As acid deposits in large parts of Sweden exceed the critical load (Kämäri *et al* 1992) and a large part of the country is naturally poor in base elements, at least a quarter of Sweden's land area is affected. The part played by forestry in soil depletion, and thereby indirectly its ability to affect the quality of runoff water, means that in the long term liming for environmental reasons needs to be widespread. Many naturally acid environments are felt to have become more acid through acidified precipitation and "sustained" forestry (Bernes 1991, Nilsson 1993). As the upper layers of the soil become acidified, acid surges during high-water periods are becoming more and more common and are persisting for longer periods. However, there are researchers who maintain that the acid conditions in lakes and watercourses in large parts of northern Sweden are not caused by air pollution (Bishop 1991, Kullberg *et al* 1993).

5. Criteria for state grants

In Sweden, state grants are given for liming of acidified surface waters which during a stable period (summer or winter) have an alkalinity of less that 0.05 meq/l. Normally, 85% of the total cost of carrying out the liming is covered by the grant. For particularly valuable projects or projects with financing difficulties, the grant may be 100% or 95% respectively. Up to and including 1991, decisions concerning an increase in grants were made by the Government and thereafter by SEPA. The total cost covers planning, procurement, control, lime, dispersion, monitoring of effects, etc.

A state grant also covers the extra costs involved in altering a waste-water treatment plant to facilitate treatment by means of lime precipitation, on the condition that the water is acidified and meets the criteria for the state grants stated above. A state grant can also be given for biological restoration in limed waters (Bergquist, Chap. 17) in the form of reintroduction of locally extinct species and restoration of colonies damaged by acidification. The grant was previously 50%, later raised to 70%, but since 1992 it has been included in the liming project grant.

6. Liming organization

6.1. Liming plans and application

The target area for liming is projected by private owners of waters, angling clubs, fish conservation associations or municipal authorities, who apply to the county administrative board for a liming grant (Fig. 2). Often, the local authority has a municipal liming plan for the water areas in question and a municipal budget for "internal activities", normally covering 15% of the costs for planned liming projects in the area. The county administrative board draws up a county plan to cover the requirements and projects in the county, and the plans are sent to SEPA in March each year. In collaboration with NBF, a central plan is prepared covering the requirements for the following budget year. An application for funds is sent at the beginning of May to the Ministry of the Environment and Natural Resources. A decision is made at the end of June and the funds granted can normally be distributed by the county administrative boards at the beginning of August.

The county administrative boards bear the overall responsibility for operations. They are responsible for environmental monitoring and liming in the region, they investigate the acidification situation and liming needs, they coordinate smaller projects and ensure that liming is carried out in places which lack financiers or interested parties.

Certain counties deal with the procurement of lime centrally for the whole county, in order to obtain a more favourable price.

The county administrative board is also responsible for liming priorities. A large lake with clear water, a turnover time of a couple of years, several species of fish, glacial relics, osprey *(Pandion haliaetus)* and an active angling club thus has a greater chance of receiving a state grant than a small, brown perch-tarn with a turnover time of 1 month, even if the latter lake would be much cheaper to lime.

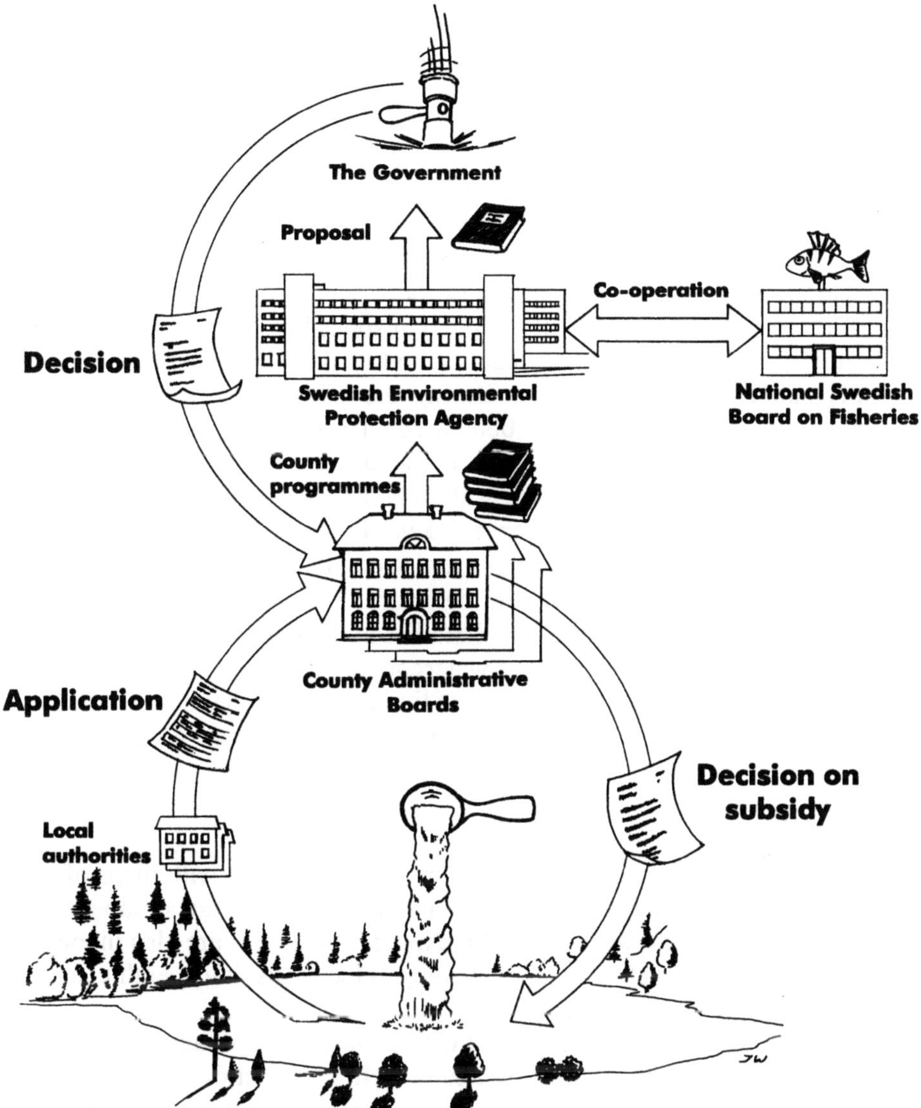

Figure 2. Application and decision chart for aquatic liming in Sweden.

6.2. Monitoring of effects

The county administrative board is responsible for the regional followup of the effects in the limed waters. The purpose of this monitoring is to optimize future liming in terms of time, dose and method. Analyses cover pH, alkalinity, colour index, calcium (or calcium + magnesium) and conductivity (SEPA 1988). Water samples are taken from the lakes twice a year and normally at a depth of 2 m. One is taken under stable conditions and the other under periods of acid surge, i.e. autumn or spring. If a doser is used, water samples should be taken six times a year, both upstream and downstream of the dosing point.

Wetland liming is also under the control of the county administrative board, with samples being taken from selected lakes twice a year or in running waters six times a year. In many cases special chemical and biological studies are instituted, the cost of which is covered by the liming grant. The county administrative board is also responsible for the implementation, evaluation and reporting of these special studies.

SEPA, in conjunction with NBF, is responsible for the countrywide monitoring of liming and for evaluation and further development of ecologically suitable liming methods. During the winter of every fifth year, samples are taken from over 5,000 Swedish lakes and a comprehensive analysis is carried out to investigate the acidification situation in Sweden (Bernes 1986, 1991). In addition, water chemistry sampling is carried out each year in nearly 200 reference lakes. In about 25 of these lakes, comprehensive biological samplings are also conducted. The reference lakes may not be limed either now or in the future.

The ISEL programme (Integrated Studies of the Effects of Liming) comprises 14 limed lakes and 7 limed watercourses. This programme aims to highlight integrated ecological effects of liming and includes a large number of water, chemical and biological analyses (Appelberg and Aldén 1992).

6.3. Staff

Consultants are often engaged to draw up a working plan with maps of the areas that can be limed, to propose a suitable dose and to act as supervisor when the treatment is carried out. Sometimes consultants are also involved in special studies, such as sample fishing, electro-fishing, benthic fauna studies or wetland inventories.

Liming matters are normally dealt with at the municipal authority by the environment and health protection board or in conjunction with the planning department. In the environmental conservation department at the county administrative board there is normally at least one person working on liming matters on a full-time basis. County administrative boards with extensive liming operations also have their own laboratories to carry out simple water analyses. At SEPA there are two people working full-time on liming issues, computer processing, information and education.

An estimate has been made of how many people in Sweden are working with liming and the cost involved (Hallin 1993). Liming involves 0.84 persons per "lime million" for production, sales, transport and dispersion. The administration within the county authorities and at SEPA involves approximately 40 people, which is equivalent to a liming grant of SEK 176 million i.e. 0.23 persons per million. Planning, procurement and monitoring of effects (SEK 100 per tonne = SEK 20 million) is equivalent to 57 man-years and 0.32 persons per million. A further 0.05 persons per million are involved in research and special studies. A total of approximately 1.5 persons per liming million are used and the total man-years can be estimated at 250-300. Of the costs of the activities, salaries etc, SEK 60-70 million i.e. more than one-third, goes back to the state in the form of payroll overheads, taxes and VAT (Hallin 1993). If the liming companies also make a profit and pay tax, the total paid by the state is even less.

6.4. Changes in the liming grant

Since the 1976/77 budget year, liming of Swedish lakes and watercourses has attracted state support. The grant was SEK 10 million per year during the first years, but this has gradually increased and in 1993 the figure stood at SEK 196 million (Henrikson and Brodin, Chap. 1). SEK 1 billion (to 1993) has been paid in state support for measures against acidification. In addition, a total of SEK 150 million has been spent by individual municipal authorities, fish conservation authorities and companies, since the state grant does not normally cover the full cost of liming.

6.5. Extent of liming

Sweden and Norway are the only countries which maintain large-scale liming operations for acidified lakes and watercourses. In Norway, several hundred surface waters are limed and there are plans to expand these operations (Hindar and Rosseland 1991, Hegna 1992). Liming of acidified lakes and watercourses is conducted on a limited scale, i.e. less than 200 waters, in a number of other countries such as Canada, Scotland, Wales, Italy, Poland, Finland, Germany, the Netherlands and the USA.

In Sweden, approximately 200,000 tonnes of lime are dispersed each year to counteract acidification of lakes and watercourses. Approximately 8,000 surface waters have been limed. The operational activity includes liming in lakes (c. 6,500 objects), doser liming in watercourses (c. 300) and liming of wetlands for both lakes and watercourses (c. 1,500) (Fig. 3). This covers more than half the acidified lake area. The limed wetland areas only make up 0.3 percent of the country's total wetland area.

The most comprehensive liming has been carried out in the most acidified parts of south-west Sweden, where approximately half the lakes have been limed at some time. Liming of our lakes and watercourses is carried out on a large scale and next to acidification it represents the biggest chemical impact on water since the melting of the inland ice.

The methods, agents and doses used and described in the following sec-

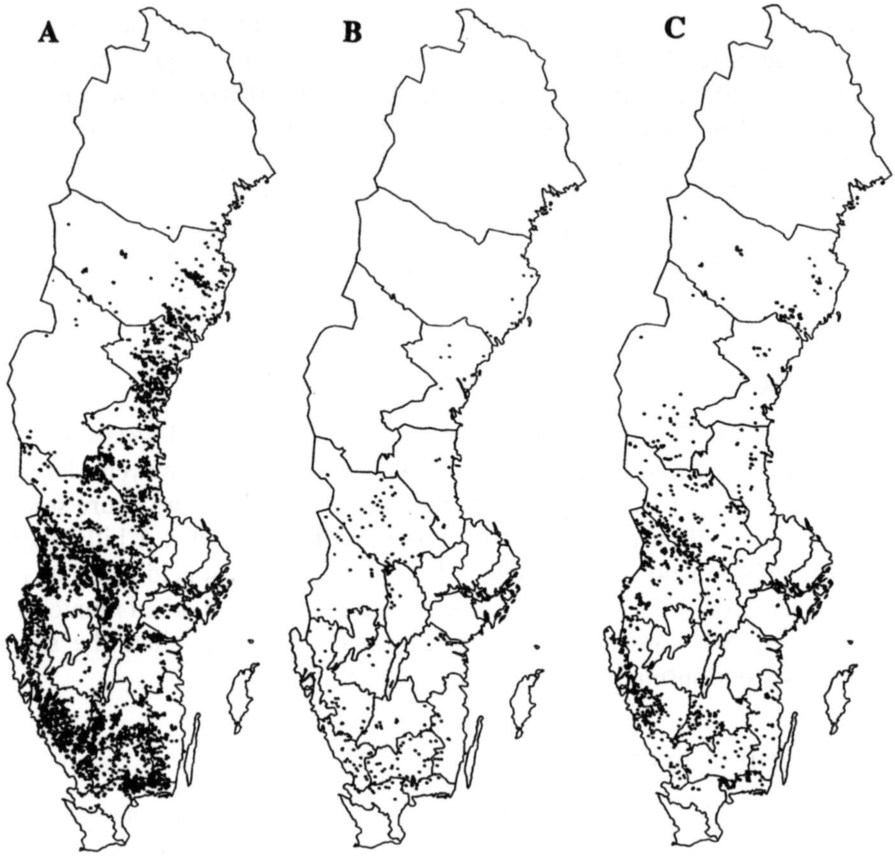

Figure 3. Lakes and watercourses in Sweden limed by means of the three major methods (1985-1993); A) lake liming (c. 6,500), B) doser liming (c. 300) and C) wetland liming (c. 1.500).

tion have been mainly applied and tested in Sweden. As described previously, however, the chemical influence of airborne deposition and the large-scale use of land produce effects that justify even greater liming activities in order to counteract acidification of soil and water in the long term.

7. Liming strategies and methods

7.1. Strategies

The strategy for dealing with acidification of surface waters depends on the resources available. If financial resources are limited, it is cheapest simply to try to save lakes, particularly those lakes that have a long turnover time.

If running waters are also to be protected, those that reach many lakes are limed and dosers are placed at strategic locations. Continued soil acidification is thus "accepted" and no effect is achieved in all the small running waters which make up the catchment system for lakes and rivers and which are often valuable nurseries for many species of fauna, including salmon.

If one wishes to save both lakes with a short turnover time and streams, liming is required upstream and/or in wetlands in the catchment area. This will mean costly operations with high lime dosing or treatment that needs to be repeated every few years. In return, the soil will be limed and the leaching of toxic metals into the watercourses will be reduced.

The strategy selected also depends on:

* what is in acute need of protection;
* what needs to be protected in the long term;
* the financial resources immediately available;
* the financial resources available in the long term;
* the benefit of the measures in relation to the cost;
* what effect the measures will have compared with not taking any action;
* the undesired side-effects that may arise.

If different sites are chosen for liming: lakes, running waters, wetlands and solid ground, various pH-increasing agents in different doses and fractions can be selected, as can different application times.

The optimal strategy is to utilize an agent at the lowest cost in order to achieve a desired effect that will last as long as possible and for undesired side-effects to be as limited and brief as possible.

7.2. Methods

Three methods - lake liming, wetland liming and doser liming - are used for spreading of more than 98% of the neutralizing agents in acidified surface waters in Sweden (Svenson, pers. comm.). A number of other liming methods to mitigate acid conditions in surface waters are available, i.e. full-scale catchment liming, road liming as an alternative to salting to reduce skidding, lime precipitation in wastewater treatment plants, ditch liming, and spreading of coarsely ground lime into watercourses. These methods can be used to complement the three major methods mentioned above.

7.2.1. Lake liming

During the first 10 years of liming operations in Sweden, emphasis was mainly on lakes and approximately 90% of the resources were used on liming directly in the lake. Such measures were generally adequate if the lakes had a turnover time of 1-2 years or longer. Even today, more than half the lakes are treated by liming directly into the water. The lime is dispersed from boats or pontoon vessels where the lime is mixed with lake water before being pumped out (Fig. 4). In the winter, the lime can also be dispersed on the ice with the aid of a tracked vehicle or snowmobile. Small lakes can also be limed successfully by spraying the lime from the lakeside across the lake surface from a bulk truck. Lakes that are poorly accessible in areas with no roads, for instance, are limed from the air by helicopter (Fig. 5).

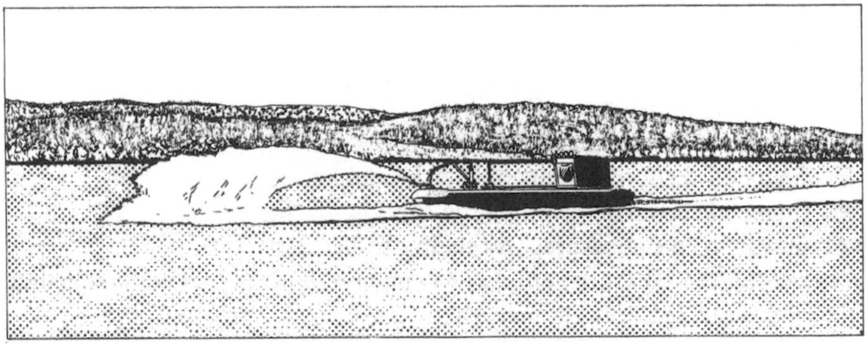

Figure 4. *Different kinds of boats are frequently used for lake liming. Drawing by J. Wihlke based on photo by A. Maslennikov.*

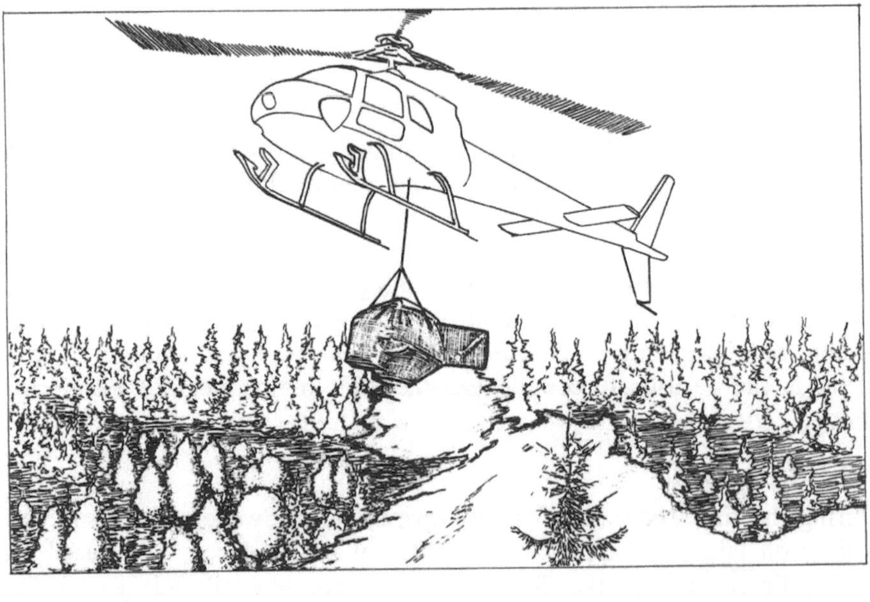

Figure 5. *Helicopters are commonly used for liming of poorly accessible lakes. Drawing by J. Wihlke.*

Dispersion on ice is the safest method of ensuring that the lime is spread in the quantities and in the locations planned. On an open lake using a pontoon or with dispersion of a lime mixture from a boat, the dispersion area may vary depending on the pressure and water content. A study has indicated that for normal dispersion on the lake surface, 0.8-2 kg/m^2 (8-20 t/ha) is needed. When dispersing from the side of a lake, the dispersion width was 8-15 m and the lime dose often 1-2 kg/m^2, i.e 10-20 t/ha (Sandberg 1992).

For normal lime doses, 5-20% of the lake area is covered. To cover the entire lake surface would take 4-6 times longer. In return, direct dissolution increased (Sandberg 1992).

The lime should thus be dispersed over a large area of the lake, although it does not necessarily need to be spread evenly over the whole surface. To achieve a slower dissolution rate and more lasting effect, it can be deposited along the shallow lake beds and hard lake beds. (Sverdrup *et al* 1986, Abrahamsson 1990b). If the lake has many promontories and is divided into several basins, the lime should be distributed so that each basin is adequately limed.

The type of lime recommended for lakes is 0-0.5 or 0-0.2 mm where 50% of the lime is finer than 50μ. Of the lime dispersed, 10-20% dissolves within a few days of being dispersed and 60-70% during the next 2-4 years (Alenäs 1986, 1988, Hultberg and Nyström 1988, Abrahamsson 1990b). In order not to create lime banks that are far too thick the lime dose should be less that 20 t/ha of lake surface (Alenäs 1986, 1988). Doses of approximately 5 t/ha have been normal in southern Sweden.

Direct liming is relatively cheap and simple to plan and implement. However, it does not prevent the influx of damaging metals from the acidified catchment area. During periods of snow-melt and high water, acid surges occur in the littoral zone of the lake which have a negative influence on the biology (Hultberg 1975, Lagerman 1983, Henrikson 1988, Hasselrot *et al* 1987, Abrahamsson 1993c). It may take several weeks before the new water becomes mixed with the "older" water from the deeper parts of the lake which, thanks to the liming, has a high pH level.

As the water in the lakes is constantly exchanged, lake liming only lasts for a limited period. The lake water neutralized by the lime eventually flows from the lake and is replaced by acid water from the catchment streams and precipitation. The duration is decided by the lime dose and the turnover time of the lake. Calculations of the duration and the lime dose have been made *inter alia* by Sverdrup *et al* (1986) and SEPA (1988). In lakes with a turnover period of less than 1 year, the duration is often equivalent to 2-3 years, while the duration in lakes with a turnover time of several years is equivalent to 4-5 turnovers (Fig. 6). Lakes with a very short turnover time can benefit from liming on the ice and from the lakeside every winter. This makes it possible to counteract the acid surges during the following spring melting period. The results and effects are discussed further in Wilander *et al* (Chap. 5).

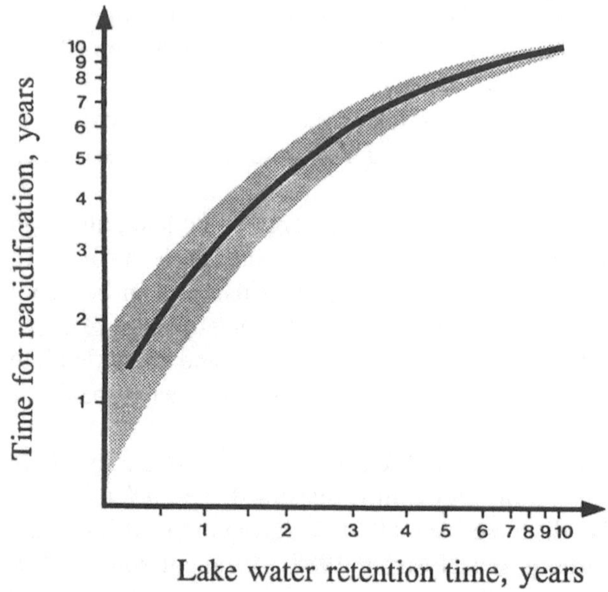

Figure 6. *Duration of neutralized conditions in limed lakes in Sweden (from SEPA 1988).*

7.2.2. Wetland liming

In water with a short turnover time - a few months - the lake must be limed at regular intervals to achieve the desired effect. Here, lake liming is often replaced or complemented by liming of wetlands which represent discharge areas, where the groundwater moves up to the surface and is transported in the upper layers of the soil to the watercourse. The chances of the lime added to the discharge areas being effective are good if the lime is dispersed on suitable soils.

Since part of the lime remains in the soil, larger quantities are required than for normal lake liming to achieve the desired effect in the watercourse. In most cases, moreover, helicopters must be used to disperse the lime, which can increase the cost. Wind drift when using helicopters for dispersion has also proved to be a significant factor; one study indicated that 70-80% of the lime landed where it should (Sandberg 1992), another stated that 50-60% of the lime landed at the desired location (Svahnberg 1992).

Liming also causes considerable changes in the wetland flora - bog moss is replaced by leaf mosses and sedges (Aronson 1990, Eriksson 1990, Fasth 1991, Rafstedt 1992, Larsson, Chap. 7). Wind drift also means that the lime could land on solid ground and damage lichens. It is thus important in wetland liming, both for ecological and financial reasons, that the lime lands where it is intended. Helicopter liming therefore should not be used in areas where there is a particular need to protect flora and fauna. However, dispersion of lime in wetlands can be done more accurately using land vehicles. A number of different types have been developed. Tracked vehicles, however, are slower and for this reason distances to the dispersion areas should not be too great.

The lime fraction is also of significance. Lake lime (0-0.5 or 0-0.2 mm) has mainly been used for wetlands. However, evaluation and development of operations are taking place on an on-going basis. Dispersion of a coarser lime than is used today, such as limestone powder of 0-1 or 0-2 mm, could reduce dust deposition and wind drift (Svahnberg 1992) and probably provide slower and more even dissolution (Abrahamsson 1993a), while at the same time it is cheaper and causes less damage to vegetation (Aronson 1992).

Effective wetland liming can be achieved with careful planning in terms of where the lime is to be deposited. The level of dissolution may in favourable cases be almost as high as in lake liming, 50-70% (Abrahamsson 1992, 1993b). Mathematical models for calculating lime requirements and the effects of wetland liming have been produced and *i.a.* been used in the USA and Canada (Gherini *et al* 1985, Warfvinge and Sverdrup 1989).

Liming of flooded (irrigation) soils initially provides rapid outward movement but uneven uptake, while the outflow (filtration) of soils provides a more even and slower outward movement, which means that the lime is utilized more efficiently (Abrahamsson 1990a, 1992). Liming of drained wetlands has normally been avoided but this could have a good effect on highly humified peatland where runoff takes place on the surface (Hasselrot and Lind 1992).

If the limed wetland surface makes up more than 1% of the runoff area, it is possible that positive effects might last at least 1 year. The lime dose should be at least 250 kg/ha of runoff area. With dispersion each year, the lime dose can be reduced to 60-110 kg/ha of runoff area (Abrahamsson 1992). The recommended dose in regions of the country is discussed in Sect. 9.2.

When pH increases in treated wetlands, a considerable proportion of the harmful metals that would otherwise leak out into the water is bound (Hasselrot *et al* 1984b, Lessmark 1987, Abrahamsson 1990a, Andersson and Eriksson 1991, Degerman 1992, Norrgren and Degerman 1993).

7.2.3. Solid ground liming

It appears from the above that, to achieve a rapid effect in the runoff water from the soil, the lime should be deposited in the discharge areas. Conversely, liming influx areas, i.e. solid ground, produces a slow effect on the runoff water and only a few percent of the added lime is initially leached out each year from the humus layer and the mineral soil, after which the effect diminishes further (Hallbäcken and Popovic 1985, Andersson and Persson 1988).

Soil liming as a method of treating streams and lakes was introduced experimentally in the 1970s, but leaching made up only a few percent each year. A summary of the short-term results has been presented (Lessmark 1987).

In order to comply with the overall goal of Swedish liming activities, viz. to restore ecosystems to their natural conditions as far possible, full-scale liming of the catchment soil is the best method in the long term. This strategy is, however, rarely applied, mainly because of the high initial costs involved and the long time needed to achieve desired effects in the water.

For long-term treatment of soil acidification, metal leaching and watercourse acidification, treatment should be carried out in the watershed area. Full-scale liming is, however, required to produce an effect on the quality of the runoff water. In return, forest soil, groundwater, streams and lakes are treated.

In a previous section the accumulated need for full-scale liming was discussed. Just to restore the level of base elements in the soil to the conditions at the beginning of the 20th century, 5 tonnes of limestone/ha would be required in southern Sweden. It is therefore hardly surprising that full-scale liming would not immediately produce a satisfactory effect on the runoff water. The soil profile down to a depth of 1.5 m in southern Sweden is acidified (Eriksson *et al* 1992) and must assimilate the slowly penetrating lime solution before sufficient effects appear in the runoff water. Total limestone utilization to the benefit of the surface water will after full-scale liming be perhaps 20% before the treatment needs to be repeated. In a transfer to large-scale solid ground liming it is necessary during an interim period, perhaps several decades, to use supplementary liming directly into the water until the profile of the forest soils has regained its buffer ability.

Many lakes in southern Sweden were strongly acidified in the 1960s and are found in areas with thin soil cover; often less than 0.5 m (Bernes 1991). However, here full-scale liming including solid ground means that the route of the lime to the watercourse is shorter and the effects are visible more quickly even if the leaching is only a few percent per year (Andersson *et al* 1982, Hasselrot *et al* 1984a, Warfvinge and Sverdrup 1988).

Liming of forests with the aid of state grants has been introduced in southern Sweden on a relatively large scale and on a trial basis. The primary aim of the treatment is to counteract nutrient depletion in the soil, which can lead to reduced forest production. Extensive liming with this aim in mind has been carried out for several years in Germany (Matzner *et al* 1985). Eventually, liming of forest soil may become important in the treatment of acidified lakes and watercourses.

In Sweden, a number of runoff areas have been limed and the effects of this on the runoff water have been studied. The lime used is primarily a mixed product of 2/3 limestone and 1/3 dolomite 0-3 mm, with a maximum of 5% coarser than 3 mm and a maximum of 30% finer than 0.25 mm, limestone powder or a mixture of lime and wood ash. The coarser material dissolves more slowly and is less damaging to the vegetation. Nevertheless, the trials indicate that lime uptake can be just as good as from finer fractions. The effects on the runoff water are moderate but positive. Lime doses of less than 2 t/ha can, however, in a strongly acidified environment, produce negative ion exchange effects in the soil and even a reduction in the pH in groundwater (Swedish Environmental Research Institute 1993).

7.2.4. Doser liming

A normally cost-effective means of treating running waters or lakes with a short turnover time consists of liming with the aid of dosers (Fig. 7). The dosers are linked directly to the watercourse and the advantages are that they can normally distribute lime in proportion to the flow, water level, water speed or pH in the watercourse and therefore lead to a certain desired downstream content. Either lake lime or other finely ground lime, dosing lime, is used as the liming agent where at least 50% of the lime is finer than 30μ. When dosers are correctly located in water that is sufficiently fast running, the effectiveness and level of dissolution are high, 70-90%.

***Figure** 7. Different kind of liming dosers can be used for liming of watercourses or lakes with short water retention time. Drawing by Kemira Ltd showing a doser device.*

The doser must be located beside a road suitable for vehicles so that it can be refilled with new lime. It should be in sufficiently fast running waters so that the lime does not settle on the bed directly below the doser. Bearing in mind the precipitation zone of toxic metals which arises in the event of an increased pH below the doser, the dosing distance should be at least several hundred metres (Dickson 1983a, Blomberg and Bjärnborg 1986, Lessmark *et al* 1986, Degerman 1992). Part of the problem is that the dosers have not always functioned when they are most needed, i.e. during high water and during cold spells.

Since liming operations began, significant developments have taken place in dosing techniques. A number of evaluations of Swedish lime dosers have been published (Tideström 1984, Sverdrup *et al* 1985, Kulinski and Dickson 1994). The total number of dosers in Sweden in continuous or temporary operation in 1993 was approximately 300. A number of different doser types are available on the market, some operated by water flow, wind power or solar cells, while others are operated by batteries or electricity. Neither diversion wells nor slurry dispensers appear to have a future and several installations have been replaced by other types.

A number of dosers are attached to an alarm to the fire station or other authority (Holmström *et al* 1987, Holmström 1988). Others are completely manual but are inspected daily by staff who live close by and who adjust the dose based on the water flow and check the level of lime in the container. Such dosers even have an important social function.

Common to all doser dispenser manufacturers is the fact that they are searching to find more operationally reliable solutions. Operational interruptions depend more on lack of supervision and service than on particular technical problems. The dosers with 20-130-tonne containers cost USD 25,000-100,000 depending on the technology employed and are in themselves of considerable value. However, with a total operating lifetime of approximately 10 years, the investment cost in most cases is soon recouped. During this period, a considerable amount of lime has been dispensed into streams and the littoral zones of lakes during periods of the year when often no other method is available.

7.2.5. Stream liming

The simple technique of spreading large amounts of coarsely ground lime into watercourses was tested at the end of the 1970s in southern Sweden. Often the aim was to increase the pH in the watercourse and to improve the immediate physical and chemical environment in spawning and nursery grounds. The chemical effects are also clear when sufficient doses are used. Approximately 1 t/ha of runoff area and 1 t/m of stream have been tested.

The effect is directly controlled by the flow: ten times the flow provides approximately ten times lower concentration in the water from dissolution.

Eventually, the lime dissolution decreases due to the accumulation of layers of algae, iron, manganese and aluminium which gradually cover the decomposition areas of the lime. Total lime utilization is therefore low (Lagerman 1983, Fleischer 1984).

7.2.6. Ditch liming

A simple and rapid method of dispersing lime consists of blowing it directly from a road down into ditches alongside the road. The method has been tested in different projects. The dose is, however, high, short-lived and covers a small area, and as road ditches are often not discharge areas in the true sense a good deal of lime is trapped in and alongside the ditch. Some is carried forward at the next high water period but the total effect is short-lived and the total exchange of lime minimal (Lagerman 1981, Lindgren 1991).

7.2.7. Road liming

In the winter, a salt-sand mixture is applied on lesser-used roads to reduce the risk of skidding. The salt, natrium chloride, increases corrosion of vehicles and, through ion exchange mechanisms in the soil, involves increased acidification of the runoff water. To test more environmentally friendly alternatives, trials have been conducted using lime in fractions of 2-5 mm as a replacement for the salt. The effect on skid prevention was the same, and on certain road surfaces it was better when lime rather than a salt-sand mixture was used. Similar experience was gained when salt was replaced by Absol, a natural product with a basic effect of 26% CaO (cost SEK 45/km) (Karlsson 1990).

By placing a line of light, coarsely crushed lime along the edge of the asphalt on the road, the road is supported, the edge of the road can be seen and the water from the road is filtered through lime gravel before it reaches the ditch. In addition to increasing the pH in the runoff water and ground, there are also visible effects with a richer and more striking roadside flora (Åkesson 1993). Road liming is thus an environmentally friendly means of sparing human lives, vehicles, the soil, vegetation and the runoff water. Calculations and experience, however, indicate that road liming can only in exceptional cases give rise to a significant increase in pH in an adjacent acidified lake or watercourse.

7.2.8. Lime precipitation in wastewater treatment plants

Wastewater treatment plants located beside acidified waters with an alkalinity of less than 0.05 meq/l and which are thus themselves entitled to a liming grant can be given a grant for rebuilding to accommodate lime precipitation. Normally, aluminium or iron salts are used as precipitation chemicals for phosphorus and organic materials. These chemicals reduce pH and alkalinity of outgoing water. With slaked lime ($Ca(OH)_2$) it is possible to achieve a good purification level and the alkalinity of the outgoing water is increased. In addition, owing to high pH, bacteria are killed and chlorination of the outgoing water can be avoided (Bergström 1977, Hultman *et al* 1982, Wikström 1984, Halland County Council 1986). Some 20 wastewater treatment plants in Sweden have received rebuilding grants in the last 10 years to facilitate purification of water by lime precipitation. However, the lime precipitation technique is considered by operating personnel as more difficult to deal with and demanding a good deal of work. In addition, a greater amount of sludge is produced, albeit with a high pH value, but in these times of agricultural changes it often involves disposal problems. In total, these factors have meant that lime precipitation in wastewater treatment plants located in an acidified environment, despite a generous state grant, has not been successful. However, the advantages for the ecosystem in the water environment are obvious (Engblom and Lingdell 1985b).

8. Liming agents and fractions

8.1. Agents

Limestone powder ($CaCO_3$) or dolomite ($CaMg(CO_3)_2$) are used in more than 95% of all liming of lakes and watercourses in Sweden. The situation is similar in Norway and the USA. The experience of past years has indicated that limestone powder and dolomite are the cheapest agents for lakes, dosers and wetlands. Limestone powder is also easily handled and the risks of harmful effects on the ecology in the event of an excessively high dosage in an acidified environment are small. The normal natural limestone contains very small amounts of harmful or undesirable elements (Table 1).

Table 1. *Maximum concentrations of metals in lime used for treatment of acidified waters in Sweden.*

Al	<20 g/kg	Co	<20 mg/kg	Cd	<0.5 mg/kg
Zn	<50 mg/kg	Cu	<20 mg/kg	Hg	<0.05 mg/kg
Cr	<50 mg/kg	V	<20 mg/kg		
Ni	<20 mg/kg	Pb	<20 mg/kg		

Limestone was formed during different geological periods through chemical eruptions or sedimentary deposition. Limestone is very common in many European countries but considerably less common and unevenly distributed throughout Sweden (Fig. 8; Bäckman 1990). Many limestone quarries have been closed in the last few decades and the transport distances to the liming areas are sometimes quite long. There are, however,

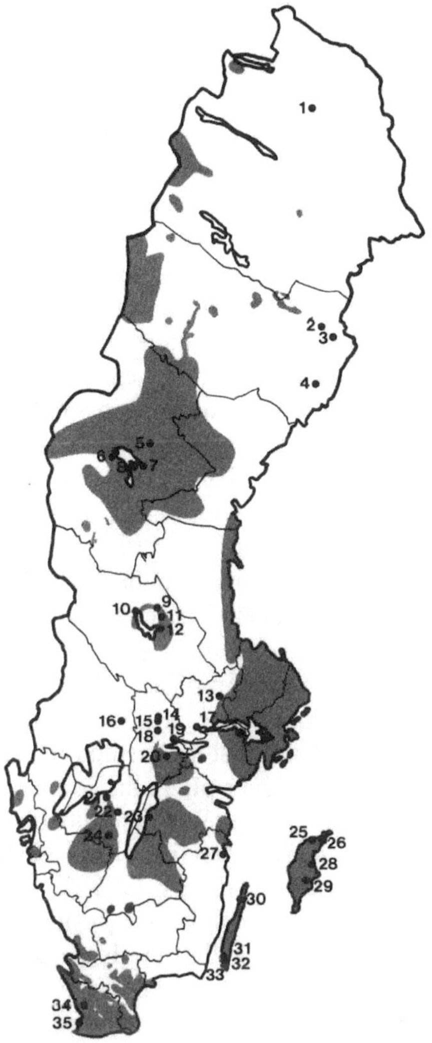

***Figure** 8. Bedrocks and/or soils rich in lime and limestone quarries (numbers) in Sweden.*

sufficient supplies of limestone in Sweden to fulfil the demands of fresh-water liming in Sweden several decades ahead.

Primary limestone was formed over 2 billion years ago and is often found in Sweden as bands in leptite formations containing iron ore. Dolomite is quarried in a few places in southern Sweden. Primary limestone normally has a high carbonate content, 85-95%. When it is ground, there are many fraction types which means that the lime powder has a high volumetric density and is thus suitable, for example, for dosers.

The largest limestone deposits normally consist of red orthoceratite limestone in southern Sweden and along the mountain chain. The carbonate content is not particularly high, 75-85%. It is suitable as environmental limestone when transport distances are not too great. Silurian limestone is quarried at a few places in southern and central Sweden. It is normally very pure with a carbonate content of around 90%. Chalk limestone is the youngest limestone and considerable reserves are to be found in southernmost Sweden. It is relatively soft with a carbonate content of 80-90%.

Sodium carbonate (Na_2CO_3) is the second most used neutralizing agent for alleviation of acidified conditions in Swedish surface waters. It is, however, added to less than 3% of all limed surface waters in Sweden. Sodium carbonate is highly soluble and produces a high pH easily. Used in the right environment, such as humic lakes with a short turnover time, the agent can be suitable. It is also used as a trace element to follow water movements and water exchange (Holmström 1984). To avoid a very high pH sodium carbonate is sometimes ploughed down into the lake sediment from where it gradually leaches out and provides a long-term effect on water quality and high utilization of base elements (Lindmark 1984, 1992a). In a number of other cases, sodium carbonate briquettes with the size of an egg have been applied on 10-30% of the lake's soft beds. They are dissolved slowly in the sediments. The sodium carbonate is considerably more expensive than lime but the dissolution rate is good even in humic lakes and irrespective of the pH which means that the method in certain cases can still possibly be competitive. The higher pH in the sediment around the sodium carbonate (pH can easily rise to over 9 in the immediate vicinity) means that the phosphorus content in the water increases which could be positive if the lake is nutritionally poor (Lindmark 1992b).

Several other neutralizing agents are available to counteract freshwater acidification. These are, however, only rarely used within the large-scale liming activity in Sweden, generally because of environmental disadvantages or significantly higher costs in comparison to limestone. Some products are worth mentioning, however.

The annual "production" of eggshells in the country is approximately 10,000 tonnes. The shell consists of porous and easily soluble lime and some membranes attached to the shell. At individual bakeries and food manufacturers, several tonnes of eggshell are produced each week. The agent has been tested in a number of lakes and in one case the effects have been carefully monitored (Dickson 1983b). The alkalinity was doubled but a large proportion of the lake's oxygen reserves was used up. The recommendation was that the eggshell membranes should be broken down on land before the eggshells are used in the water. Nevertheless, the product could be used for soil improvement rather than being dumped.

In different areas along the Swedish west coast there are shell gravel banks. This material is used sometimes as road filling and for soil preparation and has a very good liming effect. In a number of cases it has also been used for liming lakes. Mussel farming at present produces several tens of thousands of tonnes of shells per year in Denmark and Sweden which could be put to better use than simply being dumped. The liming effect is just as high as for limestone (Skoog 1993). If transport of sufficient volumes can be arranged by boat from the farming locations, the agent could be used as an alternative to limestone powder in regions such as northern Sweden where the cost of transporting by road is high.

Algal lime from *Lithothamnion calcareum* is quarried in several places, including off the coast of France. The agent has been used for a long time in waterworks and as part of biodynamic cultivation as an alternative to limestone or dolomite. It contains a number of micronutrients, including selenium, 1-2 mg/kg and 0.1-0.4% phosphorus, i.e. ten times more phosphorus than any other lime (Persson 1989, Lindström 1989). The agent has been tested in full-scale trials in southwestern Sweden (Alenäs 1986, Lindström 1992). It is easily soluble and effective, but up to now considerably more expensive than limestone. One possible area of use could be lakes that need added selenium or nutrients to reduce the content of mercury in fish (Håkanson *et al* 1990, SEPA 1991).

Spreading of wood ash on acidified forest soils can be conducted to avoid nutrient depletion in Swedish forest soils (Eriksson and Börjesson 1991, Rosén 1991, Timm 1993). In the long term, ash recirculation in forest areas can also result in positive effects in acidified surface waters, e.g. it can contribute to restore the nutrient level of the water. Ash is a residual product from the burning of fossil or bio-fuels. During burning, acid or acidifying gases are released and a surplus of base elements remains in the ash. The basic content of wood ash is equivalent to approximately 30% CaO, and it also contains 1-2% phosphorus and quantities of the macro- and micronutrients from which biological material is made up.

As a means of increasing pH in acidified water, ash can be closely compared to sodium carbonate or the washing powder of the past, potash, which was extracted from birch ash. The pH can easily reach 9-10 if the dosing is too high. The phosphorus content is also 100 times higher than in normal limestone. The agent therefore cannot be dosed directly into the water. In Sweden approximately 200,000 tonnes of ash are produced from the burning of bio-pulp and the majority is dumped.

A number of other neutralizing agents have been tested for treatment of acidified surface waters, but shown to be unsuitable. Some are them are mentioned below in order to avoid a repetition of mistakes.

Burnt lime (CaO), slaked lime ($Ca(OH)_2$) and caustic soda (NaOH) are so corrosive that they can cause injury to staff and damage to machines during dispersion and can locally induce a harmfully high pH in the water.

The silicate mineral olivine ($MgFeSiO_4$) has a high base content equivalent to approximately 67% CaO but decomposes too slowly to be of use in the field and also contains 0.2% nickel and chrome which are not desirable in the environment. The agent has been tested in laboratory trials (Eriksson 1982) and on agricultural land (Bertilsson 1980), in full lake experiments (Eriksson *et al* 1980), and in runoff areas to lakes and streams (Andersson *et al* 1982, Olsson 1987). The effects on vegetation have also been documented (Olsson 1982).

Slag limes from industry, such as T-lime, can be a very effective means of increasing the pH and were previously used to treat many lakes on the west coast of Sweden (Hultberg and Grahn 1974, Hultberg and Andersson 1982). However, in the event of excess dosing, a pH value that is too high could arise, causing an increase in aluminium levels and harm to the fish population (Degerman *et al* 1990). Slag limes are no longer used due to their often high metal content. T-lime, for example, contained approximately 0.3% chrome. Lime sludge is a lime product manufactured and used at sulphate pulp plants. There are still large quantities of lime sludge at mills that have been closed down. The agent has been tested in different areas and the pH effects have normally been good (Valdmaa and Dickson 1982). The pollutant content has, however, often been far too high for the agent to be recommended. Lime sludge with a low pollutant content, however, could be of use.

8.2. Fractions of limestone and dolomite

Limestone is crushed and ground to the desired fractions. The level of grinding is indicated by a granular curve which describes the division of fractions. The total specific surface of the lime grains determines the relative rate of dissolution. If the average diameter is reduced by one to the power of ten, the rate of dissolution increases by two to the power of ten.

Comprehensive laboratory studies and full-scale trials have been devoted to liming kinetics, dissolution and proposals for the choice of the most suitable fraction size (Sjöberg 1979, Sverdrup 1985, Warfvinge 1988). The length of time for the limestone powder to dissolve in water depends on the specific surface of the limestone grain, the pH of the water and the levels of carbon dioxide, iron, humus etc. The finer the fractions, the more rapidly the dissolution takes place and the shorter the long-term effect of lime dissolution.

Limestone is relatively difficult to dissolve (maximally, in equilibrium with the air's atmosphere, 2-2.5 meq/l when the pH is a little over 8), which means that the water's pH remains within biologically acceptable limits even directly after dispersion, provided that balanced dosing and the recommended fraction distribution are applied. The highest pH after liming is seldom more than 7-7.5 and the alkalinity 0.2-0.3 meq/l. The larger grains of limestone can take months or years to dissolve, which contributes to extending the duration of the treatment (Table 2; Dickson 1984).

Table 2*. Dissolution efficiency in lake water and pH effect for different fractions of limestone. The $CaCO_3$ dose is 20 mg/l, pH before liming is 4.6 and colour >100 Pt/l (from Dickson 1984).*

$CaCO_3$ fraction	Dissolved fraction in %		pH in lake water	
mm	3 weeks	14 weeks	3 weeks	14 weeks
0 - 0.06	83	90	7.2	7.2
0.06 - 0.2	56	90	7.1	7.2
0.2 - 0.6	40	85	6.7	7.2
0.6 - 2.0	20	43	5.7	6.6

About 80% of the limestone used in Sweden has a particle size of 0-0.2 mm with an average diameter of 30-50μ; otherwise coarser lime is used, 0-0.5 mm or 0-1 mm. When liming using a doser, the lime must dissolve quicker to provide the desired effect in the water and to ensure that sludge banks do not occur downstream. In such cases a more finely ground product is recommended, 0-0.1 mm, with an average diameter of less than 30μ. The recommended lime fractions are shown in Table 3.

Table 3*. Limits for particle size distribution for recommended lime fractions (0-0.5 and 0-0.1) in Sweden.*

Particle size 0-0.1 mm		Particle size 0-0.5 mm	
Screen (mm)	Passing screen, %	Screen (mm)	Passing screen, %
0.001	15	0.002	0
0.002	20	0.020	27
0.063	90	0.063	60
0.100	100	0.125	82
		0.250	95
		0.500	100

A finely ground limestone product can also be compressed into thicker particles, granules or pellets. Normally, the aim is to make the product easier to handle and disperse without deterioration of the dissolution process. When liming wetlands by helicopter, a coarser material is sometimes desired which will fall more heavily onto the ground and actually land where it should. Various binding agents for granulation have been tested, including ash, slaked lime and water. The granulation costs are, however, high, at least SEK 200 per tonne.

The optimal limestone powder should be both cheap and in the case of lake or wetland liming consist of fractions which last over a longer period rather than dissolving immediately. Unless the granulation costs fall considerably the optimal fractions for liming wetlands is probably a more coarsely ground natural product, 0-1 or 0-2 mm, instead of the more finely ground material used today.

8.3. Quality control of lake lime

When liming lakes and watercourses, product control is carried out where it is required that the lime should meet the conditions prescribed by SEPA (1988). The control is called P-marking and consists of rules formulated by the National Testing and Research Institute regarding sample extraction and metal analyses, fraction distribution, acid neutralization effect and moisture content (Schouenborg 1991).

9. Doses of lime and other neutralizing agents

9.1. Lime dissolution rates

When a basic element (lime, dolomite, sodium carbonate etc.) is added to a lake and sinks to the bed, direct dissolution takes place (phase 1). Down on the lake bed secondary dissolution then continues (phase 2). This is most obvious during the first year. When it has decreased and become less than the reacidification from the hydraulic loading, a dilution process remains (phase 3) (Sverdrup *et al* 1986, Abrahamsson 1990b, Sandberg 1992).

The extent of direct dissolution depends, *i.a.*, on the depth to which the lime has sunk, grain size, pH of the water and mixing conditions, dose and dispersion technique. Direct dissolution of the lime can vary from 1-2% in lakes where the lime is deposited in large depots along the lakeside to more than 50% when the dose was small and the lime spread over the deep areas of the lake (Sverdrup *et al* 1986, Sandberg 1992, Wilander *et al*, Chap. 5).

Secondary dissolution depends on the limed bed area, grain size, lime dose and water quality. With lake liming the lime dose is often in the region of 1-5 t/ha of lake area. In shallow lakes the direct dissolution is 10-20% and the remaining dissolution is thus secondary dissolution from the bed. When the lime is deposited along the lakeside, the dose is often equivalent to 10-20 t/ha. Direct dissolution is then less and much of it takes place as secondary dissolution during the next 1-2 years. The total dissolution can nevertheless be high, over 70% (Alenäs 1986, Sandberg

1992, Wilander *et al*, Chap. 5).

The dilution process is mainly affected by the turnover time of the lake. This depends on the volume of water in the lake and the size of the watershed. With a complete remixing of a dissolved chemical in the water, such as lime or sodium carbonate, the content decreases logarithmically: after 1 change of water (1 turnover period) 37% of the element remains and after 2 changes of water (2 turnover periods) 14% of the element remains.

The majority of lakes in Sweden, however, are layered or covered by ice for long periods of the year which means that the remixing is incomplete. The surface waters of the lake are therefore changed over a shorter period than in theory and the bottom water changed more slowly. For this reason reacidification due to dilution can produce acid surges down to a depth of more than 2 m in newly limed lakes during periods of high water. At the same time, the water at the lake bed is still being affected by secondary dissolution and can have an alkalinity of over 1 meq/l (Hasselrot *et al* 1987, Abrahamsson 1993c).

The actual reacidification consequently takes longer than theoretically calculated. This is due in part to the fact that the lake water is not remixed completely and partly to the fact that secondary dissolution takes place over several years. The time for reliming can, for practical purposes, be calculated with acceptable accuracy based on individual alkalinity values on a time scale.

A number of lake liming and reacidification models have been developed which can predict with reasonable accuracy lime or sodium carbonate requirements, the duration and the reacidification process (Sverdrup *et al* 1986, Abrahamsson 1990b, 1993c, Sandberg 1992, Lindmark 1992a).

Leaching of base elements from a limed wetland is determined partly by dissolution of lime and partly by leaching of exchangeable bicarbonate ions from the soil horizon. The actual dissolution of the lime can take place relatively quickly, while the leaching process takes longer. Both processes are influenced by the throughflow of water. Models which have succeeded in relative terms in replicating the actual process of measured content levels have been developed (Warfvinge 1988, Abrahamsson 1993b).

9.2. Dosage recommendations

The first dosage recommendations (NBF 1982) took into consideration water quality and turnover time but also current acid loading in the area, recalculated as limestone requirements per hectare of runoff area (70 kg/ha in southern Sweden and 15 kg/ha in northern Sweden). This sometimes entailed a certain degree of excess dosing.

The dosing now recommended (SEPA 1988) for lake liming is determined purely by the lake's turnover time and its pH prior to liming. If the pH exceeds 5.5, 10 g of limestone per m^3 in those cases where the turnover time exceeds 3 years is sufficient, while 30 g/m^3 is required if the water is changed over a period shorter than 6 months (Table 4). For lakes with a pH down to 4 approximately three times higher lime doses are required, i.e. up to 90 g/m^3 or, alternatively, liming can be carried out more often. Reliming should take place before the pH once again reaches a critical level. In return, lower doses are necessary. Doses recommended for doser liming of watercourses are shown in Table. 5.

Table 4. Recommended dose for lake liming in Sweden (SEPA 1988).

pH before liming	Alkalinity before liming, meq/l	Recommended dose g/m^3 (fraction 0-0.2/0.5 mm, 50% CaO), years			
		<0.5	0.5-0.9	1-3	>3
5.6-6.0	<0.05	30	20	15	10
5.1-5.5	0	45	30	20	15
4.6-5.0	0	60	40	30	20
4.3-4.5	0	75	50	40	25
4.0-4.2	0	90	60	50	30

Table 5. Recommended dose for liming of watercourses in Sweden (from SEPA 1988).

pH before liming	Alkalinity before liming, meq/l	Recommended dose g/m^3 (fraction 0-0.2/0.5 mm, 50% CaO)
5.6-6.0	<0.05	10
5.1-5.5	0	15
4.6-5.0	0	20
4.3-4.5	0	25
4.0-4.2	0	30

The dose for Swedish lake liming has been described by Thörnelöf and Degerman (1991) according to the formula:

dose (g/m^3) = - (LOG_{10} of retention time x 28.14) - (pH x 23.18) + 151.2.

The dose may result in a somewhat higher pH and alkalinity than that occurring under pristine conditions in the lake. The reason for this is related to the chemical goals for freshwater liming in Sweden, i.e. to prevent acid surges and to detoxify the water from harmful concentrations of metals. Hereby, suitable conditions for a restoration of a natural flora and fauna are probably provided. This national strategy and how it might be changed in the future are addressed in Henrikson and Brodin (Chap. 1). In the following, some brief comments are put forward to further elucidate the Swedish strategy on lime dosage.

It is never possible through liming to completely restore the natural chemistry in the water, as a chemical is added to the water which increases the salt content and conductivity. A dose which sometimes causes the alkalinity to rise a little higher than in nature is, however, far less harmful to the flora and fauna in the water than exposure to chronic stress at the chemical limits for survival caused by a low pH.

Precipitation of toxic metals such as aluminium, cadmium, iron and manganese takes place much more rapidly at pH 6.5-7 than at pH 5.5-6 (Stumm and Morgan 1970, Dickson 1983a, Björklund *et al* 1985, Blomberg and Bjärnborg 1986). This is particularly noticeable in cold water when the metal content levels are highest and the temperature itself is a stress factor for the biology. From a biological point of view, it is thus better to have somewhat unnaturally high pH, 6.5-7, but a more natural low aluminium content, 0.05-0.10 mg/l, than a pH of 5.5-6.0 and an aluminium level of 0.2-0.4 mg/l.

Liming of Swedish lakes and watercourses therefore normally entails a certain chemical overdosing in order to establish refuges for organisms during periods of acid surge. A certain overdosing results in improved conditions downstream since in time the beds achieve a certain lime repletion. In the event of involuntary interruptions in the operation of the lime dosers, improved chemistry can persist over a period of several days and for many kilometres downstream owing to the fact that the lime deposited earlier is dissolved (Widarsson 1988). In some watercourses the dosers are run with deliberate interruptions to make better use of the lime (Widersson, pers. comm. 1993).

If the individual is partly responsible for financing the liming project, the dosing tends to be lower and liming for fish is a priority. A study indicated that the willingness to pay decreased even when the contribution was limited to 5% of the total cost (Lindeneg and Navrud 1993).

In watercourse systems, dose calculations can be facilitated by using hydrochemical models. In the PULS model the input consists of daily temperature and precipitation figures and monthly evapotranspiration figures. The chemical part of the model is calibrated against the measured alkalinity and pH in the watercourse over a number of years and the hydrological information is obtained from current information in the region. Using model simulations, different liming treatments in lakes and running waters can be tested and mixing conditions and doser locations varied. The costs for different activities can be weighed against the effects in the form of acid surge periods or persistent pH values and their effects on the biology (Carlsson and Bergström 1989, Sanner *et al* 1993).

Dose calculation for wetland liming is much more difficult than calculation of direct liming in lakes or watercourses. Experience acquired from practical work in Sweden and mathematical models, e.g. applied in Sweden and Scotland, indicate major differences in lime requirements for different types of wetlands (Brown 1988, Tervet and Harriman 1988, Warfvinge and Sverdrup 1988). Important factors to be considered are soil chemistry, climate, and hydrological conditions within the catchment. The general lack of sufficient data on soil chemistry and hydrology weighed against the great need to counteract acidification has resulted in the development of standards for wetland liming in Sweden (SEPA 1988). The standards are based on scientific as well as practical experience.

For liming wetlands in northern Sweden a dose of 0.3 t/ha is recommended based on the whole catchment area. As the discharge areas normally only cover a small part of the catchment, the actual dose in the areas capable of being limed is considerably higher. For example, if they make up 5% of the catchment area, approximately 6 t/ha should be added. Even higher doses are required in the southern parts of the country where the watercourses are often acidic throughout the year. In this region a lime dose of 1 t/ha is thus recommended, calculated over the whole catchment area (Table 6).

The recommended dose is based on full-scale trials and continuous evaluations of wetland liming, which means that the doses can be gradually adapted to the actual results. In certain wetlands lower doses are more economical, however, dispersion must be carried out at more regular intervals than has been recommended up to now (Abrahamsson 1992). If lime is dispersed using land vehicles, and a slightly coarser lime is used than at present, the accuracy will increase since more lime lands where it is intended and the duration of the measures carried out is extended.

Table 6. *Recommended dose for liming of wetlands (from SEPA 1988). Percentages in column 3 refer to the percentage wetland area limed compared to the total catchment area. Region Götaland is in southernmost Sweden; Svealand covers southern Sweden north of Götaland; Norrland is northern Sweden.*

Region	Dose t/ha	t/ha limed area 10%	5%	2%
Götaland	1.0	10	20	50
Svealand	0.5	5	10	25
Norrland	0.3	3	6	15

10. Costs

The governmental subsidies for liming are intended to counteract anthropogenic acidification and activities are aimed at controlling the funds so that they are used optimally in terms of cost, lime uptake and effects.

The price of the product depends on its origin, base content, purity, fraction arrangement and the quantity delivered. Fine-grained and selected fractions are the most expensive as it takes time and energy to produce them. A roughly ground lime, 0-0.5 or 0-0.2 mm, costs around SEK 200, dosing lime 0-0.1 mm around SEK 250 and the extremely finely ground filler lime 0-0.005 mm around SEK 500 per tonne (excluding VAT, 1992).

If the lime is packed in large sacks of 1 tonne, the price is approximately SEK 150 more and in 25 kg sacks approximately SEK 200 more per tonne. Algal lime and sodium carbonate cost around SEK 1,500 per tonne.

The transport costs depend on the distance from the point of extraction to the area to be limed, the means of transport, the amount of lime and the time of year. The mileage charges for the lime suppliers is normally SEK 5 or 10 per tonne.

The dispersion cost depends on the quantity and method of dispersion. The most expensive method is by helicopter and using dosers with a limited annual dispersion, SEK 300 per tonne upwards. Large-scale lake liming using a boat is cheaper with a cost of around SEK 100 per tonne, whilst the cheapest is blowing directly from a bulk truck, costing approximately SEK 10 per tonne.

The time of year at which the dispersion takes place is of significance to the price. A considerable amount liming is done in late summer and autumn. During a few months, all available equipment in the country is being used. Sometimes even helicopters from abroad need to be brought in, which naturally makes the operations more expensive. With a more even distribution throughout the year, including the winter months - possibly with the exception of the breeding season - liming would be cheaper.

A few examples of costs from southern and northern Sweden are shown in Table 7. Depending on the longer transport distances, it is more expensive to lime in the northern part of the country. Often, a more fine-grained lime is needed for the dosers which is a little more expensive. The same price has been allocated to lake and wetland lime as up to now the same products have been used. Examples are taken from the actual operations in 1990. The dispersion costs for the dosers are calculated based on a depreciation period of 5 years and with an annual dosing level of 200 tonnes. Wetland liming is calculated as helicopter dispersion. As expected, the cheapest method is lake liming followed by doser liming. As a result of a lower overall level of uptake, 50%, and a high dispersion cost, wetland liming is normally the most expensive method per tonne of dissolved

Table 7. *Cost for aquatic liming in Sweden 1990. 1 SEK is approximately 0.13 US dollars (July 1994).*

Method	Material +transport SEK	Spreading SEK	Efficiency %	Cost/tonne dissolved base, SEK
Southern Sweden				
lake	200	100	70	430
doser	250	200	80	560
wetland	200	300	50	1,000
Northern Sweden				
lake	500	100	70	860
doser	550	200	80	940
wetland	500	300	50	1,600

base element (Dickson 1990). However, a somewhat coarser lime could be used in the future for wetlands to increase the dispersion accuracy and duration and to reduce damage to vegetation thereby reducing the cost slightly (Svahnberg 1992, Abrahamsson 1993a, Aronson 1992).

11. Complementary methods

As has often been stressed on various occasions, radical reductions in acid emissions is the only long-term strategy to combat acidification. Even if international measures are praiseworthy, the agreements reached today are insufficient to protect the oligotrophic Swedish waters.

Considering direct measurements in the environment, there are no satisfactory alternatives to liming, however, some forms of treatment can act as a complement to liming.

Changes in land use provide one important example. Because of the deep root system and leaf litter of the birch tree, the pH of humus and mineral soils in birch forests is around 0.5 units higher than in spruce and pine forests. The store of exchangeable base cations and the pH of the water in the soil are therefore higher (Ministry of Agriculture 1982). A general inclusion of 50% birch in suitable forests could increase the pH by 0.2-0.3 units, greatly affect the soil's pH (Lundmark 1988, Liljelund 1988, Drakenberg *et al* 1991) and achieve a certain improvement in the runoff water. The method has been successful, although up to now this has been largely a result of increased demand and a high price for birch. Forest conservation boards and forestry companies now often recommend 15-20% leaf content compared with approximately 10% during the 1970s. The soil improvement qualities of the birch should be used to a far greater extent than today as an alternative or complement to soil liming. Ash produced by burning of forest products should naturally be returned to the forest soils. Doses of 1-3 t/ha and forest generation compensate the soil depletion resulting from biomass extraction and could be done without harming sensitive soil organisms (Timm 1993).

Domestic animals that graze in forests and meadowland produce mechanical and chemical vitalization of the soil and they cause a release of base elements and nutrients through their fertilization. The domestic pig's ability to process the soil has been used for hundreds of years for soil vitalization and natural seed dispersal. The pigs also rapidly root the added lime into the mineral soils. In trials, the pH in the mineral soils rose after a month by one unit compared with an 0.1 unit increase with liming and no pigs (Dickson 1991, 1992). To use the country's forests and meadowland for animals during the vegetation period to a far greater extent than today, as with increased inclusion of birch forests, can be a cheap method of improving the condition of the soil.

Controlled burning is a quick method of transforming organically sealed nutrients in wood remnants and soil vegetation to inorganic bases: potash, soda, calcium carbonate and magnesium carbonate. The method has been used regularly in forestry, grazing and in agriculture. As the organic content is burned away and the basic salts can be lost in the runoff water, this form of soil vitalization fell out of use some decades ago. However, the method is returning and is suitable where the thickness of the soil cover is sufficient. The method entails liming and nutrient enrichment of the forest soils and the effects on the runoff water, pH, base cations and nutrients increase. Even forest fires caused increase leakage of base elements and a higher pH in acidified water which has been demonstrated in lakes on the Swedish west coast (Alenäs 1987).

The present wastewater treatment process entails low discharges of phosphorus to the lake or watercourse. This can result in a decreased alkalinity of the freshwater body due to the use of acid precipitation chemicals and nitrification of the wastewater where nitrogen has often not been removed sufficiently. A technical and chemical control ensuring that out-going water from the treatment plant is already nitrified and has an increased pH thus has a positive effect on acidification-sensitive waters both with regard to pH and oxygen conditions. Soil acidification means that phosphorus is sealed harder into the soil, and the nutrient supply to adjacent lakes and watercourses is decreased (Olsson and Pettersson 1993). Therefore, it seems less urgent to reduce the phosphorus content of wastewater discharged to nutrient-depleted and acidified lakes and watercourses (Hultman *et al* 1982).

Digging ditches in forest land is carried out to improve the conditions for forest production. Digging deeper ditches, to yet non-acidified levels, channels the water down to better soil levels and can therefore cause an increased pH in the runoff (Ramberg 1981). A number of ditching operations, however, produce pH reductions when the sulphur-rich or organic soils become aerated (Magnusson 1982, SEPA 1987). As the natural hydrological conditions are changed and conflict with other values in the natural environment, creation of forest ditches is hardly a means to be used other than for the sake of the forest itself.

Water regulation, where water from melting snow is saved in the emptied stores, can be a method to suppress acid surges further down the watercourse. In a number of watercourses the method can be effective if sufficient base elements are added from the runoff area below the adjustment point. As spring floods in the forest and in the mountains do not occur at the same time, the method can, on the contrary, strengthen acid surges in such large watercourses (Carlsson and Bergström 1989, Västernorrland County Council 1993).

12. References

Abrahamsson, I. 1990a. Ytvattenkemiska effekter från en våtmarkskalkning. [Effects on surface water chemistry of wetland liming.] - Terra-Limno Group, Mölnlycke.

Abrahamsson, I. 1990b. Lägesrapport. Vattenkemisk respons och kalkutnyttjande i kalkade sjöar. [Status report. Response of water chemistry and the utilization of lime in limed lakes.] - Terra-Limno Group, Mölnlycke.

Abrahamsson, I. 1992. Vattenkemiska resultat av våtmarkskalkning i Jönköpings län. [Results on water chemistry of wetland liming in the county of Jönköping.] - Jönköping County Council, Rep. 7/92.

Abrahamsson, I. 1993a. Uttnyttjande av kalcium i kärrmarker med kalkstensmjöl 0-1 mm. [The utilization of calcium in bog areas after liming with limestone powder 0-1 mm.] - Swedish Environm. Prot. Agency, unpubl. rep.

Abrahamsson, I. 1993b. En modell för kalkupplösning och uttransport av kalcium från kalkade våtmarker. [A model för lime dissolution and the transportation of calcium from limed wetlands.] - Terra-Limno Group, Mölnlycke.

Abrahamsson, I. 1993c. Impact of waterflows on acid-base chemistry in limed lakes. - Vatten 49: 24-33.

Ahlgren, G. 1918. Är kalkbrist orsaken till de dåliga resultaten av karpövervintringarna i Aneboda? [Is deficit of lime the cause of the unsatisfactory results of the over-wintering of carp in Aneboda?] - S. Sveriges Fiskeriför. Skrif. (1917-1919), pp. 206-212.

Åkesson, B. 1993. Blommande asfalt lockar turisterna. [Flowering asphalt enticing for the tourists.] - Land 25: 2-3.

Alenäs, I. 1986. Kalkningsprojektet Härskogen 1976-1986. [The Härskogen liming project 1976-1986.] - Swedish Environm. Res. Inst., B 846.

Alenäs, I. 1987. Effekter på ytvattenkemi inom ett brandpåverkat avrinningsområde i Ale- och Vättlefjällsområdet på svenska västkusten. [Effects on surface water chemistry in an area affected by fire in the Ale and Vättlefjäll districts on the Swedish west coast.] - Swedish Environm. Res. Inst., L87/104.

Alenäs, I. 1988. Effects of different liming agents and methods. - Acidification research in Sweden, 6: 11-12. Swedish Environm. Prot. Agency Informs.

Andersson, F. and Persson, T. (Eds.) 1988. Liming as a measure to improve soil and tree

conditions in areas affected by air pollution. - Swedish Environm. Prot. Agency 3518.
Andersson, I., Hasselrot, B. and Hultberg, H. 1982. Vattenkemiska effekter av olika sätt att applicera neutralisationsmedel i ett försurat område på västkusten. [Effects on water chemistry of various methods for application of neutralization agents in an acidified area on the west coast.] - Swedish Environm. Res. Inst., EM 0799.
Andersson, L. 1974a. Kalk- och dolomitbränning samt en översikt av Sveriges kalkindustri. [Burning of lime and dolomite and a survey of Swedish lime industry.] - R. Techn. Univ., Stockholm, FK1:5.
Andersson, L. 1974b. Svavelemission och vattenvård. [Emission of sulphur and water protection.] - Swedish Liming Organization, unpubl. rep.
Andersson, U. and Eriksson, B.-G. 1991. Biologisk och vattenkemisk undersökning i två våtmarkskalkade vattensystem. [Biological and water chemical studies in two water systems with limed wetlands.] - Jönköping County Council, Rep. 1991:2.
Appelberg, M. and Aldén, U. 1992. Integrerad uppföljning av kalkningens effekter på sjöar och vattendrag - en treårsrapport. [English summary: Integrated monitoring in limed lakes and watercourses - a three year report.] - Inform. Inst. Freshw. Res., Drottningholm (4): 1-60.
Aronson, J.-A. 1990. Våtmarkskalkning. Förändringar på miljö och vegetation. [Liming on wetlands. Changes in environment and vegetation.] - Swedish Environm. Prot. Agency 3827.
Aronson, J.-A. 1992. Vegetationsförändringar på mossor och kärlväxter efter kalkning. [Vegetational changes for mosses and vascular plants after liming.] - Consultant in Karlstad, Rep.
Bäckman, L. 1990. Krossad kalksten som halkbekämpningsmedel. Inventering av förekomster och täkter. [Crushed limestone; an agent to reduce skidding. An inventory of localities and quarries.] - Väg- och Trafikinst., Linköping, Rep. 634.
Bengtsson, B., Dickson, W. and Nyberg, P. 1980. Liming acid lakes in Sweden. - Ambio 9: 34-36.
Bergström, R. 1977. Kalk vid avloppsvattenrening. [Lime in wastewater treatment.] - In: Kalk i vattenvården. [Lime in water protection.]. - Svenska Kalkföreningens 3:e vattenvårdsdag, Stockholm, pp. 14-41.
Bernes, C. 1986. Monitor 1986 - Sura och försurade vatten. [Acid and acidified waters.] - Swedish Environm. Prot. Agency Informs.
Bernes, C. 1991. Monitor 12. Acidification and liming of Swedish freshwaters. - Swedish Environm. Prot. Agency Informs.
Bernhoff, R. 1974. Kalk förhindrar försurning av vår miljö. [Lime prevents acidification of our environment.] - Cementa Inc., Malmö, Sweden, Inf. 1974.
Bertilsson, G. 1980. Mg-Mineral (olivin, peridotit) som ingrediens i gödselmedel eller för direkt användning i jordbruket. [Mg-minerals (olivine, peridotite) as a component in fertilizers for direct agricultural use.] - Supra Co., Malmö, Rep. 1980-10-10.
Berzins, B. 1962. Kalkning av sjöar. [Liming of lakes.] - S. Sveriges Fiskeriför. Årsskr. 1959-1960: 28-35.
Bishop, K.H. 1991. Is there more to acidity in organic-rich surface waters than air pollution? An example from northern Sweden. - Vatten 47: 342-347.
Björklund, I., Haux, C., Hogstrand, C., Unger, M. and Öhrn, T. 1985. Bioackumulation i organ och förändringar av jonbalans i blodplasma hos öring vid påverkan av aluminium vid olika pH, humushalt och vattentemperatur. [Bio-accumulation in organs and changes in ion balance in the blood of trout affected by aluminium at different pH, humic content and water temperature.] - Swedish Environm. Prot. Agency 3046.
Blomberg, E. and Bjärnborg, B. 1986. Fraktionering av aluminium med jonbytesteknik. - Laboratorieförsök avseende kalkning av naturliga ytvatten. [Fractionation of aluminium

with ion-exchange technique in laboratory experiments aimed at liming of natural surface waters.] - Swedish Environm. Prot. Agency 3150.

Broberg, O. 1978. Kalkningsmedel, kalkning och effekter av kalkning. En litteraturöversikt. [English summary: Liming agents, liming and effects of liming.- A literature survey.] - Vatten 34: 104-116.

Brodin, Y.-W. 1994. Acidification - an everlasting problem, or is there hope? - Swedish Environm. Prot. Agency 4242.

Brown, D.J.A. 1988. The Loch Fleet and other catchment liming programs. - Water Air Soil Poll. 41: 409-416.

Carlsson, B. and Bergström, S. 1989. Skinnmuddselets regleringsmagasin - påverkan på alkalinitet och pH i Gideälven. [The Skinnmuddselet Reservoir - impact on alkalinity and pH in the River Gideälven.] - SMHI, Norrköping, Rep.

Degerman, E. 1992. Kalkning för djur i de rinnande vattnen. [Liming for animals in the running waters.] - In: Strategi vid kalkning av rinnande vatten. [Strategy for liming of running waters.] 12-13/5 1992, Umeå, Swedish Environm. Prot. Agency, unpubl. rep.

Degerman, E., Sjölander, E., Johlander, A., Sjöstrand, P., Höglind, K., Thorsson, L. and Carlstrand, H. 1990. Kalkning för att motverka försurningspåverkan på fisk i rinnande vatten. [English summary: Liming of acidified running waters to improve conditions for fish.] - Inform. Inst. Freshw. Res., Drottningholm (4): 27-214.

Degerman, E., Engblom, E., Lingdell, P.-E., Melin, E. and Olofsson, E. 1992. Försurning i fjällen. [English summary: Acidification in the mountains?] - Inform. Inst. Freshw. Res., Drottningholm (1).

Dickson, W. 1974. Försurning av sjöar och vattendrag. [Acidification of lakes and watercourses.] - Swedish Environm. Prot. Agency 531.

Dickson, W. 1979. Kalkning av sjöar. [English summary: Liming of lakes.] - K. Skogs-Lantbruksakad. Tidskr. Suppl. 13: 38-48.

Dickson, W. 1983a. Liming toxicity of aluminium to fish. - Vatten 39: 400-404.

Dickson, W. 1983b. Äggskal i Jungfrugöl. Preliminär rapport från ett kalkningsprojekt i Småland. [Egg shells in Lake Jungfrugöl. Preliminary report on a liming project in Småland.] - Swedish Environm. Prot. Agency, unpubl. rep.

Dickson, W. 1984. Kalkning av sjöar och vattendrag. Dosberäkningar och kalkningsmedlens metallinnehåll. Kalkfällning i reningsverk. [Liming of lakes and streams. Calculations of dosages and the metal content of liming agents. Lime precipitation in wastewater treatment plants.] - Swedish Environm. Prot. Agency 1919: 52-59.

Dickson, W. 1985. Water quality in productive forests of different atmospheric load. - Air pollution and stability of coniferous forest ecosystems. Internat. Symp., October 1984, Brno, Czechoslovakia, pp. 165-174.

Dickson, W. 1988. Försurningsläget i svenska vatten och kalkningsbehovet. [English summary: Acidification of Swedish waters and the need of lime]. - Vatten 44: 300-304.

Dickson, W. 1990. Försurning, kalkning, vattenkemi. [Acidification, liming, water chemistry.] - Nat. Board Fish. (1), pp. 12-19.

Dickson, W. 1991. Försurning och motåtgärder. [Acidification and countermeasures.] - Miljö och Framtid 7: 20-24.

Dickson, W. 1992. Grisen bot mot försurad mark. [The pig, a remedy for acidified soils.] - In: Gillberg, M. and Odén, K. (Eds.) Grisen som markberedare. [The pig for loosening of the topsoil.], Stiftelsen Ekologisk Markberedning, Rep.

Drakenberg, B., Ehnström, B.A., Liljelund, L.-E. and Österberg, K. 1991. Lövskogens naturvärden. [The natural value of the deciduous forest.] - Swedish Environm. Prot. Agency 3946.

EMEP 1992. Det Norske Meteorol. Inst., Oslo, Norway, Techn. Rep. 97.

Engblom, E. and Lingdell, P.-E. 1985a. Hur påverkar kalkdoserare bottenfaunan? [How do lime dosing devices affect the benthic fauna?]. - Swedish Environm. Prot. Agency 1994.

Engblom, E. and Lingdell, P.-E. 1985b. Hur påverkar reningsverk med olika fällningsmedel bottenfaunan? [How do wastewater plants with different precipitation agents affect the bottom fauna?]. - Swedish Environm. Prot. Agency 1998.

Eriksson, E. 1982. On the rate of dissolution of olivine in aqueous solutions. - Vatten 38: 409-415.

Eriksson, E., Johansson, S. and Nilsson, B. 1980. Kalk och olivin som sjöförbättringsmedel, erfarenheter från fullskaleförsök. [Limestone and olivine as liming agents, experience from full-scale experiments.] - Hydroconsult, Uppsala, Rep. 80-736.

Eriksson, E., Karltun, E. and Lundmark, J.-E. 1992. Acidification of forest soils in Sweden. - Ambio 21: 150-154.

Eriksson, F. 1990. Vegetationsförändringar inom kalkade områden på Fulufjället. [Vegetational changes in limed areas in the Fulufjället district.] - Nat. Board Fish., (1), pp. 34-35.

Eriksson, J. and Börjesson, P. 1991. Vedaska i skogen. En litteraturstudie. [Wood ash in the forest. A literature study.] - Vattenfall Utveckling, Stockholm, Rep.

Falkengren-Grerup, U. 1986. Soil acidification and vegetation changes in deciduous forests in southern Sweden. - Oecologia (Berlin) 70: 339-347.

Fasth, T. 1991. Våtmarkskalkning - vad händer med vegetationen? [Wetland liming - what happens to the vegetation.] - Municipalities of Jönköping and Vaggeryd, Rep.

Fleischer, S. 1984. Högvadsån. Kalkningsprojekt 1978-1981. III Vattenkemiska undersökningar. [The River Högvadsån liming project. III. Water chemistry studies.] - Halland County Council, Rep.

Gherini, S.A., Mok, L., Hudson, R.J.M., Davis, G.F., Chen, C.W. and Goldstein, R.A. 1985. The ILWAS model: formulation and application. - Water Air Soil Poll. 26: 425-460.

Grahn, O. and Hultberg, H. 1975. The neutralizing capacity of 12 different lime products used for pH-adjustment of acid water. - Vatten 2: 120-132.

Haglund, E. 1911. Redogörelse för en botanisk - torfgeologisk undersökning af fiskdammar å torfjord vid fiskeriförsöksstationen i Aneboda. [Report on a botanical and geological study of fish ponds on peat soils at the fisheries laboratory in Aneboda.] - S. Sveriges Fiskeriför. Skrif. (1906 - 1911) 8: 12-22.

Håkanson, L., Andersson, P., Andersson, T., Bengtsson, Å., Grahn, P., Johansson, J.-Å., Jönsson, C.-P., Kvarnäs, H., Lindgren, G. and Nilsson, Å. 1990. Åtgärder mot höga kvicksilverhalter i insjöfisk. Slutrapport för kvicksilverdelen av projektet: Kalkning-Kvicksilver-Cesium. [Measures to reduce mercury in lake fish. Final report from the project Liming-Mercury-Caesium.] - Swedish Environ. Prot. Agency 3818.

Halland County Council 1986. Dokumentationsrapport om kalkfällningsverken i Hallands län. [Report on wastewater treatment plants in the county of Halland using lime in precipitation.] - Halland County Council, Rep. 1986:3.

Hallbäcken, L. and Popovic, B. 1985. Markkemiska effekter av skogliga kalkningsförsök. [Effects on soil chemistry of forestry liming experiments.] - Swedish Environm. Prot. Agency 1880.

Hallin, K. 1993. Årsarbetare per en sjökalkningsmiljon i Sverige. [People/year per million Swedish kronor of lake liming in Sweden.] - Movab Co., Malmö, unpubl. rep.

Hasselrot, B. and Lind, B. 1992. Går det att kalka utdikade våtmarker? Ett försöksprojekt i Stockasjöns och Holmåsatjärnsbäckens avrinningsområde. [Is it possible to lime wetlands affected by ditch drainage. An experimental project in the catchment of the Lake Stockasjön and the Holmåsatjärnsbäcken Stream.] - Melica Consultant Co., Gothenburg.

Hasselrot, B., Andersson, I., Alenäs, I. and Hultberg, H. 1984a. Våtmarkskalkning - en metod att kalka rinnande vattendrag. [Wetland liming - a method for liming of running waters.] - Swedish Environm. Prot. Agency 1919: 65-70.

Hasselrot, B., Andersson, B.I. and Hultberg, H. 1984b. Ecosystem shifts and reintroduction of arctic char (*Salvelinus salvelinus* L.) after liming of a strongly acidified lake in southwestern Sweden. - Rep. Inst. Freshwat. Res., Drottningholm 61: 78-92.

Hasselrot, B., Andersson, B.I., Alenäs, I. and Hultberg, H. 1987. Response of limed lakes to episodic events in southwestern Sweden. - Water Air Soil Poll. 32: 341-362.

Hegna, K. (Ed.) 1992. Vassdragskalking - strategi och effekter. Referat fra FoU-seminar i 1992. [Liming of surface waters - strategies and effects. Report from scientific seminar in 1992.] - Direct. Nature Managem., Norway, DN-notat 1992-5.

Henrikson, L. 1988. Effects on water quality and benthos of acid water inflow into the limed Lake Gårdsjön. - In: Dickson, W. (Ed.) Liming of Lake Gårdsjön, an acidified lake in SW Sweden. - Swedish Environm. Prot. Agency 3426, pp. 309-327.

Hindar, A. and Rosseland, B.O. 1991. Liming strategies for Norwegian lakes. - In: Olem, H., Schreiber, R.K., Brocksen, R.W. and Porcella, D.P. (Eds.) International lake and watershed liming practices. - The Terrene Inst., Washington D.C., pp. 173-192.

Holmberg, B. 1975. Kalkning som fiskevårdande åtgärd. - Äldre erfarenheter [Liming for fishery management - earlier experiences.]. - Nat. Board Fish., unpubl. rep.

Holmström, I. 1984. Åtgärder mot försurning i Velen. Erfarenheter från utförd behandling med kalk och soda samt förslag till fortsatta åtgärder. [Measures against the acidification of Lake Velen. Experiences of lime and sodium carbonate treatment and suggestions for further measures.] - Swedish Meteorol. Hydrol. Inst., Rep. HO-14.

Holmström, I. 1988. Erfarenheter från automatisk styrning och övervakning av kalkdosering i vattendrag. [Experience of auto-regulation and monitoring of lime dosages for watercourses.] - Swedish Meteorol. Hydrol. Inst., unpubl. rep.

Holmström, I., Särén, D. and Rolandsson, V. 1987. Borås använder ny teknik för kalkning mot försurning. [The municipality of Borås uses new liming technique to counteract acidification.] - Vannet i Norden 1: 4-5.

Hultberg, H. 1975. Thermally stratified acid water in late winter - a key factor inducing self-accelerating processes which increase acidification. - Swedish Environm. Res. Inst., Rep. B:246.

Hultberg, H. and Andersson, I.B. 1982. Liming of acidified lakes: induced long-term changes. - Water Air Soil Poll. 18: 311-331.

Hultberg, H. and Grahn, O. 1974. Några kalker och kalkslaggers neutraliserande förmåga. [The neutralization capacity of some liming agents.] - Swedish Environm. Res. Inst., Rep. B:246.

Hultberg, H. and Nyström, U. 1988. The role of hydrology in treatment duration and reacidification in the limed Lake Gårdsjön. - In: Dickson, W. (Ed.) Liming of Lake Gårdsjön, an acidified lake in SW Sweden. - Swedish Environm. Prot. Agency 3426, pp. 95-134.

Hultman, B., Forsberg, L. and Engvall, R. 1982. Kan försurade sjöar och vattendrag botas med renat kommunalt avloppsvatten? [English summary: Can acidified lakes and rivers be cured by treated sewage water?] - Vatten, 38: 461-472.

Kämäri, J., Amman, M., Brodin, Y.-W., Chadwick, H.J., Henriksen, A., Hettelingh, J.-P., Kuylenstierna, J., Posch, M. and Sverdrup, H. 1992. The use of critical loads for the assessment of future alternatives for acidification. - Ambio 21: 377-386.

Karlén, A. 1979. Svavel-miljö-ekonomi. [Sulphur-environment-economy.] - PhD thesis. Kugel Tryckeri AB, Stockholm.

Karlsson, J.-Å. 1990. Halkbekämpning som miljövård? [Measures against skidding for environmental care?] - Vägverket, Falun, BD-rep. 90: 201-46.

Kulinski, I. and Dickson, W. 1994. Kalkdoserare: Erfarenheter och möjligheter. [Lime

dosers: experiences and prospects.] - Swedish Environm. Prot. Agency. (in press)
Kullberg, A., Bishop, K.H., Hargeby, A., Jansson, M. and Petersen, R.C. 1993. The ecological significance of dissolved organic carbon in acidified lakes. - Ambio 22: 331-337.
Lagerman, B. 1981. Ölen. Resultat och erfarenheter av ett kalkningsprojekt. [Experience of the Lake Ölen liming project.] - Örebro County Fishing Board, unpubl. rep.
Lagerman, B. 1983. Kalkningsprojektet Unden. [The Lake Unden liming project.] - Nat. Board Fish., unpubl. rep.
Larsson, S. and Lindquist, P. 1974. Kalkning av försurade sjöar. Del 1. Problembeskrivning samt utvärdering av kalkningen av Östra Nedsjön. [Liming acidified lakes. Sect. 1. Description of the problem and evaluation of the liming of Lake Östra Nedsjön.] - Chalmers Techn. Univ., Göteborg, Rep. 83.
Lessmark, O. 1987. Markkalkning som metod för att motverka försurning av sjöar och vattendrag. [English summary: Ground liming as a method of mitigating acidification of surface waters.] - Inform. Inst. Freshw. Res., Drottningholm (9).
Lessmark, O., Degerman, E., Johlander, A. and Sjölander, E. 1986. Effekter av kalkning på fisk omedelbart nedströms doserare. [English summary: Effects of liming on fish immediately downstream of lime dosers.] - Inform. Inst. Freshw. Res., Drottningholm (5).
Levander, T. 1975. Kalkning av sjöar. [Liming of lakes.] - Swedish Environm. Prot. Agency 607.
Liljelund, L.-E. 1988. Mer lövträd för att motverka markförsurning? [More deciduous trees to counteract soil acidification?] - K. Skogs- Lanbruksakad. Tidskr. Suppl. 22: 69-75.
Lindeneg, K. and Navrud, S. 1993. Mechanisms in environmental control II. - Nordic J. Environm. Econ. 93: 34-37.
Lindgren, G. 1991. Utvärdering av landsvägskalkning i Ljusdals kommun. [Evaluation of road liming in the municipality of Ljusdal.] - Gävleborg County Council, Rep.
Lindmark, G. 1984. Acidified lakes: ecosystem response following sediment treatment with sodium carbonate. - Verh. Internat. Verein. Limnol. 22: 772-779.
Lindmark, G. 1992a. Soda ash for mitigation in acidified waters. Long-term lake response and efficiency. - Reprint, 25th SIL Congress, Barcelona, Spain, 1992.
Lindmark, G. 1992b. Soda-briketter för "kalkning" av försurade sjöar. En sammanfattning av resultat, rapporter och artiklar angående sodabehandlingar i sjöar. [The use of soda ash briquettes in the neutralization of acidified lakes. Summary of results, reports and articles on soda ash treatments of lakes.] - Solvay Nordic AB, Stockholm.
Lindström, I., Samuelsson, H. and Wijk, S. 1993. Skogsmarkskalkning. Resultat från en fyraårig försöksperiod samt förslag till åtgärdsprogram. [Liming of forest soils. Results from a four-year trial period and suggestion for a measure programme.] - Nat. Board Forest., Rep. 1993:6.
Lindström, K. 1989. Spårämnen i algkalk med särskild hänsyn till biotillgänglighet av selen (Se) - en litteraturstudie. [Trace elements in algal lime with special reference to bioavailability of selenium (Se) - a literature study.] - Dep. Limnol., Univ. Uppsala., unpubl. rep.
Lindström, K. 1992. Sjörestaurering med algkalk eller kalksten? [Restoration of lakes by algal lime or limestone?] - Dep. Limnol., Univ. Uppsala, unpubl. rep.
Lundberg, R. 1893. Är allmännare införande af fiskuppfödning i dammar hos oss önskvärdt, och efter hvilken plan bör i så fall detta ske? [Is a more general introduction of fish breeding in ponds desirable and on what plan should this be based?] - K. Landtbruks-akad. Handl. Tidskr., 1893, pp. 316-320.
Lundin, U. and Lundh, I. 1968. Resultat av behandling av Tulltjärn med lut. [Results of treatment of Lake Tulltjärn by sodium hydoxide.] - Göteborgs och Bohus County Fishing Board, unpubl. rep.
Lundmark, J.-E. 1988. Skogsmarkens ekologi. Del 2. [The ecology of the forest soil. Part

2.] - Nat. Board Forest., Rep.
Magnusson, T. 1982. Skogs- och myrdikningens inverkan på försurningen av ytvatten. En litteraturstudie. [The impact of forest and mire drainage on surface water acidification. A literature study.] - Swedish Environm. Prot. Agency 1626.
Malmer, N. 1975. Synpunkter på kalkning av sjöar som motåtgärd mot försurning. [Views on lake liming as a counter-measure against acidification.] - Dep. Ecol. Bot., Univ. Lund, unpubl. rep.
Matzner, E., Khanna, P.K., Meiwes, K.J. and Ulrich, B. 1985. Effects of fertilization and liming on the chemical conditions and element distribution in forest soils. - Plant Soil 87: 405-415.
Ministry of Agriculture 1976. Mindre svavel - bättre miljö. [Less sulphur - better environment.] - Ministry of Agriculture Ds Jo 1976:2.
Ministry of Agriculture 1982. Acidification today and tomorrow. - Kommittén Miljö '82.
NBF (National Board of Fisheries) 1982. Råd och riktlinjer för kalkning av sjöar och vattendrag. [Guidelines for liming of lakes and streams.] - Nat. Board Fish., (1).
NBF and SEPA (National Board of Fisheries and Swedish Environmental Protection Agency) 1979. Kalkning av sjöar och vattendrag. [Liming of lakes and streams.] - Inform. Inst. Freshw. Res., Drottningholm (8).
NBF and SEPA (National Board of Fisheries and Swedish Environmental Protection Agency) 1981. Kalkning av sjöar och vattendrag 1977-1981. [Liming of lakes and streams 1977-1981.] - Inform. Inst. Freshw. Res., Drottningholm (4).
Nilsson, S.I. 1993. Acidification of Swedish oligotrophic lakes - interactions between deposition and forest growth and effects on lake-water quality. - Ambio 22: 272-276.
Norrgren, L. and Degerman, E. 1993. Effects of different water qualities on the early development of Atlantic salmon and brown trout exposed in situ. - Ambio 22: 213-218.
Olem, H., Schreiber, R.K., Brocksen, R.W. and Porcella, D.P. (Eds.) 1991. International lake and watershed liming practices. - The Terrene Inst., Washington D.C.
Olsson, B. 1982. Storskalig spridning av kalkstens- och olivinmjöl på mark: effekter på fält-och bottenskiktsvegetation. [Large scaled application of limestone and olivine powder on ground: effects on ground vegetation.] - Swedish Environm. Res. Inst., Rep. EM 811.
Olsson, H. and Pettersson, A. 1993. Oligotrophication of acidified lakes - a review of hypotheses. - Ambio 22: 312-317.
Olsson, T.I. 1987. Olivinbehandling av Gravån. Kemiska och biologiska konsekvenser. [Olivine treatment of the Gravån Stream. Chemical and biological effects.] - Dep. Ecol. Zool., Univ. Umeå., unpubl. rep.
Persson, E. 1989. Algprodukter inom jordbruk och trädgårdsnäring, husdjursnäring och miljövård. [Algal products within agriculture, horticulture, livestock business and environmental protection] - Katrineholm, unpubl. rep.
Rafstedt, T. 1992. Ekologiska effekter av våtmarkskalkning. [Ecological effects of wetland liming.] - In: Strategi vid kalkning av rinnande vatten. [Strategy for liming of running waters.] 12-13/5 1992, Umeå, Swedish Environm. Prot. Agency, Rep.
Ramberg, L. 1981. Increase in stream pH after a forest drainage. - Ambio 10: 34-35.
Rosén, K. 1988a. Skogsenergi eller fossila bränslen. En jämförelse ur försurningssynpunkt. [Energy from forest or fossil fuels. A comparison with respect to acidification.] - Swedish Environm. Prot. Agency 3521.
Rosén, K. 1988b. Effects of biomass accumulation and forestry on nitrogen in forest ecosystems. - In: Nilsson, J. and Grennfelt, P. (Eds.) Critical loads for sulphur and nitrogen. - Nordic Counc. Minist., Miljørap. 1988:15.
Rosén, K. 1991. Skörd av skogsbränslen i slutavverkning och gallring - ekologiska effekter. [Harvest of forest biomass used for fuels in final cutting and thinning - ecological

effects.] - Sveriges Lantbruksuniv., Uppsala, Rep. 5.
Sandberg, P.-E. 1992. Sammanställning av kalkningsverksamheten i vissa delar av Kopparbergs län 1979-91. [Report on the liming programme in certain parts of the county of Kopparberg, 1979-1991.] - Kopparberg County Council, Rep.
Sander, S. and Eriksson, M. 1974. Något om kalkningsprojektet vid Östra Nedsjön nov. 1972 - feb. 1974. [The Lake Östra Nedsjön liming project Nov. 1972 - Feb. 1974.] - Älvsborg County Fishing Board, Rep.
Sanner, H., Wittgren, H.B., Carlsson, B. and Holmström, I. 1993. Kalkningsplanering i vattendragssystem med PULS-modellen. [English summary: Planning of liming in water systems by use of the PULS model.] - Vatten 49: 17-23.
Schouenborg, B. 1991. Regler för P-märkning av sjökalk. [Rules for quality control of lake lime.] - SP (Statens Provningsanstalt), Rep. 1991:35.
Schmuul, R. 1970. Sjö- och fiskeriundersökningar i Härskogs-reservatet samt framtidsutsikterna för fritidsfisket inom området. [Studies of lakes and fisheries in the Härskogen Reserve and the possibilities of future recreational fishing in the area.] - Härskogsutred., Lantbruksnäm. Vänersborg, Suppl. D.
SEPA (Swedish Environmental Protection Agency) 1987. Skogs- och myrdikningens miljökonsekvenser. Slutrapport från ett projektområde. [The environmental impact of forest- and mire drainage. Final report.] - Swedish Environm. Prot. Agency 3270.
SEPA (Swedish Environmental Protection Agency) 1988. Kalkning av sjöar och vattendrag. [Liming of lakes and streams.] - Swedish Environm. Prot. Agency, Off. Guidel. 88:3.
SEPA (Swedish Environmental Protection Agency) 1991. Kvicksilver i Sverige, problem och åtgärder. [Mercury in Sweden, problems and measures.] - Swedish Environm. Prot. Agency Informs.
Sjöberg, L. 1979. Kinetics and mechanism of calcite dissolution in aqueous solution at low temperatures. - PhD thesis, Dep. Geol., Univ. Stockholm.
Skoog, A. 1993. Analysprotokoll, avseende musselskal. [Analysis record concerning mussel shells.] - Sveriges Lantbrukskem. Lab., Uppsala.
Stenberg, L. 1965. Några erfarenheter av kalkning och gödsling av fiskevatten i Värmlands län. [Some experiences of liming and fertilization of fishing waters in the county of Värmland.] - Värmland County Fishing Board, unpubl. rep.
Stumm, W. and Morgan, J.J. 1970. Aquatic chemistry. - Wiley & Sons, Intersci., New York.
Svahnberg, A. 1992. Vindavdrift och spridningsjämnhet. Undersökning av vindavdrift och spridningsjämnhet vid helikopterspridning av kalkstensmjöl på våtmarker. [Wind drift and dispersal uniformity. Study on wind drift and dispersal uniformity in helicopter liming on wetlands.] - Jönköping County Council, Rep. 8/92.
Sverdrup, H.U. 1985. Calcite dissolution kinetics and lake neutralization. - PhD thesis, Dep. Chem. Engin., Lunds Inst. Techn., Univ. Lund.
Sverdrup, H.U. and Warfvinge, P. 1988. What is left for researchers in liming? A critical review of state-of-the-art acidification mitigation. - Lake Reserv. Managem. 4: 87-97.
Sverdrup, H.U., Warfvinge, P.G. and Fraser, J. 1985. The dissolution efficiency for different stream liming methods and technologies. - Vatten 3: 155-163.
Sverdrup, H.U., Warfvinge, P.G. and Bjerle, I. 1986. A simple method to predict the time required to reacidify a limed lake. - Vatten 42: 10-15.
Swedish Environmental Research Institute 1993. Miljökonsekvensbedömning av Skogsstyrelsens förslag till program för skogsmarkskalkning. Sammanfattning och slutsatser 1993-09-29. [Environmental impact assessment of the plan for forest soil liming suggested by the National Board of Forestry. Summary and conclusions, 29.9.1993.] - Swedish Environm. Res. Inst., Rep.
Tamm, C.O. and Hallbäcken, L. 1988. Changes in soil acidity in two forest areas with dif-

ferent acid deposition: 1920s to 1980s. - Ambio 17: 56-61.
Tervet, D.J. and Harriman, R. 1988. Changes in pH and calcium after selective liming in the catchment of Loch Dee, a sensitive and rapid-turnover loch in southwest Scotland. - Aquacult. Fish. Managem. 19: 191-203.
Thörnelöf, E. and Degerman, E. 1991. Lake liming in Sweden, case studies and general experiences. - In: Olem, H., Schreiber, R.K., Brocksen, R.W. and Porcella, D.P. (Eds.) International lake and watershed liming practices. - The Terrene Inst., Washington D.C., pp. 41-60.
Tiberg, H.V. 1912. Drag ur näringslifvet i Filipstads Bergslag serskilt vid Långbanshyttan. [The trade and industry in Filipstads Bergslag, especially in Långbanshyttan.] - Wald. Zachrissons Boktryck., Gothenburg.
Tideström, H. 1984. Kalkning av rinnande vatten, utvärdering av kalkdoseringsanläggningar. [Liming of running waters, evaluation of lime doser devices.] - Swedish Environm. Prot. Agency 1873.
Timm, B. 1993. Policy för användning av träaska på skogsmark. [Policy for the use of wood ash on forest soil.] - Swedish Environm. Prot. Agency, unpubl. rep.
Valdmaa, K. and Dickson, W. 1982. Försök med mesa till sjökalkning. [Experiments with sludge in lake liming.] - Swedish Environm. Prot. Agency 1567.
Västernorrland County Council 1993. Besvär över föreläggande angående vidare åtgärder mot försurningseffekter och metallutlakning från Skinnmuddselets regleringsmagasin. [Appeal about injunction concerning further measures to counteract effects of acidification and metal discharges from the Skinnmuddselet Reservoir.] - Västernorrland County Council, unpubl, rep.
Warfvinge, P. 1988. Modeling acidification mitigation in watersheds. - PhD thesis, Dep. Chem. Engin., Lunds Inst. Techn., Univ. Lund.
Warfvinge, P. and Sverdrup, H.U. 1988. Soil liming as a measure to mitigate acid runoff. - Water Resourc. Res. 8: 701-712.
Warfvinge, P. and Sverdrup, H.U. 1989. Modeling limestone dissolution in soils. - Soil Sci. Soc. Am. J. 53: 44-51.
Widarsson, L.-E. 1988. Återförsurning vid avstängning av två kalkdoserare i Kristianstads län. [Reacidification after closing of two lime dosers in the county of Kristianstad.] - Municip. Osby & Fish. Assoc. Östra Örkened, unpubl. rep.
Wikström, U. 1984. Kalkfällning i reningsverk, ekologiska synpunkter på avloppsvattenrening. [Lime precipitation in wastewater treatment plants; wastewater treatment from an ecological point of view.] - Swedish Environm. Prot. Agency 1786: 28-31.
Wilander, A. and Ahl, T. 1972. The effects of lime treatment to a small lake in Bergslagen, Sweden. - Vatten 28: 431-445.

Chapter 5

The effects of liming on water chemistry

Anders Wilander[1], Paul Andersson[2], Hans Borg[2] and Ola Broberg[3]

[1] Swedish University of Agricultural Science, Department of Environmental Assessment, Box 7050, S - 750 07 Uppsala, Sweden
[2] Institute of Applied Environmental Research, Stockholm University, S-171 85 Solna, Sweden
[3] Jönköping County Adminstrative Board, S-551 86 Jönköping, Sweden

Summary

The consequences of liming on water chemistry are presented in terms both of the primary changes in pH, calcium and alkalinity and of secondary effects such as the impact on levels of organic materials, nutrients and trace metals. Data for the evaluation have mainly been drawn from the liming data base (KRUT), which is maintained by Swedish county councils and the Swedish Environmental Protection Agency. The condition of limed waters in Sweden and the changes which are taking place in these waters can be described in objective terms and subjected to international comparison.

On average, pH rose from 5.88 to 6.56 after liming. Only in a few cases did the pH exceed 7.5. Overdoses raised the alkalinity to levels above 0.3 meq/l in 230 lakes. Calcium concentrations increased by 0.22 meq/l (median value). The median duration of the liming effect for pH in excess of 6.0 was 1.7 years, which corresponds to 2.5 hydrological turnovers. The efficiency of lake liming measures has been evaluated by calculating the transport of calcium on the basis of decay curves and then comparing the results with the dosage. The median value for lake-liming efficiency was 63%.

Secondary changes are dependent upon the extent of the treatment and the condition of the water at the time of liming. Water with a high colour value becomes lighter, while lower colour values can involve either an increase or a decrease. Lake liming alone does not result in any improved nutrient status and there is a strong risk that the liming measures which have been undertaken will have no effect. Aluminium concentrations decline in conjunction with liming, a change which depends on the final pH and the level of organic materials. In water with a high level of organic

Liming of Acidified Surface Waters
L. Henrikson and Y.W. Brodin (Eds.)
© Springer-Verlag Berlin Heidelberg 1995

materials, there is little change in the total concentration, but there is a shift from toxic forms to more tightly bound and less toxic forms after liming.

There are indications of a decrease in the concentrations of iron and manganese after liming. As a rule, levels of other trace metals decline.

Effective liming of watercourses can be achieved using dosers or by upstream liming. The dosers cannot provide any security against acid surges when they are out of action, however.

Acid deposition affects soil processes so that the composition of the runoff changes. This means that the pre-acidification condition of lakes or watercourses cannot be restored by liming. If this is to be achieved, soil liming in the catchment area must be performed at least once. Soil liming in the discharge zone (wetlands) is more effective than in drier land, but the duration is short - approximately one year - and lake liming has a greater effect. Secondary effects such as reduced aluminium leaching and, to some extent, increased leaching of nutrients, may contribute to lake restoration.

A research programme which provides greater insight into the different processes and the links between chemical, biochemical and biological changes is essential if we are to gain a better understanding of the limed ecosystem and establish a platform for future treatment.

The objectives of liming are being achieved to a greater extent today than they were ten years ago. Treatment has become more organized so that entire water systems can be covered, rather than individual waters. Today, a larger proportion of strongly coloured waters and waters with short turnover periods are being covered than ten years ago.

1. Introduction

The first Swedish review of the effects of liming on water chemistry was presented by Dickson (1979). This was followed by a report covering an trial period (NBF, National Board of Fisheries; SEPA, Swedish Environmental Protection Agency 1981; Bengtsson *et al* 1980). Recently, reviews of liming activities in Sweden have been given by Stålhandske (1990, unpub.), Bernes (1991), and Thörnelöf and Degerman (1991), in addition to an international report by Olem (1991) and Olem *et al* (1991). The present book is the most comprehensive review on Swedish freshwater liming. This chapter summarizes current knowledge of the effects of liming on water chemistry, mainly on the basis of Swedish experience, but to some extent internationally.

The primary objective of liming is to achieve acceptable living conditions for biota which are exposed to acidification. Seen in terms of water chemistry, the objectives consist of pH values >6.0 and alkalinity >0.1 meq/l, except in cases where natural levels are lower (SEPA 1988).

Acidification lowers the buffer capacity of water and, consequently, the pH. Secondary effects include elevated levels of the toxic trace metal aluminium, which is leached from acidified soil, and reduced availability to primary producers of phosphorus, which becomes firmly bound in the soil.

1.1. Material

Nowadays, every Swedish liming project is monitored from the point of view of water quality (SEPA 1988). Each county council is responsible for data input and storage in the KRUT data base. Evaluations of liming effects are mainly founded on extracts from this data base from autumn of 1991. There are certain errors in the data base; some have been corrected, although others limit its usefulness. However, it still provides us with the only unbiased method of evaluating liming activities in Sweden in addition to the numerous comments and reports by scientists, consultants, municipalities, water authorities and county councils.

During the course of this review it was found that monitoring of the concentration and turnover of organic matter, nutrients, aluminium and other trace metals was inadequate. The present evaluation is therefore incomplete in this respect. The geographical locations of most of the lakes included in the report are shown in Fig. 1.

Figure 1. Limed lakes in southern Sweden mentioned in the text.

2. Primary effects of liming

2.1. Primary water chemistry reactions

The purpose of liming agents is to raise the pH value by lowering the concentration of hydrogen ions. Limestone ($CaCO_3$) reacts in acid water as follows:

(1) $CaCO_3\ (s) + H^+ = Ca^{2+} + HCO_3^-$

The anion of the liming agent also reacts with positively charged aluminium ions and humic acids (HA):

(2) $Al(OH)_2^+ + CO_3^{2-} + H_2O = Al(OH)_3\ (s) + HCO_3^-$
(3) $HA + CO_3^{2-} = A^- + HCO_3^-$

Through ion exchange, the cation of the liming agent can react with substances such as clay and humic acids, both in water and sediment:

(4) $HSilicate + Ca^{2+} = CaSilicate + 2\ H^+$
(5) $HA + Ca^{2+} = CaA + 2\ H^+$

Liberated hydrogen ions react with carbonate. These processes raise the concentrations of alkalinity and base cations. In limestone this mainly involves calcium, in dolomite it involves calcium and magnesium equally, and in soda it involves sodium.

The rate of the reactions described above varies. The main limiting factor is dissolution, which depends on the composition and grain size of the liming agent (Sverdrup *et al* 1986b). Calcite, for example, dissolves about 300 times faster than dolomite.

The pH value which is reached after liming is determined by the buffer capacity of the water before liming (acidity), and the amount and type of liming agent (e.g. Sverdrup 1983). Quickly dissolved and strongly alkaline liming agents such as quick lime (CaO) and soda can raise the pH extremely rapidly to around 10 in treated water (e.g. Stross and Hasler 1960), reducing the concentration of dissolved carbon dioxide. As a result of reaeration, however, it is supplied slowly until equilibrium is reached, leading to a lowering of the pH value. A solution in equilibrium with air will dissolve more limestone and the calcium concentration will be about 2 meq/l, while the pH is 8.4 (Stumm and Morgan 1981).

2.2. Acidic episodes

Organisms sensitive to acid can survive only if liming activities are effective in both time and space. Acid conditions should not prevail at any time of the year and the area protected by liming should be large enough to maintain a viable population.

Lakes are also affected by hydrological processes. During winter and spring, the inflowing water is lower in density (<4 °C) than that of the lake and therefore flows into a surficial stratum, especially in ice-covered lakes (e.g. Abrahamsson 1993). During the autumn rains, on the other hand, the inflow water may be of a higher density (being colder) than the lake water and thus flow over the bottom, as shown in Lake Gårdsjön (Hasselrot *et al* 1987). When the drainage area is acidified, these hydrological processes lead to what is known as an acidic episode (Hasselrot *et al* 1987, Henrikson 1988). Early studies in Lake Unden, southern Sweden, (NBF and SEPA 1981, Lagerman 1983) show that successful liming in tributaries results in neutralized water, which intrudes into the lake surficially beneath the ice. Liming of drainage areas and tributaries may thus help to protect biota against acidic episodes.

2.3. Lake liming

This section mainly discusses the effects of liming on calcium and magnesium, pH value and alkalinity (primary effects). In addition to limestone - mainly in the form of calcite ($CaCO_3$) - several other liming agents have been in use (NBF and SEPA 1981, Dickson and Brodin, Chap. 4). Only limestone, dolomite and soda are discussed in the following, however.

Lake liming may be performed either as whole-lake liming, in which case the liming agent is distributed over the entire lake surface, or littoral liming of shallow areas. Whole-lake activities are discussed first, with special emphasis on the immediate effect, duration and efficiency.

2.3.1. Whole-lake liming

Since most of the lakes involved in this review have been treated with limestone, it would be reasonable to monitor the results by determining calcium levels, but since analysis of calcium and magnesium (total hardness) is in common use and has little bearing on the interpretation of the results, this method is also employed in many cases.

Changes in water chemistry after liming may be interpreted either on the basis of measurements immediately after treatment or by calculating theoretical concentrations. In reality, the first alternative may lead to misinterpretation, since measurements are often made long after the treatment. The maximum effect does not appear directly after treatment, moreover, but is delayed due to slow dissolution and mixing in the water mass. Sverdrup and Warfvinge (1984) show that changes are influenced both by the dissolution rate, in certain cases, and by hydrographic conditions such as bays and thermal stratification in others.

Before the first treatment, the median pH value of Swedish lakes studied was 5.7 (Fig. 2). This rose to about 6.25 within 6 months for all the lakes concerned. After liming the pH remained above 5.5 in about 8% of all lakes, most of them with a very short turnover time - a median value of less than 6 months. The dose was small (median value below 0.1 meq/l) and the water was generally brown with a colour value of 120 (max. 350). The pH value exceeded 7.5 in twenty lakes. In nine cases this was probably attributable to the liming agent, but in eleven lakes (including one with a maximum pH value of 8.5) it was probably due to algal production during the summer. Fordham and Driscoll (1989) report pH values of 9.4 and 9.1 when a fine calcite fraction was applied to two Canadian lakes. In some of the earliest liming activities, calcium oxide and hydroxide were used, resulting in similarly high values (Wilander and Ahl 1972, Sandberg 1992).

Excessive doses should be avoided. General guidelines issued by SEPA in 1988 state that alkalinity should remain below 0.2-0.3 meq/l after liming. Concentrations above 0.3 meq/l were found in 230 lakes after liming, in about sixty they were above 0.43, with a maximum of 1.63 meq/l. One cause of overdosage may be erroneous volume calculations resulting from misinterpretation of soundings.

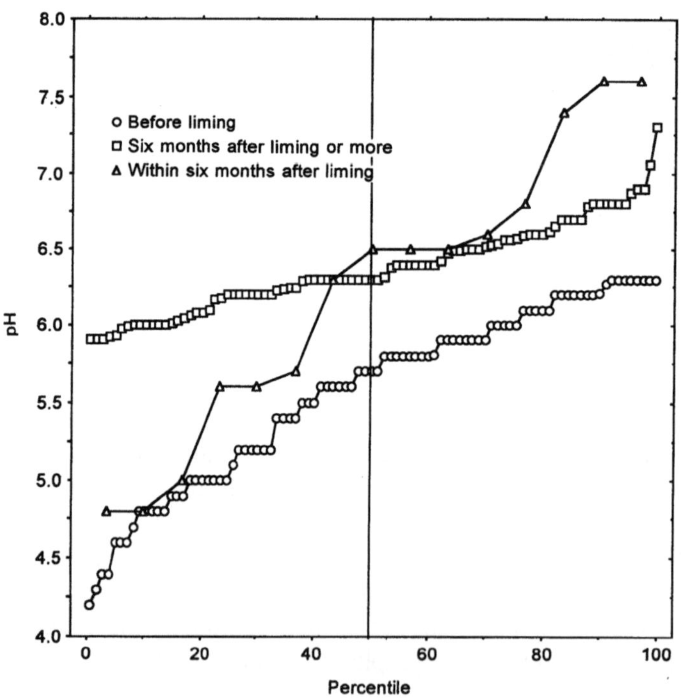

Figure 2. pH values in limed Swedish lakes before and after liming.

The addition of calcium will shift the ratio between mono- and divalent cations. A high ratio favours algae such as desmids and chrysophyceans, while a ratio <1.5 support diatoms (Reynolds 1984). Only in a few cases does the liming change the ratio to an important extent.

The duration of a liming activity may, from a water quality point of view, be evaluated against a limit value either for alkalinity or for the pH. The general guidelines issued by SEPA in 1988 state that the water should have neither a pH <6.0 nor an alkalinity <0.05 meq/l. The duration can thus, in the simplest alternative, be measured as the time which elapses between liming and the first occasion upon which observations show that the limit value is exceeded. For pH <6.0, this method gives a median duration of 1.7 years (Fig. 3).

The most important cause of "reacidification" is dilution due to the hydrological turnover of the water mass. A lake with a short turnover time is thus reacidified faster than one with a longer one. Comparisons of duration between different lakes should thus be based on the number of (theoretical) turnovers. Reacidification to an observed pH <6.0 occurs after a median of 2.5 turnovers (Fig. 4). The application of an alkalinity limit of 0.05 meq/l results in a median duration of 1.43 years and 1.91 turnovers (n = 326). A limit value <6.0 for pH thus seems somewhat stricter than one for alkalinity <0.05 meq/l.

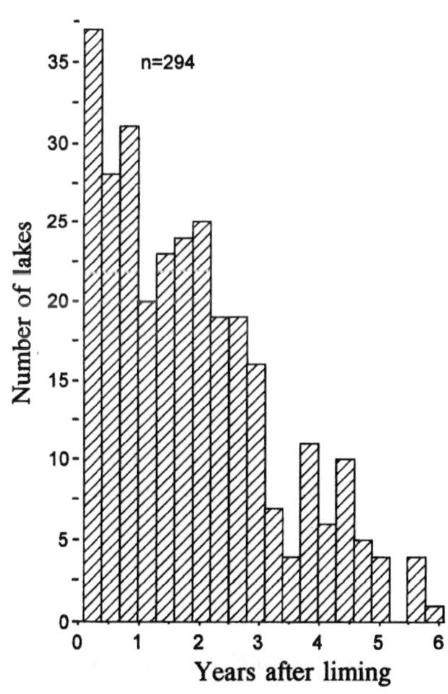

Figure 3. Duration of once-only liming of lakes (n = 294). Time required for reacidification to pH = 6.0.

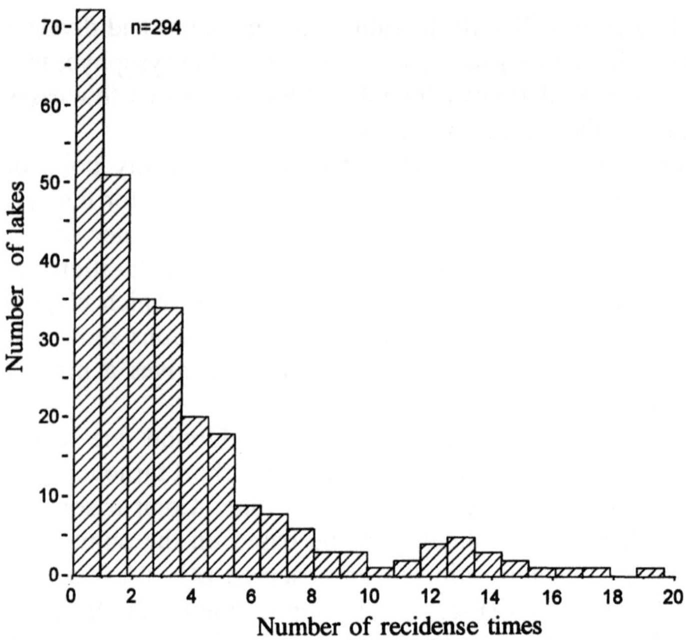

Figure 4. *Duration of once-only liming of lakes (n = 294). Number of theoretical turnovers required for reacidification to pH = 6.0.*

This simple approach provides maximum duration times, since measurements in the lakes are usually performed once each year. An alternative is to use statistically calculated reacidification curves.

In simple terms, a lake may be compared with a stirred tank with one inlet and one outlet (single-cell stirred tank reactor). Using a simple equation, Rainey (1967) described how a short-lived change in the supply of a substance results in a change in the concentrations in the lake (a decay).

(6) $$C_t = (C_o - C_a)\, e^{-b1t/T} + C_a$$

where;

C_t = concentration of substance at time = t
C_o = concentration of substance at time = 0
C_a = concentration of substance in influent
T = theoretical turnover time
b1 = constant.

This equation was used by Wilander and Ahl (1972) to follow the reacidification of Lake Ljustjärn, southern Sweden, and subsequently for Lake Gårdsjön by Hultberg and Nyström (1988) and Alenäs *et al* (1991), as well as by Adams and Brocksen (1988) on seven American lakes.

The concentration before liming (Ca) was calculated as the mean of at least two pre-liming measurements. In cases where the value was zero it was considered to be more appropriate to use a figure of -0.01 meq/l, which corresponds to a pH of 5.0 (and an insignificant concentration of aluminium). Calculations using linear regression were based on at least three measurements after liming, using a maximum of twenty-seven values. The time before the limit was exceeded can also be calculated using this equation, as can the amount of substance leaving the lake.

This is a very simple model. A much more detailed model for the calculation of reacidification was developed by Sverdrup and co-workers (Sverdrup and Warfvinge 1984, Sverdrup *et al* 1984, 1985, 1986a). This model includes the dissolution of limestone deposited on the sediment, immobilization in the sediment and stratification in the lake. The data obtained from the monitoring of most limed lakes is inadequate as a basis for such complex calculations.

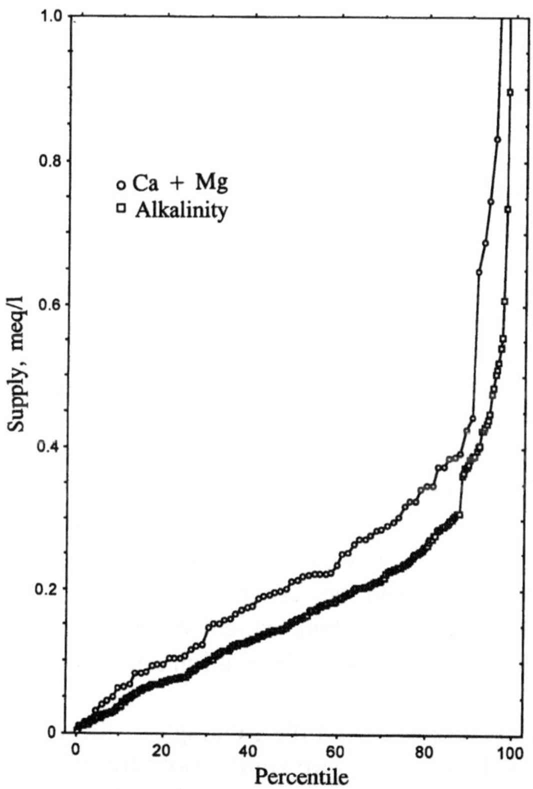

Figure 5. *The frequency distribution of the increase in calcium + magnesium and alkalinity (meq/l) after liming of Swedish lakes.*

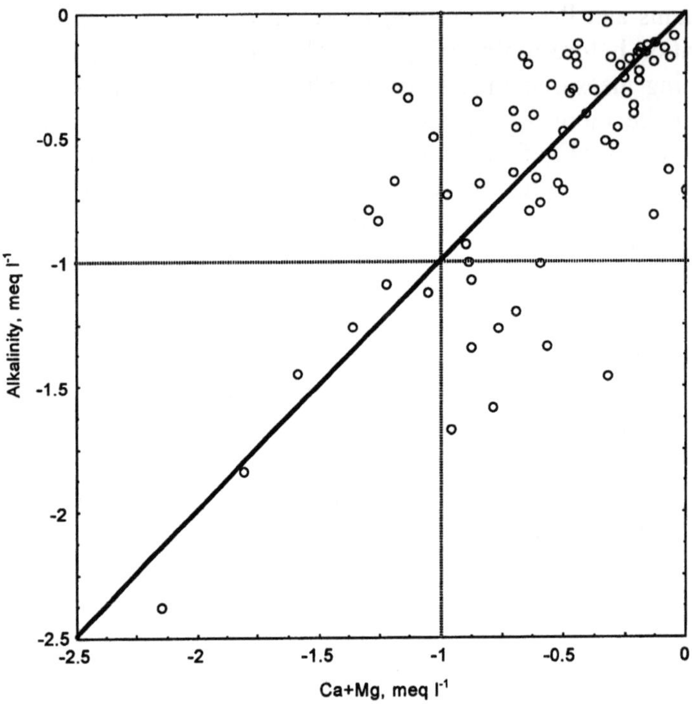

Figure 6. *Reacidification constants b1 for calcium + magnesium and alkalinity according to equation 1 (cf text).*

On the basis of these calculations, the median addition of calcium and magnesium was found to be 0.22 meq/l, while 10% of the lakes received less than 0.09 or more than 0.39 meq/l (Fig. 5). The median rise in alkalinity was 0.18 meq/l with concentrations of 0.07 and 0.4 for the 10 and 90 percentiles. The fact that cation values were higher than alkalinity values is probably attributable to the existing acidity.

Reacidification of a lake is caused by direct dilution with inflowing water and by input of acidity, which consumes the alkalinity (large-scale titration). As the cation (mainly calcium) is not affected by the acid input, the fall in alkalinity concentration should be faster than for calcium. Reacidification is described by a rate constant which varies with the turnover time (T). Deviations from purely hydraulic dilution are registered by the constant b1.

Data from only seventy-seven lakes were complete enough to allow calculation, and in most cases the values for b1 were similar both for calcium and magnesium and for alkalinity (Fig. 6), suggesting that few of the lakes had inputs of high acidity. A value of -1 implies that the reacidification is regulated by pure dilution with "clean" water. Higher values (-1, - 0) indicate that the addition of a liming agent gave a store of slowly available limestone and that the treatment thus had a good long-term effect. In some cases, however, the constant had a value less than -1, indicating that the liming agent was partially inactivated and that the liming was therefore less successful.

Equation 6 may be adapted in order to calculate the time required to reach a concentration corresponding to the limit value. The advantage, compared with the method used above, is that the constants are calculated on the basis of all available data. The duration may be calculated as the time from liming to the time when the concentration of alkalinity falls below the limit value. A limit of 0.05 meq/l gives a median duration of 4.5 turnovers ($n = 134$) with 1.2 and 12.7 turnovers for the 10 and 90 percentiles, respectively. A limit of 0.1 meq/l lowers the median duration to 2.3 turnovers.

The specific lime dose (amount of liming agent/lake volume) does not appear to influence the duration. The use of different liming agents does give rise to variation in duration, however, and a finer grain fraction gives a shorter duration (Table 1). If sedimenting limestone is inactivated at the bottom, then the use of coarser fractions should result in shorter duration times. Sverdrup and Warfvinge (1984) considered this in earlier publications but subsequently found that limestone is partially available on the lake bottom (Sverdrup *et al* 1986a).

Table 1. *Duration of limestone powder having different particle sizes when entire lakes in Sweden are limed.*

	Particle size mm	Duration 50 %	n
KM05	<0.5	5.4	24
KM02	<0.2	4.4	88
KM01	<0.1	3.2	5

Depending on the requirements, there are several indicators of the efficiency of a liming activity. The official guidelines issued by SEPA in 1988 specify an alkalinity of more than 0.05 meq/l, or that the effect of liming can no longer be traced.

Calculation of the output of calcium provides a measure of the active amount of limestone, which may be compared with the added amount. On the basis of such calculations, NBF and SEPA (1981) found that an efficiency of 50-70% is normal for fine limestone powder. Hultberg and Nyström (1988) calculated the efficiency for Lake Gårdsjön at 80%, basing their result on the best monitoring in Sweden with frequent measurements of both concentration and flow over a period of about three years. Budget calculations have also been published by Sverdrup *et al* (1984). For four Norwegian lakes, Rosseland and Hindar (1988) found efficiencies between 50-75%, while Gloss *et al* (1988) report efficiencies of 17-59% for ten lakes in Adirondack, USA.

Alternatively, the theoretical efficiency may be calculated using the integral form of Eq. (6). This provides an estimate of the dissolved limestone in the water mass which leaves the lake. This estimate gives information about the total amount of limestone and is not a measure of the amount required for a certain water quality to be maintained. Data for sixty-seven lakes give a median efficiency of 63%, with 25% of the lakes with an efficiency of less than 38% and 25% with a figure above 90%. The ratio between efficiency and liming agent dose (e.g. calcium) indicates that efficiency and dose are independent. Sverdrup and Warfvinge (1988) describe a lake with low efficiency which was limed with the highest dose using lime sludge. They consider the low efficiency to be caused by errors in dosing, however. For fifteen lakes, the calculated efficiency was more than 100% - higher than the theoretically possible value. Such results indicate shortcomings in data for which correction or verification has been impossible.

The median efficiency found here (63%) is higher than the mean (41%) for fourteen Swedish lakes which were limed on twenty-four occasions (Sverdrup and Warfvinge 1988). Their values, calculated from initial dissolution, varied between 16 and 90% as a result of differences in liming techniques, grain size and the lake pH value. According to Sverdrup and Warfvinge, higher efficiency is obtained with fine powder and low initial pH.

Humic substances are considered to influence the result of liming activities by coating the grains (e.g. Berzins 1962). An evaluation of the initial concentration of calcium (or calcium and magnesium) shows no significant correlation with colour, nor does the efficiency for seventy-six lakes with colour values ranging from 5 to 180 vary significantly with the concentration of humic substances.

The results which have been obtained do not indicate that coatings on limestone particles have any effect (Sverdrup *et al* 1984). Most probably, the speed at which fine limestone powder dissolves only permits inactivation of the small amount which dissolves more slowly. This means that the effect is so small as to be indiscernible in the available data set.

2.3.2. Littoral liming

In terms of water chemistry, there should be no difference between the liming of shallow zones (areas less than 2 m deep) and whole-lake liming. Locally, however, positive effects can be seen on the biota (Degerman *et al*, Chap. 8). In this case, thirty-six lakes with a relatively short mean residence time of 0.8 years were evaluated. The mean dose was 1.0 meq/l, with a maximum of 5.85 meq/l. With such high doses, seven lakes attained an alkalinity of more than 0.3 meq/l. The liming objectives were fulfilled in terms of a pH of 6.0 for a mean of 5.5 turnovers and alkalinity of more than 0.05 meq/l for 7 turnovers. The duration of littoral liming thus seems to be longer than that of whole-lake liming. The data set is not large enough to make it possible to distinguish between the effects of different liming agents and allow calculation of reacidification.

It was possible to calculate immediate efficiency for a few lakes only on the basis of the maximum concentration of alkalinity and the dose (cf. Sverdrup and Warfvinge 1988). If the concentration of alkalinity before liming is set to zero, the result is an immediate efficiency of 20-30% in four lakes. In one, however, with a short residence time and low dosage of limestone, it was zero.

Dolomite, which is used for liming land and lakes, dissolves more slowly than limestone. Alenäs (1986) reports an experiment in a lake limed in 1977 with 100 g/m^3 of coarse dolomite, mostly in the littoral zone. Dissolution was clearly slow, but it had no negative impact on the result. Duration was about seven years and efficiency was about 52%.

Dickson (1985) describes the situation in Lake Österämten, southern Sweden. Although this lake was mainly limed in the littoral zone with as much as 150 g/m^3, the alkalinity only reached about 0.3 meq/l. Dickson's measurements indicated a duration of at least four years (approximately five turnovers). By comparison, calculations indicate a duration of about one year and, if all limestone was dissolved directly, an efficiency of 70%. For the Norwegian lake Hovvatn, Wright (1985) found that littoral liming worked well with an efficiency of 50%. Sverdrup *et al* (1985) quote several studies which support the opinion that wave movements in the shallow zone have a positive effect on the dissolution of the liming agent and that distribution in such areas is favourable. However, they maintain that there is no proof of this and suggest that dosing should be proportional to water depth, thus making efficient, maximum direct dissolution possible. The same view has been put forward by Rosseland and Hindar (1988). Moreover, an evaluation of data from twenty lakes in central Sweden (Sandberg 1992) provides no evidence that an increased proportion of liming in the littoral zone leads to any clear increase in duration. Alenäs (1986) presents data for seven lakes near the Swedish west coast: six received doses of 8-70 g/m^3 (median approximately 35 g/m^3), with approximately 80% in the littoral zone; he reports a high median efficiency of 88% (54-95%) within eight years. Gloss *et al* (1988) report a mean contribution from the sediment of 64% of all dissolved limestone. If anything, these latter studies support Dickson's view. The conditions in the shallow zone of lakes vary so much that no general conclusion can be made, however.

2.3.3. Soda

Soda may be used as an alternative liming agent. Since it is easily soluble and raises the pH strongly at high doses, soda is injected in the sediment instead of dissolved in the water. Ion exchange takes place in the sediment:

(7) $Na_2CO_3 + H, Al\ sediment \rightarrow Na\ sediment + H^+ + Al^{3+} + CO_2$

Both H^+ and aluminium are neutralized by the soda in water and sediment:

(8) $2\ Al^{3+} + 3\ Na_2CO_3 + 3\ H_2O \rightarrow 2Al(OH)_3\ (s) + 6\ Na^+ + 3\ CO_2$

During the ongoing process of acidification, acidic ions are exchanged with sodium in the sediment. This leads to neutralization of the water.

Sediment treatment with soda (Table 2) has been performed by way of injection of soda solution in two lakes in southern Sweden and with briquettes in two other lakes (Lindmark 1984, 1990, 1991). Since the neutralization capacity is located in the soft, deepest sediment areas, the contribution of neutralized water to more surficial layers will be limited during stratification periods. However, these usually coincide with periods when the addition of acidic water is low. One disadvantage consists of the very high pH which may appear in both the sediment and the nearby water. Values of around 9.5 were found in one lake during the first summer after treatment. Such effects were not observed in the other lakes.

***Table 2.** Data on four Swedish lakes treated with soda (Lindmark 1990, 1991).*

Lake	Lilla Galtsjön	Hjärtasjön	Ålasjön	Strömsjön
Surface area, ha	14.7	20	10.3	11.7
Mean depth, m	3.2	2.7	4.8	4.4
Retention time, y	3.5	0.32	0.8	0.36
Treatment year	1980	1983	1989	1989
tonnes	40	32	52	68
eq/m^2	5.1	3.0	9.6	10.9
meq/l			2.0	2.5
Duration, years	20	4	6	5
Duration, turnovers	6	12	6	12
Efficiency (2 y), %	22	37	46	40

In one of the lakes, both pH and alkalinity values have so far been higher than the limits for liming. The duration was at least three turnovers and a prognosis indicated that it could be approximately six. Another lake, with a turnover time of four months, was reacidified after about four years, corresponding to 12 turnovers.

Na^+ was liberated during reacidification in a lake treated with soda. The efficiency may thus be evaluated on the basis of a sodium budget, but the values shown in Table 2 are based on alkalinity and thus represent an underestimate of true efficiency.

2.4. Liming of watercourses

Running water varies widely in terms of water chemistry. Acid episodes may appear during high-flow periods, mainly in springtime in the north and in autumn in the south of Sweden, while low-flow periods exhibit higher pH values. This variability makes it difficult to optimize liming activities. In addition to the variation in time, there is one in space in the form of micro-stratification with lower pH close to the sediment due to influx of acid water. This leads to acid conditions in an important area for the biota (Lingdell and Engblom 1985).

The main methods consist of liming of lakes situated upstream of the area to be protected, wetland liming and the use of dosing devices. Distribution of crushed limestone has also been tried in some waters. In several waters a variety of application methods have been used. An evaluation of the effects of liming in twenty-two running waters was performed by Degerman *et al* (1990). Table 3 shows a further analysis of these results (Degerman and Appelberg 1992). In 4 waters only the quality was above the limits. Otherwise, they came to the conclusion that in general the treatment did not have the intended result and that this could be attributed to the increased water flow during the period. They also found that the effects of liming procedures have improved with time.

Table 3. *Efficiency of different strategies for liming of 22 watercourses in Sweden. Number of occasions (%) when measured lowest pH/alkalinity during the year reached pH <5, <6 and alkalinity <0.1 meq/l, respectively (from Degerman and Appelberg 1992).*

	Wetland	Doser	Lake
pH <5.0	15.4	2.2	11.8
pH <6.0	38.5	50.6	51.6
Alk. <0.1 meq/l	46.2	57.8	53.9

In theory, lime dosers are very attractive (Dickson and Brodin, Chap. 4). The dose may be regulated for pH (Hindar 1989) or flow, which makes it possible to avoid acid episodes. The main disadvantage is that a shut-down immediately leads to a flow of acid water, resulting in damage to the biota. Fine fractions produce turbidity, which may also damage biota, although it usually stimulates it (Lingdell and Engblom 1985, Degerman *et al*, Chap. 8).

Two dosers have been installed in River Högvadsån, southwestern Sweden (Edman *et al* 1988). At one, the dose was about 55 g/m^3, and at the other it was about 13 g/m^3. The effects were very different, partially depending on upstream water quality, but mainly on the reliability of the dose feeders.

The doser in River Fylleån in the same region worked well from a water chemistry point of view (Widarsson and Johansson 1992), but may not be enough to achieve the biological objectives. Lessmark *et al* (1986) studied the effects of fifteen dosers on water chemistry and fish fauna. All of these lime feeders had shut-down periods, several being closed during the summer. During operation, the pH increased by 0.5-1.5 units immediately downstream. Toxic forms of aluminium (such as inorganically bound, labile aluminium, probably $Al(OH)^{2+}$) were present in the mixing zone. In several cases the total concentration of aluminium remained constant, although the concentrations of labile aluminium fell at pH of more than 6 and were not considered to be toxic further downstream (Lessmark *et al* 1986). Dosers function well when placed upstream from a lake which can function as a detention tank and neutralize acid surges during shut-downs.

Crushed limestone has been distributed in running water with the objective of obtaining slow but sustainable neutralization. Crushed limestone was tried on stretches of 250 and 50 m respectively in two tributaries to Lake Unden, southern Sweden (Lagerman 1983). The effect on the pH was good during low-flow periods (and initially also during high-flow periods) but after a time the contribution fell to only about 0.05 meq/l (as calcium and magnesium).

When 120 tonnes of crushed limestone were distributed along 100 m of a tributary to River Högvadsån, the alkalinity increased by less than 0.02 meq/l (Edman *et al* 1988). The stones soon became coated and further measurements showed the contribution to be 0.01 meq/l or even less, indicating an efficiency of less than 30% over a period of 8 years. The liming objectives were a pH of more than 6.0 and an alkalinity of more than 0.05 meq/l. These were only achieved during 11 and 18% of the time respectively.

The method has thus not been particularly successful and is seldom used nowadays, although it may be justified by a certain improvement in the hatching of fish roe (Degerman *et al*, Chap. 8, Dickson and Brodin,

Chap. 4).

Shore liming may be used in order to neutralize acid surficial water. Several methods were tried in the drainage area of River Stenungeån at the Swedish west coast, including wells and ditches filled with limestone near the river bed (Alenäs and Hultberg 1988). The drainage water had pH between 6.5 and 7, but it was not possible to judge the effect on the river since additional liming efforts were made upstream. A 215-m ditch which was also filled with agricultural lime yielded no obvious change in the water quality of the river (Alenäs and Hultberg 1988). Since all these activities extended along only about one third of one shore - probably less than 10% of incoming water - it is naturally unlikely that any effect could be detected.

2.5. Catchment liming

Catchment liming is performed in order to restore the base saturation on a long-term basis. Results include changes in the leaching of nutrients and aluminium and subsequently in runoff (Nihlgård *et al* 1988). Two types of land are treated with lime: dry soils, such as forest and agricultural land, and wetlands.

Lessmark (1987) reviews the effects of catchment liming for fifty-two lakes and rivers, mainly in terms of dry soils. The limestone distributed on wetlands corresponded to only about 10% of what was applied on land.

The liming of dry soils leads to ion exchange in the uppermost stratum where the calcium is mainly exchanged with H^+ and aluminium. Soils with a low base saturation require a large dose to restore the status. The movement of the calcium in the upper parts of the soil is usually vertical, which results in long transport and slow recovery of deep soil acidification and a minor effect on adjacent streams (Nihlgård *et al* 1988). Leaching from a forest soil to the groundwater during the first years after liming was thus estimated to be about 1% per year (NBF and SEPA 1981).

Andersson and Hamrin (1984) studied land liming in an area where 65% of all agricultural areas were limed with 11 t/ha (mean) or 0.8 t/ha for the whole catchment area of 10,500 ha. The effect compared with the non-limed reference stream is given in Table 4. The liming objective was thus fulfilled over a longer time than the one specified, although other factors such as limited periods of naturally non-acidified water also played a part. The efficiency of the liming for the whole area and study time was about 2%, with a variation of 1 and 5% between different areas (Lessmark 1987).

The watersheds of two lakes in southern Sweden were treated with olivine and limestone, respectively (Table 5). It is evident that watershed liming with olivine is inefficient: somewhat more than 1% over 2.5 years. The liming objectives were not achieved at all in the lake.

Table 4. Fulfillment of goals, expressed as % of time, for liming of the catchment of River Vänneån, southern Sweden, 1981-1983 (data from Andersson and Hamrin 1984).

	Fulfillment in % pH >5.5	pH >6.0	Alkalinity >0.05 meq/l
Vänneån subdivision area	26	36	37
Össjasjön	87	57	4
Grysshultasjön	49	10	10
Balkeryd[a]	35	29	28

[a] Difference upstream/downstream.

Table 5. Fulfillment of goals, expressed as % of time, for lime treatment with olivine in the catchments of Lake Lilla Holmevattnet in southern Sweden and, for reference purposes, treatment with limestone in the nearby catchment of Lake Lilla Ottervattnet (data from Andersson et al 1982).

	Agent	Tonnes	Dose t/ha	Exchange %
Lilla Holmevatten	Olivine	18	5.5	0.8
Lilla Ottervattnet	Limestone	20	6.2	4.9

	pH >6.0	Fulfillment in % pH >5.5	Alk. >0.05 meq/l
Lilla Holmevatten	0	0	0
Lilla Ottervattnet	88	100	100

For River Sundstorpsån in southern Sweden, the major part of the agricultural area, 250 ha (4.8% of the total drainage area) was limed with 3-6 t/ha (1,314 tonnes limestone) (Hasselrot and Westling 1986). The results were difficult to interpret. Hasselrot and Westling (1986) could not identify any significant effect in some areas due to lack of data before liming or acceptable conditions before treatment. Only two areas exhibited a clearly positive effect. Hasselrot and Westling (1986) report that "treatment of agricultural areas with limestone has not led to a noticeable improvement for fishery in the mainstream" and conclude that applications should cover more than 10% in order to achieve acceptable results. The efficiency was low: a calculation for one of the areas shows a transport of 7 kg calcium/ha during the first year - about 6% of the dose.

More recent studies have given similar results. When two catchments in southern Sweden were limed with doses of 0.5 t/ha limestone and 1.5 t/ha of dolomite respectively no effect could be measured during the two years following treatment (Westling and Hultberg 1990/91). Moreover, Nohrstedt (1992) found no increase in calcium concentration in lysimeter experiments with 1.0 t/ha limestone, although the soil water pH values fell by 0.25-1.5 units over a period of 3-5 years. This is most likely the result of ion exchange.

The catchment to a small lake in Norway, Lake Tjønnstrand (0.25 km^2), was limed with a dose of 3 t/ha. With two exceptions, measurements showed a pH of more than 5.5 during the first five years, with a fall during the sixth. The efficiency was 15% (Traaen 1990), which was most probably an effect of thin soil cover and higher deposition than in Sweden.

Similar results were obtained in the liming of an entire drainage area (9.0 t/ha) and solely a discharge area (16 t/ha; equal to 1.5t/ha drainage area) in Wales (Jenkins *et al* 1991). In these cases, pH values and calcium concentrations increased rapidly, while aluminium concentrations fell.

For two years, the efficiency of dry soil liming of two areas at Loch Fleet in Scotland was 4.2 and 7.8% respectively (Dalziel *et al* 1988), increasing to 10.3 and 8.5% after 3.5 years. This corresponded to 49.3 and 36.2 Keq/ha respectively (Howells *et al* 1992).

Lessmark (1987) shows in a review that the pH value rose in 59% of lakes, mainly in connection with dry soil liming, but that the value remained above 6.0 for more than one year in only 6% of the cases. The efficiency of dry soil liming was low. Brown (1988) found that the dose must be more than 1-2 t/ha of the whole catchment in order to ensure lasting results.

The studies discussed above show that, if any appreciable effect is to be obtained from dry-soil liming, the doses must be higher than what is acceptable in terms of soil preservation (cf. Brocksen *et al* 1990/91). However, low-dose liming of dry land may have a positive impact on surface waters in a long-term perspective (cf. Derome *et al* 1986, Yavitt and Newton 1990/91).

Wetland liming has become a frequent method for lake mitigation, particularly in northern Sweden. Liming of outflow areas brings about direct contact between the liming agent, the soil and the groundwater which drains to the surface water.

According to the Swedish liming register, fifty lakes with a total drainage area of 776 km^2 have been limed with a median dose of only 9 t/km^2 on wetlands. As a rule, the maximum effect does not appear until about half a year after treatment. Observations showed that the pH target was met in about 90% of cases, while the success rate for alkalinity was about 50% (Table 6). A minor increase in colour (humic substances) seems to take place after liming. The fifty lakes in the study were monitored for a maximum of four years and showed a duration of only about one year (median), independent of the chosen target (Table 7).

Table 6. The effects of wetland liming on water chemistry in 50 Swedish lakes.

		Percentile		
		10	50	90
pH	Before	5.1	6.05	6.9
	After	5.7	6.5	7.0
Alkalinity, meq/l	Before	0	0.03	0.22
	After	0.02	0.12	0.27
Ca + Mg, meq/l	Before	0.16	0.30	0.46
	After	0.19	0.28	0.54
Ca, meq/l	Before	0.13	0.22	0.47
	After	0.22	0.36	0.46
Colour	Before	20	60	180
	After	25	85	240

Table 7. Fulfillment of goals for liming (years) for 50 Swedish lakes treated by means of wetland liming.

	Percentile		
Goal	10	50	90
Alkalinity >0.05 meq/l	0.55	1.1	2.3
pH >5.5 (5 lakes)		1.0	
pH >6.0	0.7	1.1	3.2

Some examples of wetland liming are presented in the following.

Two treatments included in the study were performed using granulated dolomite (Table 8). In one case there was no effect on water chemistry (Jönsson 1990) and in the second (Grahn 1990) there was a minor impact.

The wetlands of Lake Trehörningen, southern Sweden, were limed with a dose of 0.16 t/ha (limestone) leading to fulfillment of the objective of a pH above 6.0 and a pH above 5.5 during 85 and 97% of the time during a period of three years (NBF and SEPA 1981).

Table 8. *Wetland liming with granulated dolomite in the catchments of two lakes in southern Sweden (data from Grahn 1990, Jönsson 1990).*

Lake	Dolomite tonnes	Limed area, %	Dose Catchment t/ha	Dose Limed area t/ha
Skärsjön	308	47.9	2.41	5.0
Grytsjön	307	44	1.7	3.8

Hasselrot *et al* (1984) describe results for two streams in western Sweden (Table 9). The effect in one stream decreased rapidly, falling below target after only two years. Wetland liming of a nearby area only gave pH above 6.0 for 7-9 months after treatment (Hultberg and Alenäs 1981). The efficiency for a period of 3.5 years was 37% (Andersson 1983). Additional liming activities in 1982/83 and 1983/84 raised the pH values to 7.5-7.9 and the alkalinity to above 1 meq/l (Alenäs 1987), making it clear that liming would have to be repeated in order to reach the targets. The compliance increased from 19 to 90% after additional liming.

Wetland liming of 11 selected areas in Jönköping county, southern Sweden, was evaluated by Svahnberg (1989) and later by Abrahamsson *et al* (1992). In eight larger areas the effect was good, as was the efficiency.

Liming of a mire at Loch Fleet in Scotland resulted in an efficiency of 23% within two years. This was considerably higher than that achieved by dry land liming in the same area (Dalziel *et al* 1988), although the efficiency increased to only 24.1% after 3.5 years (Howells *et al* 1992).

Table 9. *Effects of wetland liming with limestone for two streams in south-western Sweden (data from Hasselrot et al 1984).*

	Tonnes	Dose t/ha	Efficiency %	Years
Bråvattenbäcken	30	1650	35	c. 5
Nordbäcken	30	1450	25	3
	Fulfillment of goals in % of time			
	pH >5.5	pH >6.0	Alkalinity >0.05 meq/l	Years
Bråvattenbäcken	100	92	35	5
Nordbäcken	92	72	95	2
Nordbäcken	2	0	0	3-5

The duration of wetland liming is shorter than that of dry soil liming - about one year - but the efficiency is much higher, albeit considerably lower than in the case of lake liming. The explanation for the short duration may be that doses are lower than those given to lakes and that the undissolved limestone is inactivated. The location of the treatment is of vital significance (Lessmark 1987), with the best effect being obtained from relatively large doses in discharge areas (cf. Nyberg 1985).

The effects of land liming on soil processes and leaching need to be studied. This is a very complex field, as conditions in soil vary much more than in the aquatic environment (cf. Nömmik 1979).

The Swedish aquatic liming data base contains few land-liming parameter measurements other than pH value and alkalinity. It is likely that soil liming will affect both aluminium and other metal ions whose dissolution declines as pH rise. A higher pH value also mobilizes nutrients in the soil.

Lessmark (1987) reports that soil liming only lowered the aluminium concentration to about 470 μg/l. It is difficult to draw any conclusions about effects on iron and manganese since their concentrations are primarily governed by redox conditions, which are regulated by the groundwater level. Westling and Hultberg (1990/91) consequently found that soil liming had no effect on the leaching of iron or manganese.

Concentrations of total nitrogen increased after liming, while no change in phosphorus concentration was found (Lessmark 1987). Concentrations of ammonium and nitrate in runoff from two land-limed areas (Aneboda, southern Sweden, and Gårdsjön) showed no effect from the treatment (Westling and Hultberg 1990/91).

3. Secondary physical-chemical effects

The direct effects of liming on pH value and concentrations of calcium and alkalinity have been elucidated in the previous sections. These changes affect many processes in the water and it is these secondary effects which are discussed in the following.

3.1. Physical effects

Hasselrot and Hultberg (1984) showed that liming may give rise to a chemical stratification with high concentrations of dissolved limestone in the bottom water of a lake. This would appear to inhibit turnover during the year after liming. The stratification conditions later returned to normal, although concentrations of calcium and alkalinity in the hypolimnion remained high due to leaching from the bottom. Increased stratification as a result of dissolved limestone is not likely to arise under normal conditions as the increase in density will not, under normal conditions, increase significantly in comparison with variations due to temperature.

Reduced water colour brought about by liming is said to increase the heat in the hypolimnion (e.g. Yan 1983, Bukaveckas and Driscoll 1991, Bukaveckas 1991) and thus reduce the stability of the stratification. However, the studies which have been carried out do not prove that this effect is of more general importance (Olem 1991).

3.2. Effects on water chemistry

Changes in water chemistry may be caused either directly by the contents of the liming agent, such as trace elements or nutrients (Broberg 1978, Dickson 1985), or through processes in the aquatic system. Only processes in lakes are discussed in this section. Liming of land gives rise to both chemical and biological processes, which in turn affect the soil and groundwater, which reaches the surface water (Nihlgård *et al* 1988).

Changes in the concentrations of organic substances, nutrients and trace metals are reviewed in the following.

3.2.1. Organic substances and optical characteristics

A rise in pH and addition of calcium may affect the concentration and characteristics of organic, coloured substances in water. TOC (total organic carbon) or DOC (dissolved organic carbon) are seldomly recorded in the Swedish data base, therefore, indirect measurements such as colour and Secchi depth are included in the discussion.

The concentration of organic substances is a function of inputs from the drainage area, the production in the lake and elimination by physical-chemical and biological processes. Lake liming will not affect the input from land, but may increase the oxidation of organic matter in the sediment (Wright 1985, Gahnström, Chap. 6) and stimulate primary production (Larsson, Chap. 7). In Lake Gårdsjön, Broberg (1988a) found an increase in TOC concentration which was explained either by decreased photooxidation, increased sediment oxidation (and release) or a release of organic matter from *Sphagnum* mats. Alaasarela *et al* (1990) found that liming of humic waters reduced the DOC concentration by 7% within five days, but the portion of high molecular and thus potentially sedimentating organic material increased somewhat, which could lead to a larger decrease later. Marcus (1988) found no change in TOC in a statistical review of twenty limed lakes.

It is well known that the colour of humic substances increases with increased pH. When raising the pH from 4 to 7 Shapiro (1957) found an increase of 17% and Wilander *et al* (1978) for a natural water one of about 20%. A change from 5 to 7 increased the colour by 12% (Shapiro 1957). The raise in pH alone thus gives a small increase in colour.

In lakes where liming raised the pH, the colour, taken as a mean, increased (Fig. 7). The figure furthermore indicates a temporary lowering soon after the liming, caused by precipitation of humic substances and particles adsorbed on lime particles and flocculated by aluminium hydroxides (Bukaveckas and Driscoll 1991).

In lakes with a colour in excess of 150 units, the colour is reduced after liming, while in lakes with less colour it is either increased or diminished. The increase in low-colour lakes may be an effect of decreased precipitation of humic substances and more favourable conditions for phytoplankton. The solubility of aluminium hydroxides is lowest and thus the flocculating capacity highest in the pH range 5.5-6 (Stumm and Morgan 1981).

Degerman (1987) cites several findings of either increases or decreases in colour after liming. He considers the increase in clear waters to be an effect of fresh iron and manganese hydroxides, while reductions, most frequent in humic waters, are due to co-precipitation of aluminium hydroxides and possibly calcium-humic compounds. In many cases, no changes in colour are reported.

It is thus obvious that the conditions before liming, both with respect to pH value and organic substances before and pH value after liming, determine the effects on organic matter.

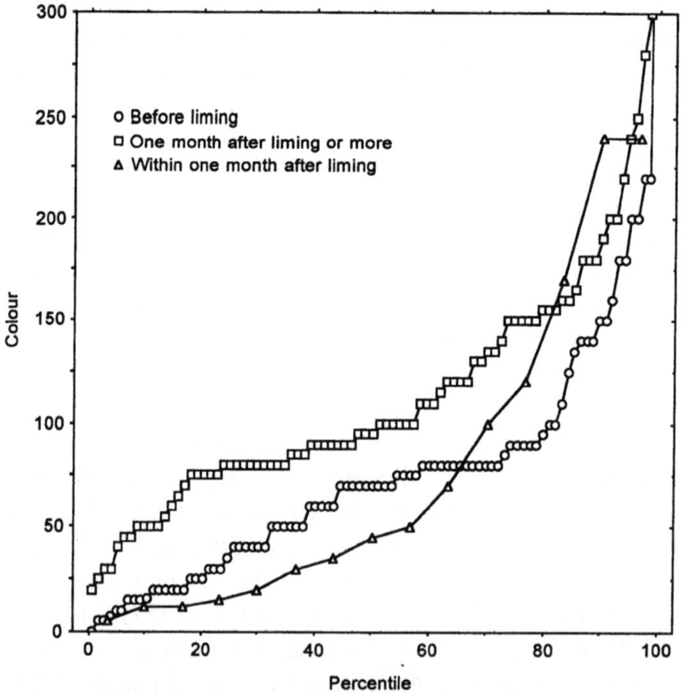

Figure 7. The frequency distribution of colour before, within one month after, and after liming of Swedish lakes.

The Secchi depth in a lake is governed by the true colour of the water and the amount and type of particles (e.g. metal hydroxides, plankton). In an early literature review Wilander and Ahl (1972) found both increases and reductions in the Secchi depth after liming. Dickson (1979) reports an immediate reduction in Secchi depth after liming of one lake in western Sweden. This also occurred in Lake Gårdsjön soon after liming, when the depth decreased from 7-13 m before to less than 3 m soon after liming, while later, during periods of medium algal biomass, it stayed at 5-7 m. In connection with an extensive algal bloom it was as low as 1.4 m (Larsson 1988). Bukaveckas and Driscoll (1991) found reduced Secchi depths in six limed lakes and increases in two reacidified lakes. In three American lakes the lowered values coincided with elevated plankton concentrations during the first year. Increased plankton biomasses are thus a common cause of reduced Secchi depths, as is ongoing precipitation of iron and aluminium hydroxides entering the lake. At strongly elevated pH values, increased Secchi depths may be explained by flocculation of humic substances (Wilander and Ahl 1972).

3.2.2. Phosphorus compounds

One early problem of acidification was identified as oligotrophication of stricken lakes (Grahn *et al* 1974, Jansson *et al* 1986). Nowadays, there is general consensus that acidification reduces the leaching of phosphorus from land (Broberg and Persson 1984, Bernes 1991). It is also clear that phosphorus is the limiting factor for production in the majority of the lakes of interest, although both phosphorus and nitrogen shortage have been proven in certain acidified lakes (Blomqvist *et al* 1989). Moreover, in other acidified waters, bacterioplankton was limited by phosphorus, nitrogen, and carbon (Bell *et al* 1993). Acidification thus changes the productivity potential in acidified lakes. These changes in inflow nutrients, both quantitative and qualitative, will not be remedied by lake liming. Lake liming alone will affect the nutrient concentrations through their presence in the liming agent and through internal chemical and biological processes.

The assumption that all limed lakes should behave in much the same way with respect to changes in nitrogen and phosphorus is unreasonable if all the basic conditions (such as turnover time, humic substances, pH values) are taken into account. There are examples of both rising and falling concentrations of both nitrogen and phosphorus compounds after liming. A study of 63 limed lakes in Sweden reveals that the phosphorus concentration increased in about half of them, while it fell in the remaining ones (Broberg 1987). A similar result was found for nitrogen. In a review of effects of liming, Bernes (1991) states that the postliming phosphorus concentration increases as a rule. However, this is contradicted by data from the Swedish data base on liming, which indicate a decrease.

A deeper knowledge of process mechanisms and the coupling between chemical, biochemical and biological changes is essential to a full understanding of how to manage the limed ecosystem.

There are several potential sources of phosphorus in the acidified lake. When and how these are mobilized and transferred to the water depends on local conditions, the liming agent and the dose, etc.

The liming agent contributes phosphorus. In regular limestone the phosphorus concentration is about 0.02-0.04%, while in Algomin (coral limestone) it is as high as 0.2%. The immediate dissolution of phosphorus is likely to be proportional to the amount of dissolved limestone. An increase of 0.2 meq calcium/l (10 g $CaCO_3/m^3$) thus results in a rise in phosphorus of 2-4 μg/l (NBF and SEPA 1981). Immediately after liming phosphorus in suspended limestone particles will also increase total-phosphorus concentrations (Broberg 1988a) which, together with phosphorus precipitated as various complexes (aluminium, iron, humic substances), will be deposited in the sediment. In the limed lake about 75% of phosphorus is organically bound and approximately 23% is NaOH-extractable (Pettersson and Olsson 1984). The latter is redox sensitive and thus released during anoxic conditions. Organically bound sediment phosphorus may be transferred to the water through biological processes. Released phosphorus will probably be resorbed if the pH of the sediment is >8 (Andersson and Gahnström 1985). When soda is used, the phosphorus increase is likely to come from the sediment (Lindmark 1982).

"*Sphagnum* mats" and "felt structures" which develop in certain acidified lakes will break down (Sangfors 1984) and disappear after liming (Lazarek 1986, Larsson, Chap. 7), releasing phosphorus. The phosphorus concentration in *Sphagnum* is low, higher amounts are found in the felt structure (cf. Table 10). Detritus is considered to accumulate in the acidified lake due to the shortage of shredders leading to a reduced breakdown (Andersson 1985, Chamier 1987). Moreover, the microbial decomposition is inhibited (Francis *et al* 1984, Gahnström, Chap. 6). Liming and recolonization ease the recycling of nutrients.

Acidification changes the composition of zooplankton fauna from cladocerans to calanoids, which are considered to have faecal pellets which decompose with difficulty, and may be even more limited as "access" is reduced by *Sphagnum* mats (Nyman 1990). This results in a decrease in phosphorus release. The effects described on phosphorus turnover are likely to disappear after successful liming.

Algal blooms have been reported some time after liming (Waters 1956, Broberg 1988b), and are associated with temporarily elevated total phosphorus concentrations, mostly particulate (Broberg 1988b). The algal species causing these blooms seem to accumulate phosphorus from the sediment during their resting stage and bring it to the free water when blooming (Boström 1986, Heyman and Lundgren 1987).

Table 10. Phosphorus content in Lake Gårdsjön, south-western Sweden, when the lake was limed for the first time. The natural phosphorus supply is 8.3 kg/year (0.027 g/m², year). Data from Andersson (1985), Broberg (1987, 1988a).

	kg	g/m^2
Sphagnum mats	1.7	0.0055
Felt structures	10.7	0.034
Surface sediments, 0-2 cm	350	1.12
Lake water	3.0	0.0096
Lime	37.4	0.11
lime momentarily dissolved	5.4	0.017

Precipitation of aluminium-phosphorus complexes has been assumed to be an important process in the reduction of diminishing phosphorus concentrations (Dickson 1980). This theory has not been verified by either mass balance studies in Lake Gårdsjön (Persson and Broberg 1985) or by regional data. The regional data set (e.g. Almer *et al* 1978) gives a nearly linear relationship between lower phosphorus concentrations and falling pH, which is not in accordance with the theoretical relation where maximum precipitation of the aluminium-phosphorus complex is found in the pH interval 5.5-6.5 (Stumm and Morgan 1981). However, acid water with high aluminium and iron concentrations entering a limed lake may produce phosphorus precipitation.

DOC has a capacity to keep nutrients such as phosphorus in solution. Immediately after liming calcium-humic complexes will be precipitated to a degree which depends on such factors as DOC, lime dose and pH (Broberg 1988b). In lakes where liming gives elevated DOC concentrations the availability of phosphorus for biota increases. Liming is generally followed by a diminished algal biomass, however, and the temporal increase in phosphorus has little effect in terms of stimulating phytoplankton (Waters 1956, Yan and Dillon 1982, Hultberg and Andersson 1982, Larsson 1988, Chap. 7). As a rule, ecosystems are very unstable the first years after liming (Evans 1989).

The negative effects of high metal concentrations on phosphorus turnover in acidified waters are reduced after liming when they are precipitated (Dillon *et al* 1979, Hasselrot *et al* 1984, Molot *et al* 1990) and also as a consequence of increased binding to organic matter.

Regeneration of organic phosphorus improves after liming when aluminium concentrations are lower and do not inhibit enzymatic hydrolysis (Jansson 1981, Brassard and Auclair 1984). The activity of phosphatases in relation to pH varies. In oligotrophic lakes it may be regulated by Ochromonadaceae algae which has an optimum pH interval of less than 6, while in eutrophic lakes with Cyanophyceae and Bacillariophyceae it lies between 7.5-8.5 (Olsson 1990, 1991). In Lake Gårdsjön, the specific phosphatase activity was highest during summer in the unlimed lake, indicating a scarcity of phosphorus under acid conditions. The changes in activity after liming indicate that the availability of phosphorus increased but it cannot be ruled out that this is caused by factors other than pH, such as changed plankton composition. The inhibition caused by aluminium at low pH is reduced in the limed lake, thus stimulating algal growth (Pettersson *et al* 1985, Törnqvist 1988).

The reduction by liming in negative effects of aluminium and other metal ions is most likely the key factor to the increased phytoplankton production found after liming even in lakes where the phosphorus concentration did not increase.

It has already been said that phosphorus concentrations in the lake-limed lake should, in the long term, decline as long as the acidification effect remains in the drainage area (Broberg and Persson 1984, Persson and Broberg 1985). This conclusion is verified by the study of Lake Gårdsjön (Fig. 8) and of nearby lakes (Hultberg and Andersson 1982). There was a falling trend with time in acid-sensitive lakes containing Arctic char *(Salvelinus alpinus)*, irrespective of whether the lakes were limed or not (Nyberg *et al* 1986). In twenty lakes studied for four years or more, no significant change in phosphorus concentration could be found (Stålhandske 1990). A more detailed study of the same data shows that the total phosphorus concentration fell in about 60% of twenty-one lakes studied for four years and in 90% of the seven lakes studied five years after liming (Stålhandske, unpubl.). No changes in concentration due to lake liming could be found in lakes in the Åva area near Stockholm or in the Delsbo area, central Sweden (Hörnström *et al* 1992a). Even though diminishing or stable phosphorus concentrations dominate the results (cf. Broberg 1987) there are examples of increased concentrations after lake liming alone. In Lake Ljustjärn, southern Sweden, the total phosphorus concentration initially rose, but fell later according to calculated dilution (Wilander and Ahl 1972). A slow increase in total phosphorus concentrations after liming has been observed in several lakes in south-western Sweden (Appelberg *et al* 1990, Hörnström *et al* 1992b), but the cause has not been discussed.

Stålhandske (unpubl.) found that the concentration in about 25% of eleven limed lakes increased, while falling in about 55%. In the class comprising land-limed lakes (twelve lakes with 9-94% of the dose) the concentration increased in as many as 75% of the cases and fell in only about 25%.

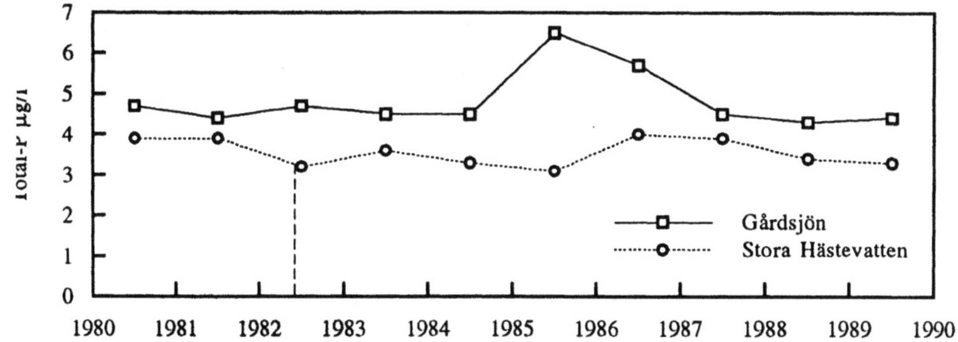

Figure 8. *Concentrations of total phosphorus in Lake Gårdsjön (solid line), before and after liming (vertical broken line), and in the nearby reference Lake Stora Hästevatten (broken line).*

Even though changes in concentrations vary between lakes, liming in a long-term perspective usually results in improved phytoplankton production (Hörnström and Ekström 1986, Larsson 1988, Chap. 7).

It is reasonable for a lowered input of phosphorus from land over a longer period of time to be reflected in the form of lower lake concentrations (Persson and Broberg 1985). A complete remedial programme should thus include catchment liming and/or fertilization with phosphorus in certain lakes (Olofsson *et al* 1988, Bergquist, Chap. 17).

3.2.3. Nitrogen compounds

Liming seems to raise and reduce the concentrations of nitrogen compounds, as in the case of phosphorus (cf. Broberg 1987). An experimental acidification in Canada using sulphuric acid did not change the nitrogen cycle in terms of the turnover of ammonium or nitrate (Schindler *et al* 1985). Several studies show little or no effect on overall concentrations of nitrogen, ammonium, and nitrate after liming (Yan *et al* 1977, Dillon *et al* 1979, Hindar and Nilssen 1983, Marcus 1988). It is, however, evident that changes in the mineralization of organic matter which take place in acidic lakes with *Sphagnum* and felt structures (cf. Larsson 1994) will, at least in the short term, influence nitrogen concentrations in the limed lake. Higher microbial activity is reflected in sudden increases in ammonium and organic nitrogen as well as in higher denitrification (Broberg 1988a). The scarcity of DOC for bacteria found in acidified lakes (Bell *et al* 1993) would diminish after liming and further promote the activity.

In lakes studied during the SBF trial period, increases in nitrogen concentrations after liming were more common that decreases (Stålhandske, unpubl.). This tendency was not found in the Swedish data base on liming - the changes in total nitrogen were small and mostly downwards instead. The increases found in DOC in many limed lakes are associated with a corresponding increase in total nitrogen, as can be seen in Lake Gårdsjön (Fig. 9; Broberg 1988a) and nearby lakes (Hultberg and Andersson 1982). Increased sediment respiration (Gahnström 1988, Chap. 6) is also likely to contribute to the organic nitrogen in the lake.

From a theoretical point of view, the conditions in acidified lakes should favour the production of N_2O during both denitrification and nitrification (Samuelsson and Klemedtsson 1991). Enhanced production of nitrous oxide during denitrification takes place at low pH and when there is a shortage of energy sources (Betlach and Tiedje 1981), as well as during nitrification at low pH and low oxygen concentrations (Johnsrud 1980, Samuelsson and Klemedtsson 1991). If nitrous oxide production is stimulated in acidified lakes, it should therefore be inhibited by liming.

The nitrate "surplus", which is commonly found in acidified lakes all year-round, changes after liming to a more normal seasonal variation, with low concentrations during the vegetation season (Hörnström *et al* 1992b). In Lake Gårdsjön, the algal bloom found after some years was accompanied by low nitrate concentrations (Broberg 1988b). It would therefore be logical to expect that, in lakes where primary production was stimulated by liming, nitrate would be consumed in relation to the production. In the Swedish data base the changes in nitrate do appear to be small, although there is a tendency for concentrations to rise.

The results indicate that the stimulating effects of liming on primary production and denitrification do not lead to a decrease in nitrate concentrations. However, the effects of processes which consume nitrate may be compensated for by simultaneous stimulation of nitrate production through nitrification and mineralization.

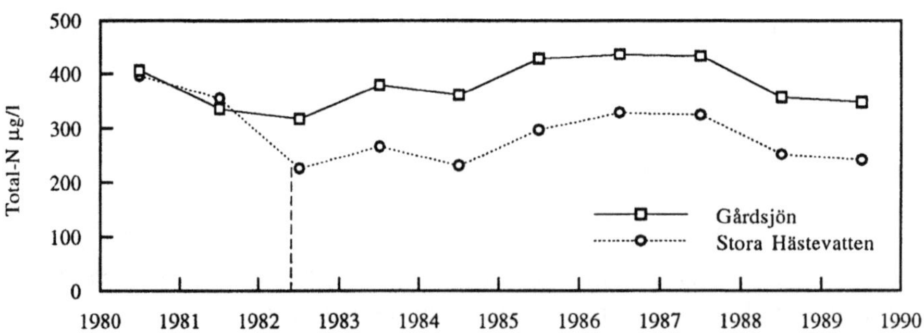

Figure 9. Concentrations of total nitrogen in Lake Gårdsjön (solid line) before and after liming (vertical broken line) and in the nearby reference Lake Stora Hästevatten (broken line).

3.2.4. Trace metals

Concentrations of trace metals and species in water are usually dependent upon pH. From a chemical point of view, higher mobility, solubility and a larger portion of free metal ions can be expected at lower pH values (Stumm and Morgan 1981). This can be observed in acidified waters (Dickson 1980, Borg 1983, Borg and Andersson 1984). Elevated concentrations have been found, particularly for aluminium, manganese, zinc, cadmium and lead, but also in certain cases for iron. These conditions should be reversed upon liming, so that concentrations fall. Lake liming promotes precipitation and sedimentation of metal compounds and catchment liming would also diminish leaching from the soil, provided that the dose is efficient (Huettl 1988).

It was noted at an early stage that liming agents, particularly waste and slag liming products, could contain trace metals in the form of high concentrations of chromium, nickel, zinc and cadmium (Broberg 1978, NBF and SEPA 1981). Although no negative effects have been observed, such agents are no longer in use (NBF and SEPA 1981). A calculation for a Finnish liming experiment with a dose of 36.4 g/m^3 in the form of limestone gave increased concentrations of lead and zinc of 0.37 and 0.58 $\mu g/l$ respectively (Weppling *et al* 1992). In general, the metals are likely to be bound relatively hard. They are therefore not dissolved in the water but transferred to the sediment directly or indirectly in the form of co-precipitation (cf. Driscoll *et al* 1989).

In the following section, the effects of liming on aluminium, iron, manganese and certain trace metals are discussed, while the effects on mercury, especially in fish, are reviewed in Meili (Chap. 16).

3.2.5. Aluminium

The toxicity of aluminium is well known from both laboratory studies (e.g. Dickinson Burrow 1977) and field observations (e.g. Dickson 1978, 1983), but the toxicity of the different forms is not clear. In acidic water the major form is cationic. The solubility is at a minimum at about pH 6, and increases at higher values, but then as an anion, $Al(OH)_4^-$ (Stumm and Morgan 1981). It is also bound to organic matter (Fig. 10) (e.g. Tipping *et al* 1988). In the pH interval 5.2-5.8, when $Al(OH)_2^+$ dominates, aluminium seems to be more toxic (e.g. Hörnström and Ekström 1983). Liming shifts the equilibrium between different forms, which seems to create an extremely toxic environment (e.g. Dickson 1983, Rosseland *et al* 1992, Degerman *et al*, Chap. 8) affecting both plants and animals - killing of fish being the most obvious result. Gill tissue is affected and at the cell level there are changes in Na-ATPase and plasma ions (Rosseland *et al* 1992).

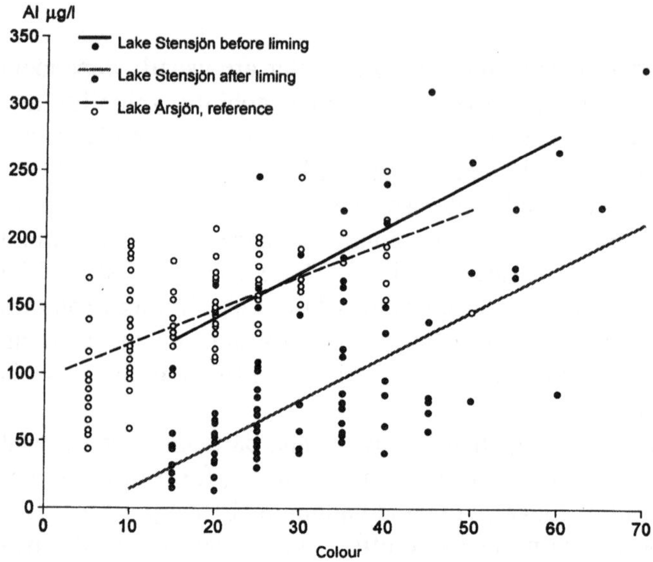

Figure 10. Aluminium concentrations and colour in Lake Stensjön (south of Stockholm), before and after liming, and in the reference Lake Årsjön.

Driscoll *et al* (1989) found that when liming operations led to high pH, the aluminium concentration remained constant for a short time when aluminate ions were produced, but as the pH values fell in the two lakes under study the aluminium concentrations also decreased.

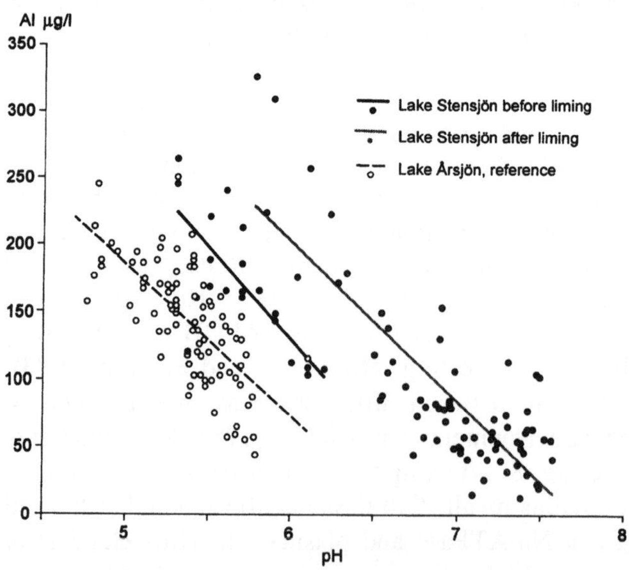

Figure 11. Aluminium concentrations and pH value in Lake Stensjön (south of Stockholm), before and after liming, and in the reference Lake Årsjön.

When the pH of a lake increases from about 5 to 6, liming usually induces rapid precipitation of aluminium (NBF and SEPA 1981). However, Niinioja *et al* (1990) found no change in aluminium concentrations after liming of two Finnish lakes, but in these cases the concentrations were already low before treatment. The negative correlation between pH and aluminium in unlimed lakes remains after liming, but generally at a lower level (Fig. 11). The liming of Lake Stensjön, near Stockholm, may have influenced the aluminium concentration directly through an increase in pH value, but also indirectly through changes in relations between organic matter and pH.

Lake liming does not reduce the input of aluminium from surrounding land areas. However, part of this aluminium is bound to humic substances during transport. The increased pH in the lake favours complex binding, thus leading to a smaller proportion in free form (e.g. Tipping and Backes 1988). In the limed Norwegian Lake Hovvatn, Wright and Skogheim (1983) found that the influent (pH 4.4) with a total aluminium level of 200 μg/l, of which approximately 100 was in the from of labile monomeric aluminium, led to a reduction of about 50 μg/l monomeric aluminium when mixed with the lake water (pH 5.4), while the concentrations of other forms (organically bound and colloidal) remained constant. Liming upstream of a fish culture, raising the pH from about 4.7 to 6.4 did not diminish the total concentration. The colloidal fraction increased from about 25 to 75%, however, and the cationic fraction fell from 200 to some tenths of a μg per litre (Dickson 1983). In the clear Lake Lysevatten in south-western Sweden, where liming increased the pH from 5.1 to 7.2, the total aluminium concentration fell from about 170 to 20 μg/l (Alenäs *et al* 1991).

In the limed lakes studied in the Swedish ISEL programme (Integrated Studies on the Effects of Liming) aluminium concentrations are low: usually <100 μg/l of acid soluble aluminium and <12 μg/l of labile inorganic aluminium (Appelberg and Aldén 1992).

Reduced concentrations are often found after catchment liming, as in the outlet of a lake in the Delsbo area, central Sweden (Fig. 12). Liming activities have led to a definite reduction in the leaching of aluminium from the soil. Similar results are reported by Andersson *et al* (1982) for limed streams on the Swedish west coast and by Rosseland and Hindar (1988) after land liming of the drainage area to Lake Tjønnstrand.

In acid, clear lakes, liming reduces the concentrations of all forms of aluminium, while in humic waters the total concentrations seem to remain constant, although the toxic, monomeric part diminishes due to increased complex binding. A permanent lowering of aluminium concentrations can only be achieved through reduction of soil acidification or land liming. Even though the toxicity of aluminium decreases after liming it can be substantial during transition, when changes between different forms occur.

In humic water a large part of the aluminium is bound in organic complexes and is thus non-toxic.

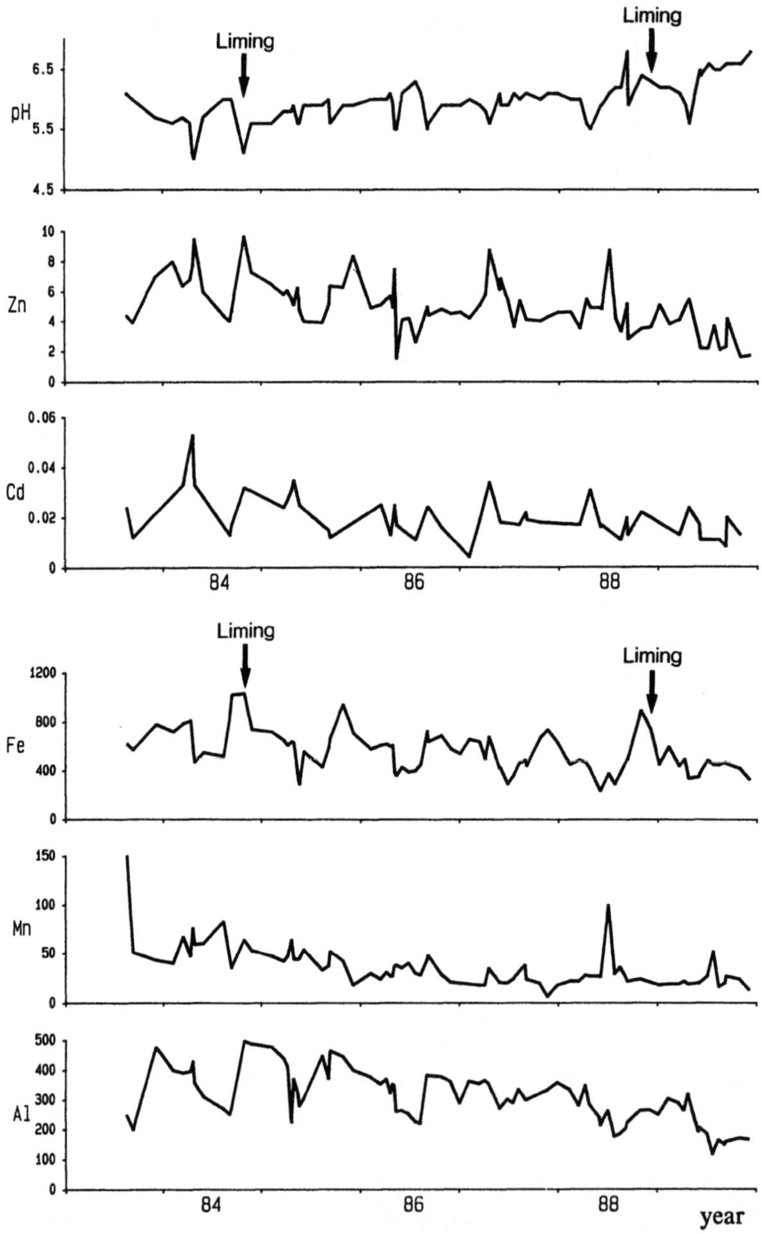

Figure 12. *pH and metal concentration (μg/l) in the outlet of the Lake Svartvallstjärn, Delsbo area (central Sweden), before and after catchment liming.*

3.2.6. Iron and manganese

The solubility of iron and manganese is largely controlled by redox conditions and pH value (Hutchinson 1957).

Reduced iron and manganese concentrations have been found after land liming in Lofsdalen in the mountainous area, central Sweden (Borg 1986) and in the Delsbo area (cf. Fig. 12). Before liming in Djursvasslan (Lofsdalen) the median concentration of iron was 700 μg/l. In connection with high flow periods, when precipitated iron hydroxides were dispersed, values as high as 10-50 mg/l were observed. After wetland liming the median concentration fell to 420 μg/l and no extreme conditions were observed. Despite this, there is no clear correlation between the pH value and the concentrations of iron and manganese in the streams studied. The flux from mires of iron and manganese is strongly regulated by redox conditions, but land liming raised pH values in outflow areas, which may increase the rate of oxidation of the mobile, reduced forms to solid hydroxides.

According to the results from the Delsbo area, there is a tendency for concentrations of iron and manganese to fall after lake liming, but there is wide variation due to changes in runoff and concentrations of humic substances. Lake turnover in spring and autumn also mixes reduced ions from the hypolimnion with the entire water mass (cf. Sanni *et al* 1983).

A review of twenty limed lakes in the USA indicated no change in concentrations of iron or manganese (Marcus 1988).

3.2.7. Other trace metals

Falls in zinc and cadmium concentrations have been observed after both land and lake liming (Fig. 12). These metal ions are relatively mobile and their occurrence is strongly dependent upon pH in soil-water systems (Bergkvist 1987, Borg and Johansson 1989), which means that liming should have a clear effect. A strong correlation between zinc and pH is evident from regional lake surveys in Sweden (Dickson 1980, Borg 1983). In addition to zinc, lead and cadmium also are affected by liming. Lead is normally bound more strongly than the other two metal ions, but despite this there may be a negative correlation with pH value and lower concentrations can be found after land liming (Fig. 13).

Driscoll *et al* (1989) studied zinc in the limed Woods Lake (USA) and found decreasing concentrations after liming even though the water was unsaturated. They consider the effect to be caused by co-precipitation. The same process may apply to other trace metals, such as cadmium.

Lake liming of Lake Hovvatn in Norway did not affect the concentrations of copper, zinc, and lead (Wright 1985). Marcus (1988) also found that liming of twenty American lakes had no effect on concentrations of cadmium, lead, and zinc, although the mean pH rose from 5.5 to 6.6.

The impact of liming on trace metals is governed by the state before liming and the range of change; small changes in pH value have a lesser effect. Reduced solubility may be counteracted by increased complex binding, for instance.

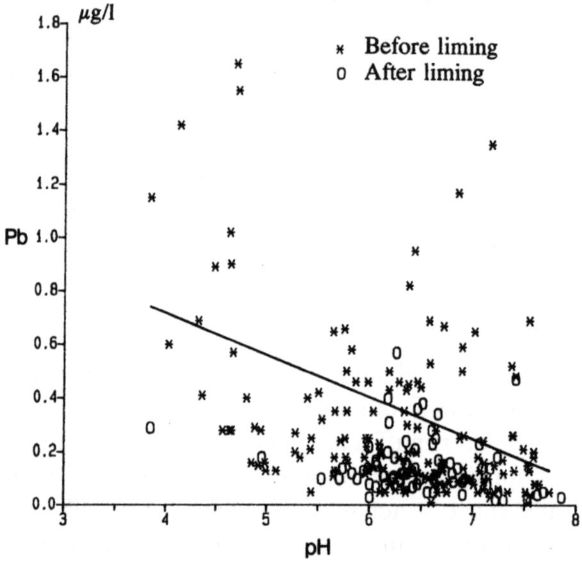

Figure 13. Lead concentrations and pH of wetland limed streams in the Lofsdalen area in the mountainous area of central Sweden.

3.3. Changes in sediment chemistry

Both sediment and free water take on a certain base neutralization capacity when acidified (BNC). A certain amount of base is thus required to raise the pH value of the sediment. Molot (1986) compiled results from different methods of measuring this capacity (lime demand) and found it to be 0.004-0.01 meq/g wet weight for acidified lake sediments.

Since liming activities change the chemistry of the free water, it would be natural to expect the sediment chemistry to change as well. Secondary processes in the water, such as precipitation of metals and increased primary production, should also affect the sediment. The sediment layers deposited each year are very thin and variations are thus difficult to trace with conventional techniques. A study of two limed lakes on the Swedish west coast thus failed to indicate any change as a result of liming (Alenäs 1986). Dickson (1983) reports elevated concentrations of aluminium and cadmium in the uppermost centimetre of the sediment after about 15 years of intermittent liming, however, and considers this to consist of sedimentation of aluminium hydroxides. Andersson and Borg (1988) found, in Lake Långsjön, south of Stockholm, that the cadmium concentration increased in the sediment while decreasing in the water. In a study of surficial sediments in nine lakes, Andersson *et al* (1991) found lower concentrations of organic matter and mercury a few years after liming. The reduction in organic matter may have been caused by microbial breakdown in the sediment (Gahnström, Chap. 6).

Dickson (1979) describes a laboratory experiment with addition of limestone or calcium chloride to a sediment sample with a relatively high pH of 6.1. The limestone addition raised the value to 7.5, but had almost no effect on concentrations of manganese, zinc, lead, and cadmium in the water. The addition of the neutral salt calcium chloride lowered the pH to 4.6, however, and increased the concentrations of all the metals analysed in the water. The reduction in pH is caused by ion exchange between calcium in the water and H^+ in the sediment. Similar results were found by Andersson *et al* (1983). In the pH interval 4-7 only about 1% of the total concentration of iron and lead is in dissolved form. Zinc and manganese behaved in a different manner, however, with a dissolved portion of about 10% at pH 4 and less than 1% at pH 7, while the portion for cadmium diminished from about 2% to below 0.1% after the pH increase.

In Lake Gårdsjön the pH in the interstitial water rose from about 5 to 6.5-7.5 soon after liming and subsequently stabilized at 6-6.5 (Gahnström 1988). After the second treatment, the pH fell immediately instead by up to one unit and later increased to pre-liming levels. This was obviously an ion exchange effect as described by Dickson (1979).

In Norway, a study of sediment reactions upon liming was made in the laboratory and in three lakes (Sanni *et al* 1983, 1984). The fact that the uptake of calcium was greater in more acid sediments confirms the ion exchange process.

Limed sediments could contribute to the ANC of the lake. Wright (1985) calculated the leaching from sediment in two Norwegian lakes at only 0.8 and 1.1% of the dose, which suggested that the buffer effect was minor. From a study of two other limed Norwegian lakes, Hongve and Abrahamsen (1984) concluded that, as generally thought, the inhibition of deposited limestone by humic substances in the sediment is just as important, since the immediate uptake in the sediment is later reversed when H^+ is exchanged leading to an outflow of calcium.

Young *et al* (1989) calculated that 16.8 and 5.5% of a fine-grain calcite reached the sediment in Woods Lake and Cranberry Lake in the USA, raising the calcium concentrations and pH values to a depth of approximately 6 cm. In general, however, this effect had disappeared after sixteen months.

Obviously, the soft sediment reacts on liming through ion exchange and elevated pH, which results in lower portions of free metal ions. Bound calcium seems to be mobilized upon reacidification and thus provides a certain, minor buffer capacity. Trace metals precipitated during liming are stored in the sediment, but the contributions are difficult to measure.

4. Effects of liming in entire aquatic systems

The effect of a single liming treatment is, as described above, finite in duration and must thus be repeated and complemented with other activities in order to maintain the liming objective.

Out of 3,708 limed objects in the Swedish liming data base (KRUT) about one third have been limed only once, approximately 70% one to three times, and 10% more than five times. Lake Svartagölen, southern Sweden, has been limed nine times since 1984 and the liming objectives have been achieved since 1987 (Fig. 14). The doses are better adapted and are less likely to exceed the upper alkalinity limits as a result.

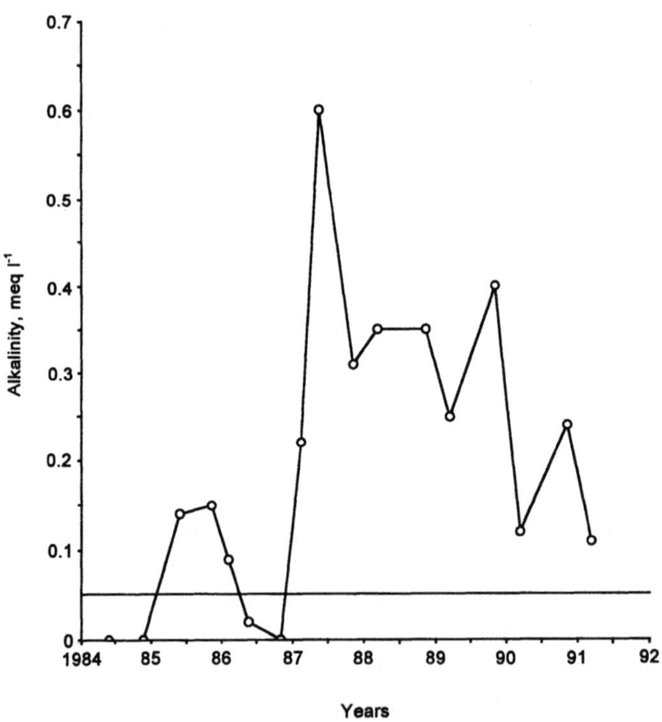

Figure 14. *An example of a multiple liming; Lake Svartagölen, southern Sweden, which has been limed nine times since 1987.*

In order to judge the effects of liming activities the results have been compared with the liming objectives (SEPA 1988). Calculations of the time of compliance were made by linear interpolation in time, for lakes with at least six observations and for the 1978-90 period (Table 11). Data were available for about 2,100 objects and reveal, for instance, that the objective of a pH value above 6 was fulfilled in 44% of the lakes during 90% of the observation time. It has not been possible to distinguish the effects of different treatments in this calculation.

Table 11. *Fulfillment of liming goals (alkalinity >0.05 meq/l, pH >6.0) for Swedish lakes during the period 1978-1990.*

	Percentile			
	10	50	90	n
Alkalinity >0.05 meq/l				
limed once	4	12	29	536
limed twice	4	16	43	552
limed >3 times	3	19	53	1034
in total	3	17	44	2122
pH >6.0				
limed once	3	10	31	536
limed twice	13	14	44	552
limed >3 times	3	20	56	1037
in total	2	17	47	2125

Obtaining the liming objective was best in objects with frequent liming. Among the twenty-two lakes which were limed more than twenty times, the goal of a pH of more than 6 was fulfilled only 30% of the time in the worst case, while in four cases the goal was obtained. In most of these cases the treatment was performed using dosers.

At the beginning of the trial period (1977-81) the liming activities usually consisted of unconnected activities. As knowledge was gained these activities have been planned to cover entire drainage areas and methods have been chosen on the basis of the conditions in the total area. All this had led to better compliance compared with ten years ago (Fig. 15).

Such activities are described in the following for two small drainage areas in western Sweden and further cases are reviewed in other chapters of the present book.

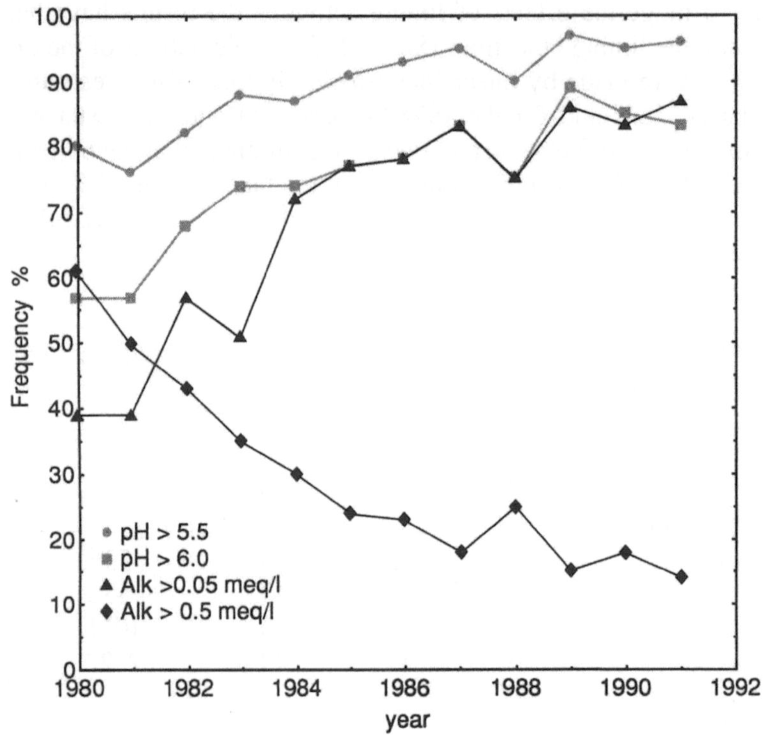

Figure 15. Attainment of goals (trends) in the 1980-91 period for limed lakes in Sweden.

4.1. River Fylleån

River Fylleån in south-western Sweden has a drainage area of 401 km^2, of which 7% is covered by lakes. Four of these lakes have short retention times. Lake liming was performed on seven lakes and a doser was installed at Ryaberg. A total of 17,470 tonnes of limestone was used during 1982-1991, with a yearly dose of about 2,000 tonnes in the last years, of which about 1,600 tonnes were applied through the doser.

The effect of the doser has been very good (Widarsson and Johansson 1992). For the sampling station further downstream beyond a small lake there has been only one observation of low pH and alkalinity. The uppermost lakes, with zero alkalinity, were limed during 1984-85 with good results (Table 12).

An idea of the total effect of the liming is provided by a comparison with the total transport in the river (Table 13). Liming activities add 730 t/year of calcium - about half of the total transport of 1,400 t/year.

Table 12. Fulfillment of goals for liming in 1984-1985 of three lakes in the River Fylleån catchment area in south-western Sweden (data from Lessmark 1990).

Lake	Fulfillment in % of time		
	pH		Alkalinity
	>5.0	>6.0	>0.05 meq/l
Fullhövden/Kvarnsjön	99	83	87
Transjön	89	73	61
Stora Svansjön	98	69	50

Table 13. Transport of calcium and alkalinity contrasted to the liming dose in the catchment of River Fylleån, south-western Sweden (cumulative).

Year	Calcium	Transport alkalinity	Lime dose
	Teq	Teq	Teq
1982			13.7
1983	80.6	20.4	38.5
1984	149.1	39.8	64.9
1985	218.6	59.9	103.5
1986	281	82.5	141.6
1987	346.3	107.5	172.1
1988	428.8	138.2	209
1989	486.9	162.3	244.1
1990	583	197.6	291.2
1991	656.8	232.3	349.4

The acid in deposition titrates some of the carbonate added, which lowers the transport of alkalinity to 25.1 Teq/year from the liming dose of 36.6 Teq/year. No transport figures are available for the period prior to liming, but a background situation may be constructed. If we adopt the simplified assumption that the transport of alkalinity was equal to the non-marine transport of calcium and magnesium in unaffected areas, the figure should be approximately 60 Teq/year. This is certainly an overestimate since acidification leads to increased weathering.

The deposition of acidifying substance may be estimated as shown in Table 14.

Table 14. *Acid deposition in the River Fylleån region in south-western Sweden (data from Granat 1990).*

	meq/m²/year	Total deposition Teq/year
H^+	70	25
SO_4^{2-} excess	100	36
NO_3^-	60	21.5
NH_4^+	70	25

A good balance exists between calculated deposition and measured transport of excess sulphate, which indicates that the estimates are reasonable. The liming treatment corresponded to 36.6 Teq/year, i.e. more than the deposition of H^+. The neutralizing effect of ammonia must be taken into account, however. If the deposition of excess sulphate and the transport of nitrate (approximately 10 Teq/year) are used as a measure of acidic load instead, then present liming activities do not correspond to the need. There is a considerable source of alkali from weathering, however, which is evident from the actual transport of base cations of approximately 97 Teq/year. Seen as a whole, River Fylleån is well managed in terms of liming.

4.2. River Enningdalsälven

River Enningdalsälven in south-western Sweden has been limed since 1980, usually annually. The total drainage area encompasses 781 km², with a lake area of 9.9%.

Since the transport of calcium and alkalinity has been measured since 1972, it is possible to evaluate the large-scale effects of liming.

Table 15. *Transport of calcium and alkalinity in River Enningdalsälven, south-western Sweden.*

	Period	Transport Teq/year
Calcium	1972-79	42
Calcium	1980-90	70
Alkalinity	1972-79	11
Alkalinity	1980-90	25
Limestone addition	1980-90	33

As can be seen from Table 15, the transport of both calcium and alkalinity increased after the start of liming. The transport of calcium increased on average 28 Teq/year and, at the same time, the liming supply amounted to 33 Teq/year, which implies a total efficiency for the whole river of about 85%.

At the river mouth, liming objectives have been fulfilled since the liming started in 1980 in terms of pH, but only since 1986 for alkalinity. The explanation for this difference may be that acid deposition in the form of sulphate decreased throughout the period.

5. Effect of liming activities on water chemistry

The prime objective of liming activities is to create acceptable conditions for the biota. Another question is that of whether or not liming restores the water chemistry. In order to answer this, reference data are necessary. Measurements for the period prior to acidification would be useful, but such data are scarce and it has not been possible to use this as a basis for evaluation in terms of a larger set. A second alternative would be to compare results for non-limed and non-acidified lakes and a third would be to apply model calculations.

5.1. Comparisons between limed and non-limed waters

Since the geographical location of a lake determines both the natural background and the extent of acidification it is risky to use data for lakes in less affected areas as a reference. Bearing this in mind, the Swedish lake inventory from 1990 can be used with caution. At the time of sampling about 70% of the limed lakes had alkalinity concentrations over 0.05 meq/l (Fig. 16; cf. Bernes 1991). Liming activities were abandoned in certain lakes where values were lower.

Calculation of the background alkalinity in limed lakes and comparison with the present reveal that several lakes have not been restored. Current alkalinity concentrations after liming are lower than the background (Fig. 17). In several cases liming activities were stopped before the samples were taken in 1990. In half of all the limed lakes the current alkalinity was equal to or lower than the calculated background, while it increased to >0.1 meq/l in 15% of the lakes and >0.2 meq/l in 5% of the lakes studied.

In limed lakes, the calcium concentrations increased somewhat in comparison with non-limed lakes where the original concentrations were low. This is an effect of liming since the changes in magnesium do not follow the same pattern. The increase is unlikely to be negative for biota. Bernes (1991) estimated that the pH after liming was restored to background levels or higher in about 3,600 previously acidified lakes.

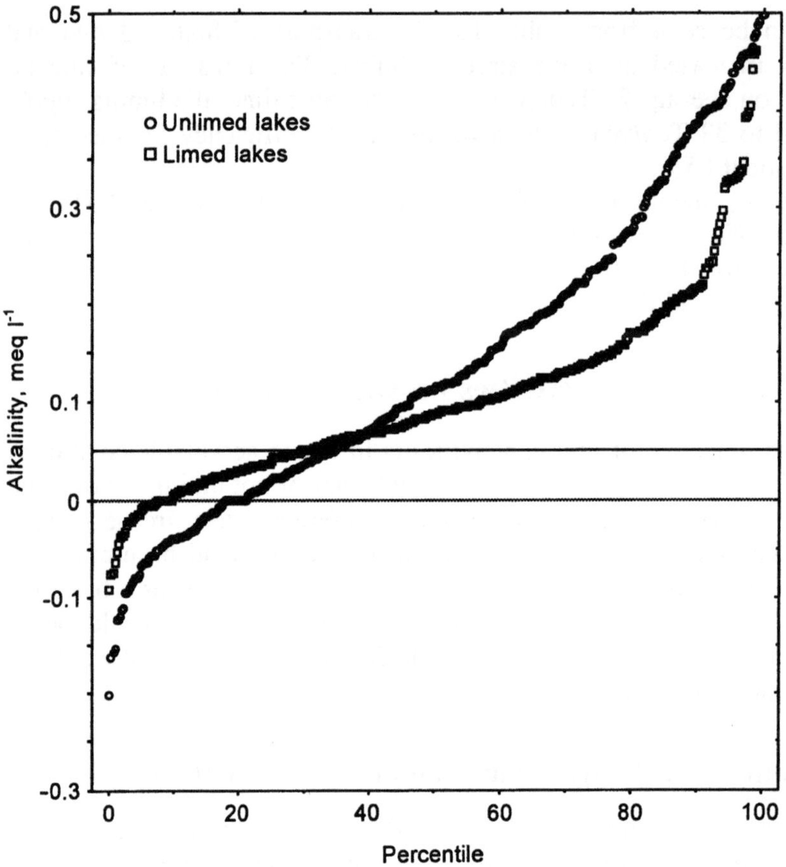

Figure 16. *The frequency distribution of alkalinity concentrations in limed and non-limed lakes in southern Sweden. Data set from 1990 the national lake survey when c. 4,000 lakes were sampled.*

Comparisons of this sort between non-limed and limed lakes in southern Sweden, for ammonia, nitrate, total nitrogen, total phosphorus and TOC indicate lower concentrations of total nitrogen and TOC in limed lakes. The data set is small, however.

The concentration of acid-soluble aluminium was lower in limed lakes than in unlimed, with values for the 90 percentile at 235 and 348 μg/l respectively. At lower concentrations the changes were small, probably because the major part of the aluminium is complex-bound and thus less influenced by changes in pH.

A comparison between metal concentrations in certain limed lakes and lakes situated in northern Sweden (and therefore less affected by acid deposition) indicates that liming may restore trace metal concentrations (Table 16).

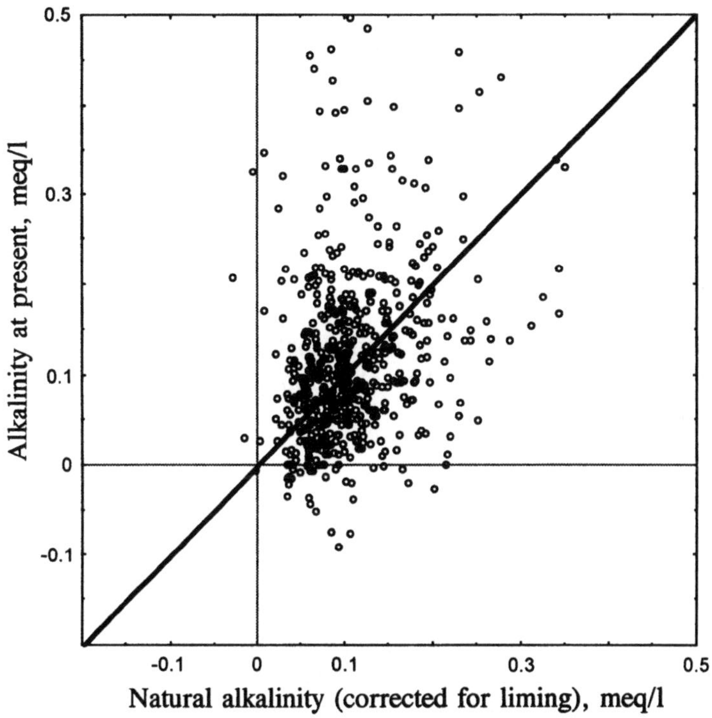

Figure 17. Comparison between measured alkalinity after liming and estimated natural alkalinity (meq/l) in Swedish lakes. Subset of data from the 1990 national lake survey.

Table 16. Concentrations of metals in limed lakes within the Swedish ISEL programme (cf. text) and non-limed forest lakes in northern Sweden (median values, μg/l) (Borg unpubl. data).

	n	Zn	Cu	Pb	Cd	Cr	Ni	Co
ISEL	14	3.8	0.39	0.26	0.018	0.17	0.31	0.05
North.	60	2.3	0.44	0.25	0.016	<0.2	<0.4	<0.2

5.2 Comparison with results from model calculations

There are several models for the calculation of acidification (e.g. Kämäri *et al* 1990). For Lake Gårdsjön Wright *et al* (1988) calculated the pre-industrial pH to be approximately 6.4 using the MAGIC mathematical model. By way of comparison it can be mentioned that the value after liming fell from 7.5 to 6.7 between 1982 and 1985, which was still higher than the "natural" value during the entire period. No models are available for the evaluation of possible restoration of organic matter, trace metal or nutrient concentrations, however.

It can be concluded from the above that pH values and alkalinity concentrations are generally restored by liming, although post-liming figures are often higher than the original ones.

6. Acknowledgements

We would like to thank Mr. B. Karlsson, who transferred and compiled all KRUT data, and Mr. G. Moberg for the collection of all these data from the county councils.

7. References

Abrahamsson, I. 1993. Impact of overflows on acid-base chemistry in limed lakes. - Vatten 49: 24-33.

Abrahamsson, I., Pettersson, L., Sandell, G. and Svahnberg, A. 1992. Vattenkemiska resultat av kalkade våtmarker i Jönköpings län. [Results on water chemistry of wetland liming in the county of Jönköping.] - Terra Limno Group, Mölnlycke.

Adams, T.B. and Brocksen, R.W. 1988. Dose-response relationships for the addition of limestone to lakes and ponds in the northeastern United States. - Water Air Soil Poll. 41: 137-163.

Alasaarela, E., Havu, J., Heikkinen, K. and Weppling, K. 1990. Neutralization of acidified watercourses. - In: Kauppi, P., Anttila, P. and Kenttämies, K. (Eds.) Acidification in Finland. Springer, Berlin Heidelberg New York, pp. 1117-1125.

Alenäs, I. 1986. Kalkningsprojektet Härskogen 1976-1986. [The Härskogen liming project 1976-1986.] - Swedish Environm. Res. Inst., B 846.

Alenäs, I. 1987. Delrapport från elfiskeundersökningar och vattenkemiska provtagningar under treårsperioden 1984-1986. [Report on electro-fishing studies and water samples during the three-year period 1984-1986.] - Swedish Environm. Res. Inst., L87/14.

Alenäs, I. and Hultberg, H. 1988. Slutrapport för projektet Stenunge å - utvärdering av olika markkalkningsåtgärder avseende vattenkemi och havsöringsbestånd. [Final report on the Stream Stenunge å project - evaluation of soil liming measures on water chemistry and sea trout population.] - Swedish Environm. Res. Inst., B 846.

Alenäs, I., Andersson, I., Hultberg, H. and Rosemarin, A. 1991. Liming and reacidification reactions of a forest lake ecosystem, Lake Lysevatten, in SW Sweden. - Water Air Soil Poll. 59: 55-77.

Almer, B., Dickson, W., Ekström, C. and Hörnström, E. 1978. Sulfur pollution and the aquatic ecosystem. - In: Nriago, J.O. (Ed.) Sulfur in the environment, 2. John Wiley & Sons, New York, pp. 271-311.

Andersson, B.I. 1985. Properties and chemical composition of surficial sediments in the acidified Lake Gårdsjön, SW Sweden. - Ecol. Bull. (Stockholm) 37: 251-263.

Andersson, G. and Gahnström, G. 1985. Effects of pH on release and sorption of dissolved substances in sediment-water microcosms. - Ecol. Bull. (Stockholm) 37: 301-310.

Andersson, G. and Hamrin, S. 1984. Effekter av markkalkning på vattenkemiska förhållanden i rinnande vatten. Vänneån. [Effects of catchment liming on water chemical conditions in running water. The Vänneån Stream.] - Swedish Environm. Prot. Agency 1865.

Andersson, I. 1983. Anråse å - resultat och erfarenheter från kalkning av en havsöringå i Stenungsunds kommun, Bohuslän 1977-1983. [The Anråse Stream - results and experience of liming of a sea trout stream in the municipality of Stenungsund, county of Bohuslän, 1977-1983.] - Swedish Environm. Res. Inst., EM 1304.

Andersson, I., Hasselrot, B. and Hultberg, H. 1982. Vattenkemiska effekter av olika sätt att applicera neutralisationsmedel i ett försurat område på västkusten. [Effects on water chemistry by various methods for application of neutralization agents in an acidified area on the west coast.] - Swedish Environm. Res. Inst., EM 0799.

Andersson, P. and Borg, H. 1988. Effects of liming on the distribution of cadmium in water, sediment, and organisms in a Swedish lake. - Can. J. Fish. Aquat. Sci. 45: 1154-1162.

Andersson, P., Borg, H. and Holm, K. 1983. Lösligheten av metaller i sediment vid olika pH och kalciumtillsats - laboratorieförsök. [The solubility of metals in sediment at different pH and calcium additions.] - Swedish Environm. Prot. Agency 1772.

Andersson, P., Borg, H. and Kärrhage, P. 1991. Kalkningseffekter på kvicksilver i fisk. [Effects of liming on mercury in fish.] - Swedish Environm. Prot. Agency 3883.

Appelberg, M. and Aldén, U. 1992. Integrerad uppföljning av kalkningens effekter på sjöar och vattendrag - en treårsrapport. [Integrated monitoring in limed lakes and streams - a three-year study.] - Inform. Inst. Freshw. Res., Drottningholm (4): 1-60.

Appelberg, M., Ekström, C. and Hörnström, E. 1990. Stora Härsjön - ett exempel på integrerad uppföljning av kalkningens effekter. [English summary: Lake Stora Härsjön - a case study for integrated monitoring of the effects of liming.] - Inform. Inst. Freshw. Res., Drottningholm (1).

Bell, R.T., Vrede, K., Stensdotter, U. and Blomqvist, P. 1993. Stimulation of the microbial foodweb in an oligotrophic, slightly acidified lake. - Limnol. Oceanog. 38: 1532-1538.

Bengtsson, B., Dickson, W. and Nyberg, P. 1980. Liming acid lakes in Sweden. - Ambio 9: 34-36.

Bergkvist, B. 1987. Soil solution chemistry and metal budgets of spruce forest ecosystems in S. Sweden. - Water Air Soil Poll. 33: 131-154.

Bernes, C. 1991. Monitor 12. Acidification and liming of Swedish freshwaters. - Swedish Environm. Prot. Agency Informs.

Berzins, B. 1962. Kalkning av sjöar. [Liming of lakes.] - Skr. S. Sverig. Fisk. Fören. Årsskrift 1956-60: 28-35.

Betlach, M.R. and Tiedje, J.M. 1981. Kinetic explanation for accumulation of nitrite, nitric oxide and nitrous oxide during bacterial denitrification. - Appl. Environ. Microbiol. 42: 1074-1084.

Blomqvist, P., Olsson, H., Olofsson, O. and Broberg, O. 1989. Enclosure experiments with low-dose additions of phosphorus and nitrogen in the acidified Lake Njupfatet, central Sweden. - Int. Rev. ges. Hydrobiol. 74: 611-631.

Borg, H. 1983. Trace metals in Swedish natural fresh waters. - Hydrobiologia 101: 27-34.

Borg, H. 1986. Metal speciation in acidified mountain streams in central Sweden. - Water Air Soil Poll. 30: 1007-1014.

Borg, H. and Andersson, P. 1984. Fractionation of trace metal in acidified fresh waters by in situ dialysis. - Verh. Internat. Verein. Limnol. 22: 725-729.

Borg, H. and Johansson, K. 1989. Metal fluxes to Swedish forest lakes. - Water Air Soil Poll. 47: 427-440.

Boström, B. 1986. The role of *Microcystis* colonies, its mucilage and associated bacteria, for nutrient fluxes from sediments to lake water - a working hypothesis. - In: Enell, M., Graneli, W. and Hansson, L.-A. (Eds.) 13th Nordic Symp., sediments. Dep. Limnol., Univ. Lund, pp. 6-8.

Brassard, P. and Auclair, J.C. 1984. Orthophosphate uptake rate constants and mediated by the 103-104 molecular weight fraction in shield lake waters. - Can. J. Fish. Aquat. Sci. 41: 166-173.

Broberg, O. 1978. Kalkningsmedel, kalkning och effekter av kalkning. En litteraturöversikt. [English summary: Liming agents, liming and effects of liming - a literature survey.] - Vatten 34: 104-116.

Broberg, O. 1987. Nutrient responses to the liming of Lake Gårdsjön. - Hydrobiologia 150: 11-24.

Broberg, O. 1988a. Liming of Lake Gårdsjön - an acidified lake in SW Sweden. Effects on: organic carbon, nitrogen and phosphorus. - In: Dickson, W. (Ed.) Liming of Lake Gårdsjön, an acidified lake in SW Sweden, pp. 135-205.

Broberg, O. 1988b. Delayed nutrient responses to the liming of Lake Gårdsjön, Sweden. - Ambio 17: 23-27.

Broberg, O. and Persson, G. 1984. External budgets for phosphorus, nitrogen and dissolved organic carbon for the acidified Lake Gårdsjön. - Arch. Hydrobiol. 99: 160-175.

Brocksen, R.W., Adams, T.B., Sverdrup, H. and Warfvinge, P. 1990/91. Terrestrial liming as a tool to mitigate acidification of Woods Lake, NY. - Water Air Soil Poll. 54: 509-527.

Brown, D.J.A. 1988. The Loch Fleet and other catchment liming programs. - Water Air Soil Poll. 41: 409-415.

Bukaveckas, P.A. 1991. Effects of lake and watershed liming on light attenuation, thermal stratification, and phytoplankton production. In: Olem, H., Schreiber, R.K., Brocksen, R.W. and Porcella, D.B. (Eds.) 1991. International lake and watershed liming practices. - The Terrene Inst., Washington D.C., pp. 129-144.

Bukaveckas, P.A. and Driscoll, C.T. 1991. Effects of whole-lake base addition on thermal stratification in three acidic Adirondack lakes. - Water Air Soil Poll. 59: 23-39.

Chamier, A.C. 1987. Effects of pH on microbial degradation on leaf litter in seven streams of the English Lake District. - Oecologia (Berlin) 71: 491-500.

Dalziel, T.R.K., Proctor, M.V. and Dickson, A. 1988. Hydrochemical budget calculation for parts of the Loch Fleet catchment before and after watershed liming. - Water Air Soil Poll. 41: 417-434.

Degerman, E. 1987. Humösa sjöar - en litteratursammanställning. [English summary: Humic lakes - a literature survey with emphasis on fish and acidification.] - Swedish Environm. Prot. Agency 3415.

Degerman, E. and Appelberg, M. 1992. The response of stream-dwelling fish to liming.- Environ. Poll. 78: 149-155.

Degerman, E., Sjölander, E., Johlander, A., Sjöstrand, P., Höglind, K., Thorsson, L. and Carlstrand, H. 1990. Kalkning för att motverka försurningspåverkan på fisk i rinnande vatten. [English summary: Liming of acidified running waters to improve conditions for fish.] - Inform. Inst. Freshw. Res., Drottningholm (4): 27-214.

Derome, J., Kukkola, M. and Mälkönen, E. 1986. Forest liming on mineral soils. Results of Finnish experiments. - Swedish Environm. Prot. Agency 3084.

Dickinson Burrow, W. 1977. Aquatic aluminium. Chemistry, toxicology and environmental prevalence. - CRC Critical Rev. Environm. Cont. 7: 167-216.

Dickson, W. 1978. Some effects of the acidification of Swedish lakes. - Verh. Internat. Verein. Limnol. 20: 851-856.

Dickson, W. 1979. Kalkning av sjöar. [English summary: Liming of lakes.]- Kungl. Skogs- och Lantbruksakad. Tidskr. Suppl. 13: 38-48.

Dickson, W. 1980. Properties of acidified water. - In: Drabløs, D. and Tollan, A. (Eds.) Proc. Int. Conf. Ecol. impact acid precip., SNSF project., Oslo-Ås, Norway, pp. 75-83.

Dickson, W. 1983. Liming toxicity of aluminium to fish. - Vatten 39: 400-404.

Dickson, W. 1985. Kalkning av sjöar och vattendrag. Dosberäkningar och kalkningsmedlens metallinnehåll. Kalkfällning i reningsverk. [Liming of lakes and streams. Calculations of dosages and the metal content of liming agents. Lime precipitation in waste water treatment plants.] In: Bogelius, A. (Ed.) Kalkning av sjöar och vattendrag. [Liming of lakes and streams.], Tällberg 1984. Swedish Environm. Prot. Agency 1919: 52-59.

Dickson, W. (Ed.) 1988. Liming of Lake Gårdsjön, an acidified lake in SW Sweden. - Swedish Environm. Prot. Agency 3426.

Dillon, P.J., Yan, N.D., Scheider, W.A. and Conroy, N. 1979. Acid lakes in Ontario, Canada: characterization, extent and responses to base and nutrient additions. - Arch. Hydrobiol. 13: 317-335.

Driscoll, C.T., Fordham, G.F., Ayling, W.A. and Oliver, L.M. 1989. Short-term changes in the chemistry of trace metals following calcium carbonate treatment of acidic lakes. - Can. J. Fish. Aquat. Sci. 46: 249-257.

Edman, G. and Fleischer, S. 1980. The river Högvadsån liming project - a presentation. - In: Drabløs, D. and Tollan, A. (Eds.) Proc. Int. Conf. Ecol. impact acid precip., SNSF project., Oslo-Ås, Norway, pp. 300-301.

Edman, G., Fleischer, S., Fritz, Ö. and Stibe, L. 1988. Högvadsån 1978-1986. [The River Högvadsån 1978-1986.] - Halland County Council, Rep. 1988:4.

Evans, R.A. 1989. Response of limnetic insect populations of two acidic, fishless lakes to liming and brook trout (*Salvelinus fontinalis*). - Can. J. Fish. Aquat. Sci. 46: 342-351.

Fleischer, S. 1984. Högvadsån. Kalkningsprojekt 1978-1981. III Vattenkemiska undersökningar. [The River Högvadsån liming project. III. Chemical studies of the water.] - Halland County Council, Rep.

Fordham, G.F. and Driscoll, C.T. 1989. Short-term changes in the acid/base chemistry of two acidic lakes following calcium carbonate treatment. - Can. J. Fish. Aquat. Sci. 46: 249-257.

Francis, A.J., Quinby, H.L. and Hendrey, G.R. 1984. Effect of lake pH on microbial decomposition of allochthonous litter. - In: Hendrey, G.R. (Ed.) Early biotic responses to advancing lake acidification. Butterworth, Boston, pp. 1-21.

Gahnström, G. 1988. The effect of liming on sediment oxygen uptake in Lake Gårdsjön. - In: Dickson, W. (Ed.) Liming of Lake Gårdsjön, an acidified lake in SW Sweden. - Swedish Environm. Prot. Agency 3426, pp. 207-243.

Gloss, S.P., Schofield, C.L. and Sherman, R.E. 1988. An evaluation of New York state liming data and the application of models from Scandinavian lakes to Adirondack lakes. - Water Air Soil Poll. 41: 241-278.

Grahn, O., Hultberg, H. and Landner, L. 1974. Oligotrophication - a self-accelerating process in lakes subjected to excessive supply of acid substance. - Ambio 3: 93-94.

Grahn, P. 1990. Kvicksilver i gädda i Örebro läns sjöar. Projekt kalkning-kvicksilver-cesium. [Mercury in pike from lakes in the county of Örebro. The project Liming-mercury-caesium.] - Örebro County Council, Rep. 1990:9.

Granat, L. 1990. Luft- och nederbördskemiska stationsnätet inom PMK. [Monitoring of the chemistry of air and deposition within PMK.] - Swedish Environm. Prot. Agency 3947.

Hasselrot, B. and Hultberg, H. 1984. Liming of acidified Swedish lakes and streams and its consequences for aquatic ecosystems. - Fisheries 9: 4-9.

Hasselrot, B. and Westling, O. 1986. Jordbruksmarkskalkning i Sundtorpsåns avrinningsområde - effekter på avrinningsvatten. [Liming of agriculture land in the River Sundstorpsån drainage area - effects on runoff water.] - Swedish Environm. Res. Inst., L86/292.

Hasselrot, B., Andersson, B.I. and Hultberg, H. 1984. Ecosystem shifts and reintroduction of artic char (*Salvelinus salvelinus* (L)) after liming of a strongly acidified lake in SW Sweden. - Rep. Inst. Freshw. Res., Drottningholm 61: 93-103.

Hasselrot, B., Andersson, B.I., Alenäs, I. and Hultberg, H. 1987. Response of limed lakes to episodic events in southwestern Sweden. - Water Air Soil Poll. 32: 341-362.

Henrikson, L. 1988. Effects on water quality and benthos of acid water inflow into the limed Lake Gårdsjön. - In: Dickson, W. (Ed.) Liming of Lake Gårdsjön, an acidified lake in SW Sweden. - Swedish Environm. Prot. Agency 3426, pp. 309-327.

Heyman, U. and Lundgren, A. 1987. Phytoplankton biomass and production in relation to phosphorus: some conclusions from field studies. - In: Persson, G. and Jansson, M. (Eds.) Phosphorus in freshwater ecosystems. Hydrobiologia 170: 211-228.

Hindar, A. 1989. Kalking av Vikedalselva. Førsøk med styring av kalkdosering etter pH målt nedstrøms kalkdoserer. [Liming of River Vikedalselva. Experiments with regulation of lime dosages using pH-measurements downstream of the doser.] - Kalking av surt vann. [Liming of acidified waters.] Norwegian Inst. Water Res., 7/89.

Hindar, A. and Nilssen, J.P. 1983. Årsrapport 1980/1981. Kalkingsprosjektet i Gjerstad, Aust-Agder. [Annual report 1980/1981. The liming project in Gjerstad, Auts-Agder.] Rep. 6/83.

Hongve, D. and Abrahamsen, H. 1984. On the efficiency of liming in acid, humus-influenced lakes. - Verh. Internat. Verein. Limnol. 22: 700-703.

Hörnström, E. and Ekström, C. 1983. pH-, närings- och aluminiumeffekter på plankton i västkustsjöar. [Effects of pH, nutrients and aluminium on plankton in Swedish west coast lakes.] - Swedish Environm. Prot. Agency 1704.

Hörnström, E. and Ekström, C. 1986. Acidification and liming effects on phyto- and zooplankton in some Swedish west coast lakes. - Swedish Environm. Prot. Agency 1864.

Hörnström, E., Ekström, C. and Andersson, P. 1992a. Tio mellansvenska sjöar. Kalkningseffekter på plankton och vattenkemi. [English summary: Ten lakes in central Sweden, liming effects on plankton and water chemistry.] - Swedish Environm. Prot. Agency 4048.

Hörnström, E., Ekström, C., Fröberg, E. and Ek, J. 1992b. Plankton and chemical/physical development in six Swedish west coast lakes, under acidic and limed conditions. - Inst. Appl. Environm. Res., Univ. Stockhom, unpubl. manuscript.

Howells, G., Dalziel, T.R.K. and Turnpenny, A.W.H. 1992. Loch Fleet: liming to restore a brown trout fishery. - Environ. Poll. 78. 131-139.

Huettl, R.F. 1988. "New type" forest declines and restabilization/revitalization strategies. - Water Air Soil Poll. 41: 95-111.

Hultberg, H. and Alenäs, I. 1981. Projekt Anråseån. Sammanfattning. [The project Anråseån Stream. Summary.] - Swedish Environm. Res. Inst., Rep.

Hultberg, H. and Andersson, I.B. 1982. Liming of acidified lakes: induced long-term changes. - Water Air, Soil Poll. 18: 311-331.

Hultberg, H. and Nyström, U. 1988. The role of hydrology in treatment duration and reacidification in the limed Lake Gårdsjön. - In: Dickson, W. (Ed.) Liming of Lake Gårdsjön, an acidified lake in SW Sweden. - Swedish Environm. Prot. Agency 3426, pp. 95-134.

Hutchinson, G.E. 1957. A treatise on limnology Vol. 1. Geography, physics and chemistry. - J. Wiley & Sons, New York.

Jansson, M. 1981. Induction of high phosphatase activity by aluminium in acid lakes. - Arch. Hydrobiol. 93: 32-44.

Jansson, M., Persson, G. and Broberg, O. 1986. Phosphorus in acidified lakes: the example of Lake Gårdsjön, Sweden. - Hydrobiologia 139: 81-96.

Jenkins, A., Waters, D. and Donald, A. 1991. An assessment of terrestrial liming strategies in upland Wales. - J. Hydrol. 124: 243-261.

Johnsrud, S.C. 1980. Heterotrophic nitrification. - In: Rosswall, T. (Ed.) Processer i kvävets kretslopp. [Processes in the turnover of nitrogen.] Swedish Environm. Prot. Agency 1213, pp. 174-180.

Jönsson, C.-P. 1990. Projekt kalkning - kvicksilver - cesium. Slutrapport från Kronobergs län. The project Liming-mercury-caesium. Final report from the county of Kronoberg.]

- Miljöfakta i Kronobergs län 1990:2.
Kämäri, J., Forsius, M. and Lepistö, A. 1990. Modelling long-term development of surface water acidification in Finland. - In: Kauppi, P., Anttila, P. and Kenttämies, K. (Eds.) Acidification in Finland. Springer, Berlin Heidelberg New York, pp. 781-810.

Lagerman, B. 1983. Kalkningsprojektet Unden. [The Lake Unden liming project.] - Samrådsgruppen för Undens kalkning 1977-1983. [The working group for the liming of Lake Unden.] - Nat. Board Fish., unpubl. rep.

Larsson, L.S. 1988. Liming effects on phytoplankton in Lake Gårdsjön 1982-1985. - In: Dickson, W. (Ed.) Liming of Lake Gårdsjön, an acidified lake in SW Sweden. - Swedish Environm. Prot. Agency 3426, pp. 245-280.

Lazarek, S. 1986. Responses of the *Lobelia*-epiphyte complex to liming of an acidified lake. - Aquat. Bot. 25: 73-81.

Lessmark, O. 1987. Markkalkning som metod för att motverka försurning av sjöar och vattendrag. [English summary: Ground liming as a method of mitigating acidification of surface waters.] - Inform. Inst. Freshw. Res., Drottningholm (9).

Lessmark, O. 1988. Utvärdering av sjökalkningar i Kalmar län. [Evaluation of lake liming in the county of Kalmar.] - The Kalmar County Council, Rep. 1988:7.

Lessmark, O. 1990. Kalkningar i Kronobergs län 1977-89. En utvärdering av sjökalkningars kemiska och biologiska effekt. [Liming in the county of Kronoberg 1977-1989. An evaluation of chemical and biological effects of lake liming.] - Miljöfakta i Kronobergs län 1990:1.

Lessmark, O., Degerman, E., Johlander, A. and Sjölander, E. 1986. Effekter av kalkning på fisk omedelbart nedströms doserare. [English summary: Effects of liming on fish immediately downstream of lime dosers.] - Inform. Inst. Freshw. Res., Drottningholm (5).

Lindmark, G. 1982. Acidified lakes: sediment treatment with sodium carbonate - a remedy? - Hydrobiologia 92: 537-547.

Lindmark, G. 1984. Acidified lakes: ecosystem response following sediment treatment with sodium carbonate. - Verh. Internat. Verein. Limnol. 22: 772-779.

Lindmark, G. 1990. Soda ash briquettes in lake sediments - a new approach in reclamation of acidified waters. - Verh. Internat. Verein. Limnol. 24: 173-178.

Lindmark, G. 1991. The use of soda ash briquettes in the neutralization of acidified lakes. - Solvay, Stockholm.

Lingdell, P.-E. and Engblom, E. 1985. Kalkdoserare - biologiska effekter. [Lime dosers - biological effects.] - In: Bogelius, A. (Ed.) Kalkning av sjöar och vattendrag. [Liming of lakes and streams.] Tällberg 1984. Swedish Environm. Prot. Agency 1919: 6-24.

Marcus, M.D. 1988. Differences in pre- and post-treatment water qualities for twenty limed lakes. - Water Air Soil Poll. 41: 279-291.

Molot, L.A. 1986. Base neutralization capacity of sediments from an acidic lake. - Water Air Soil Poll. 27: 297-304.

Molot, L.A., Dillon, P.J. and Booth, G.M. 1990. Wholelake and nearshore water chemistry in Bowland Lake, before and after treatment with $CaCO_3$. - Can. J. Fish. Aquat. Sci. 47:412-421.

NBF and SEPA (National Board of Fisheries and Swedish Environmental Protection Agency) 1981. Kalkning av sjöar och vattendrag 1977-1981. [Liming of lakes and streams 1977-81.] - Inform. Inst. Freshw. Res., Drottningholm (4).

Nihlgård, B., Nilsson, S.I. and Rosengren, U. 1988. Concentrations and fluxes of main cations and anions in soil water and run-off water. - In: Andersson, F. and Persson, T. (Eds.) Liming as a measure to improve soil and tree conditions in areas affected by air pollution. Swedish Environm. Prot. Agency 3518, pp. 40-52.

Niinioja, R., Ahtiainen, M. and Holopainen, A.-L. 1990. Liming of the acidic Lake Valealampi in eastern Finland: effects on water chemistry and phytoplankton. - In: Kauppi,

P., Anttila, P. and Kenttämies, K. (Eds.) Acidification in Finland. Springer, Berlin Heidelberg New York, pp. 1127-1143.

Nohrstedt, H.-Ö. 1992. Soil water chemistry as affected by liming and N fertilization at two Swedish coniferous forest sites. - Scand. J. For. Res. 7: 143-153.

Nömmik, H. 1979. Vilken roll kan kalken spela i framtidens skogsbruk? [English summary: The future role of liming in forestry?] - Skogs- Lantbr,-akad. Tidskr. Suppl. 13: 31-37.

Nyberg, P. 1985. Kalkning av våtmarker. - In: Grundelius, E. (Ed.) Kalkning av sjöar och vattendrag. Hagfors. [Liming of lakes and streams. Hagfors.] Swedish Environm. Prot. Agency 3167, pp. 26-33.

Nyberg, P., Degerman, E., Ekström, C. and Hörnström, E. 1986. Försurningskänsliga rödingsjöar i Syd- och Mellansverige. [English summary: Acid-sensitive Arctic char (*Salvelinus alpinus*) lakes in southern and central Sweden.] - Inform. Inst. Freshw. Res., Drottningholm (6).

Nyman, H.G. 1990. Some biological mechanisms reducing the availability of phosphorus in acid lakes. - In: Mason, B.J. (Ed.) The surface water acidification programme. Univ. Press, Cambridge, pp. 401-408.

Olem, H. 1991. Liming acidic surface waters. Lewis Publ., Chelsea, Michigan.

Olem, H., Schreiber, R.K., Brocksen, R.W. and Porcella, D.B. (Eds.) 1991. International lake and watershed liming practices. - The Terrene Inst., Washington D.C.

Olofsson, H., Blomqvist, P., Olsson, H. and Broberg, O. 1988. Restoration of the pelagic food web in acidified and limed lakes by gentle fertilization. - Limnologica (Berlin) 19: 27-35.

Olsson, H. 1990. Phosphatase activity in relation to phytoplankton composition and pH in Swedish lakes. - Freshw. Biol. 23: 353-362.

Olsson, H. 1991. Phosphatase activity in an acid, limed Swedish lake. - In: Chrost, R.J. (Ed.) Microbial enzymes in aquatic environments. Springer, Berlin Heidelberg New York, pp. 206-219.

Persson, G. and Broberg, O. 1985. Nutrient concentrations in the acidified Lake Gårdsjön: the role of transport and retention of phosphorus, nitrogen and DOC in watershed and lake. - Ecol. Bull. (Stockholm) 37: 176-192.

Pettersson, A., Hällbom, L. and Bergman, B. 1985. Physiological and structural responses of the cyanobacterium *Anabaena cylindrica* to aluminium. - Physiol. Plant. 63: 153-158.

Pettersson, K. and Olsson, H. 1984. Mobility and fractional composition of phosphorus in the sediments of oligotrophic non-acidified, acidified and limed lakes. - In: Henriksen, K. (Ed.) 12th Nordic Symposium on sediments. Rep. Bot. Inst., Univ. Aarhus, Denmark, (8): 13-25.

Rainey, R.H. 1967. Natural displacement of pollution from the Great Lakes. - Science 155: 1242-1243.

Reynolds, C.S. 1984. The ecology of freshwater phytoplankton. Univ. Press, Cambridge.

Rosseland, B.O. and Hindar, A. 1988. Liming of lakes, rivers and catchments in Norway. - Water Air Soil Poll. 41: 165-188.

Rosseland, B.O., Blakar, I.A., Bulger, A., Kroglund, F., Kvellstad, A., Lydersen, E., Oughton, D.H., Salbu, B., Staurnes, M. and Vogt, R. 1992. The mixing zone between limed and acidic river waters: complex aluminium chemistry and extreme toxicity for salmonids. - Environ. Poll. 78: 3-8.

Samuelsson, M.-O. and Klemedtsson, L. 1991. Dikväveoxidproduktion och emission från våtmarker. [English summary: Nitrous oxide production and emission from wetlands.] - Vatten 47: 317-320.

Sandberg, P.-E. 1992. Sammanställning av kalkningsverksamheten i vissa delar av Kopparbergs län 1979-91. [Evaluation of liming activities in certain parts of the county of Kopparberg, 1979-1991.] - Fiskevattenägareförb., Falun.

Sangfors, O. 1984. Produktionsbegränsande faktorer för tre akvatiska makrofyter i två försurade sjöar varav den ena genomgått kalkning. [Limiting factors for the production of three aquatic macrophytes in two acidified lakes, among which one is limed.] - Environm. Res. Group, Kil, Rep. S. 8404.

Sanni, S., Skogheim, O.K. and Hongve, D. 1983. Sediment-vann undersökelser i forbindelse med kalking av 3 vann i Gjerstad, Aust-Agder. [Sediment-water studies in connection with liming of three lakes in Gjerstad, Aust-Agder.] - The liming project, Rep. 5.83. Dep. Environm. Direct. Nat. Managem., Norway.

Sanni, S., Skogheim, O.K. and Hongve, D. 1984. Sediment-vann undersökelser i forbindelse med kalking av 3 vann i Gjerstad, Aust-Agder. Del 2. [Sediment-water studies in connection with liming of three lakes in Gjerstad, Aust-Agder. Part 2.] - The liming project, Rep. 13.84. Dep. Environm. Direct. Nat. Managem., Norway.

Schindler, D.W., Mills, K.H., Malley, D.F., Findlay, D.L., Shearer, J.A., Davies, I.J., Turner, M.A., Linsey, G.A. and Cruikshank, D.R. 1985. Long-term ecosystem stress: the effects of years of experimental acidification on a small lake. - Science 228: 1395-1401.

SEPA (Swedish Environmental Protection Agency) 1988. Kalkning av sjöar och vattendrag. [Liming of lakes and streams.] - Swedish Environm. Prot. Agency, Off. Guidel. 88:3.

Shapiro, J. 1957. Chemical and biological studies on the yellow organic acids in lake water. - Limnol. Oceanogr. 2: 161-179.

SMHI (Swedish Meteorological and Hydrological Institute) 1983. Svenskt sjöregister. - Swedish Meteorol. Hydrol. Inst.

Stålhandske, P. 1990. Försöksverksamheten 1977-88, kalkning och vattenkemiska effekter. [The trial programme 1977-88, liming and effects on water chemistry.] - National Board Fish., Rep. 1: 20-27.

Stross, R.G. and Hasler, A.D. 1960. Some lime-induced changes in lake metabolism. - Limnol. Oceanogr. 5: 265-272.

Stumm, W. and Morgan, J.J. 1981. Aquatic chemistry. - Wiley Intersci., New York.

Svahnberg, A. 1989. Våtmarkskalkning i Jönköpings län. Sammanställning av resultat. [Liming of wetlands in the county of Jönköping. A compilation of results.] - Jönköping County Council, unpubl. rep.

Sverdrup, H.U. 1983. Lake liming. - Chem. Scripta 22: 67-72.

Sverdrup, H.U. and Warfvinge, P. 1984. Integrerte massebalanser for kalkade sjöer. [Integrated mass balances for limed lakes.] - Vatten 40: 369-382.

Sverdrup, H.U. and Warfvinge, P. 1988. Lake liming in different types of acid lakes using various types of calcite powders and methods. - Water Air Soil Poll. 41: 189-222.

Sverdrup, H.U., Rasmussen, R. and Bjerle, I. 1984. A simple model for the reacidification of limed lakes, taking the simultaneous deactivation and dissolution of calcite in the sediments into account. - Chem. Scripta 24: 53-66.

Sverdrup, H.U., Warfvinge, P. and Bjerle, I. 1985. Teori og beregningseksempel for kalkningen av en sur sjö. [Theory and examples on calculation for liming of an acid lake.] - Dep. Chem. Engin., Lund Inst. Techn., Univ. Lund, Res. Rep.

Sverdrup, H.U., Warfvinge, P. and Bjerle, I. 1986a. A simple method to predict the time required to reacidify a limed lake. - Vatten 42: 10-15.

Sverdrup, H.U., Warfvinge, P. and Bjerle, I. 1986b. Experimental determination of the kinetic expression for the dissolution of K-slag, a tricalcium silicate, in acidic aqueous solution. - Vatten 42: 210-217.

Thörnelöf, E. and Degerman, E. 1991. Lake liming in Sweden, case studies and general experiences. In: Olem, H., Schreiber, R.K., Brocksen, R.W. and Porcella, D.B. (Eds.) International lake and watershed liming practices. - The Terrene Inst., Washington, pp. 41-60

Tipping, E. and Backes, C.A. 1988. Organic complexation of Al in acid waters: mode-testing by titration of a streamwater sample. - Water Res. 22. 593-595.

Tipping, E., Woof, C., Walters, P.B. and Ohnstad, M. 1988. Conditions required for the precipitation of aluminium in acidic natural waters. - Water Res. 22: 585-592.

Törnqvist, L. 1988. Studies of aluminium toxicity to green algae *(Monoraphidium dybrowskii & Stichococcus* sp.) with emphasis on phosphate metabolism. - Acta Univ. Uppsaliensis 193.

Traaen, T.S. 1990. Monitoring at Tjönnstrand. - Norwegian Inst. Water Res. Techn. Rep. for 1989. O-87116/E88460.

Waters, T.F. 1956. The effects of lime application to acid bog lakes in northern Michigan. - Trans. Am. Fish. Soc. 86: 329-344.

Weppling, K., Järvinen, M. and Rask, M. 1992. The Lake Iso Valkjärvi project - ecological studies of liming on an artificially divided acidified lake in southern Finland. - Mitteil. Hydrogr. Dienstes Österreich. 67: 62-75.

Westling, O. and Hultberg, H. 1990/91. Liming and fertilization of acid forest soil: short--term effects on runoff from small catchments. - Water Air Soil Poll. 54: 391-407.

Widarsson, L.-E. and Johansson, C. 1992. Kalkningsprojekt Fylleån. Lägesrapport för perioden september 1982 - maj 1988. [The River Fylleån liming project. Status report for the period September 1982 - May 1988. - Halland County Council, unpubl. rep.

Wilander, A. and Ahl, T. 1972. The effects of lime treatment to a small lake in Bergslagen, Sweden. - Vatten 28: 431-445.

Wilander, A., Blomquist, P. and Nylund, B. 1978. Humusämnenas roll i naturvatten. - Dep. Limnol., Univ. Uppsala, Rep. LIU 1978:10.

Wright, R.F. 1985. Chemistry of Lake Hovvatn, Norway, following liming and reacidification. - Can. J. Fish. Aquat. Sci. 42: 1103-1113.

Wright, R.F. and Skogheim, O.K. 1983. Aluminium speciation at the interface of an acid stream and a limed lake. - Vatten 39: 301-304.

Wright, R.F., Kämäri, J. and Forsius, M. 1988. Critical loads for sulphur: modelling time of response of water chemistry to changes in loading. - In: Nilsson, J. and Grennfelt, P. (Eds.) Critical loads for sulphur and nitrogen. Nordic Counc. Minist., 1988:15, pp. 201-224.

Yan, N.D. 1983. Effect of changes in pH on transparency and thermal regimes of Lohi Lake, near Sudbury, Ontario. - Can. J. Fish. Aquat. Sci. 40: 621-626.

Yan, N.D. and Dillon, P.J. 1982. Experimental neutralization of lakes near Sudbury, Ontario. - In: Studies of lakes and watersheds near Sudbury, Ontario, Chapter 6. Final limnological report. Ontario Minist. Environm., Rexdale, Ontario. Rep. SES 009/82.

Yan, N.D., Jones, J., Cave, B., Scott, L. and Powell, M. 1977. The effects of experimental elevation of lake pH on the chemistry and biology of Nelson Lake near Sudbury, Ontario. - Ontario Minist. Environm., Toronto. Techn. Rep.

Yavitt, J.B. and Newton, R.M. 1990/91. Liming effects of some chemical and biological parameters of soil (spodosols and histosols) in a hardwood forest watershed. - Water Air Soil Poll. 54: 529-544.

Young, T.C., DePinto, J.V., Rhea, J.R. and Scheffe, R.D. 1989. Calcite dose selection, treatment efficiency, and residual calcite fate after whole-lake neutralization. - Can. J. Fish. Aquat. Sci. 46: 315-322.

Chapter 6

The effects of liming on microbial activity and the decomposition of organic material

Gunnar Gahnström

Department of Ecology, University of Lund, Box 65, 221 00 Lund, Sweden

Summary

Liming of lakes and watercourses produces drastic changes in water chemistry which can be expected to affect the decomposition of organic material by influencing the density and activity of micro-organisms and detritivorous invertebrates.

Lake liming involves a short and sudden decrease in microbial activity, a "liming shock". Seen in the longer term, it can indirectly cause an increase in microbial activity and better processing of organic material if access to dead organic material improves either as a result of increased primary production or through greater availability of the accumulated organic material.

The decomposition of coarse detritus in the littoral zone of lakes and running waters is dependent upon the number and activity of detritivorous invertebrates. In most cases, the number of shredders also increases after liming, leading to faster decomposition of leaves and other coarse detritus and thus more rapid breakdown.

1. Introduction

When organic materials decompose, nutrients become available to plants once again and can be used in primary production. Many different groups of organisms take part in the process of mineralization, i.e. the decomposition of organic materials into their end products: inorganic plant nutrients, carbon dioxide and water. Micro-organisms, bacteria and fungi are mainly responsible for the mineralization process, although many invertebrates also play an important part in the first phase of decomposition. Detritus represents a source of nutrients for the organism concerned, while at the

Liming of Acidified Surface Waters
L. Henrikson and Y.W. Brodin (Eds.)
© Springer-Verlag Berlin Heidelberg 1995

same time bacteria and fungi provide an important source of food for many invertebrates, planktonic crustacaeans and insect larvae. The detritus nutrient chain is thus an important and absolutely essential part of the energy chain and the nutrient cycle.

Acidification has brought a great change in the chemical environment of freshwaters, and this has naturally had an impact on other parts of the ecosystem (Brodin, Chap. 2 and 3). Studies of the effects of acidification on microbial activity and the decomposition of organic materials have yielded somewhat contradictory results, however. Laboratory experiments have indicated that microbial activity is inhibited by higher levels of acidity (e.g. Rao and Dutka 1983, Furutani *et al* 1984), but studies of bacterial density and microbial activity in field trials and undisturbed systems have been unable to demonstrate that acidification has any significant effect (e.g. Kelly *et al* 1984, Gahnström and Andersson 1985). This is, however, contradicted by the increased accumulation of organic materials on the bottom of lakes and watercourses as a result of impaired decomposition of coarse detritus (e.g. Grahn *et al* 1974, Brock *et al* 1985, Kimmel *et al* 1985, Allard and Moreau 1986, Brezonik *et al* 1990, Kok *et al* 1990b).

Liming also causes far-reaching changes in water chemistry, which in turn has an impact on other parts of the ecosystem (Henrikson and Brodin, Chap. 1). This raises the important question: how does liming of acidified lakes and watercourses affect the density and activity of micro-organisms and detritivorous invertebrates and thus the decomposition of organic materials?

2. Bacterial density and microbial biomass

There are few studies of the effects of liming on the density or biomass of bacteria or fungi. In one experiment involving mesocosmos in the form of plastic bags containing 1,300 litres of water in the moderately humic Lake Stråken in the Aneboda area, southern Sweden, the numbers of bacterioplankton were compared in limed, acidified and unlimed mesocosmos (Gahnström *et al* 1989, 1990). The number of bacterioplankton was lower in the limed mesocosmos. Phytoplankton biomass and production also declined in the limed mesocosmos as compared with the control mesocosmos. Tranvik *et al* (in prep.) compared the number of bacteria in the water and in the sediment surface in five acidified and five limed humic lakes in southern Sweden. They found no difference in the bacterial counts between these two categories of lakes during the period in question (August-September 1990), although the phytoplankton biomass and production were roughly the same in all the lakes.

After liming of the clear waters of the acidified Lake Njupfatet in central Sweden, the bacterial density and concentration of dissolved organic carbon in the water increased, while the phytoplankton biomass decreased (Bell 1990).

Bärlocher and Petersen (1983) studied the number of aquatic hyphomycetes (fungi) at two limed and two non-limed sites in River Fylleån in south-western Sweden. They found no statistically significant difference in the number of species between the four sites, although there was a slightly greater number of species at the limed sites.

The Norwegian liming project was unable to show any long-term change in the number of bacterioplankton or in phytoplankton production as a result of liming, but both the bacterial biomass and phytoplankton production declined for a short period directly after liming (Baalsrud *et al* 1985). Liming of a humic lake in Finland led to a 50% increase in the number of bacterioplankton during the first year, but this was followed by a decline during the second year to the same level as before liming (Salonen *et al* 1990). Neither the phytoplankton biomass nor the production was affected by liming.

After liming in acidified lakes which were highly polluted with heavy metals in the Sudbury district of Canada, the number of bacteria in the water mass increased, as did the production of phytoplankton, while the number of fungi in the water mass and the number of bacteria in the sediment remained unchanged (Scheider *et al* 1975). Majumdar *et al* (1987, in Weatherley 1988) reported higher bacterial density in sediment after the liming of lakes in Pennsylvania in the USA.

The few studies available are contradictory. Some of the discrepancies can probably be explained by the choice of method used in the studies. When culturing micro-organisms, the choice of culture media is crucial to the result. It may, moreover, be difficult to distinguish living bacteria from dead particulate material or small phytoplankton when counting under the microscope, depending on how the sample has been coloured.

The number, or biomass, of micro-organisms gives no information about their activity or about how much organic materials they may process. The effects of liming on bacteria and fungi probably vary from system to system. The availability of a substrate - organic materials - affects the bacterial populations. Humic acid lakes often support a larger number of bacterioplankton than clear acid lakes (Traaen 1978). The conclusion is that liming can be expected to lead to changes in the size and activity of micro-organism populations if there is any change in the availability of substratum in the form of primary production, the amount of dissolved organic carbon or humus levels. This conclusion is applicable to most of the studies referred to above.

3. Microbial activity

There are many methods of measuring microbial turnover or decomposition activity, but two main approaches have been applied in Swedish liming studies. The first is to measure the overall activity of all the organisms involved in mineralization on the basis of carbon dioxide production, oxygen consumption or the decline in the weight of coarse detritus such as birch leaves or water lily leaves. The second is to measure a specific process by labelling a substrate such as glucose or amino acids with carbon-14, for instance.

3.1. Microbial activity in the water

The effects of liming on microbial activity in the water have been studied both in short-term mesocosmos experiments and in acidified and limed humic lakes. Both the uptake of amino acids labelled with carbon-14 and the respiration in the water mass were as great in the limed mesocosmos as in the control mesocosmos during the first summer (Gahnström *et al* 1989). Even during the second summer, when mesocosmos with sediment contact were studied, no effects from liming were registered whether on the processing of amino acids labelled with carbon-14 or on respiration (Gahnström *et al* 1990). These experiments had a duration of about one month and thus only indicated the short-term effects of liming.

Respiration in the water mass and the processing of amino acids labelled with carbon-14 were the same in the acidified as in the limed humic lakes (Tranvik *et al* 1994).

The conclusion is the same as in the previous section, i.e. that the availability of substrate is crucial to both microbial activity and to the size of the micro-organism population.

3.2. Microbial activity in the sediment

Both the immediate and the long-term effects of liming were studied in the Lake Gårdsjön project, south-western Sweden, which examined oxygen depletion in sediment from four different water depths during the period prior to liming (1979 to April 1982) and after liming in April 1982 and August 1985 to June 1986 (Gahnström 1984, 1988).

Two weeks after liming, a decline in the oxygen depletion rate in the sediment was found at all sampling points compared with the corresponding month before liming (Fig. 1). One to two months after the first liming, oxygen depletion in the sediment was once again back to the same, or higher, level as before liming. A decrease was also noted after the second liming, but only at 13 and 18 m water depth (Fig. 1b).

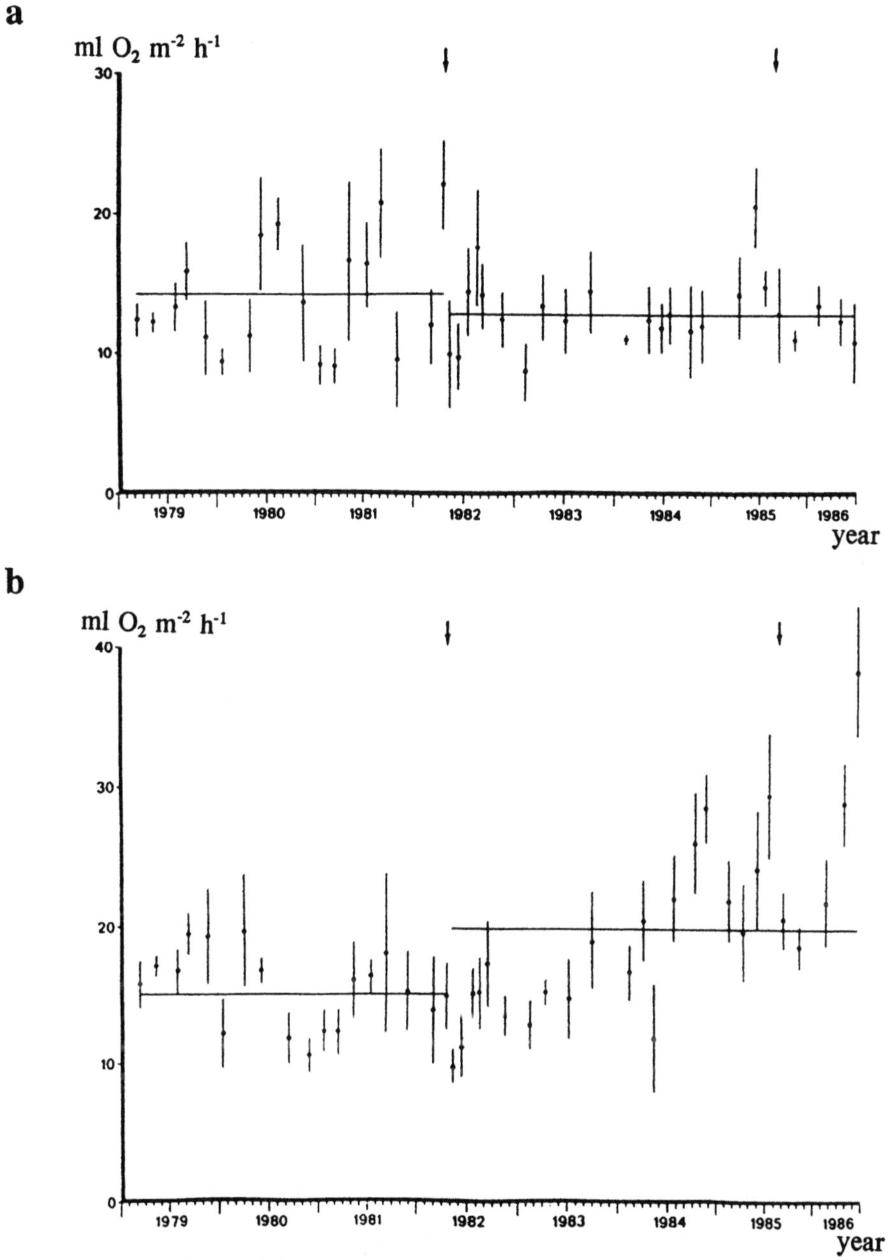

Figure 1. Oxygen depletion in the sediment (ml O_2 m^{-2} h^{-1}) at 100% oxygen saturation and +10 °C at a water depth of 2 m (a) and 18 m (b) in Lake Gårdsjön, SW Sweden. Mean values and 95% confidence intervals are shown for each sampling. The horizontal line represents the average for the periods before and after liming in April 1982. The arrows indicate when liming was performed (Gahnström 1988).

Oxygen depletion rate in the sediment increased significantly at water depths of 8, 13 and 18 m (Table 1). Oxygen depletion at 8 m increased by 25% and stabilized at this level during the following years. In the deep sediment, oxygen depletion increased during the year after liming and was 50-100% higher in 1986 than it was before liming (Fig. 1b). In the reference lakes, Stora Hästevatten and Gaffeln, there were no corresponding changes in oxygen depletion in the sediment, however (Table 1), which suggests that it was liming which, directly or indirectly, led to the increased microbial activity. Oxygen depletion in the littoral sediment at a water depth of 1-2 m did not change significantly, although the mean oxygen depletion was lower after liming than it was before liming (Fig. 1a, Table 1). Oxygen depletion in the sediment of Lake Store Finnetjenn in Norway was unaffected by liming (Baalsrud *et al* 1985).

Increased oxygen depletion and increased microbial activity, measured in terms of the processing of glucose labelled with carbon-14, was also observed in Lake Högsjön in the catchment area of River Högvadsån in south-western Sweden (Gahnström *et al* 1980).

The processing of carbon-14 labelled amino acids was the same in acidified lakes as in limed humic lakes in southern Sweden, as was respiration in the surface sediment (Tranvik *et al*, in prep).

Any reduction in microbial activity or biomass directly after liming is probably attributable to the "chemical shock" which liming involves. In Lake Gårdsjön, the pH of the water close to the bottom increased by up to four units and the pH in the interstitial water in the surface sediment in-

Table 1. *Oxygen depletion in sediment (ml O_2 m^{-2} h^{-1}) at 100% oxygen saturation and +10 °C in Lake Gårdsjön, south-western Sweden, and two nearby references - Lake Stora Hästevatten and Gaffeln - in the periods prior to liming (March 1979-March 1982) and after liming (April 1982 to June 1986).*

Sampling site	Before liming	After liming	Significance
Gårdsjön, 2 m	14.2±1.7	12.7±0.8	None
Gårdsjön, 8 m	11.6±1.0	14.4±0.7	≤0.001
Gårdsjön, 13 m	14.9±1.3	17.6±0.9	≤0.001
Gårdsjön, 18 m	15.0±1.1	19.9±1.2	≤0.001
St. Hästevatten, 2 m	13.1±3.2	12.3±1.3	None
St. Hästevatten, 9 m	11.4±1.6	11.5±0.6	None
Gaffeln, 15 m	14.0±2.1	14.6±5.0	None

creased from 5 to 6.5-7.5 (Gahnström 1988). Short-term "shocks" have also been noted in the form of reduced biomass among phytoplankton, zooplankton and benthic invertebrates (Hillbricht-Ilkowska *et al* 1977, Dillon *et al* 1979, Hultberg and Andersson 1982, Baalsrud *et al* 1985, Larsson 1988, 1994, Degerman *et al*, Chap. 8).

Oxygen depletion in the sediment varies during the year as a result of temperature and changes in the supply of dead organic materials, among other things. The increased oxygen depletion observed in Lake Gårdjön after liming was probably the result of an increase in organic materials. During the first and second years after liming, the amount of *Sphagnum* moss in Lake Gårdsjön declined from 3,000 to 10 kg (Grahn and Sangfors 1988, Larsson, Chap. 7), as did the number of benthic algae (Lazarek 1986). The production of phytoplankton increased between 1982 and 1985 from 8-10 g C m^{-2} $year^{-1}$ before liming to 15-35 g C m^{-2} $year^{-1}$ afterwards (Larsson 1988, 1991, Chap. 7).

The supply of dead organic materials thus increased after the liming of Lake Gårdsjön, which led to increased oxygen depletion in the sediment. In Lake Stora Finnetjenn the primary production did not increase after liming, which can explain why oxygen depletion in the sediment remained unchanged (Baalsrud *et al* 1985). In the humic lakes studied by Tranvik *et al* (in prep.) the primary production in the acidified lakes was as great as in the limed lakes, as was the microbial activity.

Increased oxygen depletion, particularly in the deep sediment, leads to lower oxygen levels in the hypolimnion after liming (Fig. 2; Gahnström 1988, Larsson 1988, Wilander *et al*, Chap. 5). Oxygen depletion probably also increases in the water mass after liming, which also promotes lower oxygen levels in the hypolimnion. Increased oxygen depletion in the water mass has also been noted after liming of several other lakes (Hillbricht--Ilkowska *et al* 1977, Edman *et al* 1988, Hasselrot *et al* 1984, Bukaveckas 1990, Bukaveckas and Driscoll 1991). The Secchi depth in Gårdsjön declined after liming, which meant that oxygen production via photosynthesis could not take place at the same depth in the water mass after liming as before. The oxygen supply to the hypolimnion was thus reduced after liming.

Initially, oxygen depletion and microbial activity in sediment are adversely affected by liming (liming shock), but if the supply of dead organic materials increases, the microbial activity also increases and with it oxygen depletion in the sediment. If oxygen depletion increases after liming, it leads to lower oxygen levels in the hypolimnion and possibly in a shortage of oxygen in the water close to the bottom. The increased level of microbial activity may continue if the supply of organic materials is still higher after liming.

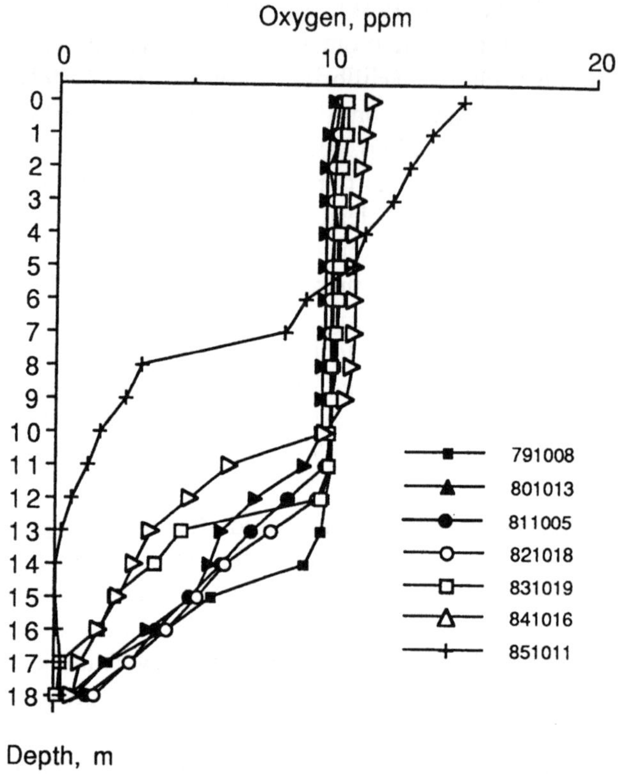

Figure 2. Oxygen levels in Lake Gårdsjön, south-western Sweden, according to a depth profile from the surface to the bottom at the end of the summer stagnation period. 1979 to 1981 were the years before liming and 1982 to 1985 were the years after liming (Gahnström 1988).

Denitrification, i.e. microbial conversion of nitrate (NO_3) to nitrogen gas (N_2) or nitrous oxide (N_2O), only occurs in environments which are devoid of oxygen or where oxygen levels are very low. After the liming of Lake Gårdsjön, the overall denitrification appears to have increased from 390 to 775 kg N $year^{-1}$, according to calculations based on the nitrogen budget of the lake (Broberg 1988). Direct measurements of the rate of denitrification in the lake sediment also show that denitrification was faster in the limed Lake Gårdsjön than in the nearby, but acidified, Lake Gaffeln (Gahnström *et al* 1993). This does not necessarily mean that denitrification is inhibited in acidified lakes, although liming may create favourable conditions for the process. Denitrification occurs only in more or less oxygen-free environments and where suitable substratum, organic materials and nitrate are available. In Lake Gårdsjön the extent of anoxic environments and the availability of organic materials increased after liming (cf. above). When organic materials are broken down, the organically bound nitrogen is released in the form of ammonia (NH_4), which can in turn be converted by micro-organisms into nitrate (nitrification). The increased availability of organic materials after liming may, via increased decomposition and nitrification, lead to an increase in the availability of nitrate, resulting in increased denitrification.

4. Decomposition of coarse detritus

In smaller watercourses, the supply of leaves and other dead material from the surrounding vegetation is extremely important as a source of energy and food for invertebrates and as substrate for bacteria and fungi. In the littoral zones of lakes, detritus from the surroundings and from water vegetation are very important as food for benthic crustaceans and insect larvae.

Grahn *et al* (1974) reported that coarse detritus is accumulated in the littoral zone of acidified lakes. The decomposition of alder leaves measured in terms of weight loss was also low in the acidified Lake Gårdsjön and in the equally acidified reference lake, Lake Hästevatten: only 20% weight loss over a period of seven months (November-May) (Andersson 1985). The weight loss of alder leaves was almost 90% in Lake Högsjön, south-western Sweden, after liming, however. In Lake Tjärnesjön, an acidified reference lake for Lake Högsjön, the weight loss amounted to about 75% (Andersson 1985). Factors other than the pH of the water would seem to be of importance in the decomposition of coarse detritus. This is also confirmed by laboratory experiments in which the addition of lime alone to water and sediment from acidified lakes failed to affect the rate of decomposition of alder leaves (Andersson 1985).

The rate of decomposition of alder leaves placed in Lake Gårdsjön six months after liming was higher than before liming: 91% from November to May compared with 27% before liming (Fig. 3). Thc following years, 1984 and 1985, the rate of decomposition of the alder leaves remained high, but it declined in the 1985-86 season to about 60%. Throughout the period studied the weight loss of alder leaves placed in the reference lake, Lake Stora Hästevatten, was between 32 and 38%.

In conjunction with the liming of River Högvadsån the decomposition of alder-leaf packs was studied in both the outflow from Lake Högsjön after liming and in the acidified reference lake, Lake Tjärnesjön (Johansson 1979). The weight loss of the leaf packs was 21% in the outflow from Lake Högsjön and 8% in the outflow from Lake Tjärnesjön during a forty-day study in April and May 1979 (Table 2). Respiration, i.e. the microbial activity in the leaf packs, was equal in both cases, but the number of detritivorous invertebrates was ten times higher in the leaf packs in the outflow from Lake Högsjön (Table 2).

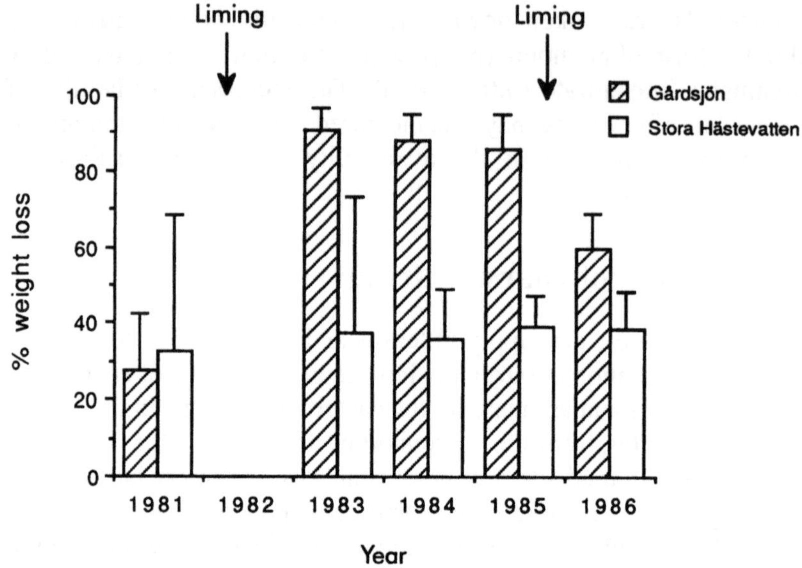

Figure 3. *Decomposition (percentage dry weight loss) of alder leaves exposed at a water depth of 1 m in Lake Gårdsjön, south-western Sweden and the unlimed reference Lake Stora Hästevatten from November to May. The columns show mean values and the vertical lines show standard deviation. The arrows show when Lake Gårdsjön was limed, in April 1982 and August 1985 (Gahnström and Andersson, in prep.).*

Table 2. *Percentage weight loss and respiration in alder-leaf packs and the number of detritivorous invertebrates per pack in the outflow from Lake Högsjön (limed) and Tjärnesjön (acidified) in south-western Sweden (Johansson 1979).*

	Högsjön	Tjärnesjön
Loss of weight, %	21.1	7.6
Respiration, ml O_2 g dry weight^{-1} h^{-1}	0.049	0.050
Number of invertebrates (n/pack)	158	13

The presence of invertebrates which fragment and consume detritus is clearly of great importance, both to the decomposition process itself and to the speed at which detritus material disappears. This was shown by Andersson (1985) in a laboratory experiment using alder leaves and water louse - *(Asellus aquaticus)*. Henrikson and Oscarson (unpubl.) studied the decomposition of birch leaves in bags with different mesh sizes (1 and 10 mm) in a total of thirty acidified, liming and unaffected lakes. After eight months the weight loss was greater in the bags with a large mesh and larger in the unaffected and limed lakes than in the acidified ones. In the small-mesh bags, the weight loss was the same in all three categories. In the large-mesh bags, shredders such as *A. aquaticus* and caddis fly larvae (Trichoptera) can fragment the leaf material, while only micro-organisms can affect decomposition in the small-mesh bags. The number of *A. aquaticus* per bag was greater in the unaffected and limed lakes (Henrikson and Oscarson, unpubl.). In Lake Gårdsjön, the number of *A. aquaticus* increased after liming in 1982, but fell again during the following years (Henrikson and Oscarson unpubl., Degerman *et al*, Chap. 8). The decline in the number of *A. aquaticus* certainly contributed to the decline in the rate of decomposition for the alder leaves in Lake Gårdsjön (Fig. 3).

The liming of Lake Store Hovvatnet in Norway had no effect on the decomposition of birch leaves in comparison with an unlimed reference lake, probably because the number of shredders was low and became no higher after liming (Fjellheim and Raddum 1988). The decomposition of water lily leaves and other water plants in laboratory experiments and in acidified and non-acidified small lakes was lower under acidified conditions, increasing in the presence of bicarbonate and plant nutrients (nitrogen and/or phosphorus) in the water, as well as in cases where the number of shredders increased (Brock *et al* 1985, Leuven and Wolfs 1988, Kok *et al* 1990a,b, Kok and van de Laar 1991, Kok and van der Velde 1991).

Most studies thus indicate that shredders play an important part in the decomposition of coarse detritus. If liming leads to an increase in the number of detritivorous invertebrates, detritus decomposition will also increase.

5. References

Allard, M. and Moreau, G. 1986. Leaf decomposition in an experimentally acidified stream channel. - Hydrobiologia 139: 109-117.

Andersson, G. 1985. Decomposition of alder leaves in acid lake water. - Ecol. Bull. (Stockholm) 37: 293-299.

Baalsrud, K., Hindar, A., Johannessen, M. and Matsow, D. (Eds.) 1985. Liming of acid water. Liming project, final report. - Dep. Environm. Direct. Nat. Managem., Norway.

Bärlocher, F. and Petersen, R.C. 1988. Aquatic hyphomycetes in an acidified and limed humic stream in Sweden. - Verh. Internat. Verein. Limnol. 23: 1279-1286.

Bell, R.T. 1990. Regulation of bacterioplankton and the microbial food web in Lake Njupfatet, Sweden. - Inst. Limnol., Univ. Uppsala, Rep. 1990, B:11.

Brezonik, P.L., Webster, K.E. and Perry, J.A. 1990. Effects of acidification on benthic community structure and benthic processes in the Little Rock Lake, Wisconsin. - Verh. Internat. Verein. Limnol. 24: 445-448.

Broberg, O. 1988. Liming of Lake Gårdsjön; an acidified lake in SW Sweden. Effects on organic carbon, nitrogen and phosphorus. - In: Dickson, W. (Ed.) Liming of Lake Gårdsjön, an acidified lake in SW Sweden. Swedish Environm. Prot. Agency 3426, pp. 135-205.

Brock, T.C.M., Boon, J.J. and Paffen, B.G.P. 1985. The effects of the season and the water chemistry on the decomposition of *Nymphaea alba* L.; weight loss and pyrolysis mass spectrometry of the particulate matter. - Aquat. Bot. 22: 197-229.

Bukaveckas, P.A. 1990. Effects of whole-lake base addition on the limnology of an acidic Adirondack lake (Woods Lake, NY, USA). - Verh. Internat. Verein. Limnol. 24: 470-473.

Bukaveckas, P.A. and Driscoll, C.T. 1991. Effects of whole-lake base addition on thermal stratification in three acidic Adirondack lakes. - Water Air Soil Poll. 59: 23-39.

Dillon, P.J., Yan, N.D., Scheider, W.A. and Conroy, N. 1979. Acidic lakes in Ontario, Canada: characterisation, extent and responses to base and nutrient additions. - Arch. Hydrobiol. Beih. Ergebn. Limnol. 13: 317-336.

Edman, G., Fleischer, S., Fritz, Ö. and Stibe, L. 1988. Högvadsån 1978-1986. Försurad - Kalkad - Pånyttfödd. [English summary: River Högvadsån 1978-1986. Acidified - limed - regenerated.] - Halland County Council 1988:4.

Fjellheim, A. and Raddum, G.G. 1988. Birch leaf processing and associated macroinvertebrates in an acidified lake subjected to liming. - Hydrobiologia 157: 89-94.

Furutani, A., Rudd, J.W.M. and Kelly, C.A. 1984. A method for measuring the response of sediment microbial communities to environmental perturbations. - Can. J. Microbiol. 30: 1408-1414.

Gahnström, G. 1984. Short-term effects of liming on sediment oxygen uptake in an acid lake. - Verh. Internat. Verein. Limnol. 22: 760-764.

Gahnström G. 1988. The effects of liming on sediment oxygen uptake in Lake Gårdsjön, Sweden. - In: Dickson, W. (Ed.) Liming of Lake Gårdsjön, an acidified lake in SW Sweden. Swedish Environm. Prot. Agency 3426, pp. 207-243.

Gahnström, G., Andersson, G. and Fleischer, S. 1980. Decomposition and exchange processes in acidified lake sediment. - In: Drabløs, D. and Tollan, A. (Eds.) Proc., Int. Conf. Ecol. impact acid precip., Norway, SNSF project, pp. 306-307.

Gahnström, G. and Andersson, G. 1985. An attempt to quantify sediment processes in the acidified Lake Gårdsjön, SW Sweden. - Ecol. Bull. (Stockholm) 37: 327-336.

Gahnström, G., Granéli, W., Tranvik, L. and Diehl, E. 1989. Effekter av försurning och kalkning på mikrobiella processer i pelagial och sediment. Experimentella studier. [Effects of acidification and liming on microbial processes in the pelagial and the sediment. Experimental studies.] - Dep. Ecol./Limnol., Univ. Lund., unpubl. rep.

Gahnström, G., Granéli, W. and Tranvik, L. 1990. Effekter av försurning och kalkning på mikrobiella processer i pelagial och sediment. Experimentella studier. [Effects of acidification and liming on microbial processes in the pelagial and the sediment. Experimental studies.] - Dep. Ecol./Limnol., Univ. Lund., unpubl. rep.

Gahnström, G., Blomqvist, P. and Fleischer, S. 1993. Are key nitrogen fluxes changed in the acidified aquatic ecosystem? - Ambio 22: 318-324.

Grahn, O., Hultberg, H. and Landner, L. 1974. Oligotrophication, a self-accelerating process in lakes subjected to excessive supply of acid substances. - Ambio 3: 93-94.

Grahn, O. and Sangfors, O. 1988. A comparative study of macrophytes in Lake Gårdsjön, during acid and limed conditions. - In: Dickson, W. (Ed.) Liming of Lake Gårdsjön, an acidified lake in SW Sweden. Swedish Environm. Prot. Agency 3426, pp. 281-308.

Hasselrot, B., Andersson, B.I. and Hultberg, H. 1984. Ecosystem shift and reintroduction of Arctic char (*Salvelinus salvelinus* (L.)) after liming of a strongly acidified lake in south-western Sweden. - Rep. Inst. Freshwat. Res., Drottningholm 61: 78-92.

Hillbricht-Ilkowska, A., Rybak, J.I., Kajak, Z., Dusoge, K., Ejsmont-Karabin, J., Spodniewska, I., Weglenska, T. and Godlewska-Lipowa, A. 1977. Effects of liming on a humic lake. - Ekol. pol. 25: 379-420.

Hultberg, H. and Andersson, B.I. 1982. Liming of acidified lakes, induced long-term changes. - Water Air Soil Poll. 18: 311-331.

Johansson, P.-A. 1979. Nedbrytning av al-löv, makrobentossammansättning samt transport av organiskt material i tre halländska bäckar. [Breakdown of alder leaves, composition of macrobenthos and transportation of organic matter in three streams in the county of Halland.] - Dep. Limnol., Univ. Lund, applic. project

Kelly, C.A., Rudd, J.W.M., Furutani, A. and Schindler, D.W. 1984. Effects of lake acidification on rates of organic matter decomposition in sediments. - Limnol. Oceanogr. 29: 687-694.

Kimmel, W.G., Murphey, D.J., Sharpe, W.E. and DeWalle, D.R. 1985. Macroinvertebrate community structure and detritus processing rates in two south-western Pennsylvania streams acidified by atmospheric deposition. - Hydrobiologia 124: 97-102.

Kok, C.J., van der Velde, G. and Landsbergen, L.M. 1990a. Production, nutrient dynamics and initial decomposition of floating leaves of *Nymphaea alba* L. and *Nuphar lutea* L. Sm. (Nymphaeaceae) in alkaline and acid waters. - Biogeochemistry 11: 235-250.

Kok, C.J., Meesters, H.W.G. and Kempers, A.J. 1990b. Decomposition rate, chemical composition and nutrient recycling of *Nymphaea alba* L. floating blade detritus as influenced by pH, alkalinity and aluminium in laboratory experiments. - Aquat. Bot. 37: 215-227.

Kok, C.J. and van de Laar, B.J. 1991. Influence of pH and buffering capacity on the decomposition of *Nymphaea alba* L. detritus in laboratory experiments; a possible explanation for the inhibition of decomposition at low alkalinity. - Verh. Internat. Verein. Limnol. 24: 2689-2692.

Kok, C.J. and van der Velde, G. 1991. The influence of selected water quality parameters on the decay rate and exoenzymatic activity of detritus of *Nymphaea alba* L. floating leaf blades in laboratory experiments. - Oecologia 88: 311-316.

Larsson, S. 1988. Effects of lime treatment on species composition and biomass of phytoplankton in Lake Gårdsjön. - In: Dickson, W. (Ed.) Liming of Lake Gårdsjön, an acidified lake in SW Sweden. Swedish Environm. Prot. Agency 3426, pp. 245-280.

Larsson, S. 1991. Gårdsjöprojektet. Effekter av kalkning och fiskisättning på växtplankton. [The Lake Gårdsjön project. Effects on phytoplankton of liming and fish introduction.] - Nat. Board Fish., Rep. 1, 1990, pp. 36-40.

Lazarek, S. 1986. Responses of the *Lobelia*-epiphyte complex to liming of an acidified lake. - Aquat. Bot. 25: 73-81.

Leuven, R.S.E.W. and Wolfs, W.J. 1988. Effects of water acidification on the decomposition of *Juncus bulbosus* L. - Aquat. Bot. 31:57-81.

Rao, S.S. and Dutka, B.J. 1983. Influence of acid precipitation on bacterial populations in lakes. - Hydrobiologia 98: 153-157.

Salonen, K., Järvinen, M., Kuoppamäki, K. and Arvola, L. 1990. Effects of liming on the chemistry and biology of a small acid humic lake. - In: Kauppi, P., Anttila, P. and Kenttämies, K. (Eds.) Acidification in Finland, Springer, Berlin Heidelberg New York, pp. 1145-1167.

Scheider, W., Adamski, J. and Paylor, M. 1975. Reclamation of acidified lakes near Sudbury, Ontario. - Ontario Minist. Environm., Rexdale, Canada, Rep.

Traaen, T.S. 1978. Bakterieplankton i innsjøer. [Bacterioplankton in lakes.] - SNSF-project: acid precipitation - effects on forest and fish. Techn. Note 41. Oslo-Ås, Norway.

Tranvik, L.T., Granéli, W. and Gahnström, G. 1994. Microbial activity in acidified and limed humic lakes. - Can. J. Fish. Aquat. Sci. (in press)

Weatherley, N.S. 1988. Liming to mitigate acidification in freshwater ecosystems; a review of the biological consequences. - Water Air Soil Poll. 39: 421-437.

Chapter 7

The effects of liming on aquatic flora

Stefan Larsson

Göteborg and Bohus County Administrative Board, S-403 40 Göteborg, Sweden

Summary

Most studies of the effects of liming on the flora of lakes and watercourses have concentrated on phytoplankton, while little attention has been devoted to epiphytic algae, macrophytes and wetland flora.

Lake liming results in an initial decline in the phytoplankton biomass, although this is generally followed by an increase after 1-2 months. The post-liming biomass may be either greater or lesser than it was during the acidification phase and a long-term decline has been observed in certain cases. New species become established and, in most lakes, the number of species increases. The composition of species changes, from initially being dominated by dinoflagellates (Dinophyceae) or blue-green algae (Cyanophyceae) to a community containing greater variety, with golden-brown algae (Chrysophyceae), green algae (Chlorophyceae) and diatoms (Bacillariophyceae). It has often been possible to observe a shift from larger to smaller plankton species. There is a suggestion that suspended calcium particles cause a release of microscopic algae from the bottom sediment, which may partially explain the unexpected algal blooms that have been observed after liming. The risk of algal blooms is probably greater, however, in cases where the acidification damage is more extensive. Few studies have been carried out on primary production among phytoplankton and no clear effects have been identified. If lakes are limed early on in the acidification phase, i.e. before extensive biological damage has occurred, less far-reaching changes are involved than if the lake has had time to become severely acidified.

Mougeotia, a filamentous green alga, which thrives under acidic conditions, disappears after liming but may become re-established if the water becomes re-acidified. Felt carpets of filamentous blue-green and green algae decline. Epiphytic growths of certain species of desmids (e.g.

Liming of Acidified Surface Waters
L. Henrikson and Y.W. Brodin (Eds.)
© Springer-Verlag Berlin Heidelberg 1995

Desmidiales) on water lobelia *(Lobelia dortmanna)* increase.

Macrophyte species usually increase after liming, but it may take time - up to several decades - for new colonies to become established. *Sphagnum* mosses decline sharply, with a consequent reduction in biomass in the most acidified lakes, and colonization of surfaces previously covered by *Sphagnum* is usually slow. Proliferation of alternate water milfoil *(Myriophyllum alterniflorum)* has been observed in a number of shallow lakes after repeated liming treatments.

Severe damage to wetland vegetation has been observed after liming. *Sphagnum* mosses are completely eliminated in many cases, and other species such as *Polytrichum* mosses usually become established in places where the *Sphagnum* has died out. Liverworts (Hepaticae) and lichens often suffer as well. The field layer is affected in the long term through an increase in the dwarf-shrub layer, among other things. The long-term biological and hydrological effects of ground-layer damage in wetlands have not been studied.

1. The conditions for flora in acidified and limed lakes

1.1. Acidification

Aquatic plants are largely dependent upon the chemical and physical conditions prevailing in the water and sediment, but are also affected by herbivorous fauna. The chemical effects of acidification, in the form of falling pH values, for instance, are thought to be the cause of the ecological changes among floras which have been observed in many inland waters during recent decades. Although this has occurred very rapidly in biological terms, the rate of change differs in the pelagic zone and in the sediment:

1) In the pelagic zone, an early reaction to changes in water chemistry can be discerned in the form of increased Secchi depth, particularly at a pH below 5.5, when the carbonate buffer capacity is exhausted. The water is buffered by aluminium compounds and humic substances instead and the pH rarely falls below 4.5.
2) The changes take place more slowly in the sediment because of buffering media in the form of mineral particles, for example, which surround the interstitial water. Many of the macrophytes such as lilies *(Nymphaea alba, Nuphar luteum)* are able to find nutrition at deeper levels via their root systems.

Natural acidification which is completely uninfluenced by anthropogenic air pollution has been in progress ever since the last ice age. This has been established through studies of phytoplankton residues in sediment (Renberg *et al* 1993). In Lake Gårdsjön, south-western Sweden, it has been shown that a shift towards more acidophilic diatoms (Bacillariophyceae) has taken place during the course of this natural acidification (Fig. 1). These studies also showed that anthropogenic acidification has increased drastically during recent decades, however.

The sediment studies are in agreement with the findings of a number of synoptic studies, suggesting far-reaching depletion of species in phytoplankton communities and changes in the abundance of macrophytes (e.g. Almer *et al* 1978, Grahn 1977). The lack of pre-industrial material upon which to base comparisons means that such findings are open to question, however.

Phosphorus is considered to be the nutrient which represents the limiting factor on pelagic production in acidified lakes (e.g. Jansson *et al* 1986). Since the lakes which have the lowest pH values are also those with the lowest phosphorus levels (Almer *et al* 1978), the logical conclusion is that acidification leads to lower production - in line with the acido-oligotrophication hypothesis (Grahn *et al* 1974). There are probably several reasons why lakes in acidified areas have low phosphorus levels. The decreased

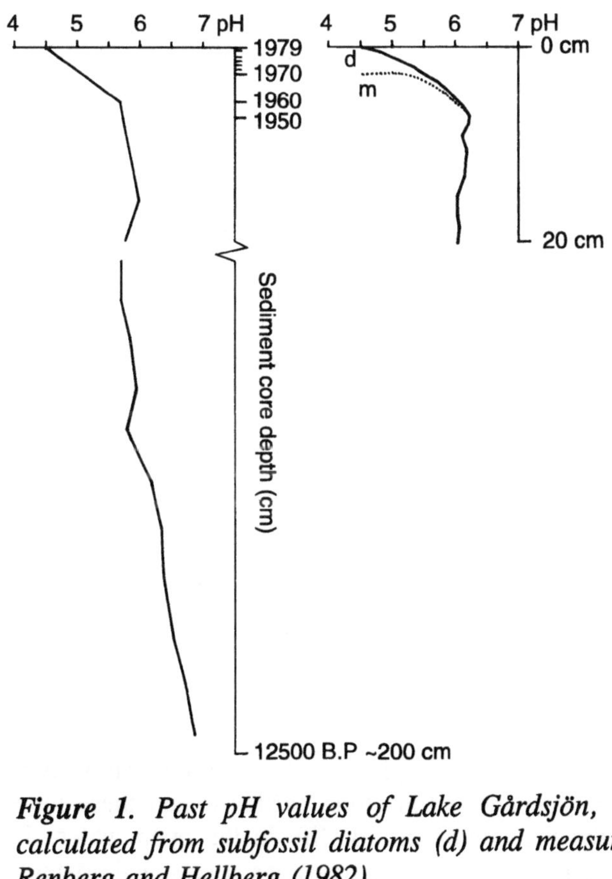

Figure 1. Past pH values of Lake Gårdsjön, south-western Sweden, calculated from subfossil diatoms (d) and measured (m). Modified from Renberg and Hellberg (1982).

mobility of phosphorus in acidified soil of the drainage area probably plays some part in this (Persson and Broberg 1985), while phosphorus can probably also be bound to metals or precipitated by aluminium so that it falls to the lake bed and thus becomes inaccessible to the plankton. Experiments in which nutrient was added to mesocosmos in Lake Njupfatet, central Sweden, were unable to demonstrate any phosphorus inhibition (Blomquist *et al* 1989). Experiments have also shown that it is the combination of nitrogen and phosphorus rather than phosphorus alone which limits production in certain types of lake such as the clear lakes in western Sweden, at least during certain parts of the year (Pettersson, unpubl.).

The fact that acidification causes a change in the nutrient balance may contribute to the elimination of groups of phytoplankton. The balance between different nitrogen fractions may be significant in that it favours certain species, as indicated in an experiment using a nitrate supplement which favoured dinoflagellates (Dinophyceae), while ammonium supplements favour *Merismopedia tenuissima* (Cyanophyceae) (Blomquist *et al* 1989).

Aluminium is particularly toxic in the pH interval of 5.1-5.8 (Nalewajko and Paul 1985) and is therefore suspected of affecting the composition of phytoplankton species in acidified waters. In laboratory experiments, adding aluminium to achieve levels typical of acidified lakes (<0.1 mg/l) produced severe damage in the form of deformation and impaired reproduction among a number of different species of algae (Hörnström *et al* 1984). The availability of humic substances in the water was shown to counteract the effect harmful effects of aluminium on several species.

Plants use carbon to build up their cells and they are therefore dependent upon a certain amount of a particular form of inorganic carbon in the water. When the pH of the water falls to <5.5, the source of carbon switches from carbonate to free carbon dioxide, at the same time as the total amount of carbon declines. This alters the conditions under which many plants are able to assimilate it. It is, however, not clear whether or not all plants can utilize carbon in the form of CO_2 via photosynthesis.

Acidification also has a number of indirect effects of a physical nature, such as changes in light conditions in the water which follow from a reduction in the shade provided by plankton. Large-scale properties such as temperature conditions may even be altered as a result of the way in which the heat-absorbing ability of the water is affected by the quantity of algae.

1.2. Liming

Certain purely mechanical effects take place in lakes immediately after liming, as a result of the suspended limestone particles, and these have physical consequences for the conditions for life in the open water. In addition to this, the addition of limestone particles has at least a short-term effect by reducing the amount of light for the benthic flora. One suggestion which has been put forward is that liming results in rapid release of microscopic algae from the bottom sediment, which may go some way in explaining the unexpected algal blooms which have been observed.

An unnaturally rapid chemical change in the aquatic environment also occurs, completely changing the conditions for vegetation. In addition to a change in the osmotic conditions and the impact this has on the process of chemical exchange through cell walls, for instance, there is a shift in the carbonic acid system. The plants which depend on carbon in the form of carbonate benefit from liming.

Aluminium and other metals have been thought to shift towards less toxic compounds as a result of the increase in pH (Wilander *et al*, Chap. 5), which can be expected to favour several plankton species in the pelagic system. Experiments using algae from Lake Gårdsjön have shown that aluminium is probably still toxic to plankton even after a lake has been limed (Pettersson unpubl.). Changes also occur in the sediment - changes in the form in which phosphorus occurs - as a result of increased mineralization after liming. This ought to affect the ability of rooted benthic vegetation to take up nutrients (Gahnström, Chap. 6).

Lake liming also has an impact on large-scale physical properties such as temperature conditions and the thermocline depth, as shown in Lake Gårdsjön (Larsson 1988, Alenäs *et al*, Chap. 14). In North American lakes like Woods Lake, liming has also resulted in a shift towards more clearly defined temperature layers (Bukaveckas 1989). The change in the thermocline depth is said to have resulted in an increase in the volume of the hypolimnion from 3 to 13% of the total volume of the lake (Bukaveckas 1991), which naturally has far-reaching biological consequences.

The changes among herbivorous fauna also have a major impact on the post-liming development of flora. This has been summarised by Degerman *et al* (Chap. 8) and probably mostly affects phytoplankton, since they are grazed upon by zooplankton.

2. Phytoplankton

2.1. Impact of acidification

Plankton netted from lakes in Risveden and Lygnersvider in western Sweden provide a clear indication that certain phytoplankton species disappeared between 1966 and 1983; at the same time, the pH fell by about one unit (Morling 1981, Morling and Willén 1990). The biomass, measured as chlorophyll α, remained unchanged, however. Using multivariate analysis, it has been possible to demonstrate clearly that the phytoplankton flora in the acidified lakes is drastically different from that in non-acidified lakes, but also that certain differences are geographical or climatic in nature (Willén *et al* 1990). One characteristic of naturally acidic lakes - at least of clear-water lakes - is the extremely impoverished species diversity among the phytoplankton communities. Johansson and Nyberg (1981) stated that there is a correlation between the number of species and the pH of a lake (Fig. 2). However, there is no correlation between total biomass and pH; sometimes the highest biomasses are to be found in the most acidic lakes (Johansson and Nyberg 1981).

There are two types of phytoplankton communities which commonly occur in acidified lakes. One of these is dominated by dinoflagellates and the other by cyanobacteria (Cyanophyceae), i.e. bacterioplankton. Typical phytoplankton communities in acidified lakes in western Sweden consist of dinoflagellates *(Peridinium inconspicuum* and *Gymnodium* cf. *uberrimum)* and golden algae (Chrysophyceae) of the genus *Dinobryon* (Almer *et al* 1978), while other lakes may be dominated by bacterioplankton of the spe-

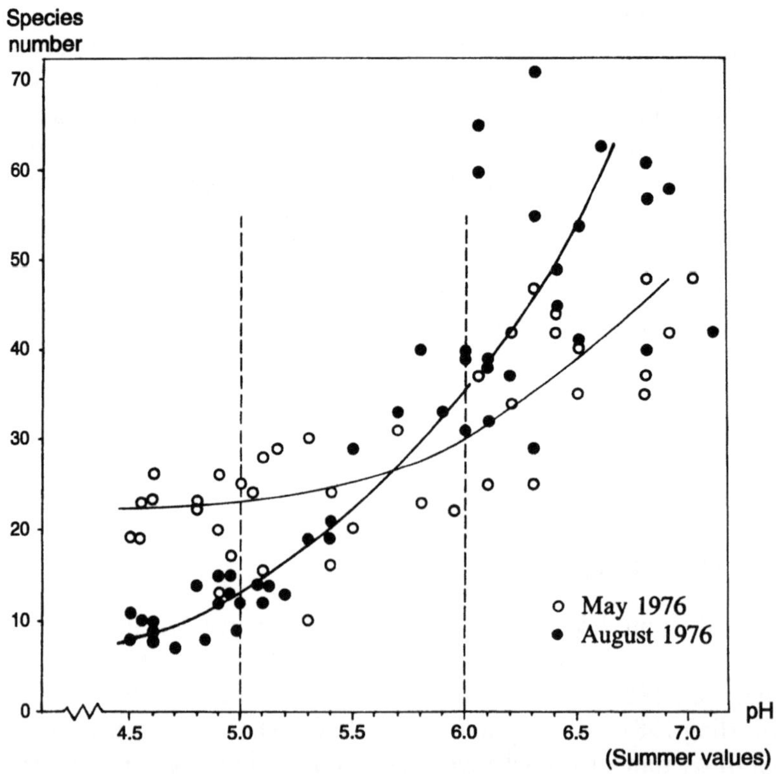

Figure 2. Number of observed phytoplankton species and lake pH in Swedish west coast lakes 1976 (from Eriksson et al 1982).

cies *Merismopedia tenuissima* (Hultberg and Andersson 1982, Andersson *et al* 1989). Bacterioplankton often dominate in certain types of acidified North American lakes (Molot *et al* 1990). According to Cronberg (1988), many acidic lakes in southern Sweden are dominated by *Gonyostomum semen*.

In studies of ecosystems in clear-water lakes in Canada, experimental acidification has provided evidence of changes among different phytoplankton groups similar to those found in synoptic studies (Findlay and Kasian 1990), which supports the hypothesis that it is acidification that causes these shifts among species. Findlay and Kasian (1991) showed, among other things, that artificial acidification with sulphuric acid to provide a pH of 5.3 meant that golden algae were replaced by dinoflagellates as the dominant group of algae. The bacterioplankton and diatoms (Bacillariophyceae) disappeared at pH 5, but the biomass remained unchanged.

The same has been demonstrated experimentally by adding sulphuric acid to mesocosmos. The dinoflagellate *Peridinium inconspicuum* replaced diatoms and green algae (Chlorophyceae) (Havens and DeCosta 1987). In these experiments, the total phytoplankton biomass increased as the pH fell. In a study in which a humic lake in Finland was artificially acidified (from pH 5.8 to 3.5), both the number of species and the biomass declined (Arvola *et al* 1986).

Little is known about the effect of acidification on primary production. The acido-oligotrophication hypothesis was put forward at an early stage (Grahn *et al* 1974), stating that acidification leads to lower production in the lakes mainly as a result of decreased availability of phosphorus for assimilation by flora as mineralization processes in the water cease. This hypothesis has not been verified by the few Scandinavian studies which have been performed and it has been questioned by Olsson and Pettersson (1993). Experiments in Canada involving the addition of acid have failed to support it, moreover - on the contrary - pelagic production increased over a six-year period after artificial acidification (Shearer *et al* 1987). In a study in which a humic lake in Finland was artificially acidified (from pH 5.8 to 3.5), production declined (Arvola *et al* 1986).

According to a synoptic study in the Sudbury region of Canada by Dillon *et al* (1979), primary production is not reduced by the fall in pH in itself, but by the availability of nutrients, particularly phosphorus. The lakes in the Sudbury study have high metal levels, however, which means that the toxic effects of such metals may have a bearing on the result and make the findings more difficult to interpret. In the Lake Gårdsjön area, however, it has been shown that the mobility of phosphorus is reduced in the surrounding soil as a result of soil acidification (Persson and Broberg 1985) and that this means that the supply of nutrients is restricted and thus indirectly reduces production in the lake.

One opinion which is generally held, however, is that the rate of biological turnover in the pelagic system declines in acidified lakes since the average size of phytoplankton in such lakes is often large. This is attributable to the dominance of large dinoflagellates with low specific production (Ramberg 1976). The acidified lakes are often dominated by dinoflagellates which have a long generation time (Malone 1980). Turnover is an expression of specific production, i.e. the production which each unit of weight of the plant can achieve. Opinions as to how the acidification process affects production conditions in lakes vary, therefore.

2.2. Number of species and composition of species after liming

The rapid changes in the post-liming physical and chemical properties of the water affect phytoplankton communities, both qualitatively through changes in the composition of species and quantitatively through changes in the density of individuals and the total biomass. Ultimately, the actual production conditions are also affected. Successive changes take place over a longer time span, moreover, in conjunction with changes in the entire food web as a result of the key role played by new generations of fish. The phytoplankton community probably attains "natural" stability only after a number of years with susequent liming (cf. Appelberg, Chap. 9).

2.2.1. Clear-water lakes

One general effect of liming is that the phytoplankton flora in severely acidified lakes, which are often dominated by dinoflagellates, becomes more diverse. The National Board of Fisheries and Swedish Environmental Protection Agency (1981) have issued a report on a number of lakes in different areas where the development of phytoplankton species was studied during the first and second years after liming. The findings show that the average number of species was more than twice that before treatment. Similar results were reported by Eriksson *et al* (1983) from three lakes in the Tiveden area, southern Sweden. The same study included Lake Stora Härsjön, south-western Sweden, where the increasing number of species was followed by a rising trophic index, indicating that the availability of nutrients improved after liming (Appelberg *et al* 1990, Alenäs *et al*, Chap. 14).

In Lake Lysevatten and in other lakes in south-western Sweden, the number of diatom species increased after liming, as did their total biomass (Hultberg and Andersson 1982). After sediment studies in Lysevatten, however, Renberg and Hellberg (1982) reported that the most common original diatom genus, *Cyclotella*, could not be found three years after liming. This genus had still not re-established itself in the lake when a new sediment study was performed twelve years later (Renberg and Hultberg 1992).

Even in the relatively large Lake Stora Holmevatten in south-western Sweden, the number of diatoms increased (including the genus *Synedra*) during the first year after liming, as did various golden algae (Hasselrot *et al* 1984). This particular genus has been observed in many lakes during the first years after liming (Larsson 1988, Bukaveckas 1991), which may be attributable to its release from epiphytic populations (cf. Sect. 4). A few years later, "new" species of golden algae also appeared in Lake Stora Holmevatten and yet another diatom species, *Cyclotella* sp., also became common. Cryptomonads constituted about half of the biomass in the autumn of 1982 and the spring of 1983, i.e. 3.5 years after the liming treatment. On the whole, however, the number of species did not increase noticeably in this lake. The main change was the shift between different groups of algae and the increase in the diversity index from 1.1 to 1.7 during the first year after treatment.

Alenäs (1986) reported an increase in species diversity in six lakes which were studied in the Härskogen district of south-western Sweden, with remarkable proliferation of the small *Cosmocladium perissum* (Desmidiales) in Lake Hornasjön during the fourth year after liming. Lake Tvärsjön, in the same area, was monitored closely both before and after liming (Hörnström *et al*, unpubl.) and it was found that the algal flora, after showing the depletion typical under acidified conditions, gradually became more natural. However, it took four years before the maximum number of species was observed, at between 43 and 60. Hörnström *et al* stated this to be a normal level of diversity in unacidified lakes.

Lake Gårdsjön is a typical nutrient-deficient clear-water lake, but differs from most other lakes in that it was totally devoid of fish when liming was undertaken, and that for experimental reasons it was kept fish-free for three years after the treatment (Alenäs *et al*, Chap. 14). A clear and successive increase in the number of species could be observed during the first three years after liming, followed by a stabilization in the fourth year (Larsson 1988). However, in the acidified reference lake, Lake Stora Hästevatten, which is in the same drainage area, the number of species was low throughout the period of the study and even displayed a shift towards more acidophilic algal flora in pace with the falling pH (Larsson, unpubl. data).

In the clear Finnish Lake Valkealampi, pH was 4.9 at the time of liming. The number of species tripled and the levels of chlorophyll and phosphorus also rose, an effect which persisted to some extent after 5.5 years (Niinioja *et al* 1990). The proportion of dinoflagellates and green algae declined after liming, while that of golden algae increased.

In Canada, positive trends have been reported after liming in Bowland Lake, Ontario, for instance (Keller *et al* 1990, Molot *et al* 1990) with a shift from dominance of cyanobacteria to a more natural composition with an increased proportion of golden algae.

In Trout Lake, USA, the development of phytoplankton was unstable during the first years after liming. Cryptomonads of the genus *Cryptomonas* sp. increased, at the expense of golden algae (Howell *et al* 1991) during the first years after liming, which meant that the proportion of edible algae available to zooplankton increased. A few years later, the proportion of green algae had increased, including desmids of the genera *Closterium* and *Spondylosium* (cf. Lake Gårdsjön above and Alenäs *et al,* Chap. 14).

A typical feature of many lakes where the impact of liming on phytoplankton has been studied is that they were severely acidified before liming treatment and that the observed impact was considerable. The larger and deeper southern Swedish lakes Stora Le (Henrikson *et al* 1989) and Unden (Lagerman 1983) were limed before acidification had progressed so far. In these cases, the change in composition and density of species was less dramatic. Norwegian studies in Lakes Langevatten and Hovvatn, which were moderately acidified (pH approximately 5.7) before they were limed, found that the fish and phytoplankton communities were still fairly diverse (Brettum and Hindar 1985). The treatment did not have any significant effect on the phytoplankton community in terms of the number and biomass of the species. The scope of the effects on phytoplankton thus depends on the degree to which the lakes were acidified when the initial treatment was performed, which means that a good strategy would be to lime in good time before the damage to the ecosystem becomes too extensive.

2.2.2. Humic lakes

Hörnström *et al* (1992) studied the effects of liming on a number of acidified lakes in mid-Sweden in the Åva area south of Stockholm and in certain lakes in the Delsbo area. In the most humic of these lakes, the number of phytoplankton species was higher than in many acidified clear-water lakes. The lakes which were most clearly damaged by acidification showed positive species trends and in several cases more trophic levels after treatment (lakes and wetlands). However, the less acidified lakes only underwent minor changes, or none at all, after liming. The most noticeable increase in the number of species was observed in the pH interval from 5 to 6. In some of the lakes, species which had previously been unknown became common, including the dinoflagellate *Peridinium goslaviense* and the green algae *Geminella* sp. As in certain lakes in western Sweden (Alenäs 1986, Larsson 1988) an episode was observed in Lake Mörtsjön during which a very small desmid, *Cosmarium pygmaeum*, proliferated (this species may possibly be identical to *Cosmocladium perissum*, which proliferated in Lakes Hornasjön and Gårdsjön (P. Holmquist, pers. comm.). There was a similarly remarkable proliferation of *Gonyostomum semen*, despite the fact that the lake in question was limed.

G. semen is a large single-cell phytoplankton from the group Raphidophyta, which is characteristic in humic lakes with moderately high phosphorus levels (Rosén 1981). It is thought to cause deterioration of the water quality by ejecting threads, trichocysts, which form a thick slime which may be toxic. People who bathe in lakes where this species blooms may complain of itching. It would also seem that such blooms are followed by depletion among zooplankton communities. *G. semen* was reported in Sweden for the first time in 1945 and has increased sharply in recent decades. In certain studies, proliferation of this large flagellate has been shown to be linked with low pH (Cronberg *et al* 1988). In more recent mesocosmos experiments (Gahnström *et al* 1990), this species has also been shown to increase if acid is added, but to decrease if lime is added.

Lake Stora Holmevatten may be described as being extremely deficient in nutrients (total phosphorus level below 5μg/l) and it is thus remarkable that in 1979, when it was still acidic, *G. semen* - a fairly demanding flagellate in terms of nutrients, one normally associated with higher phosphorus levels (Rosén 1981) - was quite dominant (Hasselrot *et al* 1984). When Lake Stora Holmevatten was limed, however, *G. semen* disappeared.

Salonen *et al* (1990) studied the effects of liming on the acidic humic Lake Pussijärvi in Finland, where the main change was that the number of bacterioplankton increased.

As an alternative to lime, soda (Na_2CO_3) has sometimes been injected. In Lake Galtsjön in southernmost Sweden, the diversity increased as a result. Like limestone, the addition of soda thus had a positive effect (Lindmark 1982). Another "positive" effect was that *G. semen*, which dominated prior to treatment, disappeared two years after the initial treatment. Instead, this lake was eventually dominated by golden algae (Lindmark 1984). In Lake Hjärtasjön, southernmost Sweden, the method has also been shown to have a positive impact on *G. semen* and other species (Lindmark 1991).

2.3. Biomass and primary production after liming

In Lake Stora Holmevatten (Hultberg and Andersson 1982) and Lake Lysevatten (Alenäs *et al* 1991), a "shock" effect was observed, with a reduction in the phytoplankton biomass immediately after liming. In Lake Gårdsjön, clear reductions in both biomass and chlorophyll were seen during the first month after liming (Larsson 1988). This shock was temporary, however, and the phytoplankton biomass recovered in both the Swedish and the North American lakes (e.g. Bukaveckas 1991) within a couple of months. The fact that this effect might be even greater in lakes treated with soda has been suggested as a reason for being restrictive with this method. This is contradicted, however, by the fact that one humic lake in southern Sweden, in which soda briquettes were distributed on the bottom, showed a clear rise in chlorophyll directly after the improvement in pH (Lindmark 1991).

In the longer term, the phytoplankton biomass probably increases (Hultberg and Andersson 1982, Alenäs 1986, Hörnström and Ekström 1986, Larsson 1988, Appelberg *et al* 1990). In Lakes Rammsjön and Ämten, southern Sweden, there was a clear increase in the phytoplankton biomass after liming, although fertilization was used to supplement the liming treatment (Andersson *et al* 1989). The diatom *Synedra* sp. constituted the majority of the biomass during the second year after liming.

Liming also resulted in an increase in the total phytoplankton biomass in certain lakes in mid-Sweden, while it declined or remained unchanged in several others (Hörnström *et al* 1992a). In Lake Galtsjön - treated with soda - the biomass increased during the first two years (Lindmark 1984).

A general trend in plankton size towards smaller species has been demonstrated in Lake Gårdsjön, for instance (Larsson 1988) and in Finnish lakes (Niinioja *et al* 1990). A similar trend was observed in Lake Galtsjön after treatment with soda (Lindmark 1984).

Woods Lake in the Adirondacks (USA) showed a significant increase in chlorophyll levels after liming, particularly in the summer, when the level was higher in the limed lake than in the control lake (Bukaveckas and Likens 1990). There was a strong positive correlation between pH and chlorophyll in this lake during the six-year study period (Bukaveckas 1991).

Drastic increases in biomass - i.e. algal blooms - have been observed in certain Swedish lakes a few years after initial liming treatments. Such lakes include Lake Gårdsjön (Larsson 1988) and Lake Hornasjön (Alenäs 1986). In Lake Gårdsjön, which was limed in 1982, the bloom commenced in the spring of 1985 and lasted throughout the summer and into late autumn. The total biomass in 1985 averaged 3.2 g wet weight per m^3, as opposed to 0.21 g wet weight per m^3 during the pre-liming years. The remarkable thing in both these lakes was that it was the same small desmid species, *Cosmocladium perissum*, which proliferated. This alga was previously unknown in clear-water algal blooms in Swedish lakes and was mainly considered to be an epiphytic species in humic waters (Thomasson, pers. comm.). In both lakes, the algal bloom was so severe and persistent that it had an impact in the form of oxygen deficiency in the deeper areas of the lakes (Henriksson, unpubl.). It is possible that the blooms may have been attributable to a surge of nutrients after the *Sphagnum* mosses had died. *C. perissum* also appeared in Lake Lysevatten a couple of years after the second liming treatment, but not to such a great extent (Andersson and Hultberg 1988). Other phytoplankton species appeared in great numbers in this lake after the second treatment, however, including the golden alga *Dinobryon sociale* var. *americanum*, with a total biomass of 12.400 g wet weight per m^3, in September 1987.

Algal blooms after liming in acidic humic lakes were noted as long ago as 1956 in Michigan in the USA (Waters 1956), when they were caused by cyanobacteria. A tendency towards massive blooms of the golden alga *Chrysochromulina* has also been observed in certain limed North American lakes (Bukaveckas 1989) and blooms were also noted in Lake Earnest after the pH had risen from 4.7 to 7.2 (Fairchild and Sherman 1990).

Primary production in the pelagic zone has only been measured in a few Swedish lakes before and after liming. The longest series of measurements is from Lake Gårdsjön, where it increased during the first four years after treatment and then fell to a level which was as low, if not lower, than it was before treatment (Larsson unpubl. data, Alenäs *et al*, Chap. 14). In the nearby Lake Lysevatten, the primary production was measured before the second treatment and a significant increase was observed (Andersson and Hultberg 1988).

In experimental mesocosmos in contact with the sediment, in the moderately humic Lake Stråken in southern Sweden, production declined after liming (Gahnström *et al* 1990). Primary production in a limed part of the River Fylleån system, south-western Sweden, was significantly lower than in an acidic reference water (Torkelsson and Åhren 1986). No pre--liming measurements have been reported in this case, however, so it is not possible to reach any conclusion as to how liming affected production.

In Lake Hecklan in mid-Sweden, where liming was supplemented by phosphorus and nitrogen, the average annual biomass increased between five and ten times (Olofsson *et al* 1988).

In Norwegian liming experiments, the production of phytoplankton declined briefly after liming (Brettum and Hindar 1985) but was unchanged over a longer period. Nor was there any change in production after liming in the clear Lake Valkealamppi in Finland (Salonen *et al* 1990). Reports from North America are available from lakes in the Sudbury region, which suffer from metal pollution. In these cases, liming led to an increase in phytoplankton production (Scheider *et al* 1975).

3. Epiphytic algae

3.1. Impact of acidification

The epiphytic flora (periphyton) in lakes and watercourses is probably severely affected by acidification (Almer *et al* 1978, Stokes 1981). In general, the benthic algal communities in acidified lakes display a high total biomass, but low diversity. In acidified watercourses, an increase in filamentous green algae has been observed along with a decrease in blue-green algae (Herrmann *et al* 1993).

A characteristic felt carpet may be found in severely acidified lakes, consisting mainly of filamentous epiphytic algae. In Lake Gårdsjön and Stora Hästevatten, this carpet consisted mainly of filamentous bacterioplankton of the *Lyngbya* and *Oscillatoria* genera, together with green algae (Lazarek 1982). There are also reports that filamentous green algae have increased dramatically as a result of acidification in watercourses in Norway (Hendrey 1976).

The filamentous green alga *Mougeotia* sp. is thought to increase in conjunction with acidification (Morling 1981, Hultberg and Andersson 1982, Alenäs *et al* 1991). Mesocosmos experiments, in which *Mougeotia viridis* increased sharply when sulphuric acid was added to reduce the pH, have also shown that it thrives at low pH (Havens and DeCosta 1987). The fact that this particular filamentous green alga is favoured by acidification is thought to be attributable to its being unusually good at utilizing carbon dioxide as its source of carbon. Large concentrations of *Mougeotia* sp. have been documented in one of the Lygnersvider lakes (Morling 1981) and in conjunction with re-acidification of previously limed lakes such as Lysevatten (Hultberg and Andersson 1982, Alenäs *et al* 1991).

The Scandinavian assumption (Alenäs *et al* 1991) that filamentous blue-green cyanobacteria benefit from acidification is contradicted by the results of artificial acidification of Canadian lakes (Fairchild and Sherman 1990), where they were discouraged, while the filamentous green algae benefited. Fairchild and Sherman (1990) stressed that changes in grazing by invertebrates influences the situation, although it would appear to be carbon dioxide that limits actual production.

Measurements in Lake Gårdsjön showed that primary production of the epiphytic complex on water lobelia *(Lobelia dortmanna)*, particularly in terms of *Eunotia* sp., was high. In the carpet between the plants, however, it was low and consisted of the green algae *Mougeotia* and *Binuclearia* (Lazarek 1985).

3.2. Number and composition of species

Less research has been done on the effects of liming on epiphytic flora in Sweden than on the effects on phytoplankton. The National Board of Fisheries and Swedish Environmental Protection Agency (1981) stated that the epiphytic communities in limed waters were no different from those in acidified waters.

In Lake Gårdsjön, however, detailed studies of certain epiphytic components were performed during the first years after treatment in 1982. Prior to the initial treatment, when the pH was about 4.7, large areas of the soft lake bed were covered by a felt carpet consisting of filamentous cyanobacteria with epiphytic acidophilic diatoms, *Eunotia veneris* and *Eunotia rhomboidea* (Lazarek 1982). Major changes could already be observed one month after liming, when the chemically more indifferent diatom *Achnanthes minutissima* was the dominant epiphyte on water lobelia (Lazarek 1986). The proportion of epiphytic desmids also increased after liming in Lake Gårdsjön, as well as in Lake Högsjön, south-western Sweden (Lazarek 1982). The felt carpet on the bed of Lake Gårdsjön gradually disintegrated after liming and disappeared completely by 1985. In several places, however, it became re-established, although the lake was no longer acidified (Sangfors 1991). The reasons for this are not clear, but one hypothesis is that the fish population which was introduced in 1985 decimated the small grazing fauna such as caddis flies (Trichoptera) and water louse *(Asellus aquaticus)*, allowing unrestricted growth of the filamentous algae. In Lake Bredvatten, south-western Sweden, the carpet which covered large areas of the lake bed during the acidified phase also disappeared completely after liming (Hultberg and Andersson 1982).

Fritzon (1984) studied epiphytes in the humic River Fylleån, which was limed using dosers. The number of species increased somewhat in the ensuing years but the diversity index was unchanged. Certain algae which are typical in acidified conditions (*Mougeotia*, for instance) declined. Other genera, such as *Synedra* (diatoms) and *Leptothrix* (iron bacteria) were not significantly affected by the treatment, however. Analyses also showed that the proportion of species typical in nutrient-deficient waters declined, indicating that the trophic level had been raised - i.e. that the nutrient situation had probably improved - while the situation upstream remained basically unchanged (Fritzon 1984). Bengtsson (1987) studied the flora in a number of limed watercourses in southern Sweden. Among the species appearing in these waters were *Batrachospermum* sp., a red alga which was also found after Lake Rävekärrs Långevatten in western Sweden had been limed (Grönlund unpubl.).

3.3. Biomass and abundance

In Lake Lysevatten, which was deliberately re-acidified after liming in the mid-1970s, the biomass of *Mougeotia* sp. was measured prior to the second liming treatment at more than 10 g dry weight per m^2. After the second treatment in 1986, the biomass declined to less than 1 g dry weight per m^2 (Alenäs *et al* 1991). Assimilation measurements showed that the optimum pH for photosynthesis was 4.8, when assimilation was five times higher that at pH 5.5 (Alenäs *et al* 1991).

In the USA, the effects of liming on epiphytic communities in a shallow lake, Lake Earnest in Pennsylvania, were studied by Fairchild and Sherman (1990). Prior to liming, the lake was severely acidified (pH 4.7) but afterwards the value rose to above 7 and the total phosphorus level doubled. *Mougeotia* sp. declined sharply from more than 99% to less than

1% of the total epiphytic biomass as a result of the treatment. The filamentous green alga *Oedogonium* increased, however, to account for about half of the biomass. In addition to the far-reaching qualitative changes, there was also a clear increase in chlorophyll levels, although the total biomass of epiphytes declined somewhat. Despite the large difference in pH before and after liming, there were no major changes in the grazing community, which meant that the differences were probably attributable to chemical and physical changes. In a series of experiments in Lake Earnest, nitrogen, phosphorus and carbon were added, either separately or in combination, and it was possible to show that nutrient supplements led to a further increase in chlorophyll levels. The main difference, however, was that liming meant that the epiphytic flora shifted from types limited by carbon to types limited by phosphorus/nitrogen.

In Bowland Lake in Ontario, Canada, *Mougeotia* sp. also declined after liming and was replaced by the green alga *Oedogonium* sp. (Jackson *et al* 1990). This lake was experimentally re-acidified, resulting in immediate re-establishment of *Mougeotia* sp. (Keller *et al* 1990).

4. Macrophytes

4.1. Impact of acidification

It was observed at an early stage in the history of acidification research that the macrophyte communities in acidic lakes in Sweden differed from those in non-acidic ones. Certain researchers linked this to acidification (Grahn 1977, Almer *et al* 1978). An invasion of *Sphagnum* mosses has been observed on the shallow beds of many acidified lakes (e.g. Grahn 1977, Alenäs *et al* 1991) and sediment studies indicated that these mosses have increased in certain lakes since the 1950s (Renberg and Wallin 1985). In Finland too, *Sphagnum* is thought to invade lakes as a result of acidification. This is the only clearly acidophilic organism found in a study of 71 lakes (Heitto 1990). The hypothesis that *Sphagnum* invades lakes as part of the process of acidification has come under sharp criticism by Vöge (1989) and others, who feel that the Swedish conclusions lack scientific foundation. *Sphagnum* invasions may mean that isoetids such as water lobelia and quillwort are driven out of competition.

Bulbous rush *(Juncus bulbosus)* and bladderwort *(Utricularia* sp.) may also be favoured by acidification. In Lake Änkatjärn on Lygnersvider, bulbous rush was studied closely by Svedäng (1990), who put forward the hypothesis that the species is competitive thanks to the fact that it can utilize the carbon dioxide which is present in high concentrations beneath the spring ice. Norwegian reports (Nilssen 1980) also believe that bulbous rush is favoured by acidification. Damage to the root system of quillwort has been reported from acidified lakes in Norway, while the parts of the plant above the sediment surface were apparently unaffected (Vöge 1989).

In the Netherlands, Roelefs (1983) felt that extreme cases of acidification may lead to the loss of all underwater vegetation, at least in acidified waters where the carbon dioxide level is below 100 μmol/l.

The general view when it comes to benthic vegetation is that the proportion of such vegetation to the total lake production increases with acidification, mainly as a result of the improved light situation on the lake bed (Howell *et al* 1991). It is not clear how the turnover for the entire benthic flora community is affected, however. In the acidified Lake Gårdsjön, for instance, the specific production of quillwort would appear to be no different from that in other lakes (Grahn 1985).

4.2. Number and composition of species

Eriksson *et al* (1982) reported that liming had a definite impact on the macrophyte population, enabling pondweeds *(Potamogeton* spp.) to re-establish themselves and reducing *Sphagnum subsecundum*. In Lake Västra Skälsjön, southern mid-Sweden, the number of species increased marginally - only one moss, *Drepanocladus trichophyllus*, which benefits from lime, recolonised certain lake-bed areas (Eriksson 1988).

In Lake Gårdsjön, new macrophyte species also re-established themselves after liming, but the increase was only small, from 17 species before liming to 19 ten years after the initial treatment (Sangfors 1991). As in many other limed lakes, the *Sphagnum* mosses disappeared in a short time. One species, intermediate bladderwort *(Utricularia intermedia)* disappeared completely during the first years after liming, but reappeared four years later. It is worth noting that a phenotype of bulbous rush previously unobserved in the lake, *Juncus bulbosus* f. *submersus*, was found on the more shallow areas in 1985, three years after the initial treatment.

Post-liming recolonization by isoetids of areas previously covered by *Sphagnum* has been reported by Hultberg and Andersson (1982). In Lake Gårdsjön, however, no particular recolonization could be observed after repeated treatments during a decade, despite the fact that the isoetids grew among the *Sphagnum* on the lake bed before liming (Sangfors 1991).

Norwegian studies of Lake Kjerrdalsvattnet showed that quillwort was favoured by liming, since the root system recovered from acidification damage and became more normal (Vöge 1989). This could be linked to an increase in the pH of the interstitial water after a certain period. Damage to the leaf surfaces of water lobelia could be observed, however, probably as a result of corrosion by limestone particles which had settled directly on them. Finnish studies have shown that the abundance of *Juncus* sp. declined drastically after the pH rose from 4.4 to 6.3 (Aulio 1987).

Colonization by species of pondweed after liming was observed both in Thrush Lake (Wright *et al* 1991) and in Woods Lake (Bukaveckas and Likens 1990), USA. In Thrush Lake, which was mildly acidic before liming, two more species appeared: lesser pondweed *(Potamogeton pusillus)* and slender naiad *(Najas flexilis)*, involving a clear increase in the macrophyte biomass in the 2-5 m depth interval (Wright *et al* 1991). *Sphagnum* cf. *platyphyllum*, which grows at a depth of about 4-10 m in this lake, was not particularly affected by liming, however, which shows that certain species of *Sphagnum* can also survive in more alkaline waters. In Woods Lake, where the pH rose somewhat more, the number of macrophyte species increased by five. The pondweed species *Potamogeton epihydrus* was the most successful of these. In total, the biomass of macrophytes declined by virtue of the disappearance of *Sphagnum* from the lake beds. In Bowland Lake, there were no measurable differences in macrophytes after liming (Jackson *et al* 1990) and the macrophyte flora was affected only marginally in other Canadian lakes (Keller *et al* 1990).

4.3. Abundance and primary production

Eriksson (1988) has provided a detailed report of changes in the range of macrophytes in five limed lakes in southern Sweden. In four of these lakes, the treatment was directly in the lake, while in the fifth, Mörtsjön, the effect was obtained from another limed lake upstream. In directly limed areas, the *Sphagnum* disappeared after the first treatment, and receded in other areas. In Västra Skälsjön, the total coverage of vegetation declined from 17.4 to 9.3%, since the *Sphagnum* coverage was reduced from 11.9 to 2.6% in 1980 and finally disappeared in 1985 (Fig. 3). Broad-leaved pondweed *(Potamogeton natans)* and alternate water milfoil *(Myriophyllum alterniflorum)* became more widespread, with denser populations. After three treatments, clear changes in the composition of macrophytes could thus be observed and new species appeared.

In Lake Lysevatten, the *Sphagnum* was also eliminated from littoral lake beds after liming, but it survived in deeper areas (3-10 m) although the growth declined from 8-10 cm/year to less than 1 cm/year (Hultberg and Andersson 1982). After subsequent liming of the same lake in 1986, the biomass fell from a total of 179 to 8 kg dry weight in the lake (Alenäs *et al* 1991). In Lake Gårdsjön, the biomass of the dominant macrophyte, *Sphagnum subsecundum*, was estimated to have declined from a total of 3,000 to 10 kg after liming (Grahn and Sangfors 1988).

The extensive decline in the *Sphagnum* population in limed lakes is probably one of the more important ecological changes which follow liming. In Lake Gårdsjön, the overall decline of *Sphagnum* on large parts of the shallow lake bed means that the total primary production of macro-

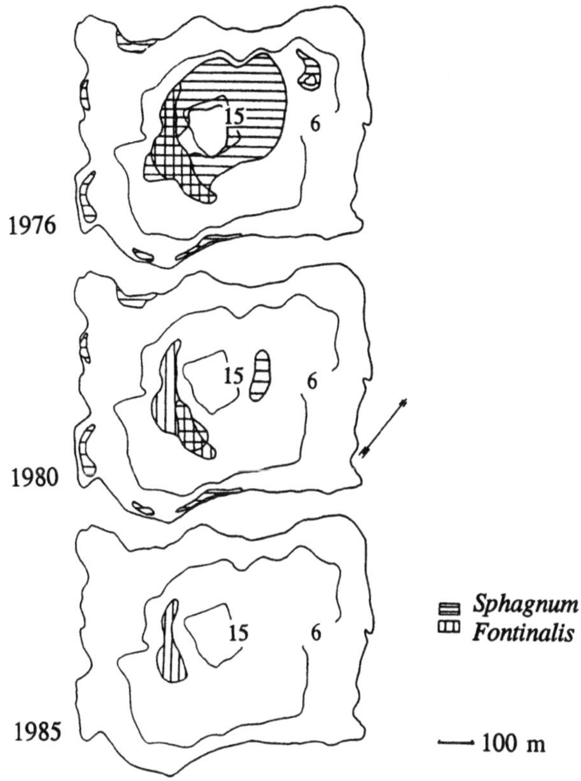

Figure 3. Distribution of Sphagnum spp. and Fontinalis antipyretica before and after liming 1976 in Lake Västra Skälsjön, southern mid-Sweden (from Eriksson et al 1983).

phytes has been estimated at 70% three years after treatment (Fig. 4).

The decline in leaf turnover among water lobelia in Lake Gårdsjön was not statistically significant, although for quillwort it was. The exceptional conditions prevailing in 1985, when the massive algal bloom lasted throughout the year, shielded the isoetids from the light and may explain the reduction in turnover. This also led to large numbers of adventitious shoots from quillwort and water lilies (Grahn and Sangfors 1988).

Liming may involve undesirable effects in the form of the proliferation of alternate water milfoil which has been observed in certain lakes in western Sweden (e.g. Grönlund 1987, Grönlund, unpubl.). Lake Rävekärrs Långevatten (maximum depth 3.5 m) has been limed on a regular basis since 1961, when the pH was 4.2. After the mid-1980s, however, the lake was more or less choked by water milfoil which had probably become established during the late 1970s (Grönlund, pers. comm.). This species grew vigorously from 1982 to 1987, mainly on the soft lake-bed areas which had been limed, which meant that in 1987 it covered approximately 42% of the entire lake surface (Fig. 5). One hypothesis is that the liming

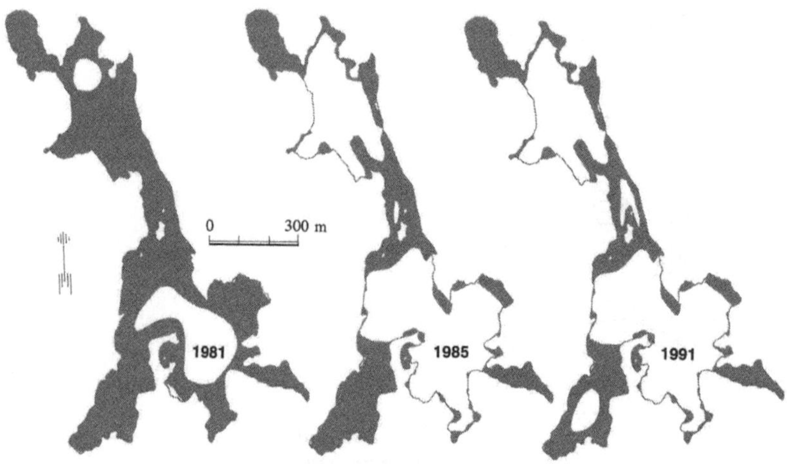

Figure 4. Coverage of macrophytes in Lake Gårdsjön, south-western Sweden, in 1981, i.e. one year before liming (pH 4.7), and three and nine years after liming, respectively (pH 6.5-7.5) (Sangfors 1991).

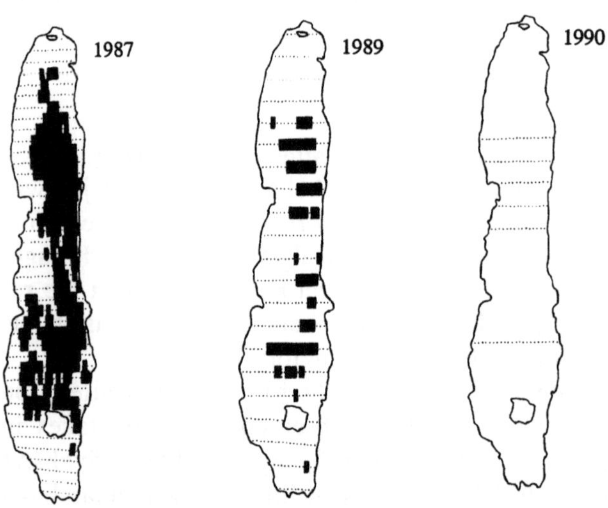

Figure 5. The development of the water milfoil Myriophyllum alterniflorum in Lake Rävekärrs Långevatten, south-western Sweden, after reacidification. The lake was regularly limed since 1961, but allowed to reacidify 1989-1990 (Grönlund unpubl. data).

of the shallow areas increased the availability of phosphorus for the root system of the water milfoil, thus promoting vigorous growth. Two other elodeids, Canadian waterweed *(Elodea canadensis)* and broad-leaved pondweed, also became more widespread. This was considered to be so undesirable that action was taken to re-acidify the lake and the alternate water milfoil disappeared. It is worth noting that large numbers of the red alga *Batrachospermum* sp. appeared during the re-acidification phase in 1989-90. Proliferation of alternate water milfoil to the detriment of leisure fishing has also been reported from other Swedish lakes (Grönlund, pers. comm.).

Soda treatment of Galtsjön had an impact on the macrophyte population similar to that observed in lakes treated with limestone: the *Sphagnum* was replaced by water lobelia and quillwort (Lindmark 1982, 1984).

5. Wetland vegetation

5.1. Impact of acidification

The Scandinavian wetlands have mainly arisen as a result of lakes becoming choked by vegetation. The peat which is thus formed provides the wetland with certain specific properties such as a capacity for ion exchange. The environment beneath the *Sphagnum* mosses is also extremely oxygen-deficient, which means that the top 30-40 cm offer special conditions for vegetation. The acidic environment is one effect of the ion exchange capability and means that seeds from vascular plants such as spruce, for example, have difficulty in taking root in the boglands.

In addition to its special chemical properties, *Sphagnum* retains water extremely well. This means that environments where it becomes established are affected in purely hydrological terms - the micro-environment is kept damp and the flow direction is affected in areas where *Sphagnum* grows.

Very little is known about the way in which anthropogenic acidification affects wetlands.

5.2. Number and composition of species

Liming of wetlands is becoming increasingly common, since it is an effective strategy for running waters (Degerman *et al*, Chap. 8, Wilander *et al*, Chap. 5). It is only recently that the effects on wetland vegetation have been studied, however. In a survey of a large amount of material from the whole of Sweden, Aronsson (1990) found that both *Sphagnum* and liverwort (Hepaticae) have been severely damaged in many limed wetland areas. In certain types of wetland, these species died out completely, requiring an extremely long time to recover. In drier places, where the

Sphagnum formed high tufts, it only died in patches, although it disappeared completely in many of the wetter sites, even at low doses. The mosses were not affected by liming, however.

In the discharge areas - i.e. those closest to the watercourses, where the groundwater comes to the surface - the lime is often distributed by virtue of the effective rise in the pH of the drainage water. As a rule, major damage to vegetation arises (Rafstedt 1991). Figure 6 shows quite clearly that even low doses can cause severe damage in certain areas. Such damage was, unexpectedly, classed as less severe in areas which had received higher doses of up to 50 t/ha.

Eriksson (1990) has examined the effects of liming on wetland vegetation on Fulufjället in mid-Sweden. This area was limed in 1977, and *Sphagnum* had died out completely after only one year. There had been some recovery after a further five years, at least in areas with good water turnover. In 1988 - more than ten years after liming - the ground layer was still seriously damaged, however, at least on sites with stagnant water. Clear damage to vegetation could still be observed twelve years after the treatment (Rafstedt 1991). In some places, the damage was less dramatic, although *Sphagnum* did decline during the first year (Eriksson 1988). After four years, it had recovered in places where it had not died completely, and

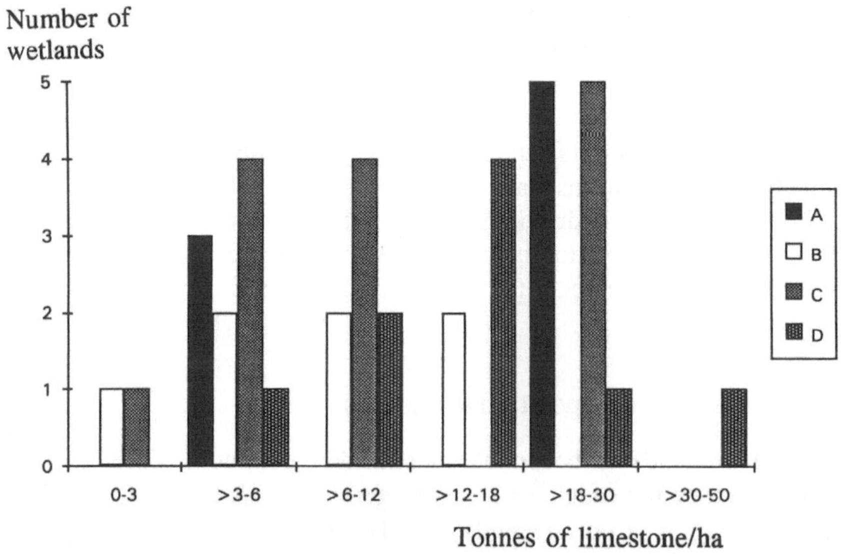

***Figure 6**. Wetland vegetation damage after liming in southern Sweden. A) no visible damage, B) visible damage in spotty patches, C) damage to continuous areas intermixed with areas without any visible damage, D) severe damage to large areas; the whole area affected (from Aronsson 1990).*

on other areas it was replaced by *Polytrichum* mosses. Reindeer mosses *(Cladonia* spp.) disappeared from the entire limed moor area, however.

Porcella *et al* (1991) studied the effects of liming on wetlands around Woods Lake in the USA and found a general transition to more neutral conditions. Great fen sedge *(Cladium mariscoides*, a sedge which is considered to be an alkaline-loving species) increased, for instance, while the oblong-leaved sundew *(Drosera intermedia)* declined, while at the same time the pH increased to 7.0 in the substrate, as opposed to 5.25 in the unlimed control areas. Damage to *Sphagnum* was still evident three years after liming around Loch Fleet in Scotland, while other mosses and vascular plants were unaffected (Clymo *et al* 1992).

5.3. Biomass and density

A number of studies on the effects of liming on the biomass and density of wetland vegetation have been performed in southern Sweden. On Herrestadsfjället, three different areas were studied one year after a treatment of 6-8 t/ha. It was observed that *Sphagnum* mosses in all three areas had almost completely died out, while at the same time certain vascular plants had increased in number (Appelquist unpubl.). Similar results were obtained in the county of Jönköping, where the dose was twice as high at 15-20 t/ha (Fasth, unpubl. data). In this case too, certain species of sedge were favoured, e.g. bottle sedge *(Carex rostrata)* and panic grass *(C. panicea)*, while *Sphagnum* mosses had completely disappeared from seven out of eight of the areas studied after four years.

5.4. Potential long-term effects

The long-term effects of wetland liming may be considerable as a result of the severe impact upon *Sphagnum* and other elements of the flora. The results depend largely upon the type of wetland to be limed. Two main trends can be discerned: (1) in the wettest wetlands, smaller tarns are formed in many cases as a result of mineralization of the *Sphagnum*, which is dissolved in the water phase in the form of humic acids. Little research has been done on the effects of this humification in terms of water chemistry and the biological consequences are unknown; (2) in many drier wetlands, the consequences will probably be quite the opposite to some extent, with an increase in higher vegetation in the form of dwarf shrub, bush and tree layers. This has already been observed in certain cases in southern Sweden (Aronsson, pers. comm.).

6. References

Alenäs, I. 1986. Kalkningsprojektet Härskogen 1976-1986. [The Härskogen liming project 1976-1986.] - Swedish Environm. Res. Inst., B 846.

Alenäs. I., Andersson, I., Hultberg, H. and Rosemarin, A. 1991. Liming and reacidification reactions of a forest lake ecosystem, Lake Lysevatten, in SW Sweden. - Water Air Soil Poll. 59: 55-77.

Almer, B., Dickson, W., Ekström, C. and Hörnström, E. 1978. Sulphur pollution and the aquatic environment. - In: Nriago, J.O. (Ed.) Sulphur in the environment. Part II. Wiley & Sons, New York, pp. 271-311.

Andersson, I. and Hultberg, H. 1988. Lake Lysevatten - a study of acidification and liming effects in a forest lake ecosystem in southwestern Sweden. - Swedish Environm. Res. Inst., B1077.

Andersson, P., Grahn, P., Hörnström, E., Nyberg, P. and Dahlquist, K. 1989. Kalkning och gödsling i Rammsjön och Ämten, Örebro län 1980-1987. Ett försök att minska kvicksilverhalten i fisk. [Liming and fertilization of Lake Rammsjön and Lake Ämten, county of Örebro, 1980-1987. An attempt to decrease mercury levels in fish.] - Swedish Environm. Prot. Agency 3584.

Appelberg, M., Ekström, C. and Hörnström, E. 1990. Stora Härsjön - ett exempel på integrerad uppföljning av kalkningens effekter. [English summary: Lake Stora Härsjön - a case study for integrated monitoring of the effects of liming.] - Inform. Inst. Freshw. Res., Drottningholm (1).

Aronsson, J.-A. 1990. Våtmarkskalkning. Förändringar på miljö och vegetation. [Wetland liming. Changes of environment and vegetation.] - Swedish Environm. Prot. Agency 3827.

Arvola, L., Salonen, K., Bergström, I., Heinänen, A. and Ojala, A. 1986. Effects of experimental acidification on phyto-, bacterio- and zooplankton enclosures of a highly humic lake. - Int. Rev. ges. Hydrobiol. Syst. Beih. 71: 737-758.

Aulio, K. 1987. Rapid decline of the mass occurrence of *Juncus bulbosus* in a deacidified freshwater. - Memo. Soc. Fauna Flora Fenn. 63: 41-44.

Bengtsson, R. 1987. Biologisk undersökning av några försurade sjöar i Kronobergs län - påväxtalger och bottenfauna. [Biological study on some acidified lakes in the county of Kronoberg - benthic algae and bottom fauna.] - Swedish Environm. Res. Inst., 11/90-586-85.

Blomquist, P., Olsson, H., Olofsson, H. and Broberg O. 1989. Enclosure experiments with low-dose additions of phosphorus and nitrogen in the acidified lake Njupfatet, central Sweden. - Int. Rev. ges. Hydrobiol. 74: 611-631.

Blomquist, P., Bell, R.T., Olofsson, H., Stensdotter, U. and Vrede, K. 1993. Pelagic ecosystem responses to nutrient additions in acidified and limed lakes in Sweden. - Ambio 22: 283-289.

Brettum, P. and Hindar, A. 1985. Effects of lime treatment on the biological system. - In: Baalsrud, K., Hindar, A., Johannessen, M. and Matsow, D. (Eds.) Liming of acid water. Liming project, final report. Dep. Environm. Direct. Nat. Managem., Norway, pp. 79-108.

Bukaveckas, P.A. 1989. Effects of calcite treatment on primary producers in acidified Adirondack lakes. II. Short-term response by phytoplankton communities. - Can. J. Fish. Aquat. Sci. 46: 352-359.

Bukaveckas, P.A. 1991. Effects of lake and watershed liming on light attenuation, thermal stratification, and phytoplankton production. - In: Olem, H., Schreiber, R.K., Brocksen, R.W. and Porcella, D.B. (Eds.) International lake and watershed liming practices. The Terrene Inst., Washington D.C., USA, pp. 129-144.

Bukaveckas, P.A. and Driscoll, C. 1991. Effects of whole-lake base addition on the optical properties of three clearwater acidic lakes. - Can. J. Fish. Aquat. Sci. 48: 1030-1040.

Bukaveckas, P.A. and Likens, G. 1990. Effects of lake and watershed liming on lake productivity and trophic interactions. - Ann. Prog. Rep. Natl. Fish Res. Center, US Fish Wildl. Serv., Kearneysville, USA.

Clymo, R.S., Foster, G.N., McKay, J., Robertson, J., Shore, R. and Skidmore, D.I. 1992. Terrestrial biology in limed catchments. - In: Howells, G. and Dalziel, T.R.K. (Eds.) Restoring acid waters: Loch Fleet 1984-1990. Elsevier Appl. Sci., London, pp. 331-361.

Cronberg, G. 1988. Bioassayförsök med *Gonyostomum semen* in situ i Östersjön, nordvästra Skåne, juli-augusti 1988 [Bio-assay trials with *Gonyostomum semen* in situ in the Baltic, north-western Skåne, July-Agust 1988]. - Research Applic.

Cronberg, G., Lindmark, G. and Björk, S. 1988. Mass development of the flagellate *Gonyostomum semen* (Raphidophyta) in Swedish forest lakes - an effect of acidification? - Hydrobiologia 161: 217-236.

Dillon, P.J., Yan, N.D., Scheider, W.A. and Conroy, N. 1979. Acid lakes in Ontario, Canada: characterization, extent and response to base and nutrient addition. - Arch. Hydrobiol. Beh. 13: 317-335.

Eriksson, F. 1988. Makrofytvegetation i kalkade sjöar. - [Macrophyte vegetation in limed lakes.] - Inform. Inst. Freshw. Res., Drottningholm (9).

Eriksson, F. 1991. Vegetationsförändringar inom kalkade områden på Fulufjället. [Vegetational changes in limed areas in the Fulufjället district.] - Nat. Board Fish., Rep. 1, 1991, pp. 34-35.

Eriksson, F., Hörnström, E., Mossberg, P. and Nyberg, P. 1982. Ekologiska effekter av kalkning i försurade sjöar och vattendrag. [English summary: Ecological effects of lime treatment in acidified surface waters.] - Inform. Inst. Freshw. Res., Drottningholm (6).

Eriksson, F., Hörnström, E., Mossberg, P. and Nyberg, P. 1983. Ecological effects of lime treatment of acidified lakes and rivers in Sweden. - Hydrobiologia 101: 145-164.

Fairchild, W.G. and Sherman, J.W. 1990. Effects of liming on nutrient limitation of epilithic algae in an acid lake. - Water Air Soil Poll. 52: 133-147.

Findlay, D.L. and Kasian, S.E.M. 1990. Phytoplankton communities of lakes experimentally acidified with sulfuric and nitric acids. - Can. J. Fish. Aquat. Sci. 47: 1378-1386.

Findlay, D.L. and Kasian, S.E.M. 1991. Response of a phytoplankton community to controlled partial recovery from experimental acidification. - Can. J. Fish. Aquat. Sci. 48: 1022-1029.

Fritzon, A. 1984. Fylleån - Biologiska effekter vid kalkning av en försurad å. - Slutrapport Fylleå-projektet. [The River Fylleån - biological effects of liming of an acidified river - final report from the the River Fylleå project.] - Dep. Limnol. Univ. Lund, unpubl. rep.

Gahnström, G., Granéli, W. and Tranvik, L. 1990. Effekter av försurning och kalkning på mikrobiella processer i pelagial och sediment. Experimentella studier. [Effects of acidification and liming on microbial processes in the pelagial and the sediment. Experimental studies.] - Dep. Ecology/Limnology, Univ. Lund, unpubl. rep.

Grahn, O. 1977. Macrophyte succession in Swedish lakes caused by deposition of airborne acid substances. - Water Air Soil Poll. 7: 295-305.

Grahn, O. 1985. Macrophyte biomass and production in Lake Gårdsjön - an acidified clearwater lake in SW Sweden. - Ecol. Bull. (Stockholm) 37: 203-212.

Grahn, O. and Sangfors, O. 1988. A comparative study of macrophytes in Lake Gårdsjön, during acid and limed conditions. - In: Dickson, W. (Ed.) Liming of Lake Gårdsjön, an acidified lake in SW Sweden. Swedish Environm. Prot. Agency 3426, pp. 281-308.

Grahn, O., Hultberg, H. and Landner, L. 1974. Oligotrophication - a self-accelerating process in lakes subjected to excessive supply of acid substances. - Ambio 3: 93-94.

Grönlund, T. 1987. Igenväxningsproblem med hårslinga - en kalkningseffekt? [Bottoms overgrown by water-milfoil - an effect of liming?] - Fiskevård (3): 28-29.

Hasselrot, B., Andersson, B.I. and Hultberg, H. 1984. Ecosysytem shifts after the re-

introduction of Arctic char (*Salvelinus salvelinus*) after liming of a strongly acidified lake in sothwestern Sweden. - Rep. Inst. Freshw. Res., Drottningholm 61: 78-92.
Havens, K.E. and DeCosta, J. 1987. Freshwater plankton community succession during experimental acidification. - Arch. Hydrobiol. 111: 37-65.
Heitto, L. 1990. Macrophytes in Finnish forest lakes and possible effects of airborne acidification. - In: Kauppi, P., Anttila, P. and Kenttämies, K. (Eds.) Acidification in Finland. Springer, Berlin Heidelberg New York, pp. 963-983.
Hendrey, G.R. 1976. Effects of pH on the growth of perihytic algae in artificial stream channels. - SNSF-project, Internat. Rep. 25. Ås, Norway.
Henrikson, L., Larsson, S., Nyman, H.G. and Oscarson, H.G. 1989. Stora Le före och efter kalkning. [Lake Stora Le before and after liming.] - Nat. Board Fish. unpubl. rep.
Herrmann, J., Degerman, E., Gerhardt, A., Johansson, C., Lingdell, P.-E. and Muniz, I. 1993. Acid-stress effects on stream biology. - Ambio 22: 298-307.
Hörnström, E. and Ekström, C. 1986. Acidification and liming effects on phyto- and zooplankton in some Swedish west coast lakes. - Swedish Environm. Prot. Agency 1863.
Hörnström, E., Ekström, C. and Duraini, M.O. 1984. Effects of pH and different levels of aluminium on lake plankton in the Swedish west coast area. - Rep. Inst. Freshw. Res., Drottningholm 61: 115-127.
Hörnström, E., Ekström, C. and Andersson, P. 1992a. Tio mellansvenska sjöar. Kalkningseffekter på plankton och vattenkemi. [English summary: Ten lakes in central Sweden, liming effects on plankton and water chemistry.] - Swedish Environm. Prot. Agency 4048.
Hörnström, E., Ekström, C., Fröberg, E. and Ek, J. 1992b. Plankton and chemical-physical development in six swedish west coast lakes, under acidic and limed conditions. - Can. J. Fish. Aquat. Sci. 50: 688-702.
Howell, E.T., Coker, G., Booth, G.M., Keller, W., Neary, B., Nicholls, K.B., Tomassini, F.D., Yan, N., Gunn, J.M. and Reitveld, H. 1991. Ecosystem responses of pH 5.9 of lake trout to whole lake liming. - In: Olem, H., Schreiber, R.K., Brocksen, R.W. and Porcella, D.B. (Eds.) International lake and watershed liming practices. The Terrene Inst., Washington D.C., USA, pp. 61-96.
Hultberg, H. and Andersson, I. 1982. Liming of acidified lakes - induced long-term changes. - Water Air Soil Poll. 18: 333-342.
Jackson, M.B., Vandermeer, E.M., Lester, N., Booth, J.A., Molot, L. and Gray, I.M. 1990. Effects of neutralization and early reacidification on filamentous algae and macrophytes in Bowland Lake. - Can. J. Fish. Aquat. Sci. 47: 432-439.
Jansson, M., Persson, G. and Broberg, O. 1986. Phosphorus in acidified lakes. The example of Lake Gårdsjön, Sweden. - Hydrobiologia 139: 81-96.
Johansson, K. and Nyberg, P. 1981. Försurning av svenska ytvatten - effekter och omfattning 1980. [Acidification of Swedish surface waters - effects and extent 1980.] - Inf. Inst. Freshw. Res., Drottningholm (6).
Keller, W., Dodge, D.P. and Booth, G.M. 1990. Experimental lake neutralization program: overview of neutralization studies in Ontario. - Can. J. Fish. Aquat. Sci. 47: 410-411.
Lagerman, B. 1983. Kalkningsprojektet Unden. [The Lake Unden liming project.] - Nat. Board Fish., unpubl. rep.
Larsson, S. 1988. Liming effects on phytoplankton in Lake Gårdsjön 1982-1985. - In: Dickson, W. (Ed.) Liming of Lake Gårdsjön, an acidified lake in SW Sweden. Swedish Environm. Prot. Agency 3426, pp. 245-290.
Larsson, S. 1991. Gårdsjöprojektet. Effekter av kalkning och fiskisättning på växtplankton. [The Lake Gårdsjön project. Effects on phytoplankton of liming and fish introduction.] - Nat. Board Fish., Rep. 1, 1990, pp. 36-40.
Lazarek, S. 1979. Periphyton in four acidified lakes of southwestern Sweden. - Res. rep. Dep. Limnol., Univ. Lund, ISSN 0348-1798.

Lazarek, S. 1981. Mikrobiella mattor i försurade sjöar. [Microbial mats in acidified lakes.] - Sv. Bot. Tidskr. 81: 225-227.

Lazarek, S. 1982a. Structure and function of a cyanophycean mat community in an acidified lake. - Can. J. Bot. 60: 2235-2240.

Lazarek, S. 1982b. Structure and productivity of epiphytic algal communities on *Lobelia dortmanna* L. in acidified and limed lakes. - Water Air Soil Poll. 18: 333-342.

Lazarek, S. 1985. Epiphytic algal production in the acidified Lake Gårdsjön, SW Sweden. - Ecol. Bull. (Stockholm) 37: 213-218.

Lazarek, S. 1986. Responses of the *Lobelia* epiphyte complex to liming. - Aquat. Bot. 25: 73-81.

Lessmark, O. 1987. Markkalkning som metod för att motverka försurning av sjöar och vattendrag. [English summary: Ground liming as a method of mitigating acidification of surface waters.] - Inform. Inst. Freshw. Res., Drottningholm (9).

Lindmark, G. 1982. Acidified lake: sediment treatment with sodium carbonate - a remedy? - Hydrobiologia 92: 537-547.

Lindmark, G. 1984. Acidified lakes: ecosystem response following sediment treatment with sodium carbonate. - Verh. Internat. Verein. Limnol. 22: 173-178.

Lindmark, G. 1991. The use of soda ash briquettes in the neutralization of acidified lakes. - Solvay, Stockholm.

Malone, T.C. 1980. Algal size. - In: Morris, I. (Ed.) The physiological ecology of phytoplankton. Blackwell, Oxford, pp. 433-463.

Molot, L., Heintsch, L. and Nicholls, K.H. 1990. Response of phytoplankton in acidic lakes in Ontario to whole-lake neutralization. - Can. J. Fish Aquat. Sci. 47: 422-431.

Morling, G. 1981. Effects of acidification on some lakes in western Sweden. - Vatten 37: 25-38.

Morling, G. and Willén, T. 1990. Acidification and phytoplankton development in some west-Swedish lakes 1966-1983. - Limnologica (Berlin) 20: 291-306.

Nalewajko, C. and Paul, B. 1985. Effects of manipulations of aluminium concentrations and pH on phosphate uptake and planktonic communities in two Precambrian shield lakes. - Can. J. Fish. Aquat. Sci. 42: 1946-1953.

NBF and SEPA (National Board of Fisheries and Swedish Environmental Protection Agency) 1981. Kalkning av sjöar och vattendrag 1977-1981. [Liming of lakes and streams 1977-81.] - Inform. Inst. Freshw. Res., Drottningholm (4).

Niinioja, R., Ahtiainen, M. and Holopainen, A.-L. 1990. Liming of the acidic lake Valkealampi in eastern Finland: effects on water chemistry and phytoplankton. - In: Kauppi, P., Anttila, P. and Kenttämies, K. (Eds.) Acidification in Finland. Springer, Berlin Heidelberg New York, pp. 1127-1143.

Nilssen, J.P. 1980. Acidification of a small watershed in southern Norway and some characteristics of acidic aquatic environments. - Int. Rev. ges. Hydrobiol. 65: 177-207.

Olofsson, H., Blomqvist, P., Olsson, H. and Broberg, O. 1988. Restoration of the pelagic food web in acidified and limed lakes by gentle fertilization. - Limnologica (Berlin) 19: 27-35.

Olsson, H. and Pettersson, A. 1993. Oligotrophication of acidified lakes - a review of hypotheses. - Ambio 22: 312-317.

Persson, G. and Broberg, O. 1985. Nutrient concentrations in the acidified Lake Gårdsjön. The role of transport and retention of phosphorus, nitrogen and DOC in watershed and lake. - Ecol. Bull. (Stockholm) 37: 176-192.

Pettersson, A. and Blomqvist, P. 1992. Bioassay for phospate demand in phytoplankton from acidified lakes: Lake Njupfatet, an example of phosphate deficiency induced by liming. - Hydrobiologia 246: 99-110.

Porcella, D.B., Fahey, T.J., Schofield, C.L., Driscoll, C.T., Newton, R.M., Raynal, D.J., Leopold, D., Yawitt, J. and DePinto, J. 1991. Limestone treatment for management of aquatic and terrestrial ecosystems. - In: Olem, H., Schreiber, R.K., Brocksen, R.W. and Porcella, D.B. (Eds.) International lake and watershed liming practices. The Terrene Inst., Washington D.C., USA, pp. 5-14.

Rafstedt, T. 1991. Effekter av kalkning (vegetation). [Effects of liming (vegetation).] - Nat. Board Fish., Rep. 1, 1990, pp. 28-33.

Ramberg, L. 1976. Relations between phytoplankton and light climate in two Swedish forest lakes. - Int. Rev. ges. Hydrobiol. 64: 749-782.

Renberg, I. and Hellberg, T. 1982. The pH history of lakes in south-western Sweden, as calculated from subfossil diatom flora of the sediments. - Ambio 11: 30-33.

Renberg, I. and Hultberg, H. 1992. A paleoecological assessment of acidification and liming effects on diatom assemblages in a Swedish lake. - Can. J. Fish. Aquat. Sci. 49: 65-72.

Renberg, I. and Wallin, J.-E. 1985. The history of acidification of Lake Gårdsjön as deduced from diatoms and *Sphagnum* leaves in the sediment. - Ecol. Bull. (Stockholm) 37: 47-52.

Renberg, I., Korsman, K. and Anderson, N.J. 1993. A temporal perspective of lake acidification in Sweden. - Ambio 22: 264-271.

Roelefs, J.G.M. 1983. Impact of acidification and eutrophication on macrophyte communities in soft waters in the Netherlands. I. Field observations. - Aquat. Bot. 17: 139-155.

Rørslett, B. and Brettum, P. 1989. The genus *Isoetes* in Scandinavia: an ecological review and perspectives. - Aquat. Bot. 35:223-261.

Rosén G. 1981. Tusen sjöar - växtplanktons miljökrav. [Thousand lakes - environmental demands of phytoplankton.] - Swedish Environm. Prot. Agency/Liber Förlag, Stockholm.

Salonen, K., Järvinen, M., Kuoppamäki, K. and Arvola, L. 1990. Effects of liming on the chemistry and biology of a small acid humic lake. - In: Kauppi, P., Anttila, P. and Kenttämies, K. (Eds.) Acidification in Finland. Springer, Berlin Heidelberg New York, pp. 1145-1167.

Sangfors, O. 1991. Inventering av makrofyter i Gårdsjön, september 1991. [Inventory of macrophytes in Lake Gårdsjön, September 1991.] - Environm. Res. Group, Kil, Rep. F91/085:2

Scheider, W., Adamski, J. and Paylor, M. 1975. Reclamation of acidified lakes near Sudbury, Ontario. - Ontario Min. Env. Rep. Techn. Rep., Toronto.

Shearer, J.A., Fee, E.J., DeBruyn, E.R. and DeClerq, D.R. 1987. Phytoplankton primary production and light attenuation responses to the experimental acidification of a small Canadian shield lake. - Can. J. Fish Aquat. Sci. 44 (Suppl. 1): 83-90.

Stokes, P.M. 1981. In: Singer, R. (Ed.), Effects of acidic precipitation on benthos. NABS, Springfield, IL, pp. 119-138.

Svedäng, M. 1990. The growth dynamics of *Juncus bulbosus* L. - a strategy to avoid competition? - Aquat. Bot. 37: 123-138.

Torkelsson, S. and Åhren, Å. 1986. Comparison of the effect of acidification and liming on a humic stream-lake system in southwest Sweden. - Dep. Limnol., Univ. Lund, Rep.

Turner, M.A., Jackson, M.B., Findlay, D.L., Graham, R.W., De Bruyn, E.R. and Vandermeer, E.M. 1987. Early responses of periphyton to experimental lake acidification. - Can. J. Fish Aquat. Sci. 44 (Suppl. 1): 135-149.

Vöge, M. 1989. Tauchuntersuchungen an Isoetiden in einigen sauren Seen Sudnorwegens. - Tuexenia 9: 29-37.

Waters, T.F. 1956. The effects of lime application to acid bog lakes in northern Michigan. - Trans. Am. Fish. Soc. 86: 329-344.

Willén, E., Hajdu, S. and Pejler, B. 1990. Summer phytoplankton in 73 nutrient-poor Swedish lakes. Classification, ordination and choice of long-term monitoring objects. - Limnologica 20: 217-227.

Wright, D., Danks, M., Popp, W., Eiler, P. and Schreiber, R.K. 1991. Biological response of Thrush Lake, Minnesota, to a protective liming treatment. - In: Olem, H., Schreiber, R.K., Brocksen, R.W. and Porcella, D.B. (Eds.) International lake and watershed liming practices. The Terrene Inst., Washington D.C., USA, pp. 111-128.

Chapter 8

The effects of liming on aquatic fauna

Erik Degerman[1], Lennart Henrikson[2], Jan Herrmann[3] and Per Nyberg[1]

[1] National Board of Fisheries, Local Authority Örebro, Trädgårdsg. 13, S - 712 12 Örebro, Sweden
[2] Municipality of Mark, S-511 80 Kinna, Sweden
[3] Institute of Natural Sciences, University of Kalmar, P.O. Box 905, S - 391 29 Kalmar, Sweden

Summary

Liming activities in Sweden have included extensive studies of fish fauna in lakes and running waters, with the result that it is now possible to make well-founded generalizations concerning the effects of treatment on this group of organisms. It is more difficult to make generalizations about lake zooplankton and benthic fauna, which have mainly been studied within the framework of research projects. Benthic faunas in running waters have almost exclusively been studied in conjunction with general monitoring of liming effects, via a large number of individual projects. Such projects have provided material to enable the effects to be described, but not the causal relationships. Little work has been done in this context on aquatic birds and only isolated studies have been performed on frogs.

A succession of changes in the animal communities commences directly after liming, which means that the observed changes in fauna vary depending on how much time has elapsed since treatment. For species with short life cycles, such changes occur relatively rapidly. Communities continue to develop over long periods after liming; in some cases for at least fifteen years.

Post-liming changes in fauna can be summarized as follows: species of fish, mayflies (Ephemeroptera), snails and zooplankton (e.g. waterfleas, Cladocera) which have died out as a result of acidification return and, in general, the number of species in these groups increases. The number of species of benthic fauna and pelagic insects may either increase, decrease or remain unchanged. One explanation in cases where the number of species does not increase is that certain acidophilic species such as non-biting

Liming of Acidified Surface Waters
L. Henrikson and Y.W. Brodin (Eds.)
© Springer-Verlag Berlin Heidelberg 1995

midges (Chironomidae) and water boatmen (Corixidae) disappear or decline to densities which are too low to be measured in practice.

Among fish and zooplankton, both the density of individuals and the biomass increase in the long term, while inconsistent results have been observed among benthic fauna. Opportunistic species, i.e. species which may appear in dense populations over a short period after liming, are to be found among several groups of benthic faunas. The density of water fowl does not appear to change immediately after liming. The declining trend in the production of young osprey *(Pandion haliaetus)* has been halted in areas where extensive liming has been performed. Improvements in the growth and condition of individual fish have been observed, together with improvements in fish and zooplankton reproduction and frog spawn survival.

Detoxification of the water is the primary reason why species which are susceptible to acidification are able to re-establish and increase in density. Such abiotic changes are the result of liming and alter the biotic conditions, i.e. competition and predation, so that in the long term they determine how individual taxa and community structure develop. For many species, the increased food supply is one important factor, while certain species may benefit from a short-term increase in the availability of a specific food resource.

The main post-liming changes occur in lakes where acidification has had a far-reaching impact on communities of organisms. Thus, protective liming provides the best strategy to preserve susceptible faunas. If liming treatment is properly performed, the fauna develops as time goes on to attain a composition which largely resembles that which existed prior to acidification - it becomes normalized. It does not, however, recover to become identical to that which originally existed. Apart from biotic factors such as predation and competition, abiotic factors such as water quality, liming strategies and recolonization determine whether the faunal community will develop to the extent that it can regain a more normal structure.

Unstable water quality in conjunction with acid surges which result from an inflow of acidified and metal-rich water into lakes or from the breakdown of dosers in running waters, for instance, may delay and obstruct such development, however. In the long run, the nutrient status of the lake is probably also of significance. The liming strategy is not crucial as long as the water quality is improved sufficiently to become reasonably stable. In certain cases, particularly in conjunction with liming of running waters, this has not been achieved and has meant that the fauna has not recovered as expected. A combination of different methods such as liming with dosers, wetland liming and liming of upstream lakes provides the best result in running waters. The possibility of local recolonization of extinct species depends on the capability of the individual species, how far from the limed water the species are to be found, and the existence of migration barriers and competition from established acidification-tolerant species. In order for locally extinct species to recover quickly, supplementary measures are sometimes required such as elimination of migration barriers.

The overwhelming proportion of post-liming changes in fauna can be considered desirable and expected, seen in the context of the effects of acidification. Certain undesirable effects have been observed in conjunction with acid surges and metal precipitation, but these can usually be remedied by adapting the liming strategy accordingly.

1. Introduction

It was, at that time, the unexplainable loss of fish populations in certain lakes in western Sweden which drew attention to the large-scale impact of acidification on fauna in the mid-1960s (Hultberg and Stenson 1970, Almer 1972). Eventually, similar losses were seen in many other Swedish lakes and among other groups of organisms (e.g. Almer *et al* 1974). The effects on fauna have provided the main indicator of aquatic acidification and the most important justification for large-scale liming operations in Sweden.

As pH declines, the number of species, e.g. of fish and invertebrates, often declines too, whether their environments are acidified as a result of natural or anthropogenic processes (Otto and Svensson 1983, Raddum and Fjellheim 1984, Muniz 1991, Appelberg *et al* 1992, 1993, Herrmann *et al* 1993). The number and composition of species (or higher taxa) are therefore often used as a measure of the degree of acidification and as a means of illustrating post-liming changes. The loss of certain species may be disguised by colonization by other species which probably did not exist previously in the lakes concerned, thus maintaining the overall number of species (e.g. Herrmann *et al* 1993). This probably applies to the water boatman *Glaenocorisa propinqua* (Henrikson and Oscarson 1978), the dragon fly *Leucorrhinia dubia* (Henrikson 1988a) and the non-biting midge *Chironomus anthracinus* (Chironomidae) (Wiederholm and Eriksson 1977, Mossberg and Nyberg 1979).

One general effect of acidification is that susceptible species decline in density or are eliminated entirely, while tolerant species are favoured and the number of individuals increases. The faunal community in the acidified waters thus takes on a different structure, which also affects certain processes, such as decomposition, in which different species have an important function (e.g. Appelberg *et al* 1993).

The main reasons for changes among fauna in conjunction with acidification consist of stress, i.e. sublethal effects, and toxicity resulting from low pH values or high levels of aluminium (e.g. Økland and Økland 1986, Elwood and Mulholland 1989, Sutcliffe and Hildrew 1989, Muniz 1991, Appelberg *et al* 1993, Herrmann *et al* 1993). It is not simply such abiotic conditions which are of importance, however. Also different types of biotic interaction, particularly in terms of predation and competition, steer developments after acidification or liming (e.g. Eriksson *et al* 1980, Appelberg *et al* 1993, Stenson *et al* 1993). The abiotic conditions set the outer limits for different species and within these limits the community structures are largely determined by biotic factors.

Fish species play a particularly important role in determining the levels of other faunas and even flora in lakes (e.g. Kerfoot and Sih 1987). It can thus be expected that post-liming changes can be related to developments among fish faunas. In running waters, the abiotic conditions, i.e. water quality and the constant movement of the water, are of greater significance than in lakes (Peckarsky 1983, Herrmann *et al* 1993). Sutcliffe and Hildrew (1989) have said that changes in water chemistry and food availability are the main factors governing the impact of acidification on benthic fauna in running waters, while fish predation is probably of lesser significance.

An overview of Swedish acidification research, including changes in the fauna, has been presented by Fleischer *et al* (1993), Appelberg *et al* (1993), Blomquist *et al* (1993), Herrmann *et al* (1993) and Stenson *et al* (1993).

Liming of lakes and watercourses represents an important part of the Swedish strategy against acidification (Henrikson and Brodin, Chap. 1). The overall goal is to preserve or re-establish a natural environment with a rich biodiversity. The reaction of the fauna to liming has been studied in various research projects and in more routine monitoring of effects. The latter has been designed in such a way that scientific evaluation is somewhat difficult.

This chapter represents an attempt to summarize experience of the post-liming development of fauna in Sweden. There is no scarcity of data when it comes to fish and it is therefore possible to draw more general conclusions. Fewer studies are available for other groups, however, making generalizations more difficult. More in-depth analyses of lake type, fish species composition, the purely chemical success of the treatment, the time after treatment and many other factors are necessary before more comprehensive conclusions can be reached. There are no Swedish studies of how mammals are affected by liming. To some extent the Swedish studies are also viewed in relation to those from other countries. Reports are divided up into lakes and running waters and then the question of whether of not the fauna regains its original structure is discussed, along with the factors which determine this, i.e. water quality, liming strategy and recolonization. More in-depth community effects and mechanisms are provided by Appelberg (Chap. 9).

International overviews of liming and the effects of liming on fauna have been presented by Weatherly (1988), Olem (1991) and Olem *et al* (1991).

2. Lakes

2.1. Fish and crayfish

2.1.1. Background - the impact of acidification

The main impact of surface water acidification is that fish fail to reproduce. The older individuals in acidified waters often remain alive for an extended period (Hultberg and Stenson 1970, Almer 1972, Appelberg 1989). Acidification causes fish death by:

1. Impaired hatching since low pH renders an enzyme (chorionase) in perivitelline fluid inactive (Runn *et al* 1977, Johansson *et al* 1977, 1981, Milbrink and Johansson 1975, Johansson and Milbrink 1976).
2. Low pH and high aluminium content cause morphological destruction of the gills of fry, together with salt loss (Johansson and Kihlström 1975, Runn 1982, Björklund *et al* 1985, Karlsson-Norrgren *et al* 1986a,b, Norrgren and Degerman 1993), and some increase in fish mortality in vulnerable phases such as salmonid smolt migration and spawning (Rosseland 1986) or newly pigmented eel fry *(Anguilla anguilla)* (Forsberg 1986).
3. Fish are probably also suffocated as a result of precipitation of metal oxides on the gills (Grahn 1980, Andersson and Nyberg 1984).

The metal uptake of fish tends to increase in conjunction with acidification (Björklund *et al* 1984, Norrgren *et al* 1991, Norrgren and Degerman 1993), while at the same time body processes - e.g. body growth and otolith formation in salmon fry (Mosegaard 1990) and the period required from fertilization to hatching (Johansson *et al* 1973) - slow down. The behaviour of the fish may also be affected, e.g. avoidance of acidified and metal-rich waters (Andersson and Nyberg 1984, Norrgren *et al* 1991, Degerman *et al* 1990). A number of biological effects of acidification on fish are indirect, resulting from changes in nutrient, competition and predation in the ecosystem (Eriksson *et al* 1980, Eriksson *et al* 1982). The fish are compelled to modify their choice of food when certain species of prey disappear (Hansson 1976, Almer and Hansson 1980). There are, however, no studies to suggest that fish populations have declined as a result of lack of food in conjunction with acidification. When fish density declines, it has far-reaching consequences for other groups of organisms such as insects (Eriksson *et al* 1980), certain mussels (Grundelius 1987) and aquatic birds (Eriksson 1987a).

Crayfish *(Astacus astacus)* and several species of fish already suffer during their most vulnerable phases at pH values just below 6, while entire populations rarely disappear until the pH falls below 5.5 (Appelberg 1987, Degerman and Nyberg 1989, Appelberg *et al* 1992).

2.1.2. Number of species

The number of fish species and the diversity of fish communities decline successively as the impact of acidification progresses. The number and diversity of species in acidified lakes have been shown to be strongly dependent upon pH, but also upon habitat diversity and nutrient level (Degerman and Nyberg 1987, Appelberg *et al* 1989a, Bergquist 1991, Appelberg and Degerman 1991). Populations of minnow *(Phoxinus phoxinus)*, burbot *(Lota lota)*, brown trout *(Salmo trutta)*, char *(Salvelinus alpinus)*, grayling *(Thymallus thymallus)*, vendace *(Coregonus albula)*, roach *(Rutilus rutilus)*, European perch *(Perca fluviatilis)*, northern pike *(Esox lucius)* and crayfish have disappeared from many acidified lakes (Almer 1972, Almer *et al* 1974, Dickson *et al* 1975, Lessmark 1976, Grahn and Hultberg 1974, Almer 1977, Andersson *et al* 1980, Almer and Hansson 1980, Johansson and Nyberg 1981, Björklund *et al* 1984, Hultberg 1985, Degerman and Nyberg 1987, 1989, Brodin 1990, Bergquist 1991). Several postliming studies have shown how populations of the above species have re-established themselves. In a study of 112 limed lakes, Degerman and Nyberg (1989) found that the average number of species increased from 3.2 before liming to 4.0 between two and four years after liming and 4.1 between five and nine years after liming. There was a statistically significant increase in fish diversity in lakes which were dominated by pelagic species (Bergquist 1991). This was thought to be attributable to the fact that pelagic species dominate in large lakes far down in water systems. Fish in these lakes thus have a greater chance of recolonizing.

2.1.3. Recruitment and number of individuals

Species disappear successively during the acidification phase and since many of the dominant species of fish, such as roach and vendace, are susceptible to acidification (Appelberg *et al* 1989a) other more tolerant species benefit in the short term. Perch are among these opportunistic species and are able to increase in number in a moderately acidified water (Degerman and Nyberg 1989). After properly performed lake-liming treatments, increased recruitment has been observed among most species: perch, roach, vendace, whitefish *(Coregonus* spp.), char, smelt *(Osmerus eperlanus)* and pike (Bengtsson *et al* 1980, Bengtsson 1982, F. Eriksson *et al* 1982, 1983, Lindström *et al* 1984, Nyberg 1984, Fraser *et al* 1985, Nyberg *et al* 1986a,b, Eriksson and Tengelin 1987, Degerman 1987, 1989, Degerman and Nyberg 1989, Andersson *et al* 1989, Lindström 1991, Degerman *et al* 1992a). This has since led to increased density of fish in the lakes, but opportunistic species such as perch may decline once again as the more susceptible, but normally dominant, species regain their position in the community (Degerman *et al* 1992a). Nyberg *et al* (1986a) showed, for example, how an increase in perch density after liming was attributable to the number of competitors and predators present. The existence of both competing and predating species involved a lower increase than simply the presence of predators - i.e. the perch were unable to increase to the same extent when facing both competition and predation.

Among many species, the expansion which takes place after liming soon declines, but the density of fish is higher than it was prior to treatment. In a study of 38 liming lakes, the overall number of individuals increased by 123% between two and four years after liming and then fell to a stable level of 85% increase (Degerman and Nyberg 1989, Degerman *et al* 1992a). The biomass, measured in terms of catch in kg per net, also increased by 40% after between two and four years in 112 of the lakes studied.

In severely acidified lakes, the condition (Lindström *et al* 1982) and growth of fish (Degerman and Nyberg 1989) tend to decline, probably because of physiological stress. Moderate acidification which eliminates competitors may benefit tolerant species, however (see the Lake Västra Skälsjön example in Nyberg, Chap. 11, and Mossberg and Nyberg 1976, Almer *et al* 1974, 1978, Eriksson *et al* 1982). The growth of perch tends to increase in acidified waters in pace with the decline in recruitment - and thereby the density - of competitors or other perch (Almer 1972, Almer *et al* 1974, Lessmark 1976, Eriksson *et al* 1982, Nyberg 1984). It follows from this that the post-liming response in terms of individual growth depends on the degree of acidification, the co-existing species and the length of time which has elapsed after liming (Alenäs 1986).

In general, post-liming growth of individuals increases among perch (Nyberg *et al* 1986a), char (Eriksson *et al* 1982, Hasselrot *et al* 1984, Nyberg 1988), and pike (Eriksson *et al* 1982). For several other species, such as whitefish, trout and roach, there are reports of good post-liming growth, but a lack of pre-liming information (Degerman 1989). Among perch, growth increases only moderately among young individuals (which live on plankton and small organisms), while the growth of piscivorous individuals increases considerably thanks to the greater numbers of the small fish which constitute their prey (Eriksson *et al* 1982, Alenäs 1986, Nyberg *et al* 1986a, Degerman and Nyberg 1989). There are also examples of perch growth declining after liming as a result of increased competition from other species which were able to recolonize the lake (Alenäs 1986, Appelberg *et al* 1990). This represents a return to the normal conditions in an unacidified lake. On the whole, the above indicates that the general increase in individual growth after liming primarily depends on the increased availability of food organisms. Reduced physiological stress is probably also of significance (Barlaup *et al* 1989).

No estimates of fish production in limed waters have been performed. The general increase in the growth of individuals which occurs after liming, in combination with increased recruitment and density of individuals, is equivalent to increased production, particularly as it has generally been observed that lakes are more extensively fished after liming (Bengtsson and Bogelius, Chap. 18). In addition, liming means that the lakes become more nutrient-rich, while the amount of carbon increases, which should mean that the environment becomes potentially more productive (Thörnelöf and Degerman 1991). Finally, the increased density of roach, which are susceptible to acidification, means that the trophic level of the lake increases as a result of the change in grazing pressure on zooplankton and thus, indirectly, on phytoplankton (Stenson 1988).

2.1.4. The structure of fish communities

After liming, primarily biotic factors regulate the fish communities within new abiotic limits which the treatment involves (Nyberg *et al* 1986a). This has been seen as a sign that the treatment has been successful (Appelberg *et al* 1989a). The response of a given species to liming must thus be seen in relation to other species, since the outcome largely depends upon interaction (Nyberg *et al* 1986a, Degerman and Nyberg 1989, Appelberg and Degerman 1991, Appelberg *et al* 1989a, 1990, 1992, Degerman *et al* 1992a). Several shifts in dominance may occur after liming, just as they may during the acidified phase (Degerman *et al* 1992a). Opportunists may take over during the acidified phase, but after liming there is a renewed shift from a perch-dominated community to a roach/pike community (Appelberg and Degerman 1991, Appelberg *et al* 1992) which supports several competitive species (Svärdson 1976). Immediately after liming, perch and ruffe *(Acerina cernua)* expand rapidly and then decline as roach and vendace increase (Degerman and Nyberg 1989). Opportunists such as perch have often been seen to increase their niches in pace with the loss of other species, only to utilize the habitat to a lesser degree after liming when the more competitive species have re-established themselves (Appelberg *et al* 1990).

In certain cases, the dominance of perch individuals has remained after liming since the previously most competitive species have been unable to recover because of a lack of recolonization opportunities (Degerman and Nyberg 1989). Perch have been able to establish very large populations due to the lack of competitors and predators (Nyberg *et al* 1986a) in lakes which have previously supported char, for instance. Char have been successfully re-introduced to these waters and have reduced the perch population (Nyberg 1988). In one of these lakes, however, trout were also re-introduced and were able to prey on young perch during the summer when char and perch are separated by temperature layers. There are thus cases where acidification has gone so far that post-liming conditions must be manipulated artificially in order to achieve normalization (Bergquist, Chap. 17). This possibly also applies to crayfish populations which have declined as a result of acidification. These often find it difficult to recover after liming, which is thought to be attributable to predation from fish. This has not been proven, however (Appelberg 1987, 1990, Nyberg *et al* 1986a, Odelström 1988, Appelberg and Odelström 1990). Crayfish populations grow slowly and a long period of time is often required after liming before the effects can be observed.

Lakes have sometimes been limed before the pH has fallen below 6 and before any changes have been proven among the fish fauna (Nyberg *et al* 1986a). After such treatments, it has not been possible to demonstrate any effects on fish communities (Degerman and Nyberg 1989) or on the growth of individuals (Eriksson *et al* 1982, Alenäs 1986). This shows that the effect of liming in acidified lakes is primarily a result of the reduced toxicity (higher pH and lower metal content) and not a matter of added calcium or phosphorus. It has also been observed that lakes which have been limed before any major changes in species dominance have occurred are able to recover more quickly (Appelberg *et al* 1992).

The effect of a liming treatment can often be observed in a fish community several years later. A change in the choice of food could be observed among perch in lakes for a period of at least five years, as the invertebrate community recovered following liming (Eriksson *et al* 1982, Hasselrot *et al* 1984, Alenäs 1986). Appelberg and Degerman (1991) noted that the fish communities slowly stabilized after liming and that the diversity increased 1-16 years after the initial treatment. In general, fish communities regain their former structure (Thörnelöf and Degerman 1991).

2.2. Zooplankton

2.2.1. Background - the impact of acidification

The main result of acidification on zooplankton communities in lakes in Europe and North America has been shown to consist of a reduction in the number of species of zooplankton, as well as in reduced densities in many cases (Almer *et al* 1978, Hobaek and Raddum 1980, Malley and Chang 1986, Yan *et al* 1988, Brett 1989a, Berge *et al* 1990, Morling and Pejler 1990, Renberg *et al* 1990, Sarvala and Halsinaho 1990). There are a number of examples of certain waterfleas, e.g. *Daphnia longispina* (Nilssen *et al* 1984) and *D. galeata mendotae* (Keller *et al* 1990b), being susceptible to low pH and raised levels of aluminium, particularly in clear-water lakes (Hörnström *et al* 1984, Havens and DeCosta 1987, Sarvala and Halsinaho 1990). The copepod *Heterocope appendiculata* is also considered to be susceptible to acidification (Nilssen *et al* 1983, Hörnström *et al* 1984). High humus levels in the water may alleviate the toxic effects, however. To judge from the differences in species composition between different acid lakes, many of the changes in species and density are secondary effects caused by changes in nutrient supply and predation and competition conditions in the zooplankton community. Generally, reduced humus levels involve a reduction in the availability of nutrients - in the form of bacteria, for instance - for filtering organisms (Johansson 1983, Francis *et al* 1984, Hessen 1985). In addition, the zooplankton community shifts towards larger species (Almer *et al* 1978, Hörnström *et al* 1984), which also involves a reduction in the nutrients available to zooplankton.

The zooplankton communities in the most acidified lakes in Scandinavia usually consist of the copepod *Eudiaptomus gracilis*, the cladoceran *Bosmina longispina* or *B. coregoni* (Almer *et al* 1978, Henrikson *et al* 1984, Nilssen *et al* 1984, Stenson and Oscarson 1985, Uimonen-Simola and Tolonen 1987, Berge *et al* 1990, Morling and Pejler 1990, Nilssen and Sandøy 1990, Rehnberg *et al* 1990, Sarvala and Halsinaho 1990, Hörnström *et al* 1992, unpubl., Lindström 1992) and certain rotifer species (Rotatoria) such as *Keratella serrulata* (Brett 1989a, Morling and Pejler 1990), *Polyarthra remata* (Almer *et al* 1978), *P. vulgata, Keratella cochlearis, Conochilus unicornis* (Morling and Pejler 1990) and *Kellicottia longispina* (Raddum *et al* 1986). In North American lakes, the corresponding communities consist of *Diaptomus minutus*, *B. longispina* and *Keratella taurocephala* (Brett 1989a,b, Schaffner 1989).

There are, however, several exceptions from this composition. Rask *et al* (1986) found that waterfleas dominated the crustacean community in a lake with a pH of 4.4 and Sarvala and Halsinaho (1990) found that the proportion of cladocerans was greatest in acid lakes since calanoid copepods were virtually non-existent in the lakes where the pH was lower than 5. They also found *Daphnia cristata* and *D. longiremis* at a pH of 4.8 in polyhumic lakes. The lack of calanoid copepods probably does not depend on the pH in itself but on competition from cladocerans, which are able to survive in these acid lakes thanks to the fact that the high humic level reduces the toxicity of the water.

The greatest change in the zooplankton community in conjunction with acidification does not occur until recruitment ceases in the fish population and the fish eventually die out. This situation allows predation-sensitive invertebrates such as the phantom midge larvae *Chaoborus* sp., the water boatman *Glaenocorisa propinqua* and diving beetles (Dytiscidae) to form dense populations (Henrikson and Oscarson 1981, Henrikson *et al* 1984, Nyberg 1984, Brett 1989a, Stenson 1990). These exercise a completely different predation pressure on the plankton community than fish, and this results in a different community structure. The depletion of the proper plankton community involves a reduction in competition and makes it possible for benthic and littoral cladoceran species such as Chydoridae and *Polyphemus pediculus* to appear pelagically (Berge *et al* 1990, Nilssen and Sandøy 1990, Nyberg unpubl., Morling and Pejler 1990, Stenson and Svensson unpubl.). In this plankton community, in which the competition from efficient, filtering cladocerans is negligible, filtering rotifers can also increase in density (Stenson 1990, Yan *et al* 1991, Nyberg unpubl.). The effect of the invertebrate predators may vary depending on the type of lake. *Chaoborus* sp. do not occur in deep clear-water lakes, for instance, even if there are no fish.

2.2.2. Species composition and diversity

The effects of a liming on a zooplankton community is very much dependent on the composition of the plankton community at the time of treatment and on the degree to which the water quality improves (Henrikson *et al* 1985). In highly acidified lakes, few new species appear as long as there are no fish (Henrikson *et al* 1984, Nyberg 1984, unpubl., Stenson 1990, Stenson and Svensson unpubl.). The same is the case if liming is performed before any reduction in species occurs (Alenäs 1986, Henrikson *et al* 1989, Hörnström *et al* 1992, Keller *et al* 1992). The fastest and most noticeable changes in species occur in acidified lakes which contain fish at the time of treatment (Hörnström *et al* 1992, 1993, Nyberg unpubl.).

Eriksson *et al* (1983) found no major short-term changes in the number of species among rotiferrs, while Hasselrot *et al* (1984) only found *Polyarthra remata* prior to liming and a further four species at the end of the first post-liming vegetation season. Of the fourteen rotifer taxa found after liming in six lakes in western Sweden, however, only *Gastropus stylifer*, *Ploesoma hudsoni*, *P. truncatum* and *Trichocera* spp. were absent from all the lakes prior to treatment (Hörnström *et al* 1993). In four lakes studied south-west of Stockholm, *Conochilus unicornis*, *Kellicottia longispina* and *Polyarthra vulgaris* were not found prior to treatment but were common afterwards (Hörnström *et al* 1992). In six lakes in the Delsbo area of central Sweden, which were less acidified at the time of liming, however, the only change in the rotifer community was that *P. truncatum* appeared in two of the lakes after liming (Hörnström *et al* 1992).

Hörnström *et al* (1993) found an average of 9.8-16.8 zooplankton species after liming in six lakes in western Sweden, compared with 4-8.5 in the same lakes before treatment. Of these, 1-4 consisted of cladocerans before liming and 3-5.5 afterwards. They also found that the composition of species in the limed lakes was normal when compared with a large number of reference lakes which were studied in 1971 and 1976. In 14 of the lakes which had an alkalinity of 0.05-0.10 meq/l in 1971, an average of 7.3 (4-10) species were found (Hörnström *et al* 1973). Stenson and Svensson (unpubl.) identified 7-11 cladoceran species 4-11 years after liming in Lake Gårdjön, south-west Sweden, compared with 3-5 before, and the diversity of the cladoceran community increased in general from about 0.1 to maximum values of over 2 the second year after re-stocking with trout. In another lake in south-west Sweden, the diversity index of the zooplankton community increased from 0.3 to 1.9 during the fourth post-liming vegetation season (Hasselrot *et al* 1984). Nyberg (unpubl.) found 4-9 crustacean taxa in ten study lakes before and 6-13 after liming. Among three lakes studied in the Fulufjäll region of central Sweden, however, only one new species appeared *(Daphnia longispina)* in one of the lakes 10-13 years after the initial treatment (Lindström 1992).

Less data is available on changes among species of copepods, partly because species in the genus *Cyclops* are rarely determined. Nyberg (unpubl.) found, however, the clearly acidification-sensitive *Heterocope appendiculata* in more lakes after liming than before, and in the seven lakes in the Delsbo region *H. appendiculata* appeared as a new species in two lakes after liming (Hörnström *et al* 1992). The species did not appear after liming, however, either in Lake Gårdjön (Stenson and Svensson unpubl.) or in the six west-coast lakes (Hörnström *et al* 1993).

All the herbivorous cladoceran species which appear in neutral and oligotrophic lakes have the potential to appear sooner or later after liming as far as water quality and nutrient conditions are concerned. Predation from fish and *Chaoborus*, in fish-free lakes can decimate susceptible species to such an extent that they cannot be traced through normal sampling. In acid lakes where the cladoceran species have been reduced, they return successively. Together with *Bosmina*, which usually also appears during the acidification stage, large populations of *Diaphanosoma brachyurum* often develop after liming (Raddum *et al* 1986, Nyberg unpubl., Stenson and Svensson unpubl.). After a further period has elapsed, *Ceriodaphnia quadrangula* and *Holopedium gibberum* appear, followed later by one or more *Daphnia* species (Hörnström *et al* 1992, 1993, Nyberg unpubl., Stenson and Svensson unpubl.). The large carnivorous species *Bythotrephes longimanus* and *Leptodora kindti* only occur in weak populations in lakes with a normal composition of fish species because of predation conditions and they are therefore only occasionally caught in sampling. Stenson and Svensson (unpubl.) found *B. longimanus* for the first time in Lake Gårdsjön seven years after liming, while *L. kindti* was not found at all. Hörnström *et al* (1993) were unable to find *B. longimanus* in one lake after liming, which was probably attributable to increased predation. *L. kindti* was found in one of the west-coast lakes before liming and in a further two afterwards. In the four lake systems studied by Nyberg (unpubl.) these species were not found at all in one, a fact with was attributed to colonization difficulties.

Rotifers, copepods and cladocerans all have dormant eggs which may lie in the sediment for a very long time. Carvalho and Wolf (1989) hatched eggs of this sort from several *Daphnia* species which may have been up to 70 years old. Herzig (1985) hatched eggs of the same age from *B. longimanus* and *L. kindti* and Hairston and De Stasio (1988) hatched rotifer eggs which could have been up to 100 years old. There is no data in the literature to show that the hatching of dormant eggs may be inhibited or stimulated by low or high pH values, however. If hatching from dormant eggs is the main reason for the appearance of new species after liming, they should appear as soon as the water quality has become satisfactory in the interstitial water of the sediment. Since this does not happen, it is more likely that new species spread to the water in question as dormant eggs with the help of the wind and, above all, aquatic birds.

Ephippium bearers, i.e. individuals with dormant eggs in brood pouches, are preferred by fish (Mellors 1975, Lessmark 1983) and the ephippia can pass through the digestive tracts of birds and fish without being damaged to any particular extent (Mellors 1975). *B. longimanus*, *L. kindti* and *H. gibberum* have, unlike most other waterfleas, exposed dormant eggs, which probably explains why they experience colonization difficulties. *H. gibberum* can withstand acidification and the appearance of the species in all lakes after liming can be attributed to the existence of sparse populations prior to liming (Nyberg unpubl.).

Some studies have documented short-lived and, in certain cases, relatively long-lived undesirable effects of liming in the form of reduced zooplankton density. Schaffner (1989) found that the post-liming density of all dominant species declined considerably in two North American lakes. About two years after treatment, the density had still not reached its previous levels (Bukaveckas 1991). Wright *et al* (1991) made similar observations. In both cases, a very easily soluble lime sludge was used and the pH increased rapidly to over 9, while the availability of food in the form of phytoplankton declined (Bukaveckas 1991). Appelberg *et al* (1990) and Hörnström *et al* (1993) noted a reduction in the occurrence of crustaceans in two lakes in south-west Sweden during the year of liming, despite the use of coarser lime fractions. In the same area, Henrikson *et al* (1985) also found that the number of crustaceans declined after liming. This was attributed to improved fish recruitment and to predation from one-summer-old perch. In most of the limed waters, however, no undesirable effects were observed (Eriksson *et al* 1983, Henrikson *et al* 1984, Raddum *et al* 1986, Hörnström *et al* 1992, Keller *et al* 1992). Nyberg (unpubl.) found that the mean density of crustaceans during the two months after liming increased by 2.5-10 times after three different liming treatments in two lakes and noted no undesirable effects after a total of 29 treatments in eight different lakes. This was thought to be attributable to the fact that moderate doses of not too easily soluble lime were used and that the treatments were performed during the spring and early summer, thus avoiding a high pH of short duration and changes in the light climate. The treatments were also performed before the phytoplankton and zooplankton populations had reached their maximum density.

Studies from Europe and North America show without doubt that, over the space of a few years, liming result in increased zooplankton density and biomass. Appelberg *et al* (1990) measured a significant post-liming increase in the zooplankton volume and Hörnström *et al* (1993) found that the total biomass, particularly the biomass of rotifers and cladocerans, was generally higher after liming. Howell *et al* (1991), Keller *et al* (1992), Nyberg (unpubl.), Raddum *et al* (1986) and Stenson and Svensson (unpubl.) noted increases in the density of zooplankton individuals after liming. Raddum *et al* (1986) measured an initial increase in the numbers of eggs among *Eudiptomus gracilis*, *Bosmina longispina* and *Holopedium gibberum*. After about two years, fertility declined, possibly as a result of a lack of nutrients as the lake became re-acidified. Lindström (1992) found that the number of eggs among *Mixodiaptomus laciniatus* decreased as the post-liming density increased. Olofsson *et al* (1988) also found an increased number of eggs among *D. longispina*, *D. longiremis* and *B. longispina*. In the lake studied, however, liming had been combined with cautious fertilization with nitrogen and phosphorus. Schaffner (1989) measured lower egg numbers in *Diaptomus minutus* after liming, however, which was considered to depend on a lack of food as a result of a sharp decline in the phytoplankton community following a change in the light climate (Bukaveckas 1991).

2.2.3. The structure of the zooplankton community

If liming leads to a neutral pH in the entire water mass, levels of toxic aluminium will decline and food supply will increase (Raddum *et al* 1986, Larsson 1988, Appelberg *et al* 1990, Hörnström *et al* 1993), which means that the zooplankton species that normally occur in oligotrophic lakes can become established (Hörnström *et al* 1993, Nyberg unpubl.). In this environment, the structure of the zooplankton community is determined by prevailing predation and competition conditions. In the extreme case, where there are no fish and the predators consist of invertebrates (Henrikson and Oscarson 1984), the community is mainly made up of copepods, rotifers and a few cladocerans (Stenson 1990, Yan *et al* 1991). *Diaphanosoma brachyurum* may, however, form dense populations during the summer months, while the invertebrate predators are undergoing a generation shift (Nyberg unpubl.). This situation is in no way representative of Swedish limed lakes, however, since they are only devoid of fish in exceptional cases.

Certain predatory invertebrates are also subject to predation and the effect they have on the zooplankton community is eliminated in limed lakes containing fish (Nyberg unpubl., Stenson and Svensson unpubl.) and where the community structure is primarily determined by the composition of the fish community (Nyberg unpubl.). In rather small and shallow lakes with little predation (pike-perch, perch, perch-char, char, trout) the proportion of rotifers and copepods is relatively small and the community is dominated by effective filtering cladocerans such as *D. longispina* (Nyberg unpubl.,

Stenson and Svensson unpubl.). In lakes with more effective zooplankton predators such as roach and vendace (Nilsson and Pejler 1973, Svärdson 1976) *D. longispina* declines sharply and is replaced by *D. cristata* (Nyberg unpubl.) and, as the predation pressure increases, also by *D. hyalina* and *D. galeata* in certain lakes (Appelberg *et al* 1990). Even other small cladocerans such as *Bosmina* spp. and *Ceriodaphnia quadrangula* may represent a significant proportion of the zooplankton community in these lakes (Hörnström *et al* 1993). Heavy predation from dense cohorts of fish may reduce the cladoceran community and make it possible for the copepods to assume a more important position in the community (Nyberg unpubl.). Apart from the fact that the two large cladocerans *L. kindti* and *B. longimanus* were not found in any lakes after liming, it is clear that the zooplankton community is normalized in terms of the composition of species and that this is determined by biotic factors, i.e. predation and competition some years after the liming treatment. No density data are available from good, unlimed reference lakes. It would, however, appear that the density of zooplankton is somewhat lower in the limed lakes studied than in unacidified, nutrient-deficient lakes.

2.3. Benthic fauna and pelagic insects

2.3.1. Background - the impact of acidification

Acidification affects the benthic fauna in lakes, both through abiotic and biotic changes. Susceptible species are found within most taxa, the most susceptible being among snails (Gastropoda), mussels (Bivalvia), amphipods and mayflies (Ephemeroptera) (e.g. Økland and Økland 1986, Bergquist *et al* 1992). These species die or suffer as a result of changes in abiotic conditions such as low pH, reduced calcium levels and increased metal levels (Økland and Økland 1986, Muniz 1991). The groups contain species with different susceptibilities to acidification (Engblom and Lingdell 1987, Bergquist *et al* 1992). One of the most susceptible of these is the mayfly *Ephemera vulgata*, while another mayfly species *Leptophlebia vespertina* can withstand very low pH. Within groups such as water boatmen (Corixidae) and backswimmers (*Notonecta* spp.) and other semi-aquatic bugs and diving beetles (Dytiscidae), however, most species can withstand acidification. For certain species such as the water louse *Asellus aquaticus* (Isopoda), the situation is less clear. It is regarded by Økland (1980) as susceptible to acidification, but still occurs in fairly high populations in certain acidified lakes (Bergquist *et al* 1992, Henrikson and Oscarson unpubl.), which suggests that factors other than pH control their occurrence. Alenäs *et al* (1992) found that *A. aquaticus* increased in number after liming, declined when the lake was re-acidified and increased again when the lake limed once more, which suggests that acidification has a negative impact.

Many species are affected by the change in predation conditions that arise when the fish species which constitute the dominant top predators are eliminated (Eriksson *et al* 1980). When this happens, taxa which can withstand acidification but are susceptible to predation, such as phantom midge larvae of *Chaoborus* spp. (Diptera), water boatmen, diving beetles and certain dragon fly (Odonata), may increase in number and take over the role of fish as top predators (Henrikson and Oscarson 1981, Henrikson 1988a). These "new" top predators will affect their prey differently than fish.

2.3.2. Composition of species and number of species

A correctly performed liming treatment reduces the toxicity of the water and makes it possible for acidification-sensitive species to become established in the lake. It may, however, take a long time for this to happen, depending on the extent to which species have been lost in nearby lakes and the ability of the species to disperse (Bergquist *et al* 1992). The mayfly *Ephemera vulgata*, which has colonized several lakes, is one example. This only took place 7-10 years after the initial treatment, however (Harnak unpubl., Mossberg and Nyberg unpubl., Lindgren unpubl.). Other mayflies which have returned after liming include *Caenis horaria*, *Cloeon dipterum*, *Heptagenia fuscogrisea* and *Centroptilum luteolum* (Hasselrot *et al* 1984, Alenäs 1986, Henrikson *et al* unpubl., Mossberg and Nyberg unpubl.). Another example may be the non-biting midge *Parakiefferiella bathophila* (Chironomidae), which was first observed after liming in four Swedish lakes (Henrikson unpubl., Mossberg and Nyberg unpubl.). In Norwegian lakes, *A. aquaticus* (Raddum *et al* 1984) and the mayfly *Siplonorus lacustris* (Fjellheim and Raddum 1992) were only observed after liming.

Snails and mussels are groups that include acidification-sensitive species which have far fewer opportunities for recolonization than those which have a phase in which they can fly or be carried by the wind or birds to limed waters. Four snails, *Radix (Lymnaea) peregra*, *Valvata cristata*, *Gyraulus albus* and *G. acronicus* and the pea mussel *Pisidium* sp. were found after liming in Swedish lakes, however (Bergquist 1980, Harnak unpubl. data).

Since many species colonize or recolonize after liming, the number of species should increase. This has been observed in many lakes (Table 1). Among the mayflies, it is clear that the number of species in general increases after liming in severely acidified lakes (Fig. 1) (Bergquist *et al* 1992). Bergquist *et al* (1992) also showed that there was a significant increase in the total number of taxa as time went on, from an average of 19 to 30 littoral taxa per lake over a ten-year period. In Lake Gårdsjön, a total of 43 taxa were found in the sublittoral and profundal zones before liming and 62 after (Henrikson unpubl.). Increases in numbers of species have also been documented in the littoral zones of lakes in northern Sweden (Söderberg unpubl.) and in western Sweden (also in the profundal zone) (Har-

Table 1. Total number of identified taxa (species, genera, etc.) among benthic fauna before and after liming. A) Lakes in central Sweden (Mossberg and Nyberg unpubl.); B) lakes in the Delsbo region (Lindgren unpubl.).

A) Lake	Number of taxa Before liming (1977)	After liming (1987)	Change, %
Mörtsjön	15	35	133
Trehörningen	15	27	80
Långsjön	25	41	64
Blanksjön	19	27	42
Västra Skälsjön	26	34	31
Stora Sirsjön	21	27	29
Iglafallssjön	26	29	12
Östra Skälsjön	28	26	-7

B) Lake	Number of taxa Before liming (1983)	After liming (1991)	Change, %
Källsjön	12	18	50
Svartvallstjärn	14	18	29
Lillvallsjön	20	23	15
Rossjön	27	24	-11
Prästvallssjön	16	13	-19
Nyvallssjön	31	19	-39
Örvallstjärn	16	9	-44
Stensjön (unlimed)	38	14	-63

nak unpubl. data). In Bowland Lake, Canada, the number of taxa increased from a total of 51 before to 66 after liming (Keller *et al* 1990b).

In several cases in Sweden, no increase in the total number of taxa has been documented after liming (Table 1). This also applies to Norwegian waters (Raddum *et al* 1986, Raddum and Fjellheim 1994). This may be attributable to the fact that certain species "disappear" and are replaced by others. In ten lakes in central Sweden, 3-11 taxa disappeared, while 10-31 came after liming (Mossberg and Nyberg unpubl.). Examples of species which may disappear include the acidophilic non-biting midges such as *Chironomus anthracinus* and *Zalutschia zalutschicola* (Mossberg and Nyberg unpubl., Lindgren unpubl.). In the littoral zone of Lake Gårdsjön, the number of taxa declined among dragonflies, water boatmen and beetles, while it increased among caddis flies and mayflies (Henrikson *et al* unpubl. data). Reductions in the number of taxa have also been observed (Table 1).

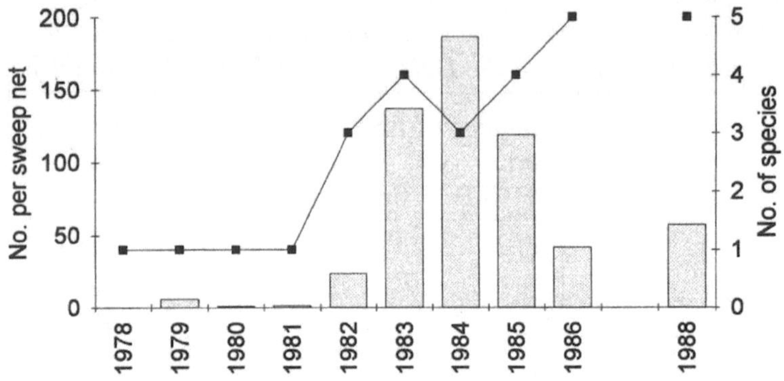

Figure 1. *Number of species (line) and abundance (bars) of mayflies (Ephemeroptera) in the littoral zone of Lake Gårdsjön before and after liming and restocking with fish (Henrikson et al unpubl. data).*

2.3.3. Abundance

Decreased, increased and unchanged densities have been measured after liming among most groups of organisms (cf. Tables 2-4 below). For certain groups, e.g. mayflies and *Asellus aquaticus*, the individuals disperse from the littoral zone down to deeper areas (Hultberg and Andersson 1982, Alenäs *et al* 1992, Henrikson unpubl.) and thus increase not only in number (Figs. 1 and 2) but also in terms of their range in the lake.

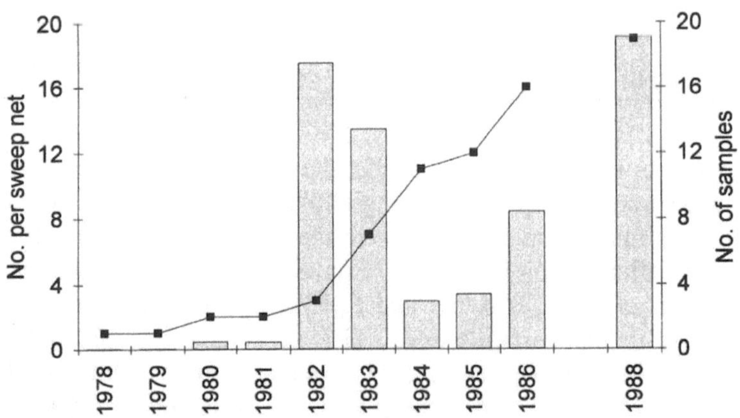

Figure 2. *Population development (bars) and occurrence (line) at 20 sample sites of water louse (Asellus aquaticus) in the littoral zone of Lake Gårdsjön before and after liming and restocking with fish (Henrikson and Oscarson unpubl.).*

Increased densities have been measured among acidification-sensitive taxa. This applies, e.g., to the pea mussel *(Pisidium* spp.) (Fig. 3) and the mayfly *Ephemera vulgata,* which declined at the same time in an unlimed reference lake (Lindgren unpubl.). In the USA, post-liming increases have been observed among snails (Wright *et al* 1991).

Crustaceans defined as "glacial relicts" - populations which became isolated after the last Ice Age - are generally sensitive to acidification (Kinsten 1986). One of these, *Mysis relicta,* increased in number after liming in Lake Unden in southern Sweden (Lagerman 1983), as well as in Trout Lake in Canada (Howell *et al* 1991). Studies on one lake in south-western Sweden, however, did not show significant changes in density among five species of glacial relicts (Henrikson *et al* 1989).

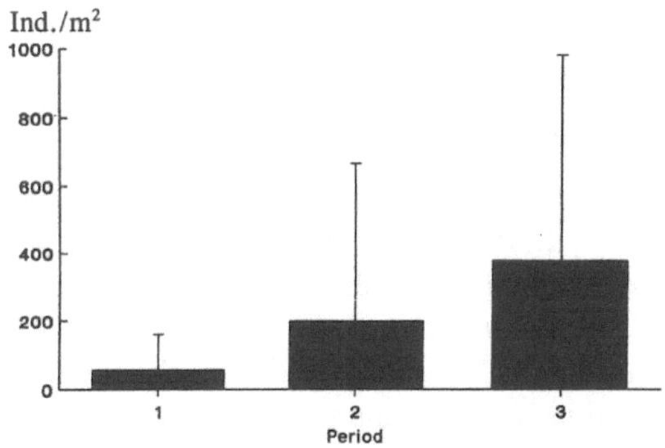

B)

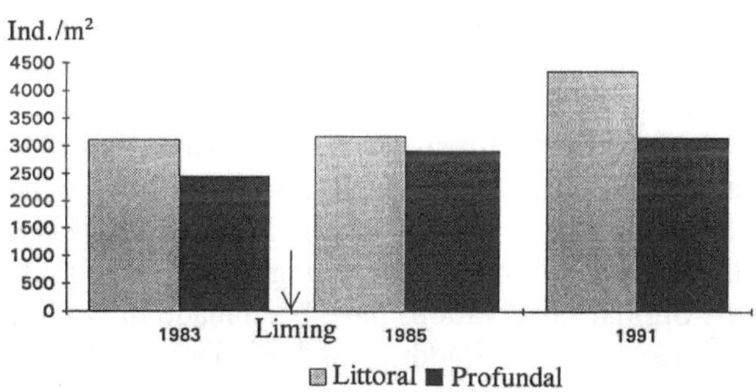

Figure 3*. Population development among pea mussels (Pisidium sp.) after liming. A) In the littoral zone in eight lakes in central Sweden (Mossberg and Nyberg unpubl.). The mean value and SE for all lakes and periods: 1 (during acidification); 2 (1-4 years after liming); 3 (5-8 years after liming). B) In the littoral zone and profundal zone in seven lakes in the Delsbo region (Lindgren unpubl.).*

There are several indications that certain species of the chironomid genera *Tanytarsus* are sensitive to acidification (Henrikson unpubl.). This group may display increased densities in the littoral zone after liming (Table 2). In two Norwegian lakes, *Tanytarsus signatus* increased sharply and become the dominant species in one of the lakes (Raddum *et al* 1986, Raddum pers. comm.). There were no clear changes among *Tanytarsus* species in certain lakes in central Sweden after liming, however (Table 2). Even acidification--tolerant chironomids such as *Heterotanytarsus apicalis* and *Pagastiella orophila* increase in many Swedish lakes after liming, which is also true for Norwegian lakes (Raddum *et al* 1984, 1986).

Species which benefit from acidification generally decline after liming. This applies to the chironomids *Chironomus anthracinus* and *Dicrotendipes* sp. in clear-water lakes and *Zalutschia zalutschicola* in humic lakes (e.g. Raddum *et al* 1986). It may, however, take a long time before these acidification indicators disappear. In one lake, it took five years before *Chironomus* disappeared (Harnak unpubl. data), possibly because of competition. In the short term, increased densities of pelagic insects, such as *Chaoborus* spp., backswimmers, water boatmen and diving beetles have been observed in Swedish, Norwegian, North American and Scottish lakes (Raddum *et al* 1984, Evans 1989, Gloss *et al* 1989, Keller *et al* 1990b, Foster 1991), possibly because of improved food availability in the form of zooplankton (Henrikson and Oscarson 1984). In the longer term, however, the development of these large free-living insects will be largely dependent upon the predation pressure from fish. In Lake Gårdsjön, the populations of phantom midge larvae, water boatmen - with the dominant *Glaenocorisa propinqua* - and diving beetles were subject to grazing when trout were introduced (Henrikson unpubl.).

Sharp increases in population density after liming, followed by declines, have been reported among the mayfly and *A. aquaticus*, for instance. In the littoral zone of Lake Gårdsjön, the mayfly reacted with a sharp increase in density, particularly the acidification-resistant *L. verspertina*, but this was followed by a marked decline (Fig. 1), although it was still found at nearly all sampling sites (Henrikson *et al* unpubl. data). Mossberg and Nyberg (unpubl.) observed that increases shortly after liming turned into significant declines in four lakes. Similar observations have been made in Norwegian lakes (Raddum *et al* 1984, 1986, Raddum and Fjellheim 1994). One explanation might be found in the improved food supply in the form of increased bacterial activity in accumulated organic material, a food resource which declines as time goes on (Eriksson et al 1983). For *A. aquaticus*, fish predation may be of importance (Mossberg and Nyberg unpubl.). This does not apply to the mayfly (Raddum and Fjellheim 1994). An increased density of *A. aquaticus* can hasten one of the first stages of the decomposition process in the limed lake, since the species is an effective shredder

(Andersson 1985, Gahnström, Chap. 6, Henrikson and Oscarson unpubl., Fjellheim and Raddum 1988).

Oligochaetes may also benefit from the improved food situation in limed lakes. Usually, they increase in population density in the littoral zone, but not so much in the sublittoral and profundal zones (Raddum *et al* 1984, Keller *et al* 1990b).

Table 2. *Changes in density among taxa of Chironomidae in Swedish lakes before and after liming; A) littoral zone, B) sublittoral (>4 m) and profundal zone. Increase +; decrease -; no change 0. References* [a] *are indicated by numbers.*

Taxa	Littoral change			Sublittoral and profundal change		
	+	-	0	+	-	0
Procladius spp.	1	1	1	1	3,9	1
Ablabesmyia spp.	1	1	1		1	1
Heterotanytarsus apicalis	1		1	3,9		1
Heterotrissocladius marcidus	1	4	1	1	1	1
Psectrocladius spp.		1,4	1	1	1,3	1
Parakiefferiella spp.	1	1	1	1,3	1	
Zalutschia spp.[b]		1,4	1	2	1	
Chironomus spp.		1	1	1,2,3,10	1	
Dicrotendipes spp.		1	1	1,2,3,10	1	1
Microtendipes spp.	1	1,4		1	1	
Pagastiella orophila	1,2		1	1,2,3	1	1
Sergentia spp.	1	1		2,3,4,10	1,2	1
Stictochironomus spp.[c]		2		2,4	9	1
Micropsectra spp.				9		
Tanytarsus spp.	1	2	1	1,2,3	9	1

[b] Including *Z. zalutschicola*
[c] Including *S. rosenschoeldi*

[a] References: 1) Mossberg and Nyberg (unpubl.): 8 lakes, central Sweden, studied 10 years after liming, cf. F. Eriksson *et al* (1982, 1983). 2) Lindgren (unpubl.): 7 lakes, central Sweden, 8 years. 3) Henrikson (unpubl.): Lake Gårdsjön, SW Sweden, 7 years. 4) Bergquist (1980): Lake Ölen, central Sweden, 4 years. 5) Alenäs (1986): Lake Stora Härsjön, SW Sweden, 10 years. 6) Hultberg and Andersson (1982): 6 lakes, western Sweden, 6 years. 7) Alenäs *et al* (1991): Lake Lysevatten, SW Sweden, 10 years. 8) Hasselrot *et al* (1984): Lake Stora Holmevatten, SW Sweden, 4 years. 9) Olofsson and Oscarson (1988): Lake Tolken, SW Sweden, 10 years. 10) Harnak (unpubl. data), Olofsson (unpubl. data): Lake Högsjön, SW Sweden, 5 years.

The densities of *Sialis* (Neuroptera) and caddis flies were observed to often decrease or remain unchanged 7-8 years after liming (Table 3). In the littoral zone of Lake Gårdsjön, the overall population density of caddis flies declined shortly after liming, particularly among Polycentropidae, which include free-living species such as *Cyrnus flavidusi* and *Holocentropus dubius* (Henriksson *et al* unpubl. data). Certain species such as *Molannodes tinctus* increased, however. Predation by fish may certainly reduce the population density of *Sialis* and large, free-living species of caddis fly, but in Lake Gårdsjön they declined before fish were introduced. This shows that other factors - competition, perhaps - are of importance.

Chironomids are prominent among the benthic fauna and include a number of species which have varying environmental demands and their susceptibility to predation by fish. Chironomids which live in sediment were observed to both increase and decrease after liming, although the decline appeared to be only temporary (Table 4). It should, however, be noted that the population density also declined in the unlimed reference lake. In Lake Gårdsjön, however, the population density was significantly lower 4-6 years after liming compared with 1-4 years before liming, which was not the case in the unlimed reference lake (Henrikson unpubl.). The Delsbo lakes in central Sweden where increases were observed were among the most acidified

Table 3*. Changes in density among animal taxa in Swedish lakes before and after liming; A) littoral zone, B) sublittoral (>4 m) and profundal zone. Increase +; decrease -; no change 0. References are indicated by numbers and listed in Table 2.*

Taxa	Littoral change			Sublittoral and profundal change		
	+	-	0	+	-	0
Oligochaeta	1,2	1	1,3	1,2		1,2,3
Ephemeroptera	1-5,10	1	1	3,5		
Odonata			1,3			
Trichoptera		1,3	1		1,3	1
Sialis lutaria		1	1,3		1,3	1
Corixidae		3			3	
Dytiscidae		3			3	
Chaoborus obscuripes		3				
Chaoborus flavicans			1	1,2,3,5	1	1
Ceratopogonidae		3			3	
Asellus aquaticus	3-8	1	1,2,9	3,5	9	1
Gastropoda	1,4,10					
Pisidium spp.	1,2		1,3,9	1		1
Sphaerium spp.				2		

prior to liming (Lindgren unpubl.). The population abundance of chironomids also increased after liming in lakes Boksjøn and Hovvatn in Norway (Raddum *et al* 1984, 1986).

Measurements of the biomass of chironomids after liming have only been performed in isolated cases, and indicated both increases and decreases (Bergquist 1980, Hultberg and Andersson 1982, Hasselrot *et al* 1984, Alenäs *et al* 1992, Henrikson unpubl.).

In most lakes, densities of chironomids in the profundal zone were lower after some time had elapsed after liming. This may be because liming leads to increased mineralization as a result of increased microbiological activity in the water mass. The "sediment rain" decreases, as does the availability of nutrients for benthic filtering organisms.

Table 4. *Mean density (individuals/m²) of sedimentary Chironomidae in the profundal zone before and after liming. A) Lakes in central Sweden (Mossberg and Nyberg unpubl.); B) lakes in the Delsbo region with water depth of sampling site indicated (Lindgren unpubl.).*

Lake	Abundance (ind/m²) 1977	1987	Change, %
A) Trehörningen	1,860	2,230	20
Långsjön	975	607	-38
Mörtsjön	2,130	871	-59
Västra Skälsjön	1,880	1,250	-34
Östra Skälsjön	1,310	1,160	-11
Stora Sirsjön	720	1,010	40
Blanksjön	422	216	-49
Iglafallssjön	1,230	977	-21
B) Källsjön (8 m)	346	361	4
Källsjön (16 m)	339	141	-58
Lillvallssjön (5 m)	620	483	-22
Lillvallssjön (9 m)	943	411	-56
Nyvallssjön (6 m)	1,030	552	-46
Nyvallssjön (12 m)	1,290	763	-41
Prästvallssjön (4 m)	502	344	-31
Prästvallssjön (8 m)	173	11	-94
Rossjön (2.5 m)	1,380	296	-79
Rossjön (5 m)	26	1,440	5438
Svartvallstjärn (4 m)	213	452	112
Svartvallstjärn (7 m)	33	282	755
Örvallstjärn (4 m)	207	304	47
Örvallstjärn (8 m)	262	852	225
Stensjön (unlimed, 4 m)	2,170	489	-77
Stensjön (unlimed, 8 m)	470	280	-40

As has been pointed out, it is not yet possible to determine with certainty any general pattern for the development of benthic fauna and individual taxa after liming. Further analyses are required. A very cautious interpretation would, however, be that oligochaetes, *A. aquaticus*, pea mussels, snails, mayfly and alder fly all react by increasing their population density after liming, while the density of phantom midges, water boatmen, diving beetles and the chironomids *Chironomus* spp., *Z. zalutschicola*, *Psectrocladius* spp., and *Tanytarsus* spp. declines. When it comes to other taxa, results are more contradictory. For chironomids, this may be attributable to the fact that it is rarely possible to study individual species, which means that shifts of species may occur within genera or species complexes.

One important observation is that changes in the benthic fauna are greatest in the waters which were most severely acidified at the time of treatment. The smallest changes were seen in lakes which were limed before they were seriously affected by acidification (Henrikson *et al* 1989, Lindgren unpubl., Mossberg and Nyberg unpubl.).

2.4. Birds

2.4.1. Background - the impact of acidification

Eriksson (1991) has summarized what is known about how birds are affected by acidification, with references to overviews for the UK by Tyler and Ormerod (1988), and North America by Longcore *et al* (1987), Hansen (1987), Diamond (1989) and Mitchell (1989).

Birds are affected both directly, through increased exposure to metals or reduced uptake of calcium, and indirectly through ecological changes caused by acidification. The latter include changes in food availability and water transparency.

Piscivorous species are naturally affected by reductions in fish populations. In western Sweden reductions in the number of osprey *(Pandion haliaetus)* and red-throated diver *(Gavia stellata)* have been documented (M.O.G. Eriksson *et al* 1983, 1988). The black-throated diver *(Gavia artica)* can compensate to some extent, however, since the transparency tends to increase as the water is acidified so that it becomes easier to discover the remaining fish (Eriksson 1985). This species can also shift to a diet of aquatic insects (e.g. Bergman and Derksen 1977). The goldeneye *(Bucephala clangula)* competes with the fish for prey (Eriksson 1979) and benefits from the disappearance of fish and from the increase in water boatmen and dragon flies, etc. Impaired breeding among dippers *(Cinclus cinclus)* in acidified watercourses in Wales has been linked with a reduction among its prey, such as caddis fly and mayfly larvae (e.g. Ormerod *et al* 1986). Preliminary results from studies of the dipper populations in western Sweden do not indicate that acidification has any effect on this species or that the food supply has changed (Åhlund unpubl. data). The grey wagtail *(Motacilla cinerea)* does not seem to be affected to any great extent by acidification in running waters (Ormerod and Tyler 1991, Johansson and Eriksson 1987).

2.4.2. *The effects of liming*

Populations of black-throated diver, goosander *(Mergus merganser)*, goldeneye, teal *(Anas crecca)* and mallard *(Anas platyrhynhcos)* were studied in eight lakes in south-western Sweden before and after liming (directly in the lake) and simultaneously in six unlimed reference lakes (Eriksson 1987a). The treatments led to increased fish population density and decreased transparency, but it was not possible to relate any changes in the bird populations to the treatment in the individual lakes. When all fourteen lakes were seen as a whole, reductions in goldeneye and mallard were observed, however, which could be related to increased fish population density and probably increased competition from fish. It was not possible to document any major liming-related changes in four other lakes which were studied in western Sweden (Svahnberg 1982, Lundgren 1982a,b).

Breeding success among black-throated diver was compared in thirteen lakes in south-west Sweden before and after liming (Eriksson 1987b). The average number of "large" chicks resulting from each attempt to breed increased from 0.30 before liming to 0.38 after, but this increase was not statistically significant. This lack of clarity may be attributable to the fact that the positive effect of increased fish population density is counteracted by a simultaneous decrease in transparency of the water or in the availability of alternative prey in the form of aquatic insects.

There are indications that the reduction trend in chick production among osprey noted by M.O.G. Eriksson *et al* (1983) in an acidified area of south-western Sweden has halted during the 1980s after extensive liming (Eriksson and Wallin unpubl.).

The size of grey wagtail broods was the same in nests in the vicinity of limed, acidified and unacidified watercourses in south-western Sweden, while the number of fully fledged young was (not significantly) lower in the acidified waters than in the limed ones (Johansson and Eriksson 1987). There was also a tendency towards a lower proportion of utilized nesting sites in the vicinity of limed (previously acidified) watercourses compared with unacidified waters.

During the trial liming period from 1977 to 1981, a certain amount of dry limestone powder was spread by tractor along shorelines and streams. It has not been possible to establish whether this had any definite impact on populations of small birds (Eriksson 1981).

The studies performed on the effects of liming on bird life have concentrated on a fairly short period after liming. In the longer term, fish populations can be expected to stabilize and conditions for birds will probably become more normal. It is also worth mentioning that the studies which are available suggest that there is no reason to fear that liming may have any undesirable impact on bird life. In conjunction with the actual dispersion of the liming agent, however, certain species such as black-throated diver, red-throated diver and osprey may be disturbed at their nesting sites. This can be avoided by not spreading lime during the nesting season. Eriksson (1991) also recommends general restrictiveness with liming in the tarns where red-throated diver nest, since the effects on shoreline vegetation - and thus on the availability of nesting sites - are not yet clearly understood.

2.5. Frogs

2.5.1. Background - the impact of acidification

In laboratory experiments, the development of spawn and tadpoles of three frog species which are found in Sweden - moor frog *(Rana arvalis)*, common frog *(Rana temporaria)* and agile frog *(Rana dalmatina)* - have been studied at different pH values and aluminium levels (Andrén *et al* 1988). At low pH, development was retarded and mortality increased among the spawn of all three species, although it was not possible to establish that aluminium had any significant effect. Tadpole mortality among moor frogs and common frogs increased, however, with both a lower pH and a higher aluminium level. The incidence of deformations among the tadpoles of both species also increased at low pH, an effect which was reinforced by an increase in aluminium. Spawn mortality differed among the three species, as did the time required for development, tadpole mortality, deformation and stress. The agile frog was the most sensitive and was affected by pH between 5 and 6. The common frog was affected below pH 5, while moor frogs could withstand pH down to 4.

This suggests that the moor frog is able to survive relatively well in acidified lakes and probably benefits from the successive loss of fish and consequent reduction in predation by fish. The increase in invertebrate predators may, however, affect the tadpole population (Henrikson 1990). A study by Andrén *et al* (1989) of moor frogs in an acidified area shows that they can possibly adapt to more acidic conditions over a period of 15 generations.

2.5.2. The effects of liming

Little research has been done ragarding the effects of liming on frogs. Andrén *et al* (unpubl.) found that there was a tendency (not significant) for moor frogs to select spawning sites with a higher pH, particularly at lower water temperatures. Liming with limestone powder directly on spawn of all three species in a laboratory experiment indicated higher survival among all species if the liming is performed at an early stage (Andrén *et al* unpubl. data). The limestone powder which came into direct contact with the spawn was not found to have any negative effects. Liming of acidified ponds in the Netherlands resulted in reduced levels of fungal infection and increased hatching success among moor frogs (Bellemakers and van Dam 1992). It would thus appear that liming is beneficial to frogs, but far too little is known as yet to enable more general conclusions to be drawn.

3. Running waters

This section describes the effects of liming on fish and benthic fauna in running waters. Few studies have been performed on birds associated with running waters; the results have been discussed above.

3.1. Fish

3.1.1. Background - the impact of acidification

Running waters in Sweden contain only a few species of fish. Electrofishing has indicated a mean of just 2.4 species per site (Degerman and Sers 1992). Trout is by far the most prevalent species. The fish fauna in watercourses is mainly regulated abiotically, but predation from pike and burbot also has a bearing on the population density of trout (Degerman and Sers 1993). Several trout populations have disappeared during acidification (Nyberg *et al* 1986b, Degerman *et al* 1990). Other species which have disappeared include Atlantic salmon *(Salmo salar)*, grayling, burbot and minnow (Degerman *et al* 1990, 1992b). The population densities of acidification-sensitive species such as eel also decrease (Degerman *et al* 1985, 1986). As pointed out in previous sections, fish populations are harmed through impaired hatching, loss of salt via the gills and possibly through direct suffocation as a result of metal precipitation.

Besides losses of fish species and individuals in acidified watercourses, the fish fauna in unacidified calcium-rich reference watercourses has remained basically unchanged during the same period (Degerman *et al* 1985, 1990). This fact, in combination with a large number of field experiments in both acidified and calcium-rich watercourses, that decreasing fish populations are linked to air pollution has been definitely confirmed (Johansson and Milbrink 1976, Andersson and Nyberg 1984, Berglund and Person 1989, Norrgren and Degerman 1993).

3.1.2. Number of species

From a compilation of data from 22 limed watercourses, it was reported that the number of species in general did not increase significantly after liming (mean 2.8-3.0), but that it did increase significantly in individual watercourses (Degerman *et al* 1990). Salmon, for instance, have returned to several watercourses after liming (Alenäs 1987, Hasselrot and Alenäs

1988, Degerman *et al* 1990) and trout have once again been observed in stretches of water where they have been absent for several years (Alenäs and Hultberg 1988, Degerman *et al* 1990). Trout have also colonized previously acidified lakes via watercourses after liming (Lindström 1991). Even short-term improvements in water chemistry after small lime doses can be beneficial. Minnows, for instance, have returned to watercourses where they were absent during the acidification phase, but along with deliberate re-acidification after liming they moved downstream again (Degerman *et al* 1990).

3.1.3. Recruitment and number of individuals

The density of individual salmonids and other species of fish decreases in acidified watercourses (Edman and Fleischer 1980, Andersson and Andersson 1984, Degerman *et al* 1985, 1986, 1987, 1990, 1992b, Sjölander 1986, Sandberg and Fogelgren 1988). A significant increase in the density of individual biomass after watercourse liming as a result of improved recruitment is often observed among trout, salmon, minnow, crayfish and bullhead *(Cottus gobio)* (Lundh 1981, Andersson *et al* 1984, Hasselrot and Hultberg 1984, Nyberg *et al* 1986a, Alenäs and Hultberg 1988, Sjöstrand 1988, 1991, Appelberg and Odelström 1990, Degerman *et al* 1990, 1992a, Degerman and Appelberg 1992). Norrgren and Degerman (1993) showed via field experiments that salmon roe hatches increased from 0% in unlimed watercourses to 90% a few hundred metres downstream from wetland lim-ing in the same watercourse. The otilith formation of yolk-sac fry also improved (Mosegaard 1990) and survival and salt balance improved in limed watercourses compared with unlimed ones (Norrgren and Degerman 1993). This shows that liming acts directly by reducing the stress caused by low pH and metals and brings both improved reproduction and survival during early stages of growth.

After migration traps were used in two rivers in south-west Sweden, it was found that the increased density of salmon and trout fry also resulted in increased smolt production and subsequently in an increased number of spawning fish (Edman *et al* 1988, Thörnelöf and Degerman 1991).

3.1.4. Growth and production of individuals

Little work has been done on determining the growth of individuals in running waters. The condition factor (the weight/length quotient) has been used instead as a measure of growth potential. The condition factor for trout fry in severely acidified watercourses has been lower than for those in non-acidified ones, while those in moderately acidified watercourses have achieved a somewhat better condition factor than in non-acidified ones (Degerman *et al* 1985, 1987). In other words, a situation similar to the one in lakes occurs: the moderate effect which eliminates competition can be favourable as long as the stress of low pH and metals does not become too great. The production of trout fry in running waters is proportional to the biomass (Mann and Penczak 1986). This means that the general increase in biomass among trout fry after liming indicates a direct increase in production, which is a factor of national importance. It has been estimated that 39-46% of salmon smolt production on the Swedish west coast is dependent upon liming (Appelberg *et al* 1989b, Bengtsson and Bogelius, Chap. 18).

3.1.5. The structure of fish communities

As in lakes, species interaction among fish, i.e. competition and predation, plays a part in the response of fish to acidification and liming. It has been observed that the Alpine bullhead *(Cottus poecilopus)* has increased in number during years when there have been mild acid surges and trout recruitment has been poor, while trout fry have regained their dominant position during years without acid surges (Degerman *et al* 1990). In waters on the Swedish west coast the density of salmon fry decreased sharply prior to liming in the 1970s and 1980s. In addition, the trout fry population expanded somewhat as a result of reduced competition from salmon, but decreased again rapidly when salmon reproduction got going again after liming (Degerman *et al* 1990, Degerman and Appelberg 1992). Lundh (1981) reported that the normal interactions between young sea trout of the same and previous years ceased during the acidification phase, but were observed once again after liming. In other words, the normal biological processes were permitted to function again in the limed watercourse.

Although liming does have a direct positive effect on fish recruitment, it may often take several years before populations regain normal numerical strength. For waters on the Swedish west coast, it has taken more than five years for the density of young salmon to return to normal (Degerman *et al* 1990). In one river, the crayfish population was still expanding 13 years after liming began (Appelberg and Odelström 1990).

3.2. Benthic fauna

3.2.1. Background - the impact of acidification

The reaction of benthic faunas in watercourses to a change in their environment, particularly in terms of acidification, has been studied in detail in Sweden. The impact of acidification is much the same as that which has been documented in lakes, i.e. reduced density or complete elimination of species within sensitive groups such as gastropods, mayflies, crustaceans, mussels and caddis flies (Økland and Økland 1986, Herrmann 1987, 1990, *et al* 1993, Elwood and Mulholland 1989, Sutcliffe and Hildrew 1989). Most stonefly species (Plecoptera), however, are fairly resistant to acidification.

The fact that different taxa react to acidification in different ways may be employed in order to assess the effects of acidification and liming. Two systems of evaluation have been applied in running waters in Sweden (e.g. Henriksson and Medin 1990, Engblom and Lingdell 1992). These include several variables, e.g. documented sensitivity to pH, density, number of species, diversity and similarity index, which provide a more reliable assessment than the reactions of individual taxa alone. These systems have also been used to shed light on the effects of liming.

3.2.2. Number of species and density

A clear increase in the number of species of benthic fauna after liming in running waters has been reported in a number of Swedish studies (Table 5). In the relatively large body of material from the Viskan water system, as well as from the tributaries of Lelången and the Magdabäcken stream, the difference before and after liming in each individual object is greater than indicated in Table 5, often by a factor of between 1.5-2 and 4. It should be noted that comparisons between the examples in Table 5 should not be made with regard to differences in water chemistry and number of species prior to liming, years after liming as well as different aims, selection of sampling sites and ability to identify species correctly. An increase in the total number of species after liming has also been documented in the USA (Herricks and Cairns 1977) and Norway (Fjellheim and Raddum 1992).

Bergquist *et al* (1992) analysed comprehensive material on limed watercourses throughout Sweden and found that the total number of benthic species increased together with the number of treatments and the time elapsing since the first treatment. The increase in species in running waters only became statistically significant after about 5 years.

Even without information about the number of species before liming, one can sometimes assume that there is an increase. In one southern Swedish river, the water chemistry before liming was so severely affected that the species which were found after liming could hardly have been present before (Medin 1991b,c). In several cases in Sweden, the benthic fauna upstream of the liming site, or in a parallel stream, has indicated that treatments have led to an increase in the number of species (Petersen *et al* 1984, Nilsson and Johansson 1985).

There are also several Swedish examples of liming not leading to any increase in the total number of benthic fauna species in watercourses (Medin 1991d,e, Medin *et al* 1993). This may be explained by colonization difficulties and acid surges.

Most species of mayfly are fairly sensitive or very sensitive to acidification. These should therefore be able to return after a successful liming treatment. Table 5 shows some cases where the number of mayfly species increased. The real increase is not always particularly great, but in several cases there has been a shift between species which substantiates an improvement in the conditions. Bergquist *et al* (1992) found that the total number

Table 5. Number of species of A) all macro-invertebrates and B) mayfly larvae (Ephemeroptera) in Swedish watercourses before and after liming. Intervals are shown where several sites or samplings are involved. The liming strategy is indicated by L for lake liming (upstream of the watercourse), W for wetland liming and D for liming with dosers.

Watercourse t. = tributaries	No of species before	after	Years after	Liming strategy	Reference[a]
A) Viskan, 13 sites	15-37	27-44	3-6	L,W,D	1,2,3
Nissan, site 1	24	43	4	W	4
Nissan, site 2	25	38	4	W	4
Lelången t., 7 sites	12-27	14-32	3-4	(L),W	4
Magdebäcken, 9 sites	8-17	16-32	1	W	5
Brunnan	9-15	24-26	1-2	W	6
Foskan	8-15	15-20	1-2	W	6
Storån	16	24	1	L	6
Nian, site 2	29	47	4	L	7
Nian, site 4	22	37	2	W	7
Prästvallsbäcken	9-17	11-31	1-8	W	8
Vasslan (strech)	14-16	30-44	3-8	W	8
B. Viskan, 13 sites	0-6	2-7	3-6	L,W,D	1,2,3
Nissan, site 1	2	5	4	W	4
Nissan, site 2	3	4	4	W	4
Lelången t., 7 sites	1-4	1-8	3-4	L,W	4
Brunnan	3-4	4-5	1-2	W	6
Foskan	0-6	3-4	1-2	W	6
Storån	4	6	1	L	6
Nian, site 4	2	5	2	W	7
Prästvallsbäcken	2-3	1-7	1-8	W	8
Vasslan (strech)	5	4-13	3-8	W	8

[a] References: 1) Henrikson and Medin (1990); 2) Ericsson and Medin (unpubl. data); 3) Ericsson and Medin (unpubl. data); 4) Medin (unpubl. data); 5) Medin and Ericsson (1993); 6) Lingdell and Engblom (1991); 7) Lindgren (1988); 8) Lingdell (unpubl. data).

of benthic faunas increased together with the number of liming treatments and the time which had elapsed since the first treatment, and also reported that the species that showed the greatest tendency to colonize were to be found among the mayflies, stoneflies and beetles (cf. Sect. 5.3).

The mayfly *Baetis rhodani* is probably relatively sensitive to acidification (Raddum and Fjellheim 1984) and has been observed as a returning species after liming (Lingdell and Engblom 1989a, 1992, Fjellheim and Raddum 1992). The sensitivity of the species has been a subject of discussion and different interpretations may depend on the fact that a moderate level of humic substances (colour index approx. 10-100 mg Pt/l) can counteract the negative influence of hydrogen ions (Lingdell and Engblom 1986, Lien *et al* 1992). In general, the total number of mayfly species increases at a given pH with an increased, but moderate, level of humic substances (Kullberg 1992, Herrmann *et al* 1993). The reason may be that the humic substances provide a certain protection against metals. At high colour (>100-200 mg Pt/l) the relatively low nutritional value of humic substances and perhaps also toxicity probably have a negative effect on the mayfly (Petersen and Persson 1987). Bergquist *et al* (1992) found that *B. rhodani* has the highest recolonization potential of any species.

The larvae of the mayfly *Leptophlebia vespertina*, *L. marginata* and *Heptagenia fuscogrisea* and the stonefly *Nemoura cinerea* can withstand very acidic water and survive at pH <4 (Bergquist *et al* 1992). After liming, with fluctuating pH and ion values, *L. vespertina* and the caddis larva *Polycentropus flavomaculatus* may proliferate, which means that these species may be seen as indicators of unstable pH conditions (Lindgren pers. comm.). The reason for this may be that the *Leptophlebia* species are fairly tolerant of low pH levels and can therefore colonize waters successfully when competition from other species is reduced. However, they compete poorly with other species which require a higher and more stable pH. The result in most cases has been a general increase followed by a rapid decrease in riffle sections of running waters, while they have remained prevalent in pool sections (Lingdell unpubl. data). Also, mayflies of the genera *Ephemera* and *Caenis* have shown a clear increase in number (Lingdell unpubl. data).

The necessity of interpreting the observed improvements cautiously cannot be overemphasized. Lingdell and Engblom (unpubl. data) have shown that the mayfly fauna in unlimed waters has gone from relatively species-rich to depletion and back to a more species-rich fauna (indicating high pH) over a 15-20 year period, yet without returning to the original species level. Henrikson and Medin (1990) also noted a clear change towards a more varied benthic fauna despite the fact that no liming was performed. The explanation may be found in the less severe acid surges of recent years and repeated recolonization from nearby watercourses. The underlying reason for this is probably climatic variation, particularly rainfall (cf. Herrmann 1991b).

Bergquist *et al* (1992) found that the number of species of caddis flies, but not stoneflies, increased along with the number of treatments and the time since the initial treatment. Several species of stoneflies and caddis flies are fairly resistant to acidification, but there are several exceptions, especially among the latter. For some of these species, particularly the stonefly *Dinocras cephalotes* and the caddis flies *Hydropsyche* spp.,

Rhyacophila spp., and *Lepidostoma hirtum*, an increase was noted after liming in several European watercourses (Nilsson and Johansson 1985, McCahon *et al* 1989, Fjellheim and Raddum 1992). Beetles generally constitute a diverse and highly varied group which has only received moderate attention. They are therefore of no great value as indicators. The riffle beetles (Elmididae) would appear to be relatively sensitive to acidification, however, and have therefore benefited from liming (Engblom *et al* 1990).

Many crustaceans are highly susceptible to falling pH. Some of them, particularly the gammarid *Gammarus pulex*, has been shown to return after liming (Lingdell and Engblom 1992). It takes some time, however, since it has a low recolonization potential since it is aquatic at all stages of its development (Bergquist *et al* 1992). A fairly neutral watercourse in Wales was acidified experimentally and limed thereafter (McCahon and Poulton 1991), after which it was found that *G. pulex* increased in number. After the addition of both acid and lime had ceased, however, the mortality rate of this organism increased sharply, possibly due to precipitation of aluminium. The precopula phase of *G. pulex*, which is an essential stage in its reproduction, lasting several days, is disrupted by acidification. This probably serious sublethal stress persists even after liming (McCahon and Poulton 1991). *A. aquaticus* also increases in numbers after liming despite the fact that it is fairly tolerant to low pH levels (Petersen *et al* 1984, Raddum *et al* 1984, Nilsson and Johansson 1985). In one stream-living crayfish, liming has been shown to prevent the increased outflow of sodium ions which occurs as the pH falls (Weatherley *et al* 1989).

Snails rarely occur in acidic watercourses (pH below about 6). They are probably not particularly good indicators of brief acid surges since the eggs are fairly resistant (Lingdell pers. comm.). The hardiest species, *Radix peregra* has benefited from liming (Petersen *et al* 1984, Lingdell and Engblom unpubl.). After liming in River Fylleån in south-western Sweden two new species appeared, *Ancylus fluviatilis* and *Physa fontinalis* (Petersen *et al* 1984). The freshwater pearl mussel (*Margaritifera margaritifera*) is susceptible to acidification and field experiments suggest that survival increases in conjunction with liming (Henrikson unpubl.).

The fact that the total number of species increases in waters after liming does not necessarily mean that the number of individuals or total production increases. Nilsson and Johansson (1985) found no general link between pH and the number of individuals. They did find, however, more individuals after liming in four out of nine Swedish watercourses studied. Raddum *et al* (1984), Lindgren (1988) and Rosemond *et al* (1992) found a definite increase in densities after liming. Elwood and Mulholland (1989) summarize by saying that the impact of acidification on total numbers, biomass and production is still unclear, which is probably also the case after liming.

Liming in River Fylleån reduced benthic fauna drift to approximately one-third (Kullberg and Petersen 1987). This represents a recovery, since several studies have shown that lowered pH increases the downstream drift of benthic fauna (Herrmann 1990).

It is questionable whether a clear rise in species number provides proof of successful liming. One reason is that a satisfying rise in production level can take place irrespective of the species number, another is that the colonizing species may be different to those originally present.

3.2.3. Functional groups

Functional groups can be discerned on the basis of the choice of food of the different benthic species, e.g. scrapers (of living attached algae), shredders (of coarse dead vegetation), sediment feeders (of finely divided dead vegetation), filter feeders (of organic particles and small organisms in the water), and predators (Malmqvist *et al* 1978). Each functional group contains species from different taxonomic groups, although certain taxa predominate in certain functional groups. Certain species may also occur in several groups. The supply of different food types may be altered by both acidification and liming, which may cause a shift between different functional groups. It has been pointed out that it is possible for species to switch functional groups if there is a shortage of food or in order to make the most of new possibilities quickly - in conjunction with liming, for instance (Weatherley 1988).

Scrapers, which are mainly species of mayfly and snails, quite often increase after liming (Nilsson and Johansson 1985). The reason for this is not necessarily the pH change in itself, but perhaps the nutrient quality of the epiphytic algae on the bottom has changed as a result of the acidification situation (Sjöström 1990). The number of filtering caddis larvae often appears to increase after liming (Lingdell pers. comm.). In a completely new stream channel, filtering blackfly (Simulidae) larvae and caddis larvae were the first to establish (Malmqvist *et al* 1991). The first colonizing functional group in a watercourse may also be scrapers, followed by sediment feeders and predators (Minshall *et al* 1983). Other factors such as the season in which the liming is performed, the reproduction capacity of the species under study and pure chance are probably just as important to the outcome of a possible (re)colonization.

As pH decreases, the rate of decomposition of leaves and other vegetation material in the watercourses also decreases, which may be beneficial to a relatively large number of species of caddis flies which play an important role as shredders (particularly the case-building Limnephilidae) (Kullberg 1992). The opposite appears to apply in conjunction with liming: this functional group declines in number (Nilsson and Johansson 1985, Kullberg and Petersen 1987). The lower rate of decomposition in conjunction with acidification can also be attributed to reduced bacterial activity, which possibly, although to a lesser extent, applies to fine detritus-rich sediment (Elwood and Mulholland 1989).

The sediment feeders appear to be relatively indifferent to acidification and associated changes. Kullberg and Petersen (1987) found that the proportion of filter feeders remained the same after liming, despite the fact that the number of food particles increased sharply. On the other hand, the quality of the particles (organic content) deteriorated. Kullberg and Petersen (1987) also found that the proportion of predators decreased after liming. Benthic predators are probably fairly tolerant to the chemical changes associated with acidification. They are often not particularly selective, but take whatever prey is available. Predator numbers may increase if liming causes increased numbers of prey.

4. Can the original fauna be restored after liming?

The above shows that the liming of acidified lakes and watercourses produces an, often desirable, change in the fauna. The question is, however, whether such changes can restore the community which existed prior to acidification (cf. Appelberg, Chap. 9, Henrikson and Brodin, Chap. 1). Since little information is available on the pre-acidification fauna, the appearance of the "original" fauna is usually unknown. It is therefore impossible to determine with any certainty whether the liming treatment has succeeded in restoring it.

In Lake Västra Skälsjön, in central Sweden, however, unique documentation is available on the composition of benthic fauna species both before and during acidification and after liming. This lake was studied as far back as 1943 by Brundin (1949) and has been monitored by Mossberg and Nyberg (1979, unpubl.) since 1973. The latter studied the similarities in the occurrence of species and density of individuals at different phases of the lake's development. They found that the fauna during the acidification phase was very similar to that immediately after liming, but quite different to the fauna 7-8 years afterwards (Fig. 4). The benthic fauna 7-8 years after liming displayed major similarities with the fauna in 1943, when the lake had not been anthropogenically acidified.

Development of chironomids in Lake Gårdsjön after liming also indicated a return to previous conditions by virtue of the sharp increase in *Psectrocladius* and *Tanytarsus* (Henrikson unpubl.). The latter genus belongs to the Tanytarsini group which, according to studies of chironomid remains in the sediment, has decreased in Lake Gårdsjön during the present century (Henrikson and Oscarson 1985).

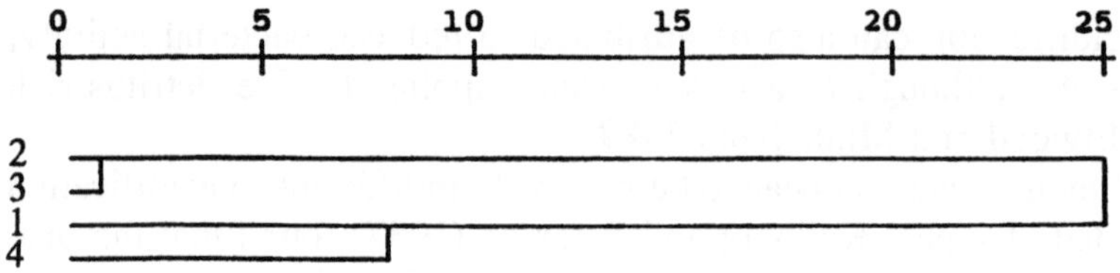

1. Pre-acidification stage
2. Acidified stage
3. Soon after liming
4. Several years after liming

Figure 4. Cluster analysis and dendrogram for densities of different benthic faunas in Lake Västra Skälsjön during different phases of acidification (Mossberg and Nyberg unpubl.).

Bergquist *et al* (1992) analysed a large body of material using cluster analysis and found that the composition of the benthic fauna in limed watercourses was significantly different from that in acidic and moderately weak alkaline (0-0.05 meq/l) reference watercourses, but not from strongly alkaline (>0.05 meq/l) ones (Fig. 5).

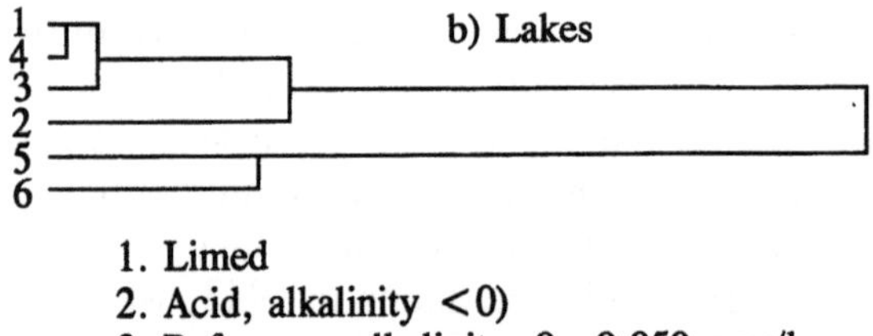

1. Limed
2. Acid, alkalinity <0)
3. Reference, alkalinity 0 - 0.050 meq/l
4. Reference, alkalinity 0.051 - 0.250 meq/l
5. Reference, alkalinity 0.251 - 0.500 meq/l
6. Reference, alkalinity >0.500 meq/l

Figure 5. Cluster dendrogram (complete linkage) based on occurrence of benthic fauna in waters of differing alkalinity. The scale indicates the distance between combined clusters. Only taxa with occurrence of >1% in limed waters have been included in the analysis; a) running waters; b) lakes (littoral zone). From Bergquist et al (1992).

Johnson *et al* (unpubl.) compared benthic faunas in the limed lakes included in the ISEL programme (Integrated Studies on the Effects of Liming) (Appelberg and Aldén 1992) with lime reference lakes, i.e. lakes at different acidification phases which were not to be limed. They found that the limed lakes had significantly lower densities of individuals than the reference lakes in the profundal zone and lower numbers of species in the sublittoral zone. They found no difference between limed lakes and reference lakes in terms of the numbers of species in the profundal zones and species diversity in the sublittoral or profundal zones.

Corresponding comparisons of littoral fauna in unlimed lakes and ISEL lakes have been performed by Appelberg and Aldén (1992). They report no differences in the density of individuals between limed lakes and neutral reference lakes, but they did find higher densities of groups susceptible to acidification in the limed lakes than in the acidified ones. The diversity in the limed lakes assumed an intermediate position between non-acidified and acidified. In a comparison of the similarities in the composition of species, it was found that the fauna in the ISEL lakes had increased over a period of time in the same way as in the unacidified reference lakes. Bergquist *et al* (1992) also showed that the littoral fauna in limed lakes displayed similarities with less affected waters. They found that the composition of species in the limed lakes was significantly different from that in the acidified and less alkaline (0-0.05 meq/l) lakes but not from more alkaline ones (>0.05 meq/l) (Fig. 5).

The zooplankton community in limed lakes displays a normal number of species and thus resembles the original conditions, except that the two large species of cladocerans, *L. kindti* and *B. longimanus* were not found in any lakes after liming. It is, however, difficult to determine the extent to which liming restores the original density of individuals among zooplankton. Nyberg (unpubl.) found mean densities of 15-40 crustaceans/l during the summer in most of the lakes studied, and in Lake Gårdsjön the corresponding density was approximately 20 individuals/l (Stenson and Svensson unpubl.). Sarvala and Halsinaho (1990) found that the mean density of crustaceans was 25 ind/l during the summer in 138 unlimed Finnish lakes. In moderately brown lakes (TOC<10 mg/l) with a pH of 6-7, the mean density was about 28 ind/l. For the year 1968, Morling and Pejler (1990) reported crayfish densities of 9-114 ind/l in 9 lakes in western Sweden which were not yet severely acidified. In four of these lakes the density was >50 ind/l. Densities after liming agreed relatively well with the Finnish findings.

Appelberg and Degerman (1991) observed that fish communities stabilized after liming and that the diversity increased over a period of 1-16 years after the initial treatment. For salmon streams on the Swedish west coast it has taken more than fifteen years before the density of salmon parr returned to normal (Degerman *et al* 1990). In River Högvadsån, the crayfish population was still expanding more than thirteen years after liming commenced (Appelberg and Odelström 1990). In general, however, fish communities return to that known prior to acidification (Thörnelöf and Degerman 1991, Appelberg and Degerman 1991).

In this context, it may also be of interest to note that liming allows normal biological processes, such as interaction between different year classes of salmon and sea trout (Lundh 1981).

The conclusion to be drawn from the above is that, as time goes on, the fauna in limed lakes and watercourses comes to resemble that prior to acidification, i.e. it normalizes. This does not, however, mean that it is identical to that before acidification - in certain cases the differences remain large.

5. Factors affecting the post-liming development of fauna

There are several reasons why various changes occur in different waters and why the fauna in limed waters is not identical to that which (probably) existed prior to acidification. Biotic factors such as predation and competition have been discussed in earlier sections; in the following, water chemistry, liming strategies and recolonization are discussed (cf. Appelberg and Aldén 1992, Appelberg, Chap. 9).

5.1. Water chemistry

The response of the fauna to liming primarily depends on how well the treatment succeeds purely in terms of water chemistry, i.e. how well the toxic effect of hydrogen ions and metals can be eliminated. If adequate detoxification of the water is achieved, species which are susceptible to acidification can become established. If the water quality does not improve sufficiently, it is quite obvious that the biological response will not meet expectations.

The water quality in most limed lakes is less stable than in unlimed ones with the same pH value, since acidified water is constantly being added. Renewed treatments must therefore be performed at regular intervals. In running waters, the maintenance of a stable water quality often involves even greater difficulties.

Problems with fish death in limed waters have mainly been associated with acid surges (in running waters) (Alenäs 1987, Degerman *et al* 1990), oxidation of sulphur and leaching of large quantities of aluminium from the soil during periods of drought (Hultberg and Johansson 1978, Alenäs 1990, Nyström 1990, Degerman *et al* 1990) and theoretically in conjunction with the outflow of acidic surface water from lakes during the winter, although this has not been confirmed by observations (Hultberg 1977, Hasselrot 1984, Abrahamsson 1993). All these causes can be predicted and can therefore be taken into account in the choice of liming strategy and method.

Acid surges can have a negative biological impact. They consist of acidic water, often rich in aluminium, from the surroundings after heavy rainfall or in conjunction with melting snow (Hasselrot *et al* 1987). In lakes, the littoral zones are particularly vulnerable, since the acidic water remains in the upper layer of the water and thus covers the shallow areas. Hultberg and Andersson (1982) considered that acid surges had caused a sharp decline in the density of *Radix peregra* in a lake in south-western Sweden. In Lake Gårdsjön, colonization by more susceptible species in the littoral zone was probably prevented by an influx of acidic and aluminium-rich water in conjunction with melting snow (Henrikson 1988b). Hasselrot *et al* (1987) also reported indirect evidence that *A. aquaticus* is affected negatively by acid surges. In running water, acid surges occur at high water or when lime dosers are inoperative. It is possible that the benthic fauna in running waters can tolerate brief acid surges, but not repeated or protracted episodes (McCahon *et al* 1989). Fish are often very sensitive and Degerman *et al* (1990) showed that even brief acid surges below pH 6 for only a few days in running water can cause fish death.

Another problem is the precipitation of metals in conjunction with liming. Degerman (1992) reported increased aluminium levels in trout immediately downstream of lime dosers. This was probably the result of acidic aluminium-rich water being limed so that the aluminium had become biologically available before having time to stabilize and before sedimentation. Similar observations were made by Lessmark *et al* (1986), who therefore recommended that lime dosers be placed at least 200 m upstream of areas in running waters containing the fish to be protected. Fish death has occurred on fish farms as a result of slow aluminium hydrolysis and precipitation in cold water after lakes immediately upstream had been limed (Dickson 1983). In another case, rainbow trout died after release within a month of liming in a severely acidified and aluminium-rich water (100-700 μg/l) (Dickson 1979). Andersson and Nyberg (1984) and Borg (1986) showed that metals other than aluminium may cause harm to the fauna in acidified waters. High concentrations and a large proportion of particulate iron and manganese during acid surge periods meant that trout in cages died in unlimed waters in northern Sweden. Herrmann *et al* (1993) reported that metal precipitation in unlimed waters may have a serious impact on mayfly larvae. Raddum *et al* (1984) and Keller *et al* (1990b) also suggested that precipitation of metals in conjunction with liming may cause a decline of certain groups of animals.

Rosseland and Hindar (1991) and Wheatherley *et al* (1991) used the concept of mixed zones, which arise when lime is added to an acidified lake or watercourse or when acidic water flows in and is mixed with neutral or limed water. They showed that the death rate among salmon and trout increased in these mixed zones since concentrations of toxic forms of aluminium increase.

Calcium is thought to protect fish against the effects of aluminium at low pH (Brown 1983). In extremely ion-deficient mountain lakes which have been limed with low doses in order to maintain the natural deficiency in electrolytes, relatively low aluminium concentrations (70 μg/l) have been assumed to cause fish death when the calcium level was around 2 mg/l (Lindström *et al* 1984). Such waters should be treated primarily by wetland liming, partly in order to minimize the effect on the water chemistry and partly to bind the metals before they reach the water.

One possibly important reason why the original fauna is not completely restored may be that liming does not always restore the trophic status of the water (Wilander *et al*, Chap. 5). Even if the internal turnover of nutrients in the water improves after liming (Gahnström 1988, Chap. 6), and food supply in the form of phytoplankton (Larsson 1988, Chap. 7, Hörnström *et al* 1993) and organic carbon (Raddum *et al* 1986, Thörnelöf and Degerman 1991, Keller *et al* 1992, Hörnström *et al* 1993) increases, the supply of phosphorus to the surface water is reduced when the catchment area is acidified (Broberg and Persson 1984, Jansson *et al* 1986, Wilander *et al*, Chap. 5). Olofsson *et al* (1988) subsequently suggested that liming should be supplemented with cautious fertilization in order to restore the production of zooplankton. Johnson *et al* (unpubl.) also found that the composition of benthic fauna in the limed lakes indicated more oligotrophic conditions than in the reference lakes. This is to be expected in the light of the fact that "sediment rain" is reduced after liming, which results in poorer food supply. It should, however, be pointed out that the reference lakes in question are richer in nutrients than the limed lakes had ever been. Lake Tolken, south-west Sweden, which was limed early in the acidification phase in 1977, has probably also undergone oligotrophication as indicated by the successive change in species composition of the chironomid fauna during the post-liming decade (Olofsson and Oscarson 1988).

In Lake Gårdsjön, unexpected oxygen depletion arose in the bottom water after liming as a result of an algal bloom, killing off the benthic fauna in the deeper areas (Larsson 1988, Henrikson unpubl.).

5.2. Liming strategies

The best liming strategy is to apply the treatment during the earliest phase of acidification, i.e. before the fauna has been affected to any great extent and the relationships between species are still intact (Henrikson *et al* 1985, 1990). Such protective liming ensures that the original populations can be preserved and that dramatic and perhaps undesirable change can be avoided. Similar views have been put forward by Canadian and Norwegian researchers (Howell *et al* 1991, Raddum and Fjellheim 1994). This is also the reason why the Swedish Environmental Protection Agency (1988) recommends that re-liming be performed at relatively high pH values, i.e. before re-acidification has progressed too far.

The liming method itself is probably not crucial to the fauna, provided the treatment succeeds in purely chemical terms and that the time required for precipitation of metals etc. is taken into account. Degerman *et al* (1990) showed, as long as the pH exceeds 6, that the method is not of any great importance. In many cases, however, it has proven to be difficult to maintain a stable water quality in running waters and in lakes with short turnover times, which means that the liming method or strategy takes on an indirect significance.

Running waters may be limed in different ways. In Sweden, three main methods have been employed: lake liming upstream of the watercourse, lime dosers in running waters and wetland or soil liming, often in drainage areas (Dickson and Brodin, Chap. 4). The question of which method is best has been the subject of considerable discussion.

Nilsson and Johansson (1985) found that the number of species among benthic faunas in running waters increased when an upstream lake was limed. Liming with dosers also led to an increase in species, but since the studies were performed in large watercourses the initial status was probably relatively good. Soil liming appeared to be less successful. It is not possible to draw any general conclusions from this study, however, since there were no data on the benthic faunas prior to liming and only three watercourses were studied for each liming method.

When liming was performed via dosers, Engblom and Lingdell (1985a) found that the longer the treatment, the greater the diversity of species downstream became (cf. Bergquist *et al* 1992). Both Engblom and Lingdell (1985b) and Henrikson *et al* (unpubl. data) claim, however, that really good results can only be obtained using dosers if they are supplemented by lake or soil liming (Fig. 6). Liming of upstream lakes alone yielded poorer results in terms of the fauna in the watercourses, but it should be pointed out that the distance from the sites studied in the watercourse to the limed lake upstream had not been taken into account. The closer the watercourse is to the limed lake, the better the effects on the fauna.

Engblom and Lingdell (1985 a,c) listed the following undesirable effects of liming by dosers: clouding of the water in conjunction with installation, and to some extent day-to-day operation, may increase the downstream drift of benthic fauna. The lime often accumulates in large banks on the bottom and these may be inactivated as a result of precipitating humic substances or algal overgrowth. Doser liming may also involve a negative change in

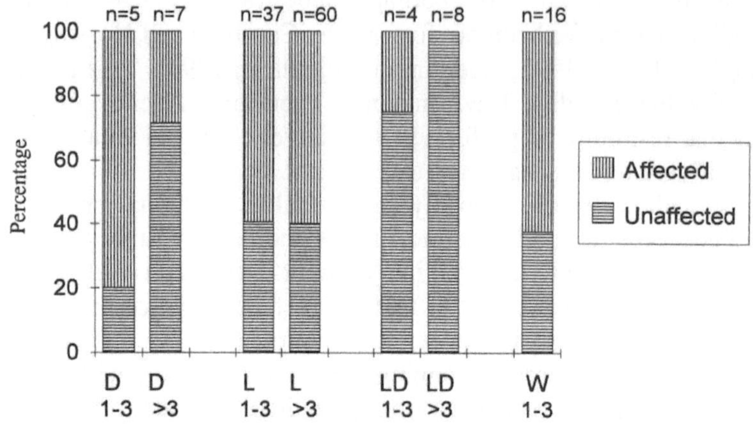

Figure 6. *Assessment of impact of acidification in running waters on the basis of composition of benthic fauna (Henrikson and Medin 1990) according to different strategies and numbers of years after liming. D = doser; L = liming of upstream lake; LD = combination of lake liming and doser; W = wetland liming; n = number of sites (Henrikson et al unpubl. data).*

the bottom structure by virtue of the reduction in physical variation. Such lime banks may well counteract drastic reductions in pH downstream if the doser is out of action. Engblom and Lingdell furthermore claimed that a doser which is out of action at a critical time is more serious than no doser at all. The reason is probably that breakdowns may lead to wide fluctuations in pH and metal concentrations and that the fauna which has developed over a long period of liming may suffer severely. Despite the technical improvements of today's dosers, they are still heavily dependent on good care.

Lessmark *et al* (1986) found high mortality among young salmon downstream of lime dosers which were out of action. Dosers have often been unreliable, but today's dosers have undergone technical improvements, as have routines for maintenance and monitoring.

In several studies from northern Sweden, it has been pointed out that wetland liming yields the best results (Ahlström and Isaksson 1990, Lindgren 1988). The reason could be that metals are fixed in the discharge and lake littoral zones, and the pH rises more slowly with a longer-term buffer capacity and a more even supply of limed water.

In some cases, lime powder was dispersed from a bulk truck directly into the stream and along the shoreline. The biological effect can hardly be expected to be adequate since the water quality probably varies far too much when bulk liming is carried out only on isolated occasions (usually spring and autumn). This was also observed in two watercourses where streams were limed in this way showing no any benthic fauna recovery

(Medin 1991f). In another watercourse, however, species which were susceptible to acidification were found (Ericsson and Medin 1992), which may be attributable to the fact that the mixing of lime with the gravel bottom had created places of refuge for such species.

Berglund and Persson (1989) demonstrated in the field that it is possible to mix crushed limestone in grayling spawning grounds in acidified waters and thereby increase the hatching rate. Large-scale trials with crushed limestone have yielded poor results in watercourses, however (National Board of Fisheries and Swedish Environmental Protection Agency 1981, Wilander *et al*, Chap. 5), with the result that this method is not recommended.

Thus different liming methods in running waters have provided various results, and combinations of methods, particularly liming of upstream lakes and dosers, appear to be best. However, in several studies, sufficient time has not passed to allow an assessment of whether imperfect restoration is due to liming methods or other reasons, e.g. low recolonization rate.

As an alternative to full-scale liming, trials have been performed with partial liming in six lakes (Andersson and Alenäs 1988) using 10% of the full dose on shallow spawning grounds. Perch, which can withstand acidification, increased significantly while roach was unable to survive since the water was still acidic. The natural fauna therefore could not be re-established. The perch in these lakes probably had high levels of mercury (Andersson and Alenäs 1988), which means that partial liming cannot be considered a viable alternative to full-scale methods.

5.3. Recolonization

One crucial factor in the development of communities of various organisms after liming is recolonization. This is the process by which species that have been eradicated as a result of environmental impact - such as acidification - return once adequate conditions have been restored (after liming). What actually happens, however, consists of several parallel processes which, biologically, may be seen as parts under the concept of dispersal biology.

New individuals are constantly dispersing from places where they have survived, such as unacidified waters, depending on the ability of the species in question and how difficult it is for them to reach their destination (e.g. the limed water). Success depends on the distance involved and the size and form of the new destination. The recolonization ability of a species is of great importance. Flying insects or zooplankton which can be carried by birds establish themselves much faster than those with entirely aquatic life cycles, such as fish, snails and mussels. In addition to such continuous colonization there is a succession of organisms which appear in the limed water for a short time and then die out. It should be pointed out that if a species is first observed after liming, this does not necessarily mean that it has colonized the water. What may be interpreted as migration may simply be that so few individuals remained at favourable sites during the acidification period that they could not be found.

The dispersal of species which are susceptible to acidification may take place from lakes or areas of watercourses, upstream or downstream, where the species has survived. Fish are able to migrate actively as long as they are not obstructed by natural or man-made barriers. Fully aquatic organisms such as crustaceans and snails can also migrate actively in running water. Insects may drift downstream during their larval stages or fly upstream as adults. They may also disperse from waters in nearby systems, but they must do so overland, either by flying or passive transport by other animals. Bergquist *et al* (1992) have claimed that colonization by benthic fauna after acidification of running waters is primarily airborne. Upstream migration takes place to a lesser extent. They considered that lake colonization by benthic fauna is mainly via the air or via bird or fish transport.

Blackfly larvae, net-spinning caddis larvae and mayflies are relatively rapid colonizers in running waters (Malmqvist *et al* 1991). Milner (1987) found, however, that mayflies and stoneflies colonized before caddis flies. Bergquist *et al* (1992) found that about one-third of the mayfly and stonefly species and a somewhat lower proportion of caddis fly species were able to recolonize. They also found that the species with the greatest propensity to recolonize were stoneflies, mayflies and beetles. Species of zooplankton with dormant eggs can be spread via birds.

Bergquist *et al* (1992) showed how benthic faunas colonize lakes and watercourses and that the number of species increases with time. They also showed that the benthic fauna does not achieve anything resembling its original composition until 5-10 years after liming. Recolonization is thus a relatively slow process, as confirmed by Henrikson *et al* (unpubl. data) in an analysis of 140 sites in limed watercourses. They found that almost half of these were still moderately or severely affected by acidification (a biological assessment based on the composition of the benthic fauna) after at least five years of liming treatments (Fig. 7). The success rate will probably improve as time goes on. The liming strategy was not taken into account (cf. above). This analysis also shows that the biological effects must be studied in order to evaluate the outcome of liming operations - chemical analyses alone are not enough. Weatherley (1988) reported that, in several cases in Europe, vigorous populations of the original fauna had still not returned 6-8 years after lime treatment in running water.

There are probably many factors which determine how well recolonization will succeed. A number of problems are involved for benthic fauna in

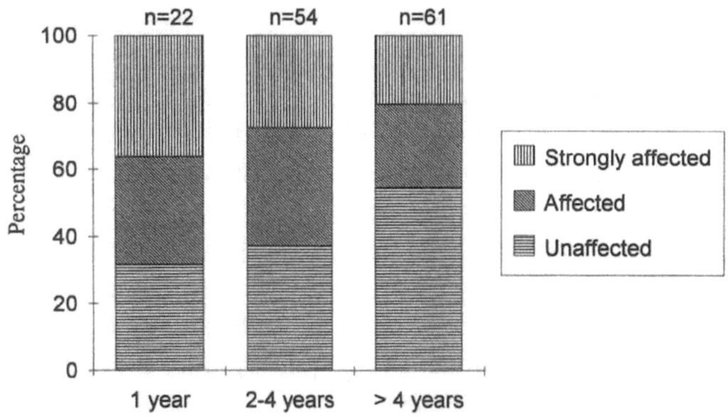

***Figure** 7. Assessment of impact of acidification in running waters on the basis of composition of benthic fauna from 140 sites (cf. Henrikson and Medin 1990) after 1, 2-4 and >4 years after liming; n = the number of sites (Henrikson et al unpubl. data).*

running waters in conjunction with liming. Acid surges are a difficult problem to overcome and have been discussed to some extent earlier. The fact that recolonization does not occur despite apparent improvements in water chemistry may be attributable to short-term acid surges which, however, not have been documented in water chemistry sampling programmes. Although acceptable mean values are obtained in terms of water chemistry, there is still considerable instability, possibly even more so than before liming. The lower the pH before treatment and the longer the delay in treatment, the more likely it is that problems of this sort will arise - with the result that only a few species may occur. Another problem is to be found in re-acidification which occurs when the lime dosage and frequency are inappropriate. This may lead to increasing metal concentrations, although levels of humic and other organic substances do not increase at the same rate. These substances may provide protection against metals, as well as having a buffer effect on the pH of the water (Herrmann *et al* 1993). Precipitation of aluminium hydroxide and other metals in the sediment may prevent recolonization by benthic fauna (Herricks and Cairns 1977, Raddum *et al* 1984). On the other hand, pH conditions may be more favourable a couple of cm below the surface of the bottom sediment (Weatherley 1988, Herrmann 1990, Lingdell pers. comm.). Moreover, it has been shown that aluminium may accumulate on or in organisms as a result of adsorption (Engblom and Lingdell 1985c, McCahon *et al* 1989). On the basis of experiments in three small artificial streams, Griffiths (1992) believes that if acid deposition were to cease - without liming - biological recovery of acidified watercourses would be possible provided potential colonizers have survived in the vicinity.

Recovery, measured in terms of the number of species returning, appears to take place more rapidly in the lower stretches of a limed watercourse, perhaps because there are a greater number of species there and it is easier to perform successful treatment. Higher up in the watercourse, species have often been killed off to a greater extent. Several species are therefore missing, which facilitates recolonization. Re-establishment of species which live only in the running waters in the lower reaches of a watercourse seem to be considerably impeded by larger lakes (Medin 1991a). The above suggests that species which can withstand acidification might be reasonably expected to colonize more quickly than those which are more susceptible (Bergquist *et al* 1992).

Another important factor in recolonization is the way in which animal communities in the acidified lakes change after liming. Weatherley (1988) suggested that an increase in the density of one or more existing species prevents other species from returning due to the competition. Venabäcken stream in central Sweden, which was probably acidified previously, is a good example. This stream once had a good population of the caddis fly *Hydropsyche* and the mayfly *Ephemera*. After liming, a community was formed in which the caddis fly *Cheumatopsyche* achieved considerable dominance and only a few of the original species were able to establish themselves (Lingdell pers. comm.). Raddum and Fjellheim (1994) put forward similar views and claimed that animal communities in severely acidified environments have a "conservational" effect and delay recolonization and the establishment of a fauna structure resembling the one which originally existed.

One way of hastening a return to a close resemblance of the original composition of species might be to facilitate recolonization or to reintroduce species, i.e. to implement supplementary measures, as discussed in detail by Bergquist (Chap. 17).

6. References

Abrahamsson, I. 1993. Impact of overflows on acid-base chemistry in limed lakes. - Vatten 49: 24-33.

Ahlström, J. and Isaksson, K.-E. 1990. Försurningsläget i Norrlands inland och fjälltrakter - Kalkning. [The acidification status of the inner parts and mountains of northern Sweden - liming.] - Swedish Environm. Prot. Agency 3781.

Alenäs, I. 1986. Kalkningsprojektet Härskogen 1976-1986. - [The Härskogen liming project 1976-1986.] - Swedish Environm. Res. Inst., B 846.

Alenäs, I. 1987. Projekt Anråse å. [The stream Anråse å project.] - Swedish Environm. Res. Inst. B 847.

Alenäs, I. 1990. Försurningsskador värst efter torka. [Most severe damage of acidification after drought.] - Municipality of Falkenberg, unpubl. rep.

Alenäs, I. and Hultberg, H. 1988. Slutrapport för projektet Stenunge å - utvärdering av olika markkalkningsåtgärder avseende vattenkemi och havsöringsbestånd. [Final report on the stream Stenunge å project - evaluation of soil liming measures on water chemistry and sea trout population.] - Swedish Environm. Res. Inst., B 904.

Alenäs, I., Hultberg, H. and Andersson, I. 1982. Kalkningsprojektet Härskogen 1976-82. [The Härskogen liming project 1976-1982.] - Swedish Environm. Res. Inst., rep.

Alenäs, I., Andersson, I.B., Hultberg, H. and Rosmarin, A. 1992. Liming and reacidification reactions of a forest lake ecosystem, Lake Lysevatten, in SW Sweden. - Water Air Soil Pollut. 59: 55-77.

Almer, B. 1972. Försurningens inverkan på fiskbestånd i västkustsjöar. [English summary: The effect of acidification on fish stocks in lakes on the west coast of Sweden.] - Inform. Inst. Freshw. Res., Drottningholm (12).

Almer, B. 1977. Ragnarök och fiskdöd - om försurningen av våra vatten. [Twilights of the gods and fish death - concerning the acidification of our waters.] - Sveriges Natur 6: 360-366.

Almer, B. and Hansson, M. 1980. Försurningseffekter i västkustsjöar. [English summary: Effects of acidification in west coast lakes.] - Inform. Inst. Freshw. Res., Drottningholm (5).

Almer, B., Dickson, W., Ekström, C., Hörnström, E. and Miller, U. 1974. Effects of acidification on Swedish lakes. - Ambio 3: 30-36.

Almer, B., Dickson, W., Ekström, C. and Hörnström, E. 1978. Sulphur pollution and the aquatic environment. - In: Nriago, J.O. (Ed.) Sulphur in the environment. Part II. Wiley & Sons, New York, pp. 271-311.

Andersson, B. and Andersson, P. 1984. The distribution of trout (*Salmo trutta* L.) in relation to pH - an inventory of small streams in Delsbo, central Sweden. - Rep. Inst. Freshw. Res., Drottningholm 61: 28-33.

Andersson, G. 1985. Decomposition of alder leaves in acid lake waters. - Ecol. Bull. (Stockholm) 37: 293-299.

Andersson, G., Dickson, W., Filipsson, O., Lindström, T. and Öhman, R. 1980. Förändringar i södra fjällområdets fiskfauna - ett samspel mellan försurning och andra faktorer. [English summary: Changes in the fish populations of the southern parts of the Swedish mountain chain - interaction between acidification and other factors.] - Inform. Inst. Freshw. Res., Drottningholm (10).

Andersson, I. and Alenäs, I. 1988. Slutrapport för projektet "Partiell kalkning - en metod att bibehålla eller åstadkomma exploaterbara fiskpopulationer i vissa försurade sjöar". [Final report on the project "partial lake liming - a method to maintain or create usable fish populations in certain acidified lakes.] - Swedish Environm. Res. Inst. B 884.

Andersson, I.B., Alenäs, I. and Hultberg, H. 1984. Liming of a small acidified river (River Anråseån) in southwestern Sweden, promoting successful reproduction of sea trout (*Salmo trutta*). - Rep. Inst. Freshw. Res., Drottningholm 61: 16-27.

Andersson, P. and Nyberg, P. 1984. Experiments with brown trout (*Salmo trutta* L.) during spring in mountain streams at low pH and elevated levels of iron, manganese and aluminium. - Rep. Inst. Freshw. Res., Drottningholm 61: 34-47.

Andersson, P., Grahn, P., Hörnström, E., Nyberg, P. and Dahlquist, K. 1989. Kalkning och gödsling i Rammsjön och Ämten, Örebro län 1980-87. Ett försök att minska kvicksilverhalten i fisk. [Liming and fertilization of Lake Rammsjön and Lake Ämten, county of Örebro, 1980-1987. An attempt to decrease mercury levels in fish.] - Swedish Environm. Prot. Agency 3584.

Andrén, C., Henrikson, L., Olsson, M. and Nilson, G. 1988. Effects of pH and aluminium on embryonic and early larval stages of Swedish brown frogs *Rana arvalis*, *R. temporaria* and *R. dalmatina*. - Holarct. Ecol. 11: 127-135.

Andrén, C., Mårdén, M. and Nilson, G. 1989. Tolerance to low pH in a population of moor frogs, *Rana arvalis*, from an acid and a neutral environment: a possible case of

rapid evolutionary response to acidification. - Oikos 56: 215-223.
Appelberg, M. 1987. Some factors regulating the crayfish *Astacus astacus* L. in acid and neutralized waters. - In: Witters, H. and Vanderborght, O. (Eds.) Ecophysiology of acid stress in aquatic organisms. Ann. Soc. R. Zool. Belg. 117, Suppl. 1, pp. 167-179.
Appelberg, M. 1989. Effektuppföljning av fisk i svenska kalkade sjöar och vattendrag. [Monitoring studies of fish in limed lakes and streams in Sweden.] - In: Kalking i vann og vassdrag. [Liming of lakes and streams.]. Direct. Nature Managem. (Norway), DN-Notat nr 4, pp. 39-44.
Appelberg, M. 1990. Population regulation of the crayfish *Astacus astacus* L. after liming an oligotrophic, low-alkaline, forest lake. - Limnologica 20: 319-327.
Appelberg, M. and Aldén, U. 1992. Integrerad uppföljning av kalkningens effekter på sjöar och vattendrag - en treårsrapport. [Integrated monitoring in limed lakes and streams - a three year study.] - Inform. Inst. Freshw. Res., Drottningholm (4): 1-60.
Appelberg, M. and Degerman, E. 1991. Development and stability of fish assemblages after lime treatment. - Can. J. Fish. Aquat. Sci. 48: 545-554.
Appelberg, M. and Odelström, T. 1990. Kräftor i sura och kalkade vatten. [English summary: Effects of acidification and liming on the freshwater crayfish *Astacus astacus*.] - Inform. Inst. Freshw. Res., Drottningholm (4): 1-25.
Appelberg, M., Degerman, E. and Nyberg, P. 1989a. Species composition and relative abundance of the fish fauna in acidified lakes in Sweden. - In: Longhurst, J.W.S. (Ed.) Acid deposition. Sources, effects and controls. British Library, Tech. Inf., pp. 143-160.
Appelberg, M., Degerman, E., Johlander, A. and Karlsson, L. 1989b. Liming increases the catches of Atlantic salmon on the west coast of Sweden. - Nordic J. Freshw. Res. 65: 44-53.
Appelberg, M., Ekström, C. and Hörnström, E. 1990. Stora Härsjön - ett exempel på integrerad uppföljning av kalkningens effekter. [English summary: Lake Stora Härsjön - a case study for integrated monitoring of the effects of liming.] - Inform. Inst. Freshw. Res., Drottningholm (1).
Appelberg, M., Degerman, E. and Norrgren, L. 1992. Effects of acidification and liming on fish in Sweden. A review. - Finnish Fish. Res. 13: 77-91.
Appelberg, M., Henrikson, B.-I., Henrikson, L. and Svedäng, M. 1993. Biotic interactions within the littoral community in Swedish forest lakes during acidification. - Ambio 22: 290-297.
Baalsrud, K., Hindar, A., Johannessen, M. and Matsow, D. (Eds.) 1985. Liming of acid water. Liming project, final report. - Dep. Environm. Direct. Nat. Managem. (Norway).
Barlaup, B.T., Åtland, Å., Raddum, G.G. and Kleiven, E. 1989. Improved growth in stunted brown trout (*Salmo trutta* L.) after reliming of Lake Hovatn, southern Norway. - Water Air Soil Poll. 47: 139-151.
Bellemakers, M.J.S. and van Dam, H. 1992. Improvement of breeding success of the moor frog (*Rana arvalis*) by liming of acid moorland pools and the consequences of liming for water chemistry and diatoms. - Environm. Poll. 78: 165-171.
Bengtsson, B. 1982. Acid rain mitigation strategies in Sweden. - AIAA 20th Aerospace Sciences Meeting, Jan 11-14, Orlando, Florida.
Bengtsson, B., Dickson, W. and Nyberg, P. 1980. Liming acid lakes in Sweden. - Ambio 9: 34-36.
Berge, F., Brodin, Y.-W., Cronberg, G., El-Daoushy, F., Høeg, H.I., Nilssen, J.-P., Renberg, I., Rippey, B., Sandøy, S., Timberlid, A. and Wik, M. 1990. Palaeolimnological changes related to acid deposition and land-use in the catchments of two Norwegian soft-water lakes. - Phil. Trans. R. Soc. London B 327: 385-389.
Berglund, I. and Persson, B.-G. 1989. Reproduktionsstrategiens betydelse för överlevnaden hos olika harrstammar i försurade och kalkade vatten. [The importance of reproductive

strategy for the survival of grayling strains in acidified and limed waters.] - Dep. Ecol. Zool., Univ. Umeå, unpubl. rep.

Bergman, R.D. and Derksen, D.V. 1977. Observations on Arctic and red-throated loons at Storkersen Point, Alaska. - Arctic 30: 40-51.

Bergquist, B.C. 1980. Undersökning av bottenfaunaförekomst, samt kvicksilverhalt i bottenfauna och sediment i sjön Ölen, Örebro län, 1976-1980. [Studies of the occurrence of benthos and the mercury content of benthos and sediment in Lake Ölen, county of Örebro.] - Dep. Limnol., Univ. Uppsala, Rep. 1980:B9.

Bergquist, B.C. 1991. Extinction and natural recolonization of fish in acidified and limed lakes. - Nordic J. Freshw. Res. 66: 50-62.

Bergquist, B.C., Engblom, E. and Lingdell, P.-E. 1992. Förekomst och kolonisation av bottenfauna i kalkade vatten. [English summary: Occurrence and colonization of benthos in limed waters.] - Inform. Inst. Freshw. Res., Drottningholm (4): 79-108.

Bernes, C. 1991. Monitor 12. Acidification and liming of Swedish freshwaters. - Swedish Environm. Prot. Agency Informs.

Bevanger, K. and Ålbu, O. 1986. Decrease in Norwegian feral mink (*Mustela vison*) population - a response to acid precipitation? - Biol. Conserv. 38: 75-78.

Björklund, I., Borg, H. and Johansson, K. 1984. Mercury in Swedish lakes - regional distribution and causes. - Ambio 13: 118-121.

Björklund, I., Haux, C., Hogstrand, C., Unger, M. and Örn, T. 1985. Bioackumulation i organ och förändringar av jonbalans i blodplasma hos öring vid påverkan av aluminium vid olika pH, humushalt och vattentemperatur. [Bioaccumulation in organs and changes of the ion balance in blood plasma in brown trout with influence of aluminium at different pH, humic content and water temperature.] - Swedish Environm. Prot. Agency 3046.

Blomqvist. P., Bell, R.T., Olofsson, H., Stensdotter, U. and Vrede, K. 1993. Pelagic ecosystem responses to nutrient additions in acidified and limed lakes in Sweden. - Ambio 22: 283-289.

Borg, H. 1986. Metals speciation in acidified mountain streams in central Sweden. - Water Air Soil Poll. 30: 1007-1014.

Brett, M.T. 1989a. Zooplankton communities and acidification processes (a review). - Water Air Soil Poll. 44: 387-414.

Brett, M.T. 1989b. The rotifer communites of acid-stressed lakes of Maine. - Hydrobiologia 186/187: 181-189.

Broberg, O. and Persson, G. 1984. External budgets for phosphorus, nitrogen and dissolved organic carbon for the acidified Lake Gårdsjön. - Arch. Hydrobiol. 99: 160-175.

Brodin, Y.-W. (Ed.) 1990. Effekter av svavel- och kvävebelastning på skogsmark, yt- och grundvatten. [English summary: Effects of sulphur and nitrogen load on forest soil, surface and groundwater.] - Swedish Environm. Prot. Agency 3762.

Brown, D.J.A. 1983. Effect of calcium and aluminium concentrations on the survival of brown trout (*Salmo trutta*) at low pH. - Bull. Environ. Contam. Toxicol. 30: 582-587.

Brundin, L. 1949. Chironomiden und andere Bodentiere der Südschweden Urgebirgseen. - Rep. Inst. Freshwat. Res., Drottningholm 32.

Bukaveckas, P.A. 1991. Effects of lake and watershed liming on light attenuation, thermal stratification, and phytoplankton production. - In: Olem, H., Schreiber, R.K., Brocksen, R.W. and Porcella, D.B. (Eds.) International lake and watershed liming practices. The Terrene Inst., Washington D.C., pp. 129-144.

Carvalho, G.R. and Wolf, H.G. 1989. Resting eggs of lake-*Daphnia*. I. Distribution, abundance and hatching of eggs collected from various depths in lake sediments. - Freshw. Biol. 22: 459-470.

Degerman, E. 1987. Humösa sjöar - en litteratursammanställning med inriktning på fisk och försurning. [English summary: Humic lakes - a literature survey with emphasis on

fish and acidification.] - Swedish Environm. Prot. Agency 3415.
Degerman, E. 1989. Redovisning av enskilda kalkningsprojekt med sjöprovfiske inom fiskeristyrelsens försöksverksamhet med kalkning av ytvatten. [Report on the test fishing in individual projects/lakes included in the program for experimental liming of surface waters administered by the Swedish Board of Fisheries.] - Inst. Freshw. Res., Drottningholm, unpubl. rep.
Degerman, E. 1992. Kalkning för djur i de rinnande vattnen. [Liming for animals in the running waters.] - In: Strategi vid kalkning av rinnande vatten. [Strategy for liming of running waters.] 12-13/5 1992, Umeå. Swedish Environm. Prot. Agency.
Degerman, E. and Appelberg, M. 1992. The response of stream-dwelling fish to liming. - Environm. Poll. 78: 149-155.
Degerman, E. and Nyberg, P. 1987. Fiskfaunans sammansättning och täthet i försurade och kalkade sjöar - en arbetsrapport. [English summary: The composition and abundance of the fish fauna in acidified and limed lakes in Sweden.] - Inform. Inst. Freshw. Res., Drottningholm (7).
Degerman, E. and Nyberg, P. 1989. Effekter av sjökalkning på fiskbestånd i sjöar. [English summary: Long-term effects of liming on fish populations in Sweden.] - Inform. Inst. Freshw. Res., Drottningholm (5).
Degerman, E. and Sers, B. 1992. Fish assemblages in Swedish streams. - Nordic J. Freshw. Res. 67:61-71.
Degerman, E. and Sers, B. 1993. A study of interactions between fish species in streams using survey data and the PCA-hyperspace technique. - Nordic J. Freshw. Res. 68.
Degerman, E., Fogelgren, J.E., Tengelin, B. and Thörnelöf, E. 1985. Förekomst och täthet av lax, öring och ål i mindre vattendrag på svenska västkusten. [English summary: Occurrence of salmonid parr and eel in relation to water quality on the west coast of Sweden.] - Inform. Inst. Freshw. Res., Drottningholm (1).
Degerman, E., Fogelgren, J.E., Tengelin, B. and Thörnelöf, E. 1986. Occurrence of salmonid parr and eel in relation to water quality on the west coast of Sweden. - Water Air Soil Poll. 30: 665-671.
Degerman, E., Lindgren, G., Lingdell, P.-E. and Nyberg, P. 1987. Kartering av strömfauna och fisk i mindre vattendrag i Norrlands inland och fjälltrakter i relation till försurning. [English summary: An inventory of benthic fauna and fish in small streams in mountainous regions of northern Sweden affected by acidification.] - Inform. Inst. Freshw. Res., Drottningholm (2).
Degerman, E., Sjölander, E., Johlander, A., Sjöstrand, P., Höglind, K., Thorsson, L. and Carlstrand, H. 1990. Kalkning för att motverka försurningpåverkan på fisk i rinnande vatten. [English summary: Liming of acidified running waters to improve conditions for fish.] - Inform. Inst. Freshw. Res., Drottningholm (4): 27-214.
Degerman, E., Appelberg, M. and Nyberg, P. 1992a. Effects of liming on the occurrence and abundance of fish populations in acidified Swedish lakes. - Hydrobiologia 230: 201-212.
Degerman, E., Engblom, E., Lingdell, P.-E., Melin, E. and Olofsson, E. 1992b. Försurning i fjällen. [English summary: Acidification in the mountains?] - Inform. Inst. Freshw. Res., Drottningholm (1).
Diamond, A.W. 1989. Impact of acid rain on aquatic birds. - Environ. Monitor. Assess. 12: 245-254.
Dickson, W. 1979. Ett exempel på metalltoxicitet vid försurning och kalkning. [An example of the toxicity of metals at acidification and liming.] - Aquannalen, Lund 1: 2-7.
Dickson, W. 1983. Liming toxicity of aluminium to fish. - Vatten 39: 400-404.
Dickson, W., Hörnström, E., Ekström, C. and Almer, B. 1975. Rödingsjöar söder om Dalälven. [English summary: Char-lakes south of River Dalälven.] - Inform. Inst. Freshw. Res., Drottningholm (7).

Edman, G. and Fleischer, S. 1980. The River Högvadsån liming project - a presentation. - In: Drabløs, D. and Tollan, A. (Eds.) Proc. Int. Conf. Ecol. impact acid precip., Norway, SNSF project., Oslo-Ås, pp. 300-301.

Edman, G., Fleischer, S., Fritz, Ö. and Stibe, L. 1988. Högvadsån 1978-1986. Försurad, kalkad, pånyttfödd. [River Högvadsån. Acidified - limed - regenerated.] - Halland County Council, Rep. 4.

Elwood, J.W. and Mulholland, P.J. 1989. Effects of acidic precipitation on stream ecosystems. - In: Adriano, D.C. and Johnson, A.H. (Eds.) Acidic precipitation, vol 2. Biological and ecological effects. Springer, Berlin Heidelberg New York, pp. 85-135.

Engblom, E. and Lingdell, P.-E. 1984. The mapping of short-term acidification with the help of biological pH indicators. - Rep. Inst. Freshw. Res., Drottningholm 61: 60-68.

Engblom, E. and Lingdell, P.-E. 1985a. Hur påverkar kalkdoserare bottenfaunan? [How do lime dosing devices affect the benthic fauna?].- Swedish Environm. Prot Agency 1994.

Engblom, E. and Lingdell, P.-E. 1985b. Hur påverkar reningsverk med olika fällningskemikalier bottenfaunan? [How do different precipitation chemicals in sewage treatment plants affect the benthic fauna?] - Swedish Environm. Prot. Agency 1998.

Engblom, E. and Lingdell, P.-E. 1985c. Synpunkter på kalkdoserare och markkalkning. [Aspects on lime dosing devices and forest soil liming.] - Limnodata Consultant Co.

Engblom, E. and Lingdell, P.-E. 1987. Vilket skydd har de vattenlevande smådjuren i landets naturskyddsområden? En studie av försurnings- och föroreningsförhållanden. [How protected are the aquatic invertebrates in national parks and nature reserves of Sweden? A study of the acidification and pollution status.] - Swedish Environm. Prot. Agency 3349.

Engblom, E. and Lingdell, P.-E. 1992. Kalkningseffekter på bottenlevande smådjur i Rödån. [Liming effects on benthic invertebrates in the River Rödån.] - Limnodata Consultant Co.

Engblom, E., Lingdell, P.-E. and Nilsson, A. 1990. Sveriges bäckbaggar (Coleoptera, Elmididae) - artbestämning, utbredning, habitatval och värde som miljöindikatorer. [English summary: The Swedish Elmididae (Coleoptera) - identification, distribution, habitat preferences, and environmental indication.] - Ent. Tidskr. 111: 105-121.

Eriksson, F., Hörnström, E., Mossberg, P. and Nyberg, P. 1982. Ekologiska effekter av kalkning i försurade sjöar och vattendrag. [English summary: Ecological effects of lime treatment in acidified surface waters.] - Inform. Inst. Freshw. Res., Drottningholm (6).

Eriksson, F., Hörnström, E., Mossberg, P. and Nyberg, P. 1983. Ecological effects of lime treatment of acidified lakes and rivers in Sweden. - Hydrobiologia 101: 145-164.

Eriksson, M.O.G. 1979. Competition between freshwater fish and goldeneyes, *Bucephala clangula* (L.), for common prey. - Oecologia (Berlin) 41: 99-107.

Eriksson, M.O.G. 1981. Undersökningar av störningseffekter på fågelfaunan i samband med kalkning. [Studies on effects of disturbance on the bird fauna in connection with liming.] - Nat. Board Fish., unpubl. rep.

Eriksson, M.O.G. 1985. Prey detectability for fish-eating birds in relation to fish density and water transparency. - Ornis Scand. 16: 1-7.

Eriksson, M.O.G. 1987a. Some effects of freshwater acidification on birds in Sweden. - ICBP Tech. Publ. 6: 1-8.

Eriksson, M.O.G. 1987b. Liming of acid lakes in Sweden: short-term effects on waterbird densities. - Wildfowl 38: 143-149.

Eriksson, M.O.G. 1991. Försurningspåverkan på fågel- och däggdjursbestånd. [English summary: Effects of acidification on bird and mammal population.] - Swedish Environm. Prot. Agency 3969.

Eriksson, M.O.G. and Tengelin, B. 1987. Short-term effects of liming on perch *Perca fluviatilis* populations in acidified lakes in south-west Sweden. - Hydrobiologia 146: 187-191.

Eriksson, M.O.G., Henrikson, L., Larsson, P., Nilsson, B.-I., Nyman, H.G., Oscarson, H.G. and Stenson, J.A.E. 1980. Predator-prey relations, important for the biotic changes in acidified lakes. - Ambio 9: 248-249.

Eriksson, M.O.G., Henrikson, L. and Oscarson, H.G. 1983. Försurning - ett framtida hot mot fiskgjusen (*Pandion haliaetus*). [English summary: Acid rain - a future danger for the osprey, *Pandion haliaetus*.] - Vår Fågelvärld 42: 293-300.

Eriksson, M.O.G., Arvidsson, B. and Johansson, I. 1988. Habitatkaraktärer hos häckningssjöar för smålom *Gavia stellata* i sydvästra Sverige. [English summary: Habitat characteristics of breeding lakes of red-throated diver, *Gavia stellata*, in south-west Sweden.] - Vår Fågelvärld 47: 122-132.

Eriksson, U. and Medin, M. 1992. Bottenfaunan på fem lokaler i Lilla Häggåns vattensystem våren 1992. [The bottom fauna at five localities in the drainage area of River Lilla Häggån during spring 1992.] - Medins Sjö- och Åbiologi Consultant Co., Mölnlycke.

Evans, R.A. 1989. Effects of liming and brook trout (*Salvelinus fontinalis*) introduction on limnetic insect populations of two acidified, fishless lakes. - Can. J. Fish. Aquat. Sci. 46: 342-352.

Fjellheim, A. and Raddum, G.G. 1992. Recovery of acid-sensitive species of Ephemeroptera, Plecoptera and Trichoptera in River Audna after liming. - Env. Poll. 78: 173-178.

Fleischer, S., Andersson, G., Brodin, Y.-W., Dickson, W., Herrmann, J. and Muniz, I. 1993. Acid water research in Sweden - knowledge for tomorrow? - Ambio 22: 258-263.

Forsberg, G. 1986. Nypigmenterade ålyngels överlevnad och födoval i en försurad sjö. [English summary: The survival and diet of elvers in an acidified lake.] - Inform. Inst. Freshw. Res., Drottningholm (8).

Foster, G.N. 1991. Aquatic beetle population changes associated with recreating a trout fishery by liming a lake catchment. - Arch. Hydrobiol. 122: 313-322.

Francis, A.J., Quinby, H.L. and Hendrey, G.R. 1984. Effect of lake pH on microbial decomposition of allochthonous litter. - In: Hendrey, G.R. (Ed.) Early biotic responses to advancing lake acidification. Butterworth, Boston, pp. 1-21.

Fraser, J.E., Saunders, W.P., von Oettingen, S.L., Britt, D.L. and Sverdrup, H.U. 1985. Ecological effects of base addition to lake ecosystems. - Internat. Sci. & Techn., Inc., Reston, Virginia.

Frick, K.G. och Herrmann, J. 1990. Aluminium accumulation in a lotic mayfly at low pH - a laboratory study. - Ecotox. Environ. Safety 19: 81-88.

Gahnström, G. 1988. The effect of liming on sediment oxygen uptake in Lake Gårdsjön, Sweden. - In: Dickson, W. (Ed.) Liming of Lake Gårdsjön - an acidified lake in SW Sweden. Swedish Environm. Prot. Agency 3426, pp. 207-243.

Gloss, S.P., Schofield, C.L., Spateholts, R. and Plonski, B.A. 1989. Survival, growth, reproduction and diet of brook trout (*Salvelinus fontinalis*) stocked into lakes after liming to mitigate acidity. - Can. J. Aquat. Sci. 46: 277-286.

Grahn, O. 1980. Fish kills in two moderately acid lakes due to high aluminium concentration. - In: Drabløs, D. and Tollan, A. (Eds.) Proc. Int. Conf. Ecol. impact acid precip., Norway, SNSF project., Oslo-Ås, pp. 310-311.

Grahn, O. and Hultberg, H. 1974. Försurningens effekter på oligotrofa sjöars ekosystem - integrerade förändringar i artsammansättning och dynamik. [The effects of acidification on oligotrophic ecosystems - integrated changes in species composition and dynamics.] - Swedish Environm. Res. Inst., B192.

Griffiths, R.W. 1992. Effects of pH on community dynamics of Chironomidae in a large river near Sudbury, Ontario. - Can. J. Fish. Aquat. Sci. 49: 76-86.

Grundelius, E. 1987. Flodpärlmusslans tillbakagång i Dalarna. [English summary: The decline of the freshwater pearl mussel in the province of Dalarna.] - Inform. Inst.

Freshw. Res., Drottningholm (4).
Hairston, N.G. and De Stasio, B.T. 1988. Rate of evolution slowed by a dormant propagule pool. - Nature 336: 239-242.
Hansen, P.W. 1987. Acid rain and waterfowl. The case for concern in North America. - Izaak Walton League of America, Minneapolis and Arlington.
Hansson, M. 1976. Biologin i en sur fjällsjö, belyst av rödingens föda. [English summary: The biology of an acid mountain lake, illustrated by the food of Arctic char] - Inform. Inst. Freshw. Res., Drottningholm (5).
Hasselrot, B. 1984. Surt vattenflöde genom en isbelagd kalkad sjö i samband med stor vattenföring. [Acid water flow through an ice-covered limed lake in connection with high water discharge.] - Swedish Environm. Res. Inst., EM 1464.
Hasselrot, B. and Alenäs, I. 1988. Kalkningsprojektet Kungsbackaån - effekter på vattenkemi och fiskpopulationer 1983-1988. [The River Kungsbackaån liming project - effects on water chemistry and fish populations 1983-1988.] - Regional Federation of Local Authorities, Gothenburg.
Hasselrot, B. and Hultberg, H. 1984. Liming of acidified lakes and streams and its consequence for aquatic ecosystems. - Fisheries 9: 4-9.
Hasselrot, B., Andersson, B.I. and Hultberg, H. 1984. Ecosystem shifts and reintroduction of Arctic char (*Salvelinus salvelinus* (L.)) after liming of a strongly acidified lake in southwestern Sweden. - Rep. Inst. Freshw. Res., Drottningholm 61: 78-92.
Hasselrot, B., Alenäs, I., Andersson, I. and Hultberg, H. 1987. Response of limed lakes to episodic acid events in southwestern Sweden. - Water Air Soil Poll. 32: 341-362.
Havens, K.E. and DeCosta, J. 1987. Freshwater plankton community succession during experimental acidification. - Arch. Hydrobiol. 111: 37-65.
Hegna, K. (Ed.) 1992. Vassdragskalking - strategi och effekter. Referat fra FoU-seminar i 1992. [Liming of freshwaters - strategies and effects. Report from a scientific seminar in 1992.] - Direct. Nat. Managem. (Norway), DN-notat 1992-5.
Henrikson, B.-I. 1988a. The absence of antipredator behaviour in the larvae of *Leucorrhinia dubia* (Odonata) and the consequences for their distribution. - Oikos 51: 179-183.
Henrikson, B.-I. 1990. Predation on amphibian eggs and tadpoles by common predators in acidified lakes. - Holarct. Ecol. 13:201-206.
Henrikson, L. 1988b. Effects on water quality and benthos of acid water inflow to the limed Lake Gårdsjön. - In: Dickson, W. (Ed.) Liming of Lake Gårdsjön - an acidified lake in SW Sweden. Swedish Environm. Prot. Agency 3426, pp. 309-327.
Henrikson, L. and Medin, M. 1990. Bottenfaunan i tjugo vattendrag i Jönköping län 1989 - En biologisk försurningsbedömning. [The bottom fauna in twenty watercourses in the county of Jönköping, a biological acidification assessment.] - Jönköping County Council, Rep. 1990:15.
Henrikson, L. and Medin, M. 1991. Bottenfaunan 1990, En biologisk försurningsbedömning. [The bottom fauna 1990, a biological acidification assessment.] - Jönköping County Council, Rep. 1991:4.
Henrikson, L. and Oscarson, H.G. 1978. Fish predation limiting the abundance and distribution of *Glaenocoris propinqua* (Hemiptera). - Oikos 31: 102-105.
Henrikson, L. and Oscarson, H.G. 1981. Corixids (Hemiptera-Heteroptera), the new top predators in acidified lakes. - Verh. Internat. Verein. Limnol. 21: 1616-1620.
Henrikson, L. and Oscarson, H.G. 1984. Lime influence on macro-invertebrate zooplankton predators. - Rep. Inst. Freshw. Res., Drottningholm 61: 93-103.
Henrikson, L. and Oscarson, H.G. 1985. History of Lake Gårdsjön: the development of chironomids. - Ecol. Bull. (Stockholm) 37: 58-63.
Henrikson, L., Oscarson, H.G. and Stenson, J.A.E. 1984. Development of the crustacean zooplankton community after lime treatment of the fishless Lake Gårdsjön, Sweden. -

Rep. Inst. Freshw. Res., Drottningholm 61: 104-114.
Henrikson, L., Nyman, H.G., Oscarson, H.G. and Stenson, J.A.E. 1985. Changes in the zooplankton community after lime treatment of an acidified lakes. - Verh. Internat. Verein. Limnol. 22: 3008-3013.
Henrikson, L., Larsson, S., Nyman, H.G. and Oscarson, H.G. 1989. Stora Le före och efter kalkning. [Lake Stora Le before and after liming.] - Nat. Board Fish., unpubl. rep.
Henrikson, L., Larsson, S., Nyman, H.G. and Oscarson, H.G. 1990. Are we liming the lakes for the benefit of an undisturbed ecosystem or for a high pH-level? - Internat. Conf. Acidic deposition, its nature and impact, Glasgow, September 1990, Conf. Abstracts, p. 607.
Herricks, E.E. and Cairns, J. 1977. The effects of lime neutralization of acid mine drainage on stream ecology. - 32nd Annual Purdue Industrial Waste Conf., May 1977.
Herrmann, J. 1987. Aluminium impact on freshwater invertebrates at low pH - a review. - In: Landner, L. (Ed.) Speciation of metals in water, sediment and soil systems. Lecture Notes in Earth Sciences 11. Springer, Berlin Heidelberg New York, pp. 157-175.
Herrmann, J. 1990. Physiological, food chain and ecological effects among benthic invertebrates exposed to low pH and associated metal concentrations. - In: Mason, B.J. (Ed.) The surface waters acidification programme. Univ. Press, Cambridge, pp. 383-396.
Herrmann, J. 1991a. Bottom fauna as indicators of organic pollution in running waters. - In: Meyer, Z. (Ed.) Proc. 1st Int. Sem. Environment protection - regional problems, Zalesie/Szczecin, Poland, pp. 7-11.
Herrmann, J. 1991b. Bioindication of nutrient and organic pollution by benthic stream invertebrates - pitfalls and problems. - Verh. Internat. Verein. Limnol. 24: 3079.
Herrmann, J., Degerman, E., Gerhardt, A., Johansson, C., Lingdell, P.-E. and Muniz, I.P. 1993. Acid-stress effects on stream biology. - Ambio 22: 298-307.
Herzig, A. 1985. Restings eggs - a significant stage in the life cycle of crustaceans *Leptodora kindti* and *Bythotrephes longimanus*. - Verh. Internat. Verein. Limnol. 22: 3088-3098.
Hessen, D.O. 1985. The relation between bacterial carbon and dissolved humic compounds in oligotrophic lakes. - FEMS (Fed. Eur. Microbiol. Soc.) Microbiol. Ecol. 31: 215-223.
Hobaek, A. and Raddum, G.G. 1980. Zooplankton communities in acidified lakes in south Norway. - SNSF-project,. Internal Rep. 75/80. Ås, Norway.
Hörnström, E., Ekström, C., Miller, U. and Dickson, W. 1973. Försurningens inverkan på västkustsjöar. [English summary: Effects of acidification on lakes in the Swedish west coast region. - Rep. Inst. Freshw. Res., Drottningholm (4).
Hörnström, E., Ekström, C. and Duraini, M.O. 1984. Effect of pH and different levels of aluminium on lake plankton. - Rep. Inst. Freshw. Res., Drottningholm 61: 115-127.
Hörnström, E., Ekström, C. and Andersson, P. 1992. Tio mellansvenska sjöar. Kalkningseffekter på plankton och vattenkemi. [English summary: Ten lakes in central Sweden, liming effects on plankton and water chemistry.] - Swedish Environm. Prot. Agency 4048.
Hörnström, E., Ek, J., Ekström, C. and Fröberg, E. 1993. Physical-chemical conditions, plankton, and bottom fauna in Lake Tvärsjön and other lakes 1973-87, under acidic and limed conditions. - Swedish Environm. Prot. Agency, unpubl. rep.
Howell, E.T., Coker, G., Booth, G.M., Keller, W., Neary, B., Nicholls, K.B., Tomassini, F.D., Yan, N., Gunn, J.M. and Reitveld, H. 1991. Ecosystem responses of pH 5.9 lake trout lake to whole lake liming. - In: Olem, H., Schreiber, R.K., Brocksen, R.W. and Porcella, D.B. (Eds.) International lake and watershed liming practices. The Terrene Inst., Washington, D.C., USA, pp. 61-96.
Howells, G. and Dalziel, T.R.K. (Eds.) 1992. Restoring acid waters: Loch Fleet 1984-1990. - Elsevier Applied Science, London.

Hultberg, H. 1977. Thermally stratified acid water in the late winter - a key factor inducing self-accelerating processes which increase acidification. - Water Air Soil Poll. 7: 279-294.

Hultberg, H. 1985. Changes in fish populations and water chemistry in Lake Gårdsjön and neighbouring lakes during the last century. - Ecol. Bull. 37: 64-72.

Hultberg, H. and Andersson, B.I. 1982. Liming of acidified lakes: induced long-term changes. - Water Air Soil Poll. 18: 311-331.

Hultberg, H. and Johansson, S. 1978. Rapport rörande orsakerna till metall- och syrabelastningen av grundvattentillrinningen till delar av Stenunge å i Stenungsunds kommun. [Report on the causes of metal and acid load by groundwater discharge to parts of the stream Stenunge å in the municipality of Stenungsund.] - Swedish Environm. Res. Inst., unpubl. rep.

Hultberg, H. and Stenson, J. 1970. Försurningens effekter på fiskfaunan i två bohuslänska sjöar. [The effects of acidification on the fish fauna in two lakes in the county of Bohuslän.] - Fauna och flora 1: 11.

Jansson, M., Persson, G. and Broberg, O. 1986. Phosphorus in acidified lakes: the example of Lake Gårdsjön, Sweden. - Hydrobiologia 139: 81-96.

Jernelöv, A., Nagell, B. and Svensson, A. 1981. Adaptations to an acid environment in *Chironomus riparius* (Diptera, Chironomidae) from Smoking Hills, N.W.T. , Canada. - Holarct. Ecol. 4: 116-119.

Johansson, J.-Å. 1983. Seasonal development of bacterioplankton in two forest lakes in central Sweden. - Hydrobiologia 101: 71-88.

Johansson, K. and Nyberg, P. 1981. Försurning av svenska ytvatten - effekter och omfattning 1980. [English summary: Acidification of surface waters in Sweden - effects and extent 1980.] - Inform. Inst. Freshw. Res., Drottningholm (6).

Johansson, N. and Kihlström, J. 1975. Pikes (*Esox lucius* L.) shown to be affected by low pH values during first weeks after hatching. - Env. Res. 9: 12.

Johansson, N. and Milbrink, G. 1976. Some effects of acidified water on the early development of roach *Rutilus rutilus* L. and perch *Perca fluviatilis* L. - Water Res. Bull. 12: 39-48.

Johansson, N., Kihlström, J.E. and Wahlberg, A. 1973. Low pH values shown to affect developing fish eggs (*Brachydanio rerio* Ham.-Buch.). - Ambio 2: 42-43.

Johansson, N., Runn, P. and Milbrink, G. 1977. Early development of three salmonid species in acidified water. - Zoon 5: 127-132.

Johansson, N., Runn, P. and Sothell, M. 1981. Perivitelline pH of salmonid eggs in relation to ambient pH. - Water Res. Bull. 17: 994-999.

Johansson, O. and Eriksson, M.O.G. 1987. Häckningsframgång och utbredning hos forsärla *Motacilla cinerea* i sydvästra Sverige; en jämförelse mellan sura och icke sura vattendrag. [English summary: Breeding performance and distribution of grey wagtail, *Motacilla cinerea*, in southwest Sweden, a comparison between acidic and circumneutral streams.] - Vår Fågelvärld 46: 365-374.

Johlander, A. and Sjöstrand, P. 1990. Kompletterande kalkningsuppföljning i Högvadsån 1988. [Supplementary monitoring studies of the liming in River Högvadsån 1988.] - Nat. Board Fish., Reg. Fish. Off., Jönköping, unpubl. rep.

Johnson, R.K., Eriksson, L. and Wiederholm, T. 1993. Comparison of profundal macroinvertebrate communities of lime- and limed (IKEU) reference lakes. - Statens Lantbruksuniv., Uppsala, unpubl. rep.

Karlsson-Norrgren, L., Björklund, I., Ljungberg, O. and Runn, P. 1986a. Acid water and aluminium exposure; experimentally induced gill lesions in brown trout, *Salmo trutta* L. - J. Fish. Dis. 9: 11-25.

Karlsson-Norrgren, L., Dickson, W., Ljungberg, O. and Runn, P. 1986b. Acid water and aluminium accumulation in farmed brown trout, *Salmo trutta* L. - J. Fish. Dis. 9: 1-9.

Kerfoot, W.C. and Sih, A. 1987. Predation - direct and indirect impact on aquatic communities. - Univ. Press, New England, London.
Keller, W. and Yan, N.D. 1991. Recovery of crustacean zooplankton species richness in Sudbury area lakes following water quality improvements. - Can. J. Fish. Aquat. Sci. 48: 1635-1644.
Keller, W., Dodge, D.P. and Booth, G.M. 1990a. Experimental lake neutralization program: overview of neutralization studies in Ontario. - Can. J. Fish. Aquat. Sci. 47: 410-411.
Keller, W., Molot, L.A., Griffiths, R.W. and Yan, N.D. 1990b. Changes in the zoobenthos community of acidified Bowland lake after whole-lake neutralization and lake trout (*Salvelinus namaycush*) reintroduction. - Can. J. Fish. Aquat. Sci. 47: 440-445.
Keller, W., Yan, N.D., Howell, T., Molot, L.A. and Taylor, W.D. 1992. Changes in zooplankton during the experimental neutralization and early re-acidification of Bowland Lake near Sudbury, Ontario. - Can. J. Fish. Aquat. Sci. 49: 52-62.
Kinsten, B. 1986. Förekomsten av glacialrelikter i mellansverige med betoning på försurningseffekter. [English summary: The occurrence of glacial relict crustaceans in central Sweden, with emphasis on the effects of acidification.] - Inform. Inst. Freshw. Res., Drottningholm (11).
Kullberg, A. 1992. Benthic macroinvertebrate community structure in 20 streams of varying pH and humic content. - Env. Poll. 78: 103-106.
Kullberg, A. and Petersen, R.C. 1987. Dissolved organic carbon, seston and macroinvertebrate drift in an acidified and limed humic stream. - Freshw. Biol. 17: 553-564.
Lagerman, B. 1983. Kalkningsprojektet Unden. [The Lake Unden liming project.] - Nat. Board. Fish., unpubl. rep.
Larsson, L.S. 1988. Liming effects on phytoplankton in Lake Gårdsjön 1982-1985. - In: Dickson, W. (Ed.) Liming of Lake Gårdsjön - an acidified lake in SW Sweden. Swedish Environm. Prot. Agency 3426, pp. 245-280.
Lessmark, O. 1976. Försurningens inverkan på fiskfaunan i några småländska sjöar. [English summary: The effect of acidification on the fish fauna of some lakes in the province of Småland.] - Inform. Inst. Freshw. Res., Drottningholm (7).
Lessmark, O. 1983. Competition between perch (*Perca fluviatilis*) and roach (*Rutilus rutilus*) in south Swedish lakes. - PhD thesis, Dep. Limnol., Univ. Lund.
Lessmark, O., Degerman, E., Johlander, A. and Sjölander, E. 1986. Effekter av kalkning på fisk omedelbart nedströms doserare. [English summary: Effects of liming on fish immediately downstream of lime dosers.] - Inform. Inst. Freshw. Res., Drottningholm (5).
Lien, L., Raddum, G.G. and Fjellheim, A. 1992. Critical loads of acidity to freshwater - fish and invertebrates. - NIVA (Norwegian Institute for Water Research), Rep. 0-89185, 3.
Lindgren, G. 1988. De bottenlevande djuren i Nianån - en studie av kalkningseffekter på bottenfaunan. [Liming effects on the benthic animals in the stream Nian.] - Gävleborg County Council, unpubl. rep.
Lindström, T. 1991. Fiskbeståndsutvecklingen på Fulufjäll, synpunkter på materialet som underlag för långtidsövervakning år 1958-1990. [The development of fish populations in the Fulufjäll district, consideration of the data basis for monitoring 1958-1990.] - Inst. Freshw. Res., Drottningholm, unpubl. rep.
Lindström, T. 1992. Zooplankton på Fulufjället 1976-91. [English summary: Zooplankton in the Fulufjäll district, 1976-1991, the southern high mountain.] - Inform. Inst. Freshw. Res., Drottningholm (2): 35-68.
Lindström, T., Dickson, W., Hanson, M. and Andersson, G. 1982. Dålig kondition hos röding i ett surt område - en effekt av näringsbrist eller fysiologisk stress. [English summary: Low condition factor in fish of acidified lakes - an effect of insufficient food

or physiological stress?] - Inform. Inst. Freshw. Res., Drottningholm (5).

Lindström, T., Dickson, W. and Andersson, G. 1984. Reclaiming acid high mountain lakes by liming: a progress report. - Rep. Inst. Freshw. Res., Drottningholm 61: 128-137.

Lingdell, P.-E. and Engblom, E. 1986. Värdet av dagsländan *Baetis rhodani* som vattenkvalitetsindikator. [The mayfly nymph *Baetis rhodani* as a valuable water quality indicator.] - In: Proc. Biologiske metoder i vassdragsovervikningen. [Biological methods in water monitoring.], Norwegian Association of Limnology, pp. 48-58.

Lingdell, P.-E. and Engblom, E. 1989a. Kalkningseffekter i Foskan och Bunnan. En studie av bottenfauna augusti 1988. [Liming effects on the benthic fauna in the Rivers Foskan and Bunnan, August 1988.] - Kopparberg County Council, Rep. 1989:2.

Lingdell, P.-E. and Engblom, E. 1989b. Försurningssituationen i några jämtländska sjöar och vattendrag. [The acidification status in some lakes and streams in the county of Jämtland.] - Limnodata Consultant Co., Stockholm.

Lingdell, P.-E. and Engblom, E. 1991. Försurningssituationen i några sjöar och vattendrag i Kopparbergs län. En studie av bottenfaunan 1969 till 1989. [The acidification status in lakes and watercourses in the county of Kopparberg. A study of the bottom fauna 1969-1989.] - Kopparberg County Council, Rep. 1991:1.

Lingdell, P.-E. and Engblom, E. 1992. Försurningssituationen i några Vätterbäckar. En studie av bottenfauna oktober 1991. [The acidification status in some tributaries to Lake Vättern. A study of the benthic fauna, October 1991.] - Skaraborg County Council, Rep. 3/92.

Longcore, J.R., Ross, R.K. and Fisher, K.L. 1987. Wildlife resources at risk through acidification of wetlands. - Trans. N. A. Wildl. & Nat. Res. Conf. 52: 608-618.

Lundgren, S. 1982a. Ornitologisk undersökning av Visen och Opperhalen 1982. [Ornitological study of Lake Visen and Opperhalen.] - Nat. Board Fish., unpubl. rep.

Lundgren, S. 1982b. Undersökning av storlommens reproduktion i Malensjöarna år 1982. [Studies of the reproduction of black-throated diver in the Malensjöarna lakes, 1982.] - Nat. Board. Fish., unpubl. rep.

Lundh, I. 1981. Kalkningseffekter på öringbestånd i Tjöstelsrödsån. [English summary: Effects of liming on the sea trout population in River Tjöstelsrödsån.] - Inform. Inst. Freshw. Res., Drottningholm (7).

Malley, D.F. and Chang, P.S.S. 1986. Increase in the abundance of Cladocera at pH 5.1 in experimentally-acidified lake 223, Experimental Lakes Area, Ontario. - Water Air Soil Poll. 30: 629-638.

Malmqvist, B., Nilsson, L.M. and Svensson, B.S. 1976. Höstens lövfall - energikälla för bäckens djursamhälle [Autumn leaves - energy source for stream benthos]. - Fauna och Flora 71: 284-292.

Malmqvist, B., Nilsson, L.M. and Svensson, B.S. 1978. Dynamics of detritus in a small stream in southern Sweden and its influence on the distribution of the bottom animal communities. - Oikos 31: 3-16.

Malmqvist, B., Rundle, S., Brönmark, C. and Erlandsson, A. 1991. Invertebrate colonization of a new, man-made stream in southern Sweden. - Freshw. Biol. 26: 307-324.

Mann, R.H.K. and Penczak, T. 1986. Fish production in rivers: a review. - Pol. Arch. Hydrobiol. 33(3/4): 233-247.

McCahon, C.P. and Poulton, M.J. 1991. Lethal and sub-lethal effects of acid, aluminium and lime on *Gammarus pulex* during repeated simulated episodes in a Welsh stream. - Freshw. Biol. 25:169-178.

McCahon, C.P., Brown, A.F., Poulton, M.J. and Pascoe, D. 1989. Effects of acid, aluminium and lime additions on fish and invertebrates in a chronically acidic Welsh stream. - Water Air Soil Poll. 45: 345-359.

Medin, M. 1991a. Biologisk bedömning av försurningssituationen i tillflöden till Lelången våren 1991. [Biological assessment of the acidification status in the tributaries to Lake Lelången in the spring 1991.] - Medins Sjö- och Åbiologi Consultant Co., Mölnlycke, Rep.

Medin, M. 1991b. Bottenfaunan vid tre lokaler i Högvadsån våren 1991, en biologisk bedömning av miljön. [The bottom fauna at three localities in the River Högvadsån during spring 1991, a biological assessment of the environment.] - Medins Sjö- och Åbiologi Consultant Co., Mölnlycke, Rep.

Medin, M. 1991c. Bottenfaunan på fem lokaler i Suseåns avrinningsområde våren 1991. [The bottom fauna at five localities in the River Suseån catchment area during spring 1991.] - Medins Sjö- och Åbiologi Consultant Co., Mölnlycke, Rep.

Medin, M. 1991d. Bottenfaunan på sex lokaler i Vänneåns avrinningsområde hösten 1990. [The bottom fauna at six localities in the River Vänneån catchment area during spring 1991.] - Medins Sjö- och Åbiologi Consultant Co., Mölnlycke, Rep.

Medin, M. 1991e. Bottenfaunan på tjugofyra lokaler i Alsteråns avrinningsområde våren 1991. [The bottom fauna at twenty-four localities in the River Alsterån catchment area during spring 1991.] - Medins Sjö- och Åbiologi Consultant Co., Mölnlycke, Rep.

Medin, M. 1991f. Bottenfaunaundersökning på tolv lokaler i Emmaboda kommun våren 1991. [The bottom fauna at twelve localities in the municipality of Emmaboda during spring 1991.] - Medins Sjö- och Åbiologi Consultant Co., Mölnlycke, Rep.

Medin, M. and Ericson, U. 1993. Återkolonisering och återutsättning av bottenfauna i Silen sjöarna 1993. [Recolonization and restocking of bottom fauna in the Silen lakes, 1993.] - Medins Sjö- och Åbiologi Consultant Co., Mölnlycke, Rep.

Medin, M., Ericson, U. and Nilsson, C. 1993. Studier av bottenfaunan i försurade och kalkade vattendrag. [Studies on the bottom fauna in acidified and limed watercourses.] - Värmland County Council, Rep. 1993:19.

Mellors, W.E. 1975. Selective predation on ephippial *Daphnia* and the resistance of ephippial eggs to digestion. - Ecology 56: 974-980.

Milbrink, G. and Johansson, N. 1975. Some effects of acidification on roe of roach (*Rutilus rutilus* L.) and perch (*Perca fluviatilis* L.) - with special reference to the Åvaå Lake system in eastern Sweden. - Rep. Inst. Freshw. Res., Drottningholm 54: 52-62.

Milner, A. 1987. Colonization and ecological development of new streams in Glacier Bay National Park, Alaska. - Freshw. Biol. 18: 53-70.

Minshall, G.W., Andrews, D.A. and Manuel-Faler, C.Y. 1983. Application of island biogeographic theory to streams: macroinvertebrate recolonization of the Teton River, Idaho. - In: Barnes, J.R. and Minshall, G.W. (Ed.) Stream ecology, application and testing of general ecological theory. Plenum Press, London, pp. 279-297.

Mitchell, B.A. 1989. Acid rain and birds: how much proof is needed? - Am. Birds 43: 234-241.

Morling, G. and Pejler, B. 1990. Acidification and zooplankton development in some west Swedish lakes 1966-1983. - Limnologica (Berlin) 20: 307-318.

Mosegaard, H. 1990. Otolitanalys av subletala försurningseffekter på lax och öring i rinnande vatten. [Otolithe analysis of sublethal effects of acidification on salmon and trout in running waters.] - Inst. Limnol., Uppsala Univ., research applic.

Mossberg, P. and Nyberg, P. 1976. Försurningseffekter på bottendjur och fisk i Västra Skälsjön. [English summary: Effects of acidification on bottom fauna and fish in Lake Västra Skälsjön.] - Inform. Inst. Freshw. Res., Drottningholm (9).

Mossberg, P. and Nyberg, P. 1979. Bottom fauna of small acid forest lakes. - Rep. Inst. Freshw. Res., Drottningholm, 58: 77-87.

Muniz, I.P. 1991. Freshwater acidification: its effects on species and communities of freshwater microbes, plants and animals. - Proceed. R. Soc. Edinburgh 97B: 227-254.

NBF and SEPA (National Board of Fisheries and Swedish Environmental Protection Agency) 1981. Kalkning av sjöar och vattendrag 1977-1981. [Liming of lakes and streams

1977-81.] - Inform. Inst. Freshw. Res., Drottningholm (4).

Nilssen, J.P. and Sandøy, S. 1990. Recent lake acidification and cladoceran dynamics: surface sediment and core analyses from lakes in Norway, Scotland and Sweden. - Phil. Trans. R. Soc. London B 327: 299-309.

Nilssen, J.P., Østerdahl, T. and Potts, W.T.W. 1983. Physiologi til zooplankton under forsurning og kalking: et pilotstudium med radioisotoper. [Acidification and liming effects on zooplankton physiology: a pilot study using radio-isotopes.] - The liming project, Rep. 7/83: 1-31. Dep. Environm. (Norway).

Nilssen, J.P., Østerdahl, T. and Potts, W.T.W. 1984. Species replacements in acidified lakes: physiology, predation or competition? - Rep. Inst. Freshw. Res., Drottningholm 61: 148-153.

Nilsson, A.N. and Johansson, A. 1985. En jämförelse av bottenfaunan i några kalkade och okalkade vattendrag; med tonvikt på kalkningsmetodik. [English summary: A comparative study of the benthos of limed and non-limed streams; effects of different liming methods.] - Inform. Inst. Freshw. Res., Drottningholm (11).

Nilsson, N.-A. and Pejler, B. 1973. On the relation between fish fauna and zooplankton composition in north Swedish lakes. - Rep. Inst. Freshw. Res., Drottningholm 53: 51-77.

Norrgren, L. and Degerman, E. 1993. Effects of different water qualities on the early development of Atlantic salmon and brown trout exposed in situ. - Ambio 22: 213-218.

Norrgren, L., Wicklund-Glynn, A. and Malmborg, O. 1991. Accumulation and effects of aluminium in the minnow (*Phoxinus phoxinus* L.) at different pH levels. - J. Fish Biol. 39: 833-847.

Nyberg, P. 1984a. Effects of liming on fisheries. - Phil. Trans. R. Soc. London B 305: 549-560.

Nyberg, P. 1984b. Impact of *Chaoborus* predation on planktonic crustacean communities in some acidified and limed forest lakes in Sweden. - Rep. Inst. Freshw. Res., Drottningholm, 61: 154-166.

Nyberg, P. 1988. Reclamation of acidified Arctic char (*Salvelinus alpinus* L.) lakes in Sweden by means of liming. - Verh. Internat. Verein. Limnol. 23: 1732-1742.

Nyberg, P. and Thörnelöf, E. 1988. Operational liming of surface waters in Sweden. - Water Air Soil Poll. 41: 3-16.

Nyberg, P., Appelberg, M. and Degerman, E. 1986a. Effects of liming on crayfish and fish in Sweden. - Water Air Soil Poll. 31: 669-687.

Nyberg, P., Degerman, E., Ekström, C. and Hörnström, E. 1986b. Försurningskänsliga rödingsjöar i Syd- och Mellansverige. [English summary: Acid-sensitive Arctic char (*Salvelinus alpinus*) lakes in southern and central Sweden.] - Inform. Inst. Freshw. Res., Drottningholm (6).

Nyström, U. 1990. Extremt höga aluminiumhalter efter torrsommaren 1990. [Extremely high aluminium concentrations following the dry summer in 1990.] - Swedish Environm. Res. Inst., unpubl. rep.

Odelström, T. 1988. The food choice of the crayfish *Astacus astacus* L. in relation to environmental conditions. - Acta Univ. Ups., Comprehensive summaries of Uppsala Dissertations from the faculty of Science 165.

Økland, K.A. 1980. Mussels and crustaceans: studies of 1,000 lakes in Norway. - In: Drabløs, D. and Tollan, A. (Eds.) Proc. Int. Conf. Ecol. impact acid precip., Norway, SNSF project., Oslo-Ås, pp. 324-325.

Økland, J. and Økland, K.A. 1986. The effects of acid deposition in benthic animals in lakes and streams. - Experientia 42: 471-486.

Olem, H. 1990. Liming acidic surface waters. - In: Acidic deposition: state of science and technology. - National Acid Precipitation Assessment Program, Washington, USA, Rep. 15.

Olem, H. 1991. Liming acidic surface waters. - Lewis Publ., Chelsea, Michigan, USA.

Olem, H., Schreiber, R.K., Brocksen, R.W. and Porcella, D.B. (Eds.) 1991. International lake and watershed liming practices. The Terrene Inst., Washington, D.C., USA.

Olofsson, H., Blomqvist, P., Olsson, H. and Broberg, O. 1988. Restoration of the pelagic food web in acidified and limed lakes by gentle fertilization. - Limnologica (Berlin) 19: 27-35.

Olofsson, J. and Oscarson, H.G. 1988. Undersökningar av kalkningseffekter på bottenfaunan i sjön Tolken, Marks kommun, Älvsborgs län. [Studies of the effects of liming on the bottom fauna in Lake Tolken, municipality of Mark, county of Älvsborg.] - Municipality of Mark, Rep. 1988:6.

Ormerod, S.J. and Tyler, S.J. 1991. The influence of stream acidity and riparian land use on the feeding ecology of grey wagtails *Motacilla cinerea* in Wales. - Ibis 133: 53-91.

Ormerod, S.J., Allison, N., Hudson, D. and Tyler, S.J. 1986. The distribution of breeding dippers (*Cinclus cinclus* (L.); Aves) in relation to stream acidity in upland Wales. - Freshw. Biol. 16: 501-507.

Ormerod, S.J., Weatherley, N.S., Merrett, W.J., Gee, A.S. and Whitehead, P.G. 1990. Restoring acidified streams in Upland Wales: a modelling comparison of the chemical and biological effects of liming sand reduced sulphate deposition. - Environm. Poll. 64: 67-85.

Otto, C. and Svensson, B. 1983. Properties of acid brown water streams in south Sweden. - Arch. Hydrobiol. 99:15-36.

Parson, J.D. 1968. The effects of acid stripmine effluents on the ecology of a stream. - Arch. Hydrobiol. 65: 25-50.

Peckarsky, B.L. 1983. Biotic interactions or abiotic limitations? A model of lotic community structure. - In: Fontaine, T.D. and Bartell, S.M. (Eds.) The dynamics of lotic ecosystems. Ann Arbor Sci. Publ., Michigan, pp. 303-323.

Petersen, R.C. and Persson, U. 1987. Comparison of the biological effects of humic materials under acidified conditions. - Sci. Total Envir. 62: 387-398.

Petersen, R.C., Kullberg, A., Persson, U. and Fritzon, A. 1984. Fylleån - Biologiska effekter vid kalkning av en försurad å. [River Fylleån - biological effects of liming in an acidified river.] - Dep. Limnology, Univ. Lund, Rep.

Porcella, D.B. 1989. Lake Acidification Mitigation Project (LAMP): an overview of an ecosystem pertubation experiment. - Can. J. Fish. Aquat. Sci. 46: 246-248.

Raddum, G.G. and Fjellheim, A. 1984. Acidification and early warning organisms in freshwater in western Norway. - Verh. Internat. Verein, Limnol. 22: 1973-1980.

Raddum, G.G. and Fjellheim, A. 1994. Invertebrate community changes caused by reduced acidification. - In: Steinberg, C.E.W. and Wright, R.F. (Eds) Acidification of freshwater ecosystems: implications for the future. J. Wiley & Sons, pp. 345-354.

Raddum, G.G., Hagenlund, G. and Halvorsen, G. 1984. Effects of lime treatment on the benthos of Lake Søndre Boksjø. - Rep. Inst. Freshw. Res., Drottningholm, 61: 167-176.

Raddum, G.G., Brettum, P., Matzow, D., Nilssen, J.-P., Skov, A., Sveälv, T. and Wright, R.F. 1986. Liming the acid Lake Hovvatn, Norway: a whole-ecosystem study. - Water Air Soil Poll. 31: 721-763.

Rask, M. 1991. Iso Valkjärvi research: an introduction to a multidisciplinary lake liming study. - Finnish Fish. Res. 12: 25-34.

Rask, M., Heinänen, A., Salonen, K., Arvola, L., Bergström, I., Liukkonen, M. and Ojala, A. 1986. The limnology of a small, naturally acidic, highly humic forest lake. - Arch. Hydrobiol. 106: 351-371.

Renberg, I., Brodin, Y.-W., Cronberg, G., El-Daoushy, F., Oldfield, F., Rippey, B., Sandøy, S., Wallin, J.-E. and Wik, M. 1990. Recent acidification and biological changes in Lilla Öresjön, southwest Sweden, and the relation to atmospheric pollution and land-use history. - Phil. Trans. R. Soc. London B 327: 391-396.

Rosemond, A.D., Reice, S.R., Elwood, J.W. and Mulholland, P.J. 1992. The effects of stream acidity on benthic invertebrate communities in the south-eastern United States. - Freshw. Biol. 27: 193-209.
Rosseland, B.O. 1986. Biological effects of acidification on tertiary consumers. Fish population responses. - Water Air Soil Poll. 30: 451-460.
Rosseland, B.O. and Hindar, A. 1991. Mixing zones, a fishery management? - In: Olem, H., Schreiber, R.K., Brocksen, R.W. and Porcella, D.B. (Eds.) 1991. International lake and watershed liming practices. The Terrene Inst., Washington, D.C., USA., pp. 161-173.
Runn, P. 1982. Effects of low pH on the development of early stages of fish. - PhD thesis, Univ. Uppsala, Acta Univ. Ups., p. 647.
Runn, P., Johansson, N. and Milbrink, G. 1977. Some effects of low pH on the hatchability of eggs of perch, *Perca fluviatilis* L. - Zoon 5: 115-125.
Sandberg, P.-E. and Fogelgren, J.-E. 1988. Inventering av fiskbestånd i 167 rinnande vattendrag i Kopparbergs län. [Inventory of fish populations in 167 streams in the county of Kopparberg.] - Fiskevattenägareförbundet, Falun.
Sarvala, J. and Halsinaho, S. 1990. Crustacean zooplankton of Finnish forest lakes in relation to acidity and other environmental factors. - In: Kauppi, P., Anttila, P. and Kenttämies, K. (Eds.) Acidification in Finland. Springer, Berlin Heidelberg New York, pp. 1009-1027.
Schaffner, W.R. 1989. Effects of neutralization and the addition of brook trout (*Salvelinus fontinalis*) on the limnetic zooplankton communities of two acidic lakes. - Can. J. Fish. Aquat. Sci. 46: 295-305.
SEPA (Swedish Environmental Protection Agency) 1988. Kalkning av sjöar och vattendrag. [Liming of lakes and streams.] - Swedish Environm. Prot. Agency, Official guidelines 88:3.
Sjölander, E. 1986. Strömfiske/-ar och försurning ur västernorrländskt perspektiv. [The fisheries/fishes and acidification viewed from the county of Västernorrland.] - Västernorrland County Council, Rep. 4.
Sjöstrand, P. 1988. Sammanställning av uppföljande elfiskeriundersökningar 1988 inom Fylleåns kalkningsprojekt. [Report on follow-up electro-fishing studies in 1988 within the River Fylleån liming project.] - Nat. Board Fish., Reg. Fish. Off., Jönköping, Rep.
Sjöstrand, P. 1991. Sammanställning av uppföljande undersökningar 1990 inom Högvadsåns kalkningsprojekt. [Report on follow-up studies in 1988 within the River Högvadsån liming project.] - Nat. Board Fish., Reg. Fish. Off., Jönköping, Rep.
Sjöström, P. 1990. Food and feeding of *Baetis rhodani* (Ephemeroptera) in acid environments. - In: Mason, B.J. (Ed.) The surface waters acidification programme. Univ. Press, Cambridge, pp. 427-428.
Stenson, J.A.E. 1973. On predation and *Holopedium gibberum* (Zaddach) distribution. - Limnol. Oceanogr. 18: 1005-1010.
Stenson, J.A.E. 1988. Fauna structure and water quality. - Aqua Fennica 18: 179-184.
Stenson, J.A.E. 1990. Creating conditions for changes in prey community structure by *Chaoborus* spp. in a lake in Sweden. - Hydrobiologia 198: 205-214.
Stenson, J.A.E. and Eriksson, M.O.G. 1989. Ecological mechanisms important for the biotic changes in acidified lakes in Scandinavia. - Arch. Environ. Contam. Toxicol. 18: 201-206.
Stenson, J.A.E. and Oscarson, H.G. 1985. Crustacean zooplankton in the acidified Lake Gårdsjön. - Ecol. Bull., (Stockholm) 37: 224-231.
Stenson, J.A.E., Svensson, J.-E. and Cronberg, G. 1993. Changes and interactions in the pelagic community in acidified lakes in Sweden. - Ambio 22: 277-282.
Sutcliffe, D.W. and Hildrew, A.G. 1989. Invertebrate communities in acid streams. - In:

Morris, R., Taylor, E.W., Brown, D.J.A. and Brown, J.A. (Eds.) Acid toxicity and aquatic animals. Univ. Press, Cambridge, pp. 13-29.

Svahnberg, A. 1982. Ornitologisk undersökning av Tolken 1982. [Ornithological study of the Lake Tolken in 1982.] - Nat. Board Fish., unpubl. rep.

Svärdson, G. 1976. Interspecific population dominance in fish communities of Scandinavian lakes. - Rep. Inst. Freshw. Res., Drottningholm 55: 144-171.

Thörnelöf, E. and Degerman, E. 1991. Lake liming in Sweden - case studies and general experiences. - In: Olem, H., Schreiber, R.K., Brocksen, R.W. and Porcella, D.B. (Eds.) International lake and watershed liming practices. The Terrene Inst., Washington, D.C., USA, pp. 41-60.

Tyler, S.J. and Ormerod, S.J. 1988. Effects on birds. - In: Ashmore, M.R., Bell, J.N.B. and Garretty, C. (Eds.) Acid rain and Britain's natural ecosystems. Imperial College, Centre for Environmental Technology, pp. 97-109.

Uimonen-Simola, P. and Tolonen, K. 1987. Effect of recent acidification on Cladocera in small clear-water lakes studied by means of sedimentary remains. - Hydrobiologia 145: 343-351.

Vanni, M.J. 1988. Freshwater zooplankton community structure: introduction of large invertebrate predators and large herbivores to a small-species community. - Can. J. Fish. Aquat. Sci. 45: 1758-1770.

Ward, A. 1991. Kalkning i Nissan - hur har vi lyckats? [The liming of River Nissan - did we succeed?] - Jönköping County Council, Rep. 1991:5.

Weatherley, N.S. 1988. Liming to mitigate acidification in freshwater ecosystems: a review of the biological consequences. - Water Air Soil Poll. 39: 421-437.

Weatherley, N.S. and Ormerod, S.J. 1992. The biological response of acidic streams to catchment liming compared to the changes predicted from stream chemistry. - J. Environm. Managem. 34: 105-115.

Weatherley, N.S., Thomas, S.P. and Ormerod, S.J. 1989. Chemical and biological effects of acid, aluminium and lime additions to a Welsh hill stream. - Environm. Poll. 56: 283-297.

Weatherley, N.S., Rutt, G.P., Thomas, S.P. and Ormerod, S.J. 1991. Liming acid streams: aluminium toxicity to fish in mixing zones. - Water Air Soil Poll. 55: 345-354.

Wiederholm, T. and Eriksson, L. 1977. Benthos of an acid lake. - Oikos 29: 261-267.

Wren, C.D., Stokes, P.M. and Fischer, K.L. 1986. Mercury levels in Ontario mink and otter relative food levels and environmental acidification. - Can. J. Zool. 64: 2854-2859.

Wright, D., Danks, M., Popp, W., Eiler, P. and Schreiber, R.K. 1991. Biological response of Thrush Lake, Minnesota, to a protective liming treatment. - In: Olem, H., Schreiber, R.K., Brocksen, R.W. and Porcella, D.B. (Eds.) International lake and watershed liming practices. The Terrene Inst., Washington, D.C., USA, pp. 111-128.

Yan, N.D., Keller, W., Pitblado, J.R. and Mackie, G.L. 1988. *Daphnia-Holopedium* relationships in Canadian Shield lakes ranging in acidity. - Verh. Internat. Verein. Limnol. 23: 252-257.

Yan, N.D., Keller, W., MacIsaac, H.J. and McEachern, L.J. 1991. Regulation of zooplankton community structure in an acidic lake by *Chaoborus*. - Ecol. Appl. 1: 52-65.

Chapter 9

The impact of liming on aquatic communities

Magnus Appelberg

Swedish National Board of Fisheries Institute of Freshwater Research, Fish Monitoring Group, S - 178 93 Drottningholm, Sweden.

Summary

The far-reaching changes in the number and composition of plant and animal species which normally occur in conjunction with the liming of lakes and watercourses involve major changes in the structure and function of communities of organisms and the complexity of food webs. This is primarily attributable to changes in water chemistry, but interaction within and among species is of crucial importance to the post-liming development of a community. The short-term effects of liming often differ from those of the long-term effects. Normally, individual species expand rapidly for a short time, since competition and predation are reduced and there is a surplus of nutrients. In certain cases, this may have undesirable effects such as blooms of certain phytoplankton species, sharp increases in macrophytes or abnormal expansion of fish populations. The long-term impact, on the other hand, is to a great extent determined by biotic factors such as competition and predation within the new post-liming abiotic limits which emerge.

Despite the fact that, in most cases, the flora and fauna in limed lakes and watercourses resemble those which are to be found in unacidified waters, most surveys suggest that communities in limed waters will never fully return to their pre-acidification states. The post-liming production capacity of the water is crucial to both the species composition and the length of the food chain. In general, the complexity of the food web in limed water will be reduced as a result of a continued decline in the supply of inorganic nutrients, which may in turn be attributable to decreased mobility of phosphorus in the soil profile of the acidified catchment area.

In severely acidified lakes where fish populations have completely disappeared, the number of trophic levels has been reduced from four to three: primary producers (phytoplankton, macrophytes and epiphytic algae), herbivores, and invertebrate secondary consumers (predators). After liming,

Liming of Acidified Surface Waters
L. Henrikson and Y.W. Brodin (Eds.)
© Springer-Verlag Berlin Heidelberg 1995

the number of trophic levels usually returns to four when fish resume their position as the highest level in the community.

The change in the length of the food chain affects the underlying trophic levels from the top downwards, i.e. predation has an impact on lower levels. In the pelagic food chain, the composition of zooplankton is affected in such a way that the smaller forms of plankton tend to dominate, which in turn alters the grazing pressure on the phytoplankton and has a significant impact on the composition of phytoplankton species. Even in the littoral food web, a shift from larger to smaller invertebrates takes place. A "bottom-up" effect has been demonstrated in a number of cases, with changes at lower trophic levels affecting higher levels after liming. The outcome of a liming treatment thus depends largely on whether the predator or the prey is the most affected by the change in water chemistry.

A limed freshwater usually constitutes a more variable environment than an unacidified water, with greater fluctuations in the acidification-dependent variables over a given period of time. This can be expected to result in communities where the average food chain is shorter than in unaffected waters. While local disturbances affect the rate at which communities develop, both predation and the disturbances themselves may affect competition between different species. Species diversity may therefore be greater in limed waters, since less competitive species have a better chance to survive.

1. Introduction

When acidified lakes and watercourses are limed, extensive changes normally occur in the number and composition of species of plants and animals, as well as in the density and biomass at all trophic levels. This is primarily attributable to the change in water chemistry but, as has been emphasized in other chapters of this book (Larsson, Chap. 7, Degerman *et al*, Chap. 8), organisms also interact with one another both within and between the different trophic levels. Such interactions primarily consist of competition within and between species and predation. It is the overall result of the abiotic and biotic effects that ultimately determines the structure and function of the post-liming community.

If it is to be possible to describe the effects of liming on entire communities, comprehensive knowledge is required regarding the way in which abiotic factors influence individual organisms, or classes of organisms, and how these organisms interact in a developing community. Studies of the latter have not normally been taken into account when monitoring the effects of liming - the main interest has focused on the direct impact of changes in water chemistry on individual species or groups of organisms.

A more community-oriented approach has been applied in individual studies, however. Stenson *et al* (1978) showed experimentally that a reduction in the fish population (roach, *Rutilus rutilus*) had a decisive impact on the lower flora and fauna. It was possible to relate these findings to acidification since the change in the fish population resembled changes which had been observed in acid lakes. Field experiments highlighted the role of fish communities as a controlling factor in acidified lakes -"top-down control" of lower trophic levels. Subsequently, Eriksson *et al* (1980) studied the role and impact of invertebrate predators in acidified lakes, especially their impact on zooplankton communities in lakes where fish had died out. Hultberg and Andersson (1982), Hasselrot *et al* (1984), Nyberg (1984) and Henrikson *et al* (1984) also adopted a more community-oriented approach in describing the effects of liming within and between different trophic levels. Trophic interactions have also been included in comprehensive studies of Lake Gårdsjön (e.g. Andersson and Olsson 1985, Dickson 1988) and in the study of the Norwegian Lake Hovvatn (Raddum *et al* 1986).

Studies of entire communities, or subcommunities, have also been performed in conjunction with fertilization trials in limed waters (Andersson *et al* 1989, Olofsson *et al* 1988, Blomqvist *et al* 1992). The latest of these studies was not intended as an analysis of the effects of liming, however, but of how nutrient additives affected the structure and dynamics of the post-liming pelagic ecosystem in acidified lakes.

Reports on the structure and long-term development of fish communities in lakes after liming include Appelberg *et al* (1989, 1990), Appelberg and Degerman (1991) and Degerman *et al* (1992). An integrated monitoring project on the effects of liming, ISEL (Integrated Studies of the Effects of Liming), also describe the links between and within the structures of communities in limed lakes and watercourses (Appelberg and Aldén 1992).

There have been fewer studies of communities in limed running waters, for understandable reasons. The organisms in running water are more dependent on variations in their surroundings. Degerman *et al* (1990) concluded, however, that competition existed between brown trout *(Salmo trutta)* and bullhead *(Cottus gobio)* in several limed rivers in western Sweden. Degerman and Appelberg (1992) have also shown that interaction between Atlantic salmon *(Salmo salar)* and brown trout represents a structuring factor in fish communities in limed running waters. The lack of studies on changes in communities in limed running waters does, however, mean that the remainder of this chapter must be restricted to a discussion of plant and animal communities in lakes.

2. Communities of organisms

A community may be defined as a group of populations which exist together in time and space (Begon *et al* 1990). This definition differs from the normal interpretation of the word "ecosystem", which includes both the communities and their surroundings. However, the community consists of more than simply the sum of its member species; it is the sum of these species and the interaction among them. In order to determine the characteristics of the community, it must be studied as a whole. The purpose of this chapter is an attempt to identify and describe repeated patterns in the common characteristics of communities in limed waters. The community may be described with the help of the following terms: number of species and species diversity (the proportions between the species), succession (the way in which the composition of species develops) and climax, which describes the stage at which the community has attained a more or less stable equilibrium. Energy and material flows through the communities may also be used for descriptive purposes. The factors which are assumed to influence the communities may be classed as abiotic (acidification, liming, climatic changes, etc.) or biotic (predation, competition, parasites and disease).

The long-term development of communities in limed waters is of particular interest, as is the question whether the community will ultimately resemble that which existed prior to acidification. It is therefore important to try to describe the structure and stability after liming. Has the community any prospect of regaining its original status and, if so, to what extent will it be stable? In order to answer these questions, the complexity of the food web and the length of the food chains must also be described.

3. Short-term and long-term effects of liming

The immediate changes which a community undergoes after liming take place during a period when it is out of balance but is striving to achieve a greater degree of stability. The factors affecting unbalanced communities are often quite different from those which control communities in equilibrium (Chesson 1986, Begon *et al* 1990). The short-term effects of liming may therefore differ considerably from the long-term ones. Naturally, short- and long-term effects must be viewed in the light of the types of organism under study: two years represent many generations of phytoplankton, while half a century may be required for the more long-lived fish species to achieve the same number of generations. Since the changes in the higher trophic levels - which often consist of fish - also affect the lower ones, however, the development of fish communities is often crucial.

The short-term effect usually entails a rapid expansion of individual species during a brief period, since the levels of competition and predation are low and there is a surplus of nutrients. In certain cases, this might lead to undesirable developments, such as blooms of certain species of phytoplankton (Svensson *et al*, Chap. 10, Larsson 1988), rapid development of macrophytes such as alternate water milfoil *(Myriophyllum alterniflorum*; Larsson, Chap. 7) or rapid expansion of certain fish populations (Degerman *et al* 1992). The long-term effect, however, is largely determined by biotic factors such as competition and predation within the new abiotic limits which have been set as a result of liming.

The fact that the long-term effects involve a community which is completely different from the one which develops immediately after liming can be exemplified by the rapid increase in diatoms and their successive replacement by chrysophytes (Eriksson *et al* 1983). A succession of species has also been noted when *Sphagnum* mosses disappear in conjunction with liming. After expanding immediately after liming, pondweed *(Potamogeton natans)* and alternate water milfoil are successively replaced by various other species (Eriksson 1988). The post-liming development of zooplankton is largely determined by changes in the fish community. A clear succession of zooplankton has been observed in several previously fish-free lakes where species vulnerable to predators are successively replaced by less vulnerable species once the fish return (Degerman *et al*, Chap. 8). Populations of water louse *(Asellus aquaticus)* frequently increase immediately after liming, subsequently returning to "normal" densities (Eriksson *et al* 1983, Henrikson and Oscarson, unpubl.).

The development of fish communities after liming is highly time-dependent (Appelberg and Degerman 1991, Degerman *et al* 1992). Short-term effects often include rapid growth of opportunistic species such as European perch *(Perca fluviatilis)*. As more competitive species such as roach, bream *(Abramis brama)* and vendace *(Coregonus albula)* recolonize the lake, the first species decreases in number. This normally means that the diversity of fish communities increases as time elapses after liming.

This shows that the development of plant and animal communities in limed waters varies over time, mainly as a result of the fact that the biotic interactions depend on the development of the most long-lived species (usually fish) and their ability to recolonize. So-called cascading effects, i.e. the fact that the effect of a change at one trophic level in the ecosystem influences several other levels, can thus be expected to occur for a considerable period after liming (cf. Svensson *et al*, Chap. 10, Appelberg, Chap. 12). The changes which result from liming will therefore persist until the most long-lived species (usually fish) have attained a stable level. In many cases, this appears to require more than ten to fifteen years.

The conclusion to be drawn from the studies of the effects of liming which have been presented to date is that such effects should not be seen too short-sightedly. In many cases, development is slow, so it will probably take many years - even decades - before community structures stabilize. Unfortunately, many studies have focused on much shorter periods and this has probably led to misinterpretation of the results of liming in certain cases. The rapid growth in populations of certain species during the initial years after liming has e.g. been seen as a lasting and unintentional effect.

4. Post-liming abiotic factors

When following up the effects of liming, it is often difficult to distinguish between direct effects and other effects which are a secondary result of changes in water chemistry. The direct effects of water quality changes after liming vary among different species. Reduced toxicity, as a result of higher pH, and lower concentrations of aluminium affect most classes of organisms (Larsson, Chap. 7, Degerman *et al*, Chap. 8), but other changes in water chemistry may also play an important part in the development of certain organisms.

The rapid growth of certain plants in acidified waters has been assumed to be the result of a shift in the carbon dioxide equilibrium, resulting in higher levels of carbon dioxide in the water (Svedäng 1992). The plants which exclusively use carbon dioxide as their carbon source are favoured by acidification, while species which also use carbonic acid suffer. *Sphagnum* mosses, bulbous rush *(Juncus bulbosus)* and epiphytic algae *Mougeoutia* sp., which use carbon dioxide exclusively, have at times been highly abundant in acidified lakes, but have declined rapidly after liming (Hultberg and Andersson 1982, Eriksson 1988, Svedäng 1992, Alenäs *et al* 1991). The succession of plants after liming thus depends largely on the fact that plants which require carbon dioxide are at a disadvantage as a result of the change in water chemistry, while those which also can use hydrogen carbonate are favoured. Although subsequent recolonization by isoetids, for instance, is subject to competition, it is mainly abiotic factors which allow species to colonize and recolonize.

In addition to changes in water chemistry, liming also has an impact on the physical conditions in the lake. Perhaps the two factors most in evidence are changes in light conditions and the heterogeneity of habitats. After liming, transparency tends to decline in the clearest lakes, while the clarity of acid, but humic, lakes may improve as a result of precipitation of humic particles (Degerman 1987, Degerman *et al* 1992). A reduction in transparency means that the rooted vegetation on the bottom of the lake declines while the conditions for the production of phytoplankton change (Larsson, Chap. 7). The change in transparency also affects competition between aquatic birds (Eriksson 1987, Eriksson *et al* 1988). A decreased transparency after liming has been shown to affect the ability of the water to absorb energy from sunlight. In Lake Gårdsjön the depth of the epilimnion decreased from 10 m before liming in 1979 to 5 m in 1985 (Larsson 1988), which was expected to result in a higher water temperature after liming.

A change in macrophytic vegetation, as a result of changes in either water chemistry or the light climate, can have consequences for the composition and density of the benthic fauna. When *Sphagnum* moss disappears after liming (Grahn 1977, Eriksson 1988) and is replaced by other rooted species, it brings a change in the micro-habitat of the benthic fauna.

Depending on which types of plant that replaces the *Sphagnum* moss, the result may be either favourable or unfavourable to organisms at a higher trophic level. In general, the prey's (benthic fauna and fish) chances of avoiding predators are greater when the macrophytic vegetation increases. This has an impact on the entire community (Persson 1986). Diehl (1988) showed experimentally that the vegetation was of crucial importance to hunting efficiency among predatory fish. The reduced distribution of *Sphagnum* moss may also affect the species composition of the benthic fauna. Henrikson (1993) showed that several (benthic fauna) taxa were more numerous in *Sphagnum* moss than on vegetation-free lake bottoms. For dragonfly larvae *Leucorrhinia dubia*, the *Sphagnum* moss provides a refuge from predatory fish. Nyman (1990) showed that *Sphagnum* moss could function as a sediment trap in which the accumulated organic material could be utilized by the benthic organisms.

5. Acidification and community development

The effects of a change in the habitat, whether caused by acidification or liming, vary depending on the stability of the community. The more stable the community, the more effectively it can withstand stress (Odum 1981). The short-term effects of low-level stress in a stable community may therefore be relatively severe, while the long-term effects may still be small. This is because the stable community is subject to stabilizing factors which imbue it with a high degree of elasticity and an ability to regain equilibrium. In the young, unstable community, however, only a small amount of stress may cause major changes, since it is strongly affected by destabilizing factors (low elasticity).

The degree of acidification determines how great the change after liming will be. The more acid the water before liming, the more it will change after liming. This is also what has been observed at all trophic levels in a number of studies of the effects of liming (see Larsson, Chap. 7, Degerman *et al*, Chap. 8). Nyberg *et al* (1986) showed that the greatest post-liming changes in a fish community were found when the pH before liming was below 5, for instance, while the effects of liming on moderately acidified waters (pH >5.5) were small (see also Degerman *et al*, Chap. 8). This is to some extent attributable to the fact that the change in the habitat parameters is greatest when severely acidified communities are limed. Another reason is that the communities in recently acidified waters are often unstable, which means that they are less resistant to the change which liming involves. The effects of liming on a moderately acidified community in which the structure is still more or less intact have often been shown to be relatively limited (Fraser *et al* 1985, Nyberg *et al* 1986).

Apart from the species which are not favoured by the change in water chemistry after liming (such as *Sphagnum* moss and some species of benthic fauna which are favoured by acidification), the species remaining in the acidified community are the ones to dominate initially in the limed water. If species which have died out as a result of acidification have no opportunity to recolonize after liming (Bergquist, Chap. 17), communities will not return to pre-acidification levels.

6. Productivity and community structure

The productivity of a lake represents one of the most fundamental factors in the post-liming development of the community (e.g. Wilander *et al*, Chap. 5, Gahnström, Chap. 6, Larsson, Chap. 7). Productivity is primarily controlled by the supply of inorganic nutrients from the drainage area, but it is also affected by the internal turnover of inorganic nutrients. It has been suggested that acidification reduces the supply of inorganic nutrients by decreasing the mobility of phosphorus in the soil profile of the drainage area (Broberg and Persson 1984, Wilander *et al*, Chap. 5), although this hypothesis has been questioned more recently (Olsson and Pettersson 1993).

Despite the importance of inorganic nutrients in the post-liming course of events, surprisingly few studies have focused on the flow of energy and materials through the communities. Such studies have mainly been linked with the Lake Gårdsjön project (Andersson and Olsson 1985, Dickson 1988) or with studies of fertilization as a supplementary post-liming measure (Olofsson *et al* 1988, Blomqvist *et al* 1992). Since Lake Gårdsjön has had no natural fish population for several decades, however, it is difficult to apply the findings generally to waters which support fish communities. In these two studies, moreover, the main interest focused on the pelagic ecosystem, while it is most probably the littoral system which dominates production in most limed lakes (Appelberg *et al* 1993a). It would be reasonable to suppose that liming affects the turnover of inorganic nutrients. Since the density of fragmenting benthos in the littoral zone increases, so does the decomposition of detritus (Andersson 1985, Gahnström, Chap. 6, Henrikson and Oscarson, unpubl.). This could involve increased turnover of inorganic nutrients in the longer term, since the availability of organic material for microbial decomposition increases (Gahnström, Chap. 6).

The importance of post-liming inorganic nutrient levels to phytoplankton communities was noted by Eriksson *et al* (1983), who found that in addition to the impact of increased pH on the composition of phytoplankton species after liming, variations in the supply of inorganic nutrients were of crucial importance. Blomqvist *et al* (1992) showed experimentally that the supply of inorganic nutrients (phosphorus and nitrogen) was the prime factor in determining the biomass of phytoplankton, not the liming treatment.

Results reported from the ISEL programme also highlight the fact that productivity is crucial to the composition of species at higher trophic levels in limed waters. In general, the communities in the limed waters under study were less productive than in equivalent neutral lakes. This was found when the benthic faunas were used as a trophic indicator (Johnson *et al* 1992). The same authors also showed that the number of individuals in the profundal benthic fauna communities in the limed lakes in the ISEL programme was significantly lower than in the neutral reference lakes. Despite this, no differences in the diversity of species were found between the different types of lake. The littoral benthic fauna also indicated that the limed lakes included in the programme were less productive than the corresponding unacidified reference lakes (Lingdell and Engblom 1992).

The productivity of a water is reflected at all levels in the community. This is shown by the fact that there is a co-variation in the number and diversity of species between different classes of organisms and different trophic levels. In the ISEL programme, a positive correlation was found between the number of phyto- and zooplankton species in the fourteen lakes in the study. The number of species of phytoplankton and zooplankton also increased with the number of benthic fauna taxa in the littoral zone and the diversity of the pelagic fish communities (Fig. 1). This co-variation was largely attributable to the post-liming productivity of the lake; the more productive the lake, the higher the number and diversity of species.

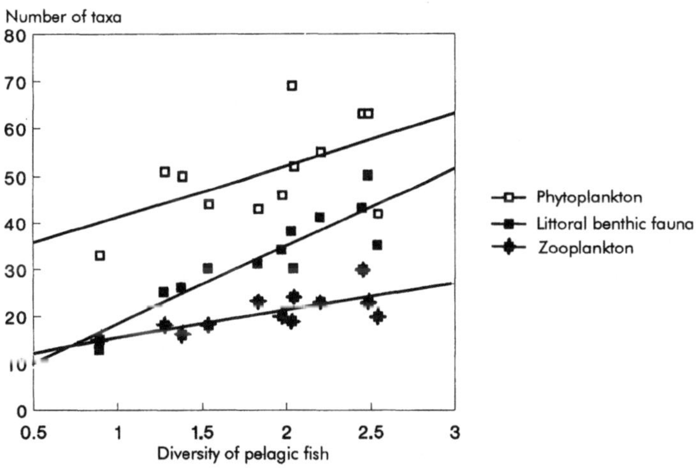

Figure 1*. Relationships between trophic levels in limed waters. Number of taxa of phytoplankton, zooplankton and littoral benthic fauna versus diversity in the pelagic fish community in the fourteen lakes included in the Swedish ISEL programme for integrated studies of the effects of liming. Species diversity in the pelagic fish community increases in pace with increased levels of nutrients. From Appelberg and Aldén (1992).*

7. Biotic factors: competition as a controlling factor

Competition within and among species is one of the most significant biotic factors in the post-liming development of communities. The more stable the environment, the more important competition becomes as a structuring factor (Wiens 1977), which means that it attains increasing importance as time elapses after liming. In communities which are *dominance controlled*, i.e. those in which the structure is primarily determined by a few highly competitive species, one can expect a short-term increase in opportunistic uncompetitive species directly after liming. Such species are successively replaced by more competitive ones. This is a more or less general development in most ecosystems which are subject to some form of perturbation. In the *donor-controlled* communities, i.e. communities which are governed by species occurrence, different species are assumed to have an equal opportunity to colonize and compete on equal terms. In such communities, therefore, a "vacancy" may be filled by any species (Begon *et al* 1990). In Swedish limed lakes where the recruitment of most organisms is strongly seasonal, one may assume that dominance is the most common controlling factor.

One of the most striking examples of the significance of competition in a community controlled by dominance is the succession of fish species after liming. In waters where acidification has not completely eliminated the fish populations, northern pike *(Esox lucius)* and European perch usually remain. These lakes, which in many cases also support coregonids such as whitefish (*Coregonus* sp.), vendace or cyprinids (e.g. roach and bream), often undergo a clear succession after liming in which the communities consist predominantly of European perch for a period of ten to fifteen years. During this time the coregonids and cyprinids gradually take over. One example of this can be found in Stora Härsjön in south-western Sweden (Appelberg *et al* 1990, Appelberg, Chap. 12). The European perch population in this lake increased rapidly after liming, but once the vendace had re-established itself the perch decreased in density. The trend continued when roach began to recolonize the lake.

In cases where coregonids and cyprinids have been reduced or eliminated, perch tend to increase sharply in number after liming (Nyberg *et al* 1986, Degerman *et al* 1992). In Lake Upprämen, central Sweden, the density of European perch increased by more than tenfold during the first five years after liming and then decreased to more normal levels seven years later (Institute of Freshwater Research, unpubl.). In this case, the stabilization of the European perch population was attributable to the fact that the food resources were probably overexploited, resulting in a high level of intraspecific competition. A more comprehensive analysis of the development of fish populations in limed waters indicates that this trend is fairly common (Appelberg and Degerman 1991, Appelberg *et al* 1993b).

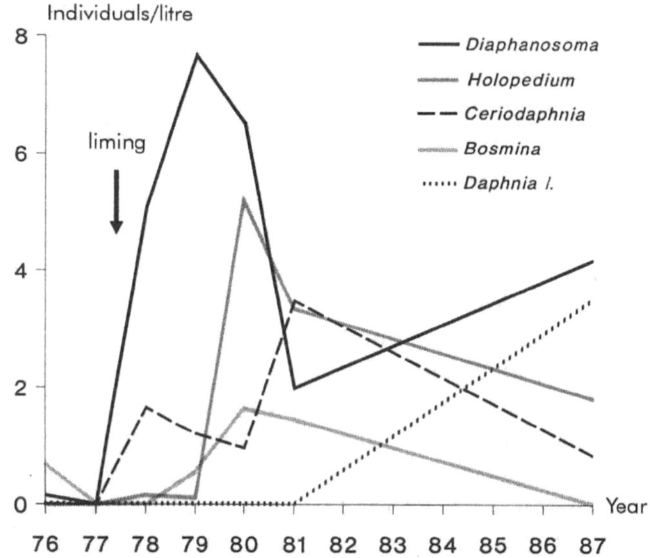

Figure 2. Post-liming development of planktonic waterfleas in Lake Trehörningen, in the River Åvaån water system south of Stockholm. The figures are from samples in July and August (from Nyberg, unpubl.).

If the competition between species is low before liming, populations of some species may grow rapidly afterwards. Several studies have shown that the planktonic waterfleas, *Diphanosoma brachyurum*, for instance, may develop into very dense populations shortly after liming (Fig. 2) (Raddum *et al* 1986, Nyberg unpubl., Stenson *et al* 1993). This species was then successively replaced by several other plankton species, probably as a result of increasing competition.

8. "Bottom-up" effects

Alongside competition, predation is the most important biotic factor controlling communities in limed waters. A predator-prey relationship is often complex and is affected by changes in density of both predator and prey. Several of the studies which have been performed in conjunction with monitoring of liming effects have shown that changes at lower trophic levels affect higher levels from below. It has, for instance, been shown that the predator-prey relationship has changed drastically after liming, particularly in ecosystems which have been subjected to severe acidification stress prior to liming.

Changes in the composition of phytoplankton affect the food supply of filtering zooplankton after liming. Eriksson *et al* (1983) reached the conclusion that the increased number of filterable phytoplankton which has been observed after liming was one of the most common reasons for the fact that the proportion of phytoplankton-consuming zooplankton increased in density. Similar results were noted in the experimental studies of nutrient supply performed in the limed Lake Hecklan in southern Sweden (Olofsson *et al* 1988). In this experiment, nutrients were added in order to change the phytoplankton biomass without promoting undesirable (inedible from the point of view of the zooplankton) phytoplankton species. The increase in the phytoplankton biomass also resulted in increased density and biomass among the herbivorous zooplankton.

The density of the planktonic phantom midge larvae *Chaborus flavicans* and *C. obscuripes*, both of which are effective zooplankton predators, normally increases in lakes which are still devoid of fish after liming. This has been observed in the case studies in Lakes Gårdsjön and Blanksjön in southern Sweden and was considered to be a direct consequence of increased food supply, in this case Rotatoria. These populations could increase since the density of the competing waterfleas was low (Henrikson *et al* 1985, Nyberg 1984, Stenson 1990). An increase in the food supply as a result of a higher level of bacterial activity - and thus an increase in the decomposition of allochthonous material - has also been seen as one of the explanations as to why the water louse *(Asellus aquaticus)* been found to increase in number after liming (Eriksson *et al* 1983). An increase in the supply of food for piscivorous fish is also one of the main explanations as to why large fish grow more rapidly after liming. Nyberg *et al* (1986) showed that the growth of large European perch individuals increased in pace with the recruitment of their prey after liming.

9. "Top-down" effects

The "top-down" effect of predators on lower trophic levels is nowadays considered to be one of the most significant factors in structuring communities in lakes. In one study, Stenson *et al* (1978) reduced the fish biomass in an oligotrophic lake on the Swedish west coast, which sharply affected the trophic level of the lakes and produced a dramatic change in the structure of the communities. Eriksson *et al* (1980) also showed that the shift from fish predators to invertebrate predators, which dominates in severely acidified lakes where fish have died out, had a decisive impact on the structure of zooplankton communities (see the example in Degerman *et al*, Chap. 8). This modified "top-down" effect has also been seen as a contributory factor in the reduced productivity of the lake, as a result of reduced turnover of inorganic nutrients.

In general, it is assumed that selective predation on competitively dominant food organisms favours diversity in a community (Begon *et al* 1990).

This also applies if the predators are generalists since they feed on the species which are highest in density. Such generalists are in a position to influence lower trophic levels by switching to other food items when the density of their prey falls too low. Follow-up of liming effects has yielded several examples of the effect of predators on lower trophic levels. Henrikson *et al* (1985) found that planktonic crustacean communities changed more rapidly after liming when fish remained in the system than otherwise. This was mainly attributable to the fact that the fish influenced the competition between the species of zooplankton by feeding on the most competitive ones. This has been confirmed more recently by Nyberg (unpubl.), who has shown that the number of species of planktonic crustaceans after liming increased in pace with the density of planktivorous fish.

One clear example is to be found in the predation of *Chaborus* larvae which can be reduced sharply as a result of predation by fish. This leads in turn to a reduction in the pressure from the larvae on zooplankton, while the pressure exerted by fish on the plankton increases. The result in this shift between predators alters the composition of the zooplankton community drastically (Nyberg 1984, unpubl., Stenson 1990, Stenson *et al* 1993). After re-establishment of fish in a limed water, the zooplankton predator *Glaenocorisa propinqua* (a water boatman) declines, thus resulting in reduced pressure on the plankton community. This provides a further example of how the invertebrate predators which can be regarded as generalists affect lower trophic levels (Henrikson and Oscarson 1984, Henrikson unpubl.).

The significance of predation for lower trophic levels has also been demonstrated in lakes which have been recolonized by vendace after liming. Since this species is an effective zooplankton predator, the composition of zooplankton species shifts to forms which are less susceptible to predation as the vendace increase in density (Appelberg *et al* 1990, Degerman *et al*, Chap. 8, Appelberg, Chap. 12). Predation by fish has also been shown to affect benthic fauna after liming. The density of crayfish *(Astacus astacus)*, which are susceptible to acidification, is supposed to be controlled by predation in many limed lakes, particularly by perch (Appelberg and Odelström 1986, 1988, Appelberg 1990, 1992). This is probably one reason why the species often does not recover after liming.

Predation also seems to be the structuring factor among populations of prey, since their defence mechanisms against the predators are affected (e.g. Kerfoot and Sih 1987). In a study of four limed lakes, Appelberg and Odelström (1986) showed that the choice of habitat and growth among crayfish were highly dependent on the number of predatory fish in the lake. If there were a large number of predators, the activity of the crayfish declined, which in turn led to a reduced growth rate and smaller size upon sexual maturity. Henrikson (1988, 1993) showed the indirect effects of predatory fish on dragonfly larvae *(Leucorrhinia dubia)* in acid/limed lakes. This species, which utilizes *Sphagnum* moss both as a hiding place and as a feeding habitat, changed colour according to its habitat in order to reduce the risk of discovery by predators.

A link between predators and lower trophic levels has been demonstrated in both unacidified lakes (e.g. Persson *et al* 1988) and limed waters. In the Stora Härsjön case study, there was a very strong positive correlation between the total phosphorus level and the density of planktivorous fish after liming (Fig. 3). The increase in density of planktivorous fish in the lake probably contributed to an increase in the turnover and availability of inorganic nutrients, which in turn resulted in a higher production level. The trophic index of this lake, based on the composition of phytoplankton, increased during the period under study.

It would be reasonable to assume that the development of communities in limed waters is structured through both "bottom-up" and "top-down" effects. Fish play a fundamental role in the development of communities after liming, through their ability to affect lower trophic levels, while the development of fish populations after liming is highly dependent upon the production capacity of the lake. In a review of the effects of liming on

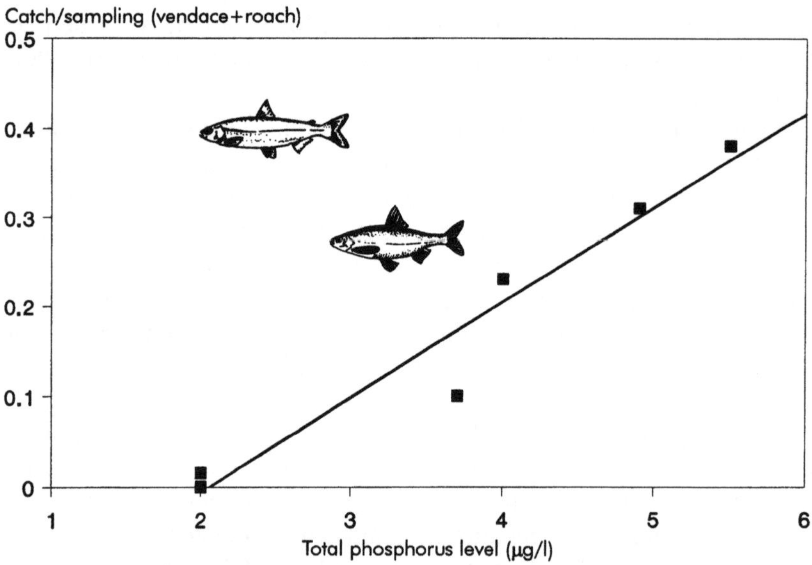

Figure 3. The relationship between total levels of phosphorus and catches of zooplanktivorous fish in Lake Stora Härsjön, SW Sweden, following liming in 1978. Both the phosphorus levels and the density of zooplanktivorous fish have increased successively over a ten-year period after the initial liming treatment. From Appelberg et al (1990).

organisms, Fraser *et al* (1985) divided the limed lakes into different categories depending on whether or not fish still remained in the pre-liming system. A classification of this sort is not relevant to Swedish conditions, however, since fish are usually introduced after liming, albeit not as the original fauna but in the form of salmonids.

10. Predation and changeable environments

In many contexts, the number of species in a community provides an important measure of the development and status of that community. Ultimately, the number of species indicates biodiversity and an increased biodiversity is in itself one of the objectives of liming. Within most classes of organisms, the number of species and species diversity decline in pace with an increase in acidification (Brodin, Chap. 3, Larsson, Chap. 7, Degerman *et al*, Chap. 8). After liming, the improved water quality usually leads to an increased number of species at all trophic levels: macrophytes (Hultberg and Andersson 1982, Eriksson *et al* 1983, Eriksson 1988), phytoplankton (Hörnström and Ekström 1986), zooplankton (Hörnström and Ekström 1986, Nyberg unpubl., Stenson *et al* 1993), benthic fauna (Lingdell and Engblom 1992, Bergquist *et al* 1992, Mossberg and Nyberg unpubl.) and fish (Degerman and Nyberg 1989, Appelberg and Degerman 1991).

A certain amount of predation, or a moderate level of variation in the environment, may have a favourable effect on the development of the number of species and species diversity. The intermediate disturbance hypothesis states that the maximum diversity in a community is achieved when the environment varies at an "intermediate level" (Connell 1978). The background to this is that a variable environment in the form of repeated liming pulses, for example, reduces the risk of any species being eliminated through competition, since the dominant species are reduced by the variation. A similar effect is obtained if the dominant species are reduced through predation. This in turn leads to increased diversity. A stable environment after liming does not automatically result in increased diversity, therefore. A dominance-controlled community, in which certain species are competitively superior, means that the species which colonize the water at an early stage do not have to be the ones which dominate when the community has reached its climax (Fig. 4; Begon *et al* 1990). A variable environment can thus be assumed to favour species with a high rate of development but with a low optimum density.

It has been found that succession at several trophic levels is largely controlled by competition. In the zooplankton community it is controlled by planktonic crustaceans through a combination of predation and competition. Nyberg (unpubl.) suggested that the increased density of the waterflea *Daphnia cristata* after liming was mainly attributable to the fact that

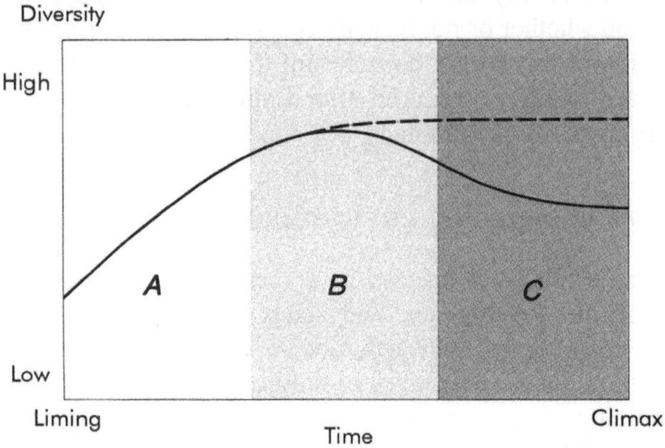

Figure 4. Post-liming trends in species diversity in a dominance-controlled community are mainly determined by interspecies competition: A) colonization by opportunistic species with a rapid rate of development but low optimum density; B) colonization by new, more dominant species with lower rate of development and higher optimum density, outcompeting the pioneer species to some extent; C) the most competitive species eventually dominate the community and outcompete the subdominant species. Unbroken line: communities which are only affected to a minor extent by disturbances of predation. Broken line: communities which are vulnerable to disturbances of predation. From Begon et al (1990).

D. longispina had been reduced by predatory fish. The continued decline in the dominant planktonic crustacean made it possible for other, less dominant species to increase in density (cf. Degerman *et al*, Chap. 8).

An example of the way in which stabilization of the environment may eventually reduce diversity after liming is to be found in the development of zooplankton in Lake Stora Härsjön (Appelberg, Chap. 12). Before liming, 7-10 taxa were noted. Three years after liming the number peaked at 20 taxa, but subsequently declined to about 15 taxa (Appelberg *et al* 1990). In this case, as in many others, the change in predatory pressure from fish affected the way in which the zooplankton communities developed.

11. Parasites and disease: an unknown factor?

The significance of parasites and disease as structuring factors is probably considerably underestimated, both in studies of acidification and in liming follow-up studies. Canadian studies have shown that the degree to which a species of crayfish was infected by *Thelohania contejeani* increased in conjunction with experimental acidification of a lake (France and Graham 1985). Similar results have also been obtained in experiments on signal crayfish *(Pacifastacus leniusculus)* and the crayfish plague *(Aphanomyces astaci)* (Appelberg, Odelström and Persson, unpubl. data).

Bellemakers and van Damm (1992) showed that liming of acidified waters in the Netherlands led to only 5% of spawn of the frog *Rana arvalis* being infected by fungus, against 95% in comparable acidified and unlimed waters. Little is known, however, about the positive or negative effects of liming on diseases and parasites. Such effects would probably depend on the susceptibility of the parasite to acidification. Since the condition of the host organism often improves after liming, it can be assumed that their resistance to parasites and disease increases in general.

12. Stability of post-liming communities

Few studies have looked into the stability of plant and animal communities after liming, despite the fact that the ultimate aim of liming is to "restore" ecosystems which have been damaged by acidification. Many authors have observed instability in different areas of the community, but have stated that there is often no way of measuring stability. It can be assumed, however, that aquatic communities in waters which have been severely acidified over a longer period of time have developed a certain level of stability, and that communities which are more or less unaffected by acidification will display a relatively high degree of stability. Waters which have only been acidified for shorter periods of time or which are still in an accelerating phase of acidification may, however, be assumed to contain unstable communities which are in a process of transformation. Liming can therefore be expected to affect these three types of community in different ways.

A number of studies indicate that fewer changes take place in a community as time elapses after liming. Hörnström *et al* (1992) showed that the main changes in the number of phytoplankton species took place in the first two or three years after liming and that only minor changes could be observed thereafter. Changes in the number of zooplankton species were also greatest during the first three to five years, after which only minor changes took place (Stenson 1990, Appelberg, Chap. 12, Nyberg, Chap. 11, Svensson *et al*, Chap. 10, Nyberg unpubl.). These results may indicate that the instability of the communities is greatest during the years immediately after the initial liming treatment and that they subsequently become more stable.

The stability of the communities is largely dependent upon the effect of liming treatment on water chemistry. Repeated "surges" in the form of reliming, with intermediate acid periods, counteract any tendency of communities towards equilibrium. If reacidification occurs, a return to the pre-liming situation can be expected.

As has been pointed out above, changes at lower trophic levels will depend on changes at higher levels. The stability of zooplankton communities is influenced by developments in fish communities, for instance. In a study of the long-term development of fish communities in 77 lakes, Appelberg and Degerman (1991) showed that their diversity increased as time elapsed after liming, along with the pH and the heterogeneity of the habitat. However, the instability of the communities (measured as Euclidean distance in terms of changes in the proportions of species) was higher in diverse communities in comparison with less diverse communities. This means that the stability declined during the first three years after liming, followed by gradual stabilization during the following twelve years (Fig. 5). It was not possible to determine how "stable" the communities were after that time. The results of the case study of Lake Stora Härsjön, for instance, showed that changes caused by liming may occur more than fifteen years after the initial treatment (Appelberg, Chap. 12). The development of the fish communities in these 77 lakes supports the hypothesis that complex and diverse communities are less stable than simple ones and contradicts the old view that complex systems are the most stable ones (cf. Begon *et al* 1990).

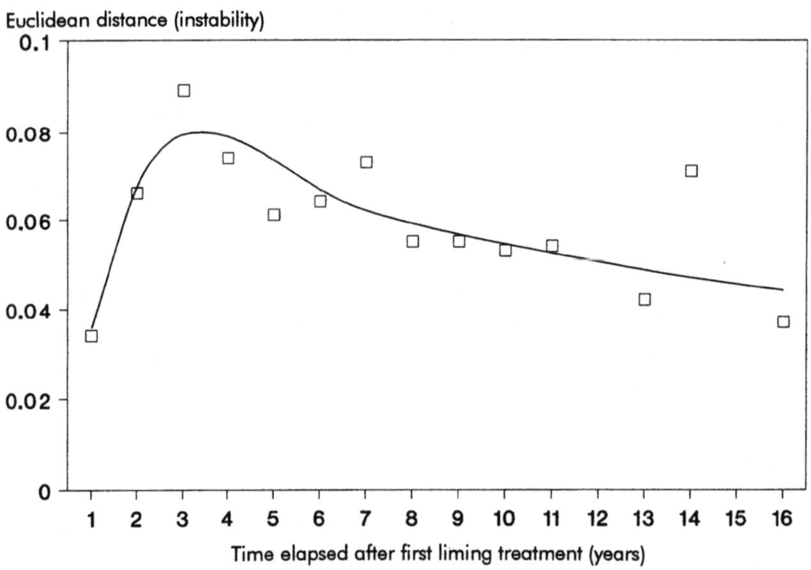

Figure 5. *Post-liming trends in the stability of fish communities in 77 limed Swedish lakes. The stability, computed in terms of the change in proportions of fish species sampled on two occasions, versus the time elapsed since the first liming treatment. The higher the value of the Euclidean distance, the lower the stability. From Appelberg and Degerman (1991).*

13. Trophic levels and food chains

It is often impossible to determine the number of trophic levels in an ecosystem exactly, since many species are omnivores or change their diet and behaviour during the course of their life cycles. However, it has often been found useful to view the food web in terms of simple chains in which each link represents a trophic level (e.g. Oksanen *et al* 1981, Pimm and Kitching 1987, Schoener 1989, Persson *et al* 1992) in order to compare the complexity of communities in different waters. A food chain of this sort can be used to describe the relationships between the trophic levels in a community.

The length of the food chain, or the number of trophic levels, has long been considered to be dependent upon the amount of energy supplied to the food web, or to the productivity of the system (Fretwell 1977, Oksanen *et al* 1981, Persson *et al* 1992). Other factors such as the heterogeneity of the habitat (Persson *et al* 1992) and the degree of disturbance (Pimm and Kitching 1987) have also been shown to affect the length of the food chain. In general, a reduction in the production capability of the water, reduced heterogeneity and increased environmental variation have been shown to result in a shorter food chain.

In severely acidified lakes in which fish have died out, the number of trophic levels is reduced from four to three. These levels consist of the primary producers (phytoplankton, rooted vegetation and epiphytic algae), herbivores, and invertebrate secondary consumers (predators) (Henrikson *et al* 1980, Appelberg *et al* 1993a). Although this shortening of the food chain is primarily attributable to the toxic effects of low pH and high metal concentrations, it affects the dynamics of the community. After liming, which allows recolonization of the water by fish, the number of trophic levels normally returns to four when the predatory fish which belong to the highest trophic level re-appear in the community (Fig. 6).

This change has a far-reaching impact on the structure and dynamics of the community. On the basis of relationships between trophic levels as described by Hairston *et al* (1960) (the HSS hypothesis), it can be assumed that the competitiveness of the highest trophic level, which consists of invertebrate predators in the acid system and of piscivorous fish in the limed system, is limited in a community which is approaching equilibrium, since there is no higher trophic level to prey upon it. This trophic level is below its optimum density, or carrying capacity, and is therefore affected if the production platform changes, i.e. if the primary production either decreases or increases. In Lake Gårdsjön the density of corixids increased after liming, probably as a result of increased production in the underlying trophic level (Henrikson and Oscarson 1984). This in turn means that the second highest trophic level, which consists of herbivorous invertebrates in the acidified system and of benthivorous and planktivorous fish and inverte-

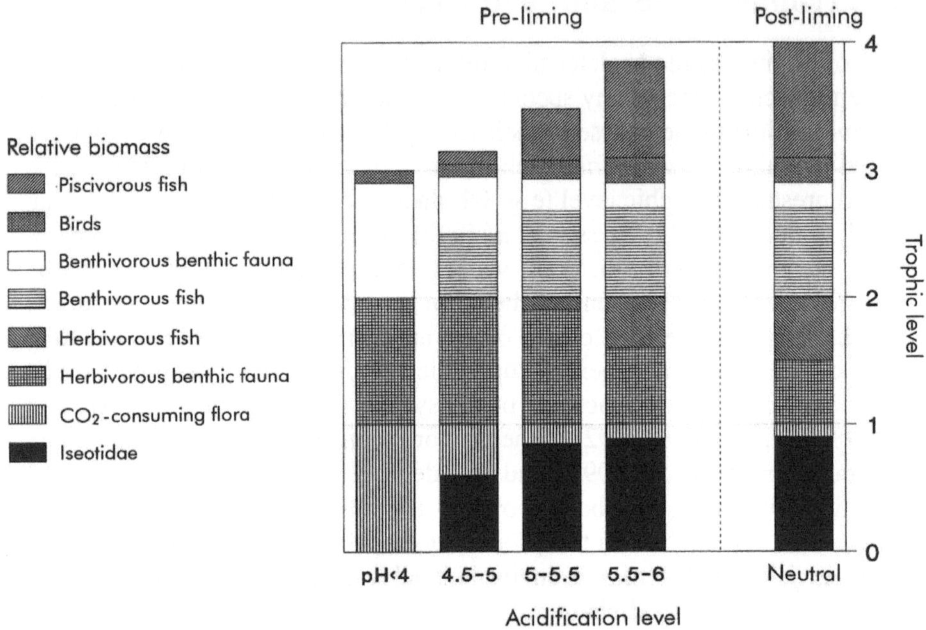

Figure 6. *Schematic representation of the length of the food chain (number of trophic levels) for different degrees of acidification before and after liming. The relative biomass for different groups of organisms is shown for each trophic level.*

brates in the limed system, should be limited by predation. This change means that the benthivorous invertebrates which function as top predators in the acid system are in turn affected by predatory fish in the limed system. This has also been verified in a number of monitoring studies on the effects of liming (see above, Degerman *et al*, Chap. 8).

The change in the length of the food chain also has an impact on the underlying trophic levels. The composition of zooplankton in the pelagic food chain is affected in that the smaller forms of plankton gain increased dominance, which increases the pressure on phytoplankton and has a considerable impact on their composition. A shift from large invertebrates to smaller forms also takes place in the littoral food web.

These theoretical links between the trophic levels may be considered to apply in cases where the communities are approaching some form of equilibrium, or climax. In most studies on the effects of liming, however, no such equilibrium has been achieved and the findings have been interpreted in different ways.

Persson *et al* (1992) argued that discrepancies between the predicted length of the food chain and the production level in the water were also influenced by other factors in the environment, such as the heterogeneity of the habitat. Studies of acidification and liming also show that other abiotic factors, such as the direct toxic effects of low pH or high metal concentrations, also affect the length of the food chain. This was suggested by Odum (1981), who described the communities in favourable environments as better organized than those in variable environments. Regardless of which of the above factors is considered to limit the length of the food chain, both the restricted production model and the disturbance model provide the same predictions concerning regulation of the community in a given trophic structure. In changeable surroundings, the trophic structure is affected more by environmental disturbances than by limitations in food. The result of a change in water chemistry is thus highly dependent on whether it is the predator or the prey which is most affected by the change.

14. Can acidified communities be restored?

Comparisons between communities in limed water and oligotrophic, unacidified waters often yield considerable similarities (Bergquist *et al* 1992, Hörnström *et al* 1993). Mossberg and Nyberg (unpubl.) have shown that the non-biting midge fauna in Lake Västra Skälsjön was much the same between five and seven years after liming as it was prior to acidification forty years previously (cf. Degerman *et al*, Chap. 8, Nyberg, Chap. 11). Bergquist *et al* (1992) showed that the benthic fauna in limed waters showed a greater similarity to the fauna in unacidified waters than in acidified ones. Despite the fact that limed lakes and watercourses have, in most cases, developed flora and fauna which resemble that of unacidified waters, most factors suggest that the communities in limed waters will not, in the long term, regain their pre-acidification status (cf. Henrikson and Brodin, Chap. 1). There are some possible explanations for this:

1) Since the complexity of the food web and the length of the food chain depend on the production capability of the water, the decreased mobility of phosphorus in the soil profile of the drainage area of the limed water will, in general, cause a reduction in the complexity of the food web. When the community in the limed water approaches climax, it should therefore have a shorter average food chain than communities in unacidified areas. This applies when phosphorus is the main factor limiting primary production.
2) The post-liming development of the community depends on the colonization opportunities for the species concerned (Larsson, Chap. 7, Degerman *et al*, Chap. 8, Bergquist, Chap. 17). The recolonization opportunities for various species of fish are of the greatest significance, since they are expected to affect the remainder of the community by way of a "top-down" effect. In cases where fish cannot recolonize a severely acidified and fishless water after liming, the community will still contain

only three trophic levels in the future. The development of the species in this type of community will be quite different from that where fish are the top predators.

3) A limed water normally represents a more variable environment than an unacidified water, partly because of the surges in water chemistry which arise because liming treatments take place at shorter or longer intervals, and partly because the water chemistry in acidified areas is subjected to a great deal of variation via deposition. In both cases, the risk of major variations in the acidification-dependent variables increases, and this can also be expected to have a bearing on the development of the community. Assuming that all the communities are striving to achieve a climax or a more or less stable equilibrium, those in limed waters should show a relatively slow development at a low level of organization. Such communities should thus be less complex and efficient than the ones in unacidified waters. In theory, this should also result in communities with a shorter average food chain than in unacidified waters.
4) At the same time as a higher level of local disturbances affects the rate at which the community develops, such disturbances may reduce the competitive relationship between species in dominance-controlled communities. This means that species diversity may even be higher in limed waters, since less competitive species are in a position to avoid being outcompeted.

15. Acknowledgements

I would like to express my thanks to Gunnar Andersson, Jan Herrmann, Stellan Hamrin, Ulrika Aldén and Lennart Henrikson for checking my manuscript, and also to the Swedish Environmental Protection Agency for their financial support.

16. References

Alenäs, I., Hultberg, H. and Rosmarin, A. 1991. Liming and re-acidification reactions of a forest lake ecosystem, Lake Lysevatten, in SW Sweden. - Water Air Soil Poll. 59: 55-77.

Andersson, F. and Olsson, B. (Eds.) 1985. Lake Gårdsjön - an acid forest lake and its catchment. - Ecol. Bull. 37 (Stockholm).

Andersson, G. 1985. Decomposition of alder leaves in acid lake water. - Ecol. Bull. (Stockholm) 37: 293-299.

Andersson, P., Grahn, P., Hörnström, E., Nyberg, P. and Dahlquist, K. 1989. Kalkning och gödsling i Rammsjön och Ämten, Örebro län 1980-87. Ett försök att minska kvicksilverhalten i fisk. [Liming and fertilization of Lake Rammsjön and Lake Ämten, county of Örebro, 1980-1987. An attempt to decrease mercury levels in fish.] - Swedish Environm. Prot. Agency 3584.

Appelberg, M. 1990. Population regulation of the crayfish *Astacus astacus* L. after liming an oligotrophic, low-alkaline, forest lake. - Limnologica. 20: 319-327.

Appelberg, M. 1992. Liming as a measure to restore crayfish populations in acidified lakes. - Finnish Fish. Res. 14: 93-105.

Appelberg, M. and Aldén, U. 1992. Integrerad uppföljning av kalkningens effekter på sjöar och vattendrag - en treårsrapport. [Integrated monitoring in limed lakes and streams - a three-year study.] - Inform. Inst. Freshw. Res., Drottningholm (4): 1-60.

Appelberg, M. and Degerman, E. 1991. Development and stability of fish assemblages after lime treatment. - Can. J. Fish. Aquat. Sci. 48: 545-554.

Appelberg, M. and Odelström, T. 1986. Habitat distribution, growth and abundance of the crayfish *Astacus astacus* L. in the littoral zone of four neutralised lakes. - In: Appelberg, M. The crayfish *Astacus astacus* in acid and neutralised environments. PhD thesis, Dep. Limnol., Univ. Uppsala.

Appelberg, M. and Odelström, T. 1988. Interaction between European Perch (*Perca fluviatilis*) and juvenile *Pacifastacus leniusculus* (Dana) in a pond experiment. - In: Goeldlin, P. (Ed.) Freshwater crayfish VII. Musée Zool. Cant. Lausanne, Switzerland, pp. 37-45.

Appelberg, M., Degerman, E. and Nyberg, P. 1989. Species composition and relative abundance of the fish fauna in acidified lakes in Sweden. - In: Longhurst, J.W.S. (Ed.) Acid deposition. Sources, effects and controls. British Lib., Tech. Inf., pp. 143-160.

Appelberg, M., Ekström, C. and Hörnström, E. 1990. Stora Härsjön - ett exempel på integrerad uppföljning av kalkningens effekter. [English summary: Lake Stora Härsjön - a case study on integrated monitoring of the effects of liming.] - Inform. Inst. Freshw. Res., Drottningholm (1).

Appelberg, M., Henrikson, B.-I., Henrikson, L. and Svedäng, M. 1993a. Species succession and the importance of biotic interactions within the littoral community during acidification. - Ambio 22: 290-297.

Appelberg, M., Degerman, E. and Norrgren, L. 1993b. Effects of acidification and liming on fish in Sweden. A review. - Finnish Fish. Res. 13: 77-91.

Begon, M., Harper, J.L. and Townsend, T.W. 1990. Ecology: individuals, populations and communities. - Blackwell Scient. Publ., London.

Bellemakers, M.J.S. and van Damm, H. 1992. Improvement of breeding success of the moor frog (*Rana arvalis*) by liming of acid moorland pools and the consequences of liming for water chemistry and diatoms. - Environm. Poll. 78: 165-171.

Bergquist, B., Engblom, E. and Lingdell, P.-E. 1992. Förekomst och kolonisation av bottenfauna i kalkade vatten. [English summary: Occurrence and colonization of benthos in limed waters.] - Inform. Inst. Freshw. Res., Drottningholm (4): 79-108.

Blomqvist, P., Olsson, H., Olofsson, H. and Broberg, O. 1992. Enclosure experiments with low dose additions of phosphorus and nitrogen in the acidified Lake Njupfatet, central Sweden. - Int. Rev. ges. Hydrobiol. 74: 611-631.

Broberg, O. and Persson, G. 1981. External budgets for phosphorus, nitrogen and dissolved organic carbon for the acidified Lake Gårdsjön. - Arch. Hydrobiol. 99: 160-175.

Chesson, P.L. 1986. Environmental variation and the coexistence of species. - In: Diamond, J. and Case, T.J. (Eds.) Community ecology. Harper and Row, New York, pp. 240-256.

Connell, J.H. 1978. Diversity in tropical rainforests and coral reefs. - Science 199: 1302-1310.

Degerman, E. 1987. Humösa sjöar - en litteratursammanställning med inriktning på fisk och försurning. [English summary: Humic lakes - a literature survey with emphasis on fish and acidification.] - Swedish Environm. Prot. Agency 3415.

Degerman, E. and Appelberg, M. 1992. The response of stream-dwelling fish to liming. - Environm. Poll. 78: 149-155.

Degerman, E. and Nyberg, P. 1989. Effekter av sjökalkning på fiskbestånd i sjöar. [English summary: Long-term effects of liming on fish populations in Sweden.] - Inform. Inst. Freshw. Res., Drottningholm (5).

Degerman, E., Sjölander, E., Johlander, A., Sjöstrand, P., Höglind, K., Thorsson, L. and Carlstrand, H. 1990. Kalkning för att motverka försurningpåverkan på fisk i rinnande vatten. [English summary: Liming of acidified running waters to improve conditions for fish.] - Inform. Inst. Freshw. Res., Drottningholm (4): 27-214.

Degerman, E., Appelberg, M. and Nyberg, P. 1992. Effects of liming on the occurrence and abundance of fish populations in acidified Swedish lakes. - Hydrobiologia 230: 201-212.

Dickson, W. (Ed.) 1988. Liming of Lake Gårdsjön, an acidified lake in SW Sweden. - Swedish Environm. Prot. Agency 3426.

Diehl, S. 1988. Foraging efficiency of three freshwater fish: effects of structural complexity and light. - Oikos 53: 207-214.

Eriksson, F. 1988. Makrofytvegetation i kalkade sjöar. - [Macrophyte vegetation in limed lakes.] - Inform. Inst. Freshw. Res., Drottningholm (9).

Eriksson, F., Hörnström, E., Mossberg, P. and Nyberg, P. 1983. Ecological effects of lime treatment of acidified lakes and rivers in Sweden. - Hydrobiologia 101: 154-164.

Eriksson, M.O.G. 1987. Liming of acid lakes in Sweden: short-term effects on water bird density. - Wildfowl 38: 143-149.

Eriksson, M.O.G., Arvidsson, B. and Johansson, I. 1988. Habitatkaraktärer hos häckningssjöar för smålom *Gavia stellata* i sydvästra Sverige. [Habitat characteristics for lakes in south-western Sweden where breeding of red-throated diver *(Gavia stellata)* takes place.]. - Vår Fågelvärld 42: 293-300.

Eriksson, M.O.G., Henrikson, L., Nilsson, L., Nilsson, B.-I., Nyman, H.G., Oscarson, H.G., Stenson, J.A.E. and Larsson, P.K. 1980. Predator-prey relations, important for the biotic changes in acidified lakes. - Ambio 9: 248-249.

France, R.L. and Graham, L. 1985. Increased microsporidian parasitism of the crayfish *Orconectes virilis* in an experimentally acidified lake, - Water Air Soil Poll. 26: 129-136.

Fraser, J.E., Saunders, W.P., von Oettingen, S.L., Britt, D.L. and Sverdrup, H.U. 1985. Ecological effects of base addition to lake ecosystem. - General Research Cooperation, 7655 Old Springhouse Rd, McLean, V., Canada.

Fretwell, S.D. 1977. The regulation of plant communities by the food chains exploiting them. - Persp. Biol. Med. 20: 169-185.

Gilinsky, E. 1984. The role of fish predation and spatial heterogeneity in determining benthic community structure. - Ecology 65: 455-468.

Grahn, O. 1977. Macrophyte succession in Swedish lakes caused by deposition of airborne acid substances. - Water Air Soil Poll. 7: 295-305.

Hairston, N.G., Smith, F.E. and Slobodkin, L.B. 1960. Community structure, population control and competition. - Am. Nat. 94: 421-425.

Hasselrot, B., Andersson, I. and Hultberg, H. 1984. Ecosystem shifts and reintroduction of Arctic char (*Salvelinus salvelinus* (L.)) after liming of a strongly acidified lake in south-western Sweden. - Rep. Inst. Freshw. Res., Drottningholm 61: 78-92.

Henrikson, B.-I. 1988. The absence of antipredator behaviour in the larvae of *Leucorrhinia dubia* (Odonata) and the consequences for their distribution. - Oikos 51: 179-183.

Henrikson, B.-I. 1993. Invertebrates in *Sphagnum* mosses in acidified lakes and the colour adaptation and substrate preference in the dragonfly *Leucorrhinia dubia* (Odonata, Anisoptera). - Ecography 16: 143-153.

Henrikson, L. and Oscarson, H.G. 1984. Lime influence on macro-invertebrate zooplankton predators. - Rep. Inst. Freshw. Res., Drottningholm 61: 93-103.

Henrikson, L., Nyman, H.G., Oscarson, H.G. and Stenson, J.A.E. 1980. Trophic changes without changes in external loadings. - Hydrobiologia 68: 257-263.

Henrikson, L., Oscarson, H.G. and Stenson, J.A.E. 1984. Development of the crustacean community after lime treatment of the fishless Lake Gårdsjön, Sweden. - Rep. Inst. Freshw. Res., Drottningholm 61: 104-114.

Henrikson, L., Nyman, H.G., Oscarson, H.G. and Stenson, J.A.E. 1985. Changes in the zooplankton community after lime treatment of an acidified lake. - Verh. Internat. Verein. Limnol. 22: 3008-3013.

Hörnström, E. and Ekström, C. 1986. Acidification and liming effects on phyto- and zooplankton in some Swedish west coast lakes. - Swedish Environm. Prot. Agency 1863.

Hörnström, E., Ekström, C. and Andersson, P. 1992. Tio mellansvenska sjöar. Kalkningseffekter på plankton och vattenkemi. [English summary: Ten lakes in central Sweden, effects of liming on plankton and water chemistry.] - Swedish Environm. Prot. Agency 4048.

Hörnström, E., Ekström, C., Fröberg, E. and Ek, J. 1993. Plankton and chemical-physical development in six Swedish west coast lakes under acidic and limed conditions. - Can. J. Fish. Aquat. Sci. 50: 688-702.

Hultberg, H. and Andersson, I. 1982. Liming in acidified lakes: induced long-term changes. - Water Air Soil Poll. 18: 311-331.

Johnson, R.K., Wiederholm, T. and Eriksson, L. 1992. Comparison of profundal macroinvertebrate communities of lime- and limed (IKEU) - reference lakes. - Swedish Environm. Prot. Agency, unpubl. rep.

Kerfoot, W.C. and Sih, A. 1987. Predation. Direct and indirect impacts on aquatic communities. - Univ. Press, New England, London.

Larsson, S. 1988. Liming effects on phytoplankton in Lake Gårdsjön 1982-1985. - In: Dickson, W. (Ed.) Liming of Lake Gårdsjön, an acidified lake in SW Sweden. - Swedish Environm. Prot. Agency 3426, pp. 245-290.

Lingdell, P.-E. and Engblom, E. 1992. Underlag till utvärdering av den integrerade kalkningseffektuppföljningen (IKEU). [Basic data for analysis of the Integrated Studies of the Effects of Liming (ISEL).] - Limnodata Consultant Co., Gunnilbo.

Nyberg, P. 1984. Impact of *Chaoborus* predation on planktonic crustacean communities in some acidified and limed forest lakes in Sweden. - Rep. Inst. Freshw. Res., Drottningholm 61: 154-166.

Nyberg, P., Appelberg, M. and Degerman, E. 1986. Effects of liming on crayfish and fish in Sweden. - Water Air Soil Poll. 31: 669-687.

Nyman, H.G. 1990. Some biological mechanisms reducing the availability of phosphorus in acid lakes. - In: Mason, B.J. (Ed.) The surface water acidification programme. Univ. Press, Cambridge, pp. 401-408.

Odum, E.P. 1981. The effects of stress on the trajectory of ecological succession. - In: Barret, G.W. and Rosenberg, R. (Eds.) Stress effects on natural ecosystems. John Wiley and Sons Ltd., New York, pp. 43-48.

Oksanen, L., Fretwell, S.D., Arruda, J. and Niemelä, P. 1981. Exploiting ecosystems in gradients of primary productivity. - Am. Nat. 118: 240-261.

Olofsson, H., Blomqvist, P., Olsson, H. and Broberg, O. 1988. Restoration of the pelagic food web in acidified and limed lakes by gentle fertilisation. - Limnologica (Berlin) 19: 27-35.

Olsson, H. and Pettersson, A. 1993. Oligotrophication of acidified lakes - a review of hypotheses. - Ambio 22: 312-317.

Persson, L. 1986. Effects of reduced interspecific competition on resource utilisation in perch (*Perca fluviatilis*). - Ecology 67: 355-364.

Persson, L., Andersson, G., Hamrin, S.F. and Johansson, L. 1988. Predatory regulation and primary production along the productivity gradient of temperate lake ecosystems - In: Carpenter, S.R. (Ed.) Complex interactions in lake communities. Springer,

Berlin Heidelberg New York, pp. 45-59.

Persson, L., Diehl, S., Johansson, L., Andersson, G. and Hamrin, S.F. 1992. Trophic interactions in temperate ecosystems: a test of food chain theory. - Am. Nat. 140: 59-84.

Pimm, S.L. and Kitching, R.L. 1987. The determinants of food chain lengths. - Oikos 50: 302-307.

Raddum, G.G., Brettum, P., Matzow, D., Nilssen, J.P., Skov, A., Sveälv, T. and Wright, R.F. 1986. Liming the acid Lake Hovvatn, Norway: a whole-ecosystem study. - Water Air Soil Poll. 31: 721-763.

Schoener, T.W. 1989. Food webs from the small to the large. - Ecology 70: 1559-1589.

Stenson, J.A.E. 1990. Creating conditions for changes in prey community structure by *Chaoborus* spp. in a lake in Sweden. - Hydrobiologia 198: 205-214.

Stenson, J.A.E., Bohlin, T., Henrikson, L., Nilsson, B.-I., Nyman, H.G. and Oscarsson, H.G. 1978. Effects of fish removal from a small lake. - Verh. Internat. Verein. Limnol. 20: 794-801.

Stenson, J.A.E., Svensson, J.-E. and Cronberg, G. 1993. Changes and interactions in the pelagic community in Swedish acidified lakes. - Ambio 22: 277-282.

Svedäng, M.U. 1992. Carbon dioxide as a factor regulating the growth dynamics of *Juncus bulbosus* (L.) - Aquat. Bot. 42: 231--240.

Wiens, J.A. 1977. On competition and variable environments. - Am. Sci. 65: 590-597.

Chapter 10

Liming strategies and effects: the Lake Gårdsjön case study

Jan-Erik Svensson[1], Lennart Henrikson[2], Stefan Larsson[3] and Anders Wilander[4]

[1] Department of Zoology, University of Göteborg, S-413 90 Göteborg, Sweden
[2] Municipality of Mark, S-511 80 Kinna, Sweden
[3] Göteborg and Bohus County Administrative Board, S-403 40 Göteborg, Sweden
[4] Swedish University of Agricultural Science, Department of Environmental Assessment, Box 7050, S-750 07 Uppsala, Sweden

1. Documentation

Lake Gårdsjön and its catchment area have been studied since the early 1970s and have, along with other lakes in the vicinity, been the target of the greatest individual Swedish effort in the field of acidification and liming research. The majority of studies in the lake itself were concluded in the early 1990s, but some still continue to a limited extent today.

The findings of this research have been reported in about 80 scientific articles up to 1992. The conditions in and around the lake during the acidification phase have been presented in 38 articles in a book edited by Andersson and Olsson (1985). The short-term effects of liming have been presented in another book edited by Dickson (1988). This book discusses hydrological and biological aspects, as well as water chemistry (Broberg 1988a, Gahnström 1988, Grahn and Sangfors 1988, Henrikson 1988, Hultberg and Nyström 1988, Johansson and Nilsson 1988, Johansson *et al* 1988, Larsson 1988). Studies have also been published on water chemistry (Hultberg and Grennfelt 1986, Broberg 1987, 1988b, Hasselrot *et al* 1987, Andersson *et al* 1992), and biology (Lazarek 1982, 1986, Henrikson and Oscarson 1984, Henrikson *et al* 1984, Stenson 1990, Sangfors 1991, Stenson *et al* 1993, Stenson and Svensson 1994, Henrikson unpubl., Svensson and Stenson unpubl.).

In the present context it is not possible to provide a comprehensive overview of the Lake Gårdsjön studies. We have therefore chosen two areas - water chemistry and, particularly, biological changes and mecha-

Liming of Acidified Surface Waters
L. Henrikson and Y.W. Brodin (Eds.)
© Springer-Verlag Berlin Heidelberg 1995

nisms in the pelagic zone. The effects of liming in Lake Gårdsjön are also discussed in several other chapters in this book: water chemistry (Wilander *et al*, Chap. 5), microbial processes (Gahnström, Chap. 6), macrophytes, phytoplankton and benthic algae (Larsson, Chap. 7) and zooplankton and benthic fauna (Degerman *et al*, Chap. 8).

2. Geographical orientation

Lake Gårdsjön is situated in the River Anråsån drainage basin in south-western Sweden, 60 km north of Göteborg (Fig. 1). It is noticeably affected by its location, only 15 km from the sea. There is a high level of acid deposition. During the 1979-81 period, sulphur deposition amounted to about 30 kg/ha and year (160 meq/m^2 and year) (Grennfelt *et al* 1985), but has declined to something over 10 kg/ha and year (>60 meq/m^2 and year) (Lövblad *et al* 1992).

Figure 1. *The geographic location of Lake Gårdsjön.*

The bedrock in the drainage basin is dominated by acidic rock types such as gneiss and granite, covered by a very thin moraine. A relatively large proportion of the drainage basin consists of open rock. The vegetation mostly consists of coniferous forest of varying age. Because of the nutrient-poor catchment area and the high percentage of lakes (15%), the lake has oligotrophic, clear water. Table 1 presents some other information on the lake. The traits of the drainage basin and the high deposition imply that Lake Gårdsjön represents an unusually severely affected lake.

Table 1. *Morphological and hydrological data on Lake Gårdsjön. Runoff data from 1979-1985.*

Drainage area (ha)	211	Mean depth (m)	4.9
Lake area (ha)	31.2	Precipitation (mm/year)	1,100
Lake volume (m^3)	1.5 x 10^6	Runoff (mm/year)	611
Max. depth (m)	18.5	Retention time (years)	1.6

3. Acidification status

Before Lake Gårdsjön was acidified, it appears to have been a typical west-Swedish oligotrophic forest lake. An inventory of the fish population in 1949 showed European perch *(Perca fluviatilis)*, northern pike *(Esox lucius)*, roach *(Rutilus rutilus)*, rudd *(Carassius carassius)*, tench *(Tinca tinca)* and eel *(Anguilla anguilla)* (Hultberg 1985). The pH was 6.25 and the transparency was 7 m.

Renberg and Wallin (1985) documented the pH trends in Lake Gårdsjön using paleolimnological methods, mainly on the basis of studies of diatom shells in sediment core samples, and found that the pH had fallen from above 6 to below 5 during the 1950s and 1960s. This means that Lake Gårdsjön suffered severely from acidification. Despite the fact that sulphate deposition has declined in recent years, the concentrations in the water have not declined to a corresponding extent.

Fish species disappeared successively, and when the research project began in 1978, Lake Gårdsjön was completely devoid of fish. The top predators in the pelagic environment consisted of phantom midge larvae *(Chaoborus* spp.) and water boatmen (Corixidae) - organisms which are susceptible to fish predation but can withstand acidic and metal-rich waters. Transparency was considerable. Measurements showed it to be up to 13 m, allowing *Sphagnum* mosses to increase to such an extent that they could even be found in the deepest areas of the lake (18 m). Isoetids *(Lobelia dortmanna, Isoetes* spp.) receded, however.

Species richness of phytoplankton and zooplankton were very low. Large species dominated among the phytoplankton, the most important being dinoflagellates (Dinophyceae) such as *Peridinium inconspicuum* and *Gymnodinium uberrimum* (Fig. 2). At times, colonial golden algae (Chrysophyceae) were also common, including *Dinobryon sociale* var. *americanum*. Biomass and primary production of phytoplankton were low. The mean summer biomass (June-August) in the epilimnion was 0.19 in 1980 and 0.17 mg/l (wet weight) in 1981. Primary production in terms of area was determined during the same time period as 57 and 52 mg C/m^2 and day respectively. The planktonic crustaceans were dominated by *Eudiaptomus gracilis* (Copepoda), although *Bosmina coregoni* (Cladocera) was also common. Dominant rotifers were *Ascomorpha ecaudis*, *Keratella cochlearis*, *Ploesoma triacanthum* and *Polyarthra remata*. During the final year of acidification, 1981, a total of 4 cladoceran species and 14 rotifer species were identified.

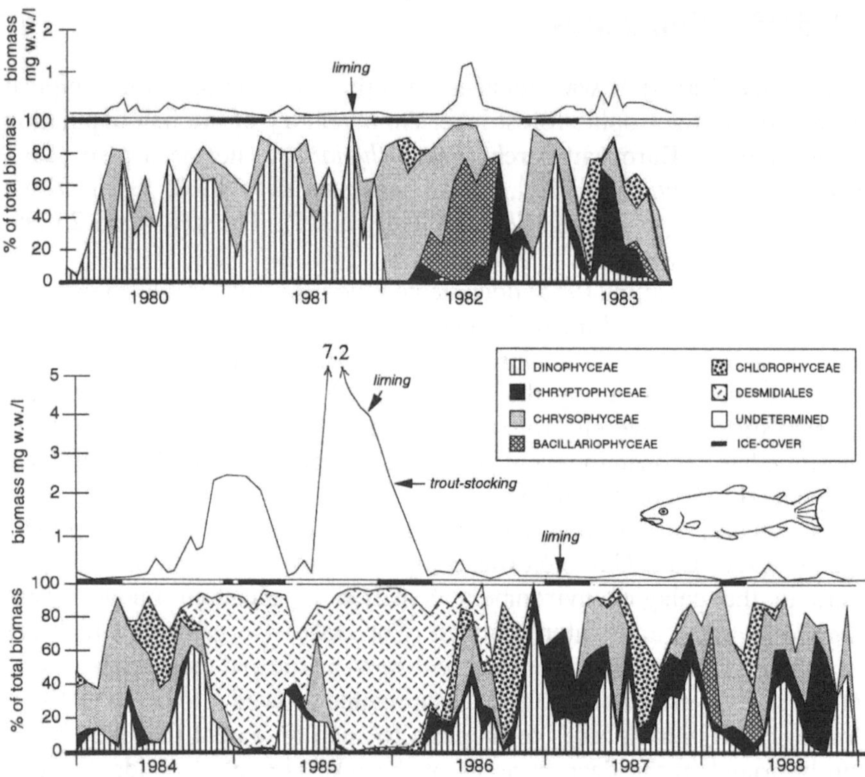

***Figure 2.** The development of dominant phytoplankton groups in Lake Gårsjön before and after liming.*

3.1. Interactions in the pelagic zone

Unfavourable abiotic conditions were one important reason for the structure of the pelagic community during the acidified period. Many of the organisms which are susceptible to high metal concentrations and/or low pH were absent, which probably provides the main explanation for the low diversity. Fish, many species of cladocerans, e.g. *Daphnia*, and many species of phytoplankton are susceptible to high aluminium concentrations (Hörnström *et al* 1984, Brett 1989). Within the limited abiotic framework, biotic factors probably also played a part in determining the structure of the community. Factors such as predation, competition and availability of food had varying degrees of impact on the organisms which succeeded in living in Lake Gårdsjön while it was acidified. It was possible to distinguish both "bottom-up" effects, i.e. characteristics among lower trophic levels which affect higher levels, and the converse, i.e. "top-down" effects (see Appelberg, Chap. 9).

Some of the important organisms and interactions in the pelagic zone of Lake Gårdsjön during the acidified phase are illustrated in Fig. 3.

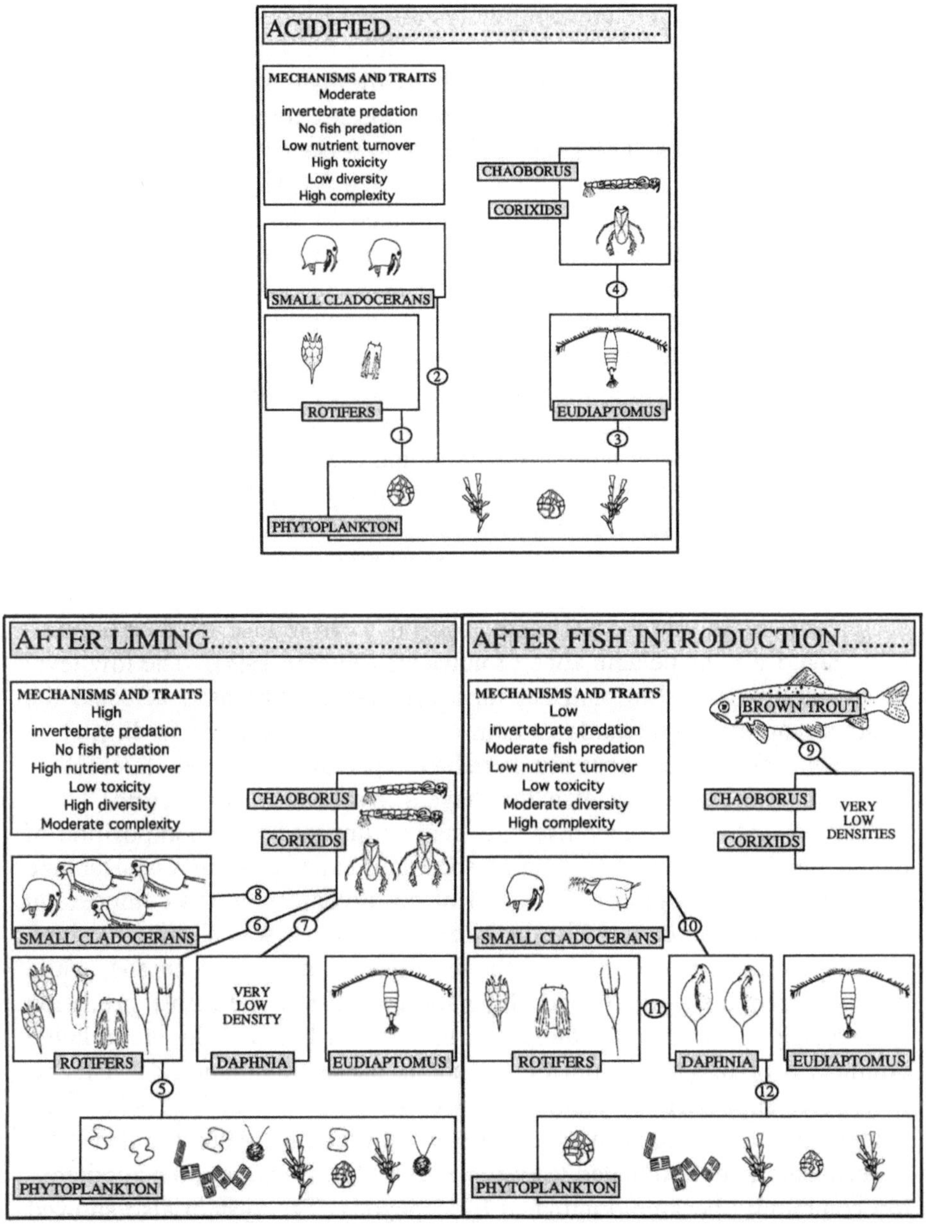

Figure 3. *Characteristics, mechanisms and interactions among pelagic organisms in Lake Gårdsjön during the acidification phase, after liming and after re-introduction of fish (cf. comments on interactions in the text).*

Since the phytoplankton community suffered from species depletion and low productivity, there were few opportunities for a diverse and abundant zooplankton community to develop. There are many specialists among the rotifers, but many of the most important species forage mainly on small phytoplankton species. The same applies to most cladocerans (Sterner 1989). The dominance of large phytoplankton in Lake Gårdsjön during the acidification phase may therefore have contributed to the depletion of species and the low densities of rotifers and cladocerans (illustrated by interactions (1) and (2) in Fig. 3, a bottom-up effect). The copepod *E. gracilis*, which is the dominant species in the zooplankton community, differs from rotifers and cladocerans in terms of its preferred diet. It grazes on relatively large species of phytoplankton and is able to live on *P. inconspicuum* and *D. sociale* var. *americanum*, which were common species in Lake Gårdsjön during the acidification phase (Svensson and Stenson 1991).

The low diversity and production in the pelagic system may have been self-perpetuating to some extent. The relative dominance of *E. gracilis* and the low densities of rotifers and cladocerans may have contributed to a low rate of nutrient turnover. By producing rather insoluble fecal pellets which accumulate rapidly in the sediment, copepods can reduce nutrient accessibility and drain the pelagic zone of nutrients (Nyman 1990). The low level of availability and slow nutrient turnover may in turn have affected the structure of the phytoplankton community (interaction (3) in Fig. 3, a top-down effect) and possibly benefited species which are tolerant to starvation, i.e. large species which have a low level of production.

The pelagic insects may also have had an impact on the composition of the zooplankton community. In a mesocosmos experiment in a neutral lake aimed at shedding light on the effect of top predators on the composition of the zooplankton community in Lake Gårdsjön, it was shown that both *Chaoborus* and corixids reduced the densities of many cladocerans, while copepods remained unaffected (Nyman *et al* 1985). Selective predation from pelagic insects may therefore have played a part in the relative dominance of *E. gracilis* during the acidic phase (interaction 4). This is, moreover, supported by studies carried out in a neutral lake where the pelagic insects also took over the role of top predators when the fish were experimentally exterminated (Stenson *et al* 1978). Here, too, the zooplankton community became dominated by copepods. The system also showed many other similarities with the acidified lakes in western Sweden (Eriksson *et al* 1980). Among other things, a dinoflagellate-dominated phytoplankton community developed.

The structuring mechanisms, selective feeding and nutrient recirculation among zooplankton and selective predation among pelagic insects have all been documented in different experiments associated with the Lake Gårdsjön study. They therefore enjoy qualitative support, but it should be pointed out that their importance in the acidified lake has not been quantified. Besides, other structuring mechanisms which have not been studied in closer detail - such as selective toxicity and competition among both phytoplankton and zooplankton - may also have been significant.

4. Liming strategy and re-introduction of fish

Lake Gårdsjön was first limed on 14 April 1982 using limestone powder ($CaCO_3$, grain size 0-0.5 mm). The lakes situated upstream remained unlimed and some were used as reference lakes. All the lime, 100 tonnes (23.3 mg Ca/l), was spread directly in the lake, about 80% being dispersed over the shallow areas. This was done in order to reduce the effects of acid runoff from land and tributaries. The lime was mixed with lake water and spread as slurry from a pontoon. Subsequent treatments were performed on 20 August 1985 (36.5 tonnes); 15 November 1986 (100 tonnes); 8 November 1989 (55 tonnes) and 6 December 1992 (56.1 tonnes).

Apart from studying the effects of liming, the Lake Gårdsjön project had the purpose of identifying the role of biological regulation mechanisms such as predation, competition and changes in food resources in the post-liming development of the pelagic system. The lake was therefore allowed to remain free of fish during the first few years after treatment. It was not until September 1985 that brown trout *(Salmo trutta)* were released (1,800 individuals, body length 15-17 cm). This was the only species of fish in the lake until June 1990, when 40,000 rainbow trout fry *(Oncorhynchus mykiss*, 3-4 cm) and 2,000 brown trout fry (4-5 cm) were released. In June 1991, a further 9,000 rainbow trout fry (4-5 cm) were released. Neither the brown trout nor the rainbow trout reproduced in Lake Gårdsjön.

5. The effects of liming

5.1. Water chemistry

The liming treatment in the spring of 1982 resulted in a rapid increase in pH to about 7.6 and alkalinity rose to 0.55 meq/l (Fig.4). The concentration of calcium rose from about 0.088 to 0.685 meq/l at the same time. The difference in the increase in calcium (0.597) and alkalinity (0.55) was attributable to the existing acidity in the water, which was estimated at 0.055 meq/l (Hultberg and Nyström 1988).

In 1982, about 60% of the limestone dissolved immediately. A further 20% dissolved from areas close to the shore during the ensuing years. This first treatment had a duration of just over three years. The alkalinity declined gradually and in conjunction with the 1985 thaw a figure of 0.05 meq/l was obtained.

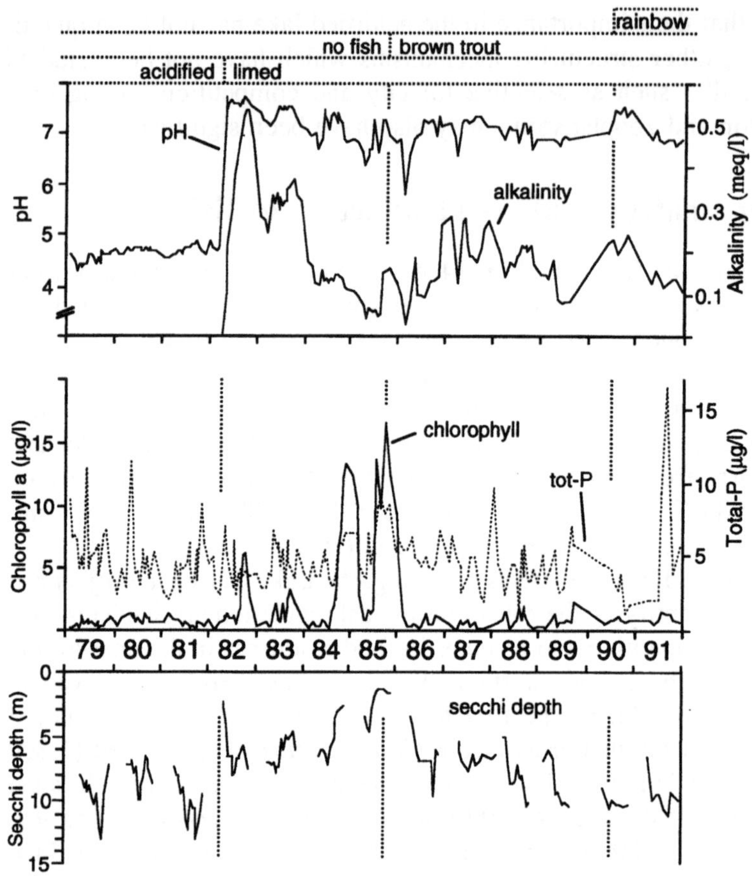

Figure 4. *Development of pH, alkalinity, phosphorus concentration and chlorophyll concentration in the epilimnion of Lake Gårdsjön before and after liming and re-introduction of fish.*

The treatment did not completely prevent the impact of acid surges. During the late winter of 1986, for instance, acidic water was seen to form layers beneath the ice with a pH of 4.2-4.7 in the surface water and pH 5.0-6.4 in deeper water. This probably meant that susceptible species such as the water louse *Asellus aquaticus* were unable to colonize the most exposed shallow areas. This type of acid surge occurs during thaw, when all the water running into the lake has a temperature of below 4°C and thus a lower density than the somewhat warmer lake water (Abrahamsson 1993). Cold acidic water may also form layers during the autumn when the density of the water entering the lake is higher than the much warmer lake water (higher than 4°C). This means that it forms a layer over the bottom, which was the case in November 1991. In such situations, the effect may be dampened to some extent by the capacity of the sediment to neutralize acid.

In addition to the obvious primary changes, significant changes took place in transparency, total aluminium, nitrate, and dissolved and particulate organic carbon (Table 2). Since the catchment area had not been limed and the supply of leached aluminium had probably therefore not changed, the decline in aluminium has been caused by precipitation and sedimentation. The reason why total aluminium concentrations did not fall more sharply may be that about half of the aluminium was bound in a non-toxic organic form which is not precipitated. The decrease in nitrate and the increase in the organic carbon fractions reflect different biological changes in the lake. The production of phytoplankton was higher than before the first years after liming, which meant that the production of particulate organic carbon (POC) increased, as the assimilation of nitrate probably also did. Moreover, the impaired light climate after liming probably also meant that the photochemical oxidation of dissolved organic carbon (DOC) declined. The decomposition of the *Sphagnum* and algal carpets which existed prior to liming may also have contributed to the dissolving of organic material (Broberg 1988a).

The transparency decreased after liming. Between 1984 and 1985, it was less than 1.5 m at times. After 1985, it increased again and was at the same level as during the acid phase by the final year of the project. This development largely follows that of the phytoplankton biomass. The chlorophyll concentrations (Fig. 4) also increased after liming and decreased after the lake had been restocked with fish.

Neither liming nor the new stocks of fish affected the phosphorus concentrations in the pelagic zone of the lake to any significant degree (Fig. 4, Table 2). In conjunction with heavy algal blooms in 1984 and 1985, however, the phosphorus concentrations were somewhat higher and at one point, during the last year of the project, an exceptionally high level was measured (cf. Wilander *et al*, Chap. 5). The 1985 algal bloom also caused increased oxygen depletion, resulting in a noticeable oxygen deficiency which had a negative impact on the benthic fauna.

***Table** 2. Physical-chemical data of the surface water of Lake Gårdsjön during acidified (1978-1982, means) and limed (1982-1992) conditions.*

	Acidified	Limed		Acidified	Limed
pH	4.66	7.17	Org-N (μg/l)	237	240
Alkalinity (meq/l)	0	0.201	Total-P (μg/l)	5	4.8
Transparency (m)	9.1	6.6	Total-Al (μg/l)	299	179
Ca (meq/l)	0.088	0.375	DOC (mg/l)	2.33	4.33
SO_4 (meq/l)	0.208	0.200	POC (mg/l)	0.45	0.66
NO_3-N (μg/l)	99	85			

5.2. Pelagic organisms

Liming had an immediate "shock" effect on the phytoplankton community. The biomass, production and number of species declined, but subsequently recovered quickly. Only a few months after treatment, new phytoplankton species appeared, while at the same time biomass and primary production increased and the earlier dominance of dinoflagellates came to an end. During the limed, fish-free period, primary production was considerably higher than before liming (Fig. 5). During the autumn bloom of 1982, the diatom *Synedra nana* dominated and biomass figures of over 2 mg/l (wet weight) were measured. During 1983 and 1984, various new species of golden algae, green algae (Chlorophyceae) and cryptophyceans dominated (Fig. 2). In the autumn of 1984, the trend of increasing diversity changed and a very dense population of the small desmid *Cosmocladium perissum* arose. This species then dominated throughout 1985 and attained very high densities at times. During the summer of 1985, the mean biomass of the phytoplankton community in the epilimnion was 4.0 mg/l, which is 20-25 times higher than during the acidification phase. The proportion of small species in the phytoplankton biomass increased after liming.

After 1985, phytoplankton biomass declined sharply, as did primary production. This decline is mainly attributable to the decline in the *Cosmocladium* bloom during the autumn of 1985. It also coincided with the re-introduction of fish. From 1986-1988, biomass (Fig. 4) and primary production (Fig. 5) remained at the same level as during the acidification phase. Once again, the proportion of dinoflagellates increased, although they failed to recover the same dominance as under acidic conditions. Cryptophyceans and diatoms were common at times. The composition of the phytoplankton community after 1988 is not known in detail, but it can be assumed from chlorophyll concentrations (Fig. 4) that the biomass still remained low.

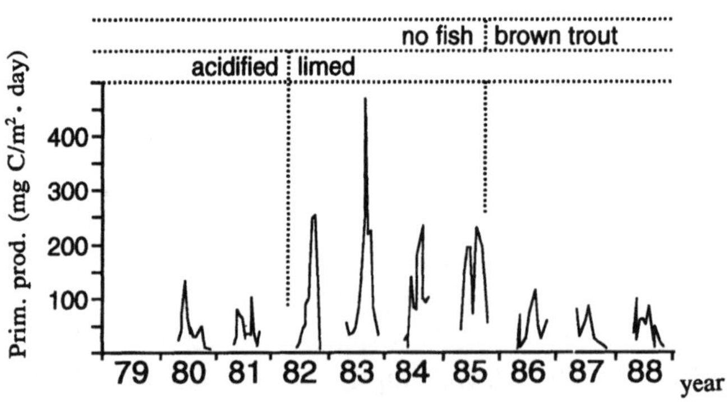

Figure 5. Changes in primary production in Lake Gårdsjön before and after liming and re-introduction of fish.

The number of species increased after liming (Fig. 6), as did the diversity of the zooplankton community. Several new species of rotifers had already appeared during the summer of 1982 in considerable densities, including *Asplanchna priodonta*, *Conochilus unicornis* and *Polyarthra vulgaris*. During the fourth year after treatment, 24 species of rotifers were identified, compared to 14 during the last year of acidification. The biomass of the rotatoria increased considerably (Fig. 7), as did their share of the total zooplankton biomass. The crustacean community was not affected by the treatment quite as noticeably as the rotifer community, however. The copepod *E. gracilis* continued to dominate, but several new species were identified. In 1985, eleven species of cladocerans were found, compared to four the year before liming. These new species usually appeared in low densities, but one of them, *Diaphanosoma brachyurum*, increased in density only a month or two after treatment and was the dominant species during the summers of 1983 and 1984 (Fig. 7).

After restocking with fish, the number of species of rotifers and cladocerans declined (Fig. 6). During 1991, twenty species of rotifers and nine species of cladocerans were identified. Most of the rotifers declined in density, however, although the rotifer biomass was still considerably higher than during the acidification phase. Unlike the rotifers, most of the immigrating cladocerans attained their highest densities during the years after fish were re-introduced. This applies to *Bosmina longirostris*, *Bythotrephes longimanus*, *Ceriodaphnia quadrangula*, *Daphnia longispina*, *Holopedium gibberum* and *Polyphemus pediculus*. The biomass of cladocerans thus increased considerably. The crustacean communities stabilized

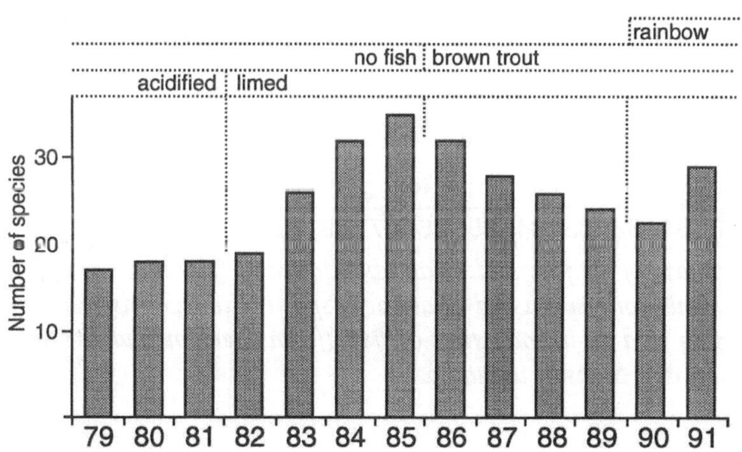

Figure 6. *Numbers of identified species of zooplankton (rotifers and cladocerans) in summer samples from Lake Gårdsjön before and after liming and re-introduction of fish.*

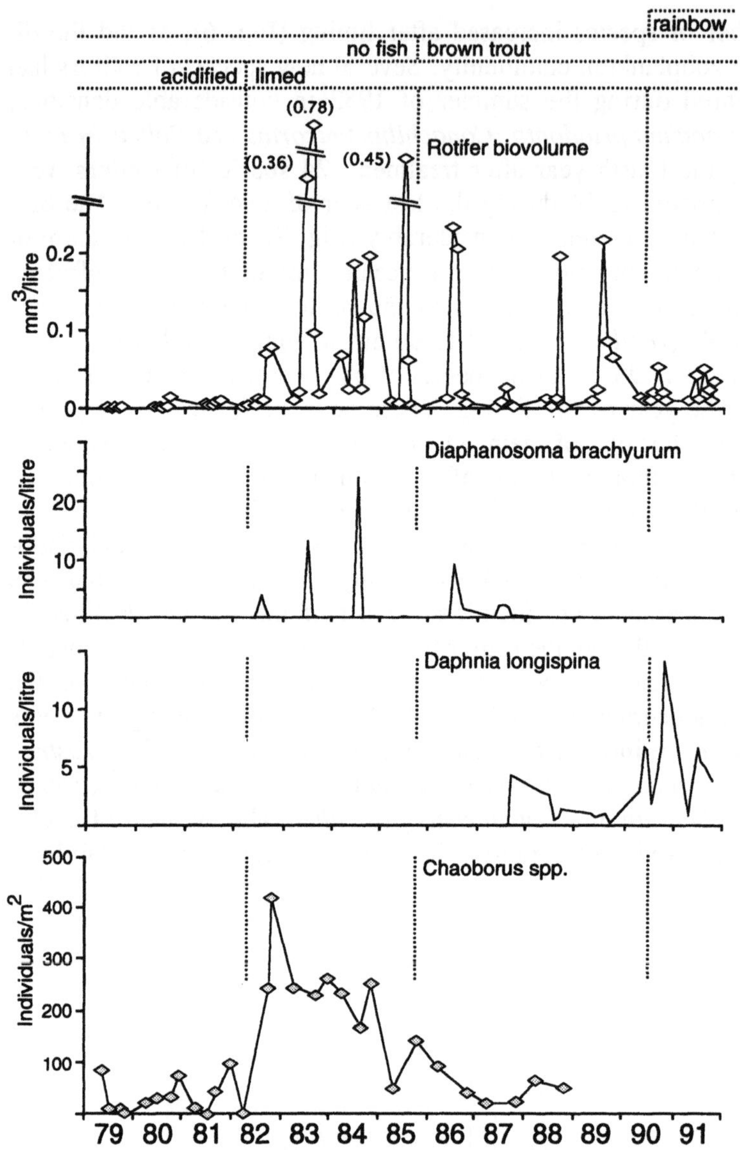

Figure 7. Biomass of rotifers and mean abundance of Diaphanosoma brachyurum, Daphnia longispina and Chaoborus spp. in Lake Gårdsjön before and after liming and re-introduction of fish (from Stenson and Svensson 1994, Svensson and Stenson unpubl.).

after the autumn of 1987, with the large cladoceran *Daphnia longispina* dominating (Fig. 7) together with the copepod *E. gracilis*. At the same time, the density of other herbivorous cladocerans decreased.

Pelagic insects, *Chaoborus* (Fig. 7) and corixids, increased considerably in density after liming. After fish were re-introduced, due to predation, these densities declined to the same levels as during the acidification phase.

On the whole, the pelagic system was characterized by considerable instability during the limed and fish-free period and for some years after fish were re-introduced. The treatment resulted in a considerable increase in diversity and production and a succession of different species followed, both among phytoplankton and zooplankton. It was not until some years after fish were re-introduced that the system appeared to have stabilized, albeit at a lower level of diversity and production.

5.2.1. Post-liming interactions in the pelagic zone

Liming changed the abiotic properties of the pelagic environment. It meant, for instance, that aluminium concentrations declined and the pH rose. This made it possible for susceptible species of phytoplankton and zooplankton to re-establish. The following is an attempt to summarize the way in which biological mechanisms may have contributed to the development of the system within the extended abiotic framework. This summary represents a qualitative model. The importance of interactions identified here has not been quantified, and other biotic and abiotic mechanisms may naturally have been important. The interactions and properties which have been identified as characterizing the pelagic system of Lake Gårdsjön after liming and re-introduction of fish are illustrated in Fig. 3.

The post-liming structure of the phytoplankton community was probably regulated by competition. The phosphorus concentrations did not change to any significant extent during the first post-liming years, but the increase in chlorophyll concentrations and primary production suggest that the availability of nutrients improved considerably. Small phytoplankton, which were able to benefit quickly from the improved food supply, were probably favoured. They represent an important source of food for many of the rotifers which dominated after liming. Thanks to the higher level of production and diversity among the phytoplankton, therefore, many rotifer populations were able to increase in density (interaction (5) in Fig. 3).

The post-liming increase in *Chaoborus* and corixids is probably also related to the improved food supply, e.g. in the form of an increase in the number of rotifers. These may be important to the survival and growth of the youngest *Chaoborus* which in turn may represent an important element in the diet of corixids (interaction 6).

The pelagic insects probably had a major influence on the way in which the cladoceran community developed. Their selective predation, combined with high densities, limited the growth of many cladoceran populations (7 and 8). One exception was *D. brachyurum*. This species is not as susceptible to predation from invertebrates since, unlike other herbivorous cladocerans, it has a well-developed flight response when attacked (Drenner and McComas 1980).

5.2.2. Interactions in the pelagic zone after re-introduction of fish

A new top predator was introduced by virtue of the re-introduction of fish, while at the same time the number of trophic levels increased from three to four. The community which subsequently developed differed in many ways from the one which had previously emerged when the lake was devoid of fish. The re-introduction of fish had an impact on all trophic levels and also affected some of the abiotic properties of the system.

During the first weeks, brown trout fed largely on the pelagic insects (9) - *Chaoborus* and corixids represented >95% of the prey identified in trout stomachs in October 1985. By virtue of this reduction in the number of pelagic insects, the competition between different species of cladocerans increased. During the initial years, the density of many small species increased. The larger and more effective *D. longispina* then took over the role of dominant herbivore, whereas the less competitive species again declined (10). Competition from *D. longispina* was probably also an important reason why many rotifers decreased in density (11).

At the same time, *Daphnia* exercised an intense grazing pressure on the phytoplankton community (12), which was probably the most important reason for the low phytoplankton biomass during this period. As *Daphnia* dominance increased, the size composition of the herbivores shifted from smaller to larger species, which had an impact on nutrient turnover in the pelagic zone. The larger cladocerans have a much slower turnover than the small rotifers. Structural changes in the phytoplankton community, such as a greater proportion of large dinoflagellates, may also have contributed to the lower nutrient circulation. The light climate in the lake improved as a result of the low phytoplankton densities - the considerable transparency during the later years of the project may therefore be attributed to biological regulation mechanisms.

The purpose of introducing rainbow trout, which feed on planktonic crustaceans to some extent, was to try to reduce the density of *D. longispina* and once again achieve higher production in the phytoplankton community. This was unsuccessful, however. Analyses of stomach contents showed that *D. longispina* may well have been the most important source of food for the rainbow trout which were released in the lake, but the density still failed to decrease, probably because the density of the rainbow trout - and thus predation - was too low.

6. Conclusions

The initial liming treatment led to an immediate and unnaturally sharp increase in alkalinity and pH to levels well above natural ones. Once this had settled and better-adapted treatments had been performed, more natural conditions were obtained. Despite some overdosing, it has not been possible to eliminate acid surges. Lake Gårdsjön was originally an oligotrophic lake, and this has not changed since the treatments.

Abiotic factors imposed strict limits on the fauna and flora in Lake Gårdsjön during the acidification phase, which means that the prime reasons for the post-liming changes were also abiotic. The "detoxification" of the water made it possible for species which are susceptible to acidification to establish. Different biological regulation mechanisms affected the structure of communities during both the acidic and the limed phases, however, and these ultimately led to the stabilization of the pelagic system.

The importance of different biological mechanisms probably altered when fish were re-introduced. During the limed, fish-free period, the zooplankton community was regulated by predation. The high predation pressure on herbivorous crustaceans had a bearing on the structure of the phytoplankton community and the way in which it was largely determined by competition. When fish were re-introduced, the direct effect was a reduction of pelagic insects. As a result, the zooplankton community became increasingly regulated by competitive interactions, which in turn probably meant that the phytoplankton community was limited by an effective herbivore, i.e. by predation. This is in contrast to the situation during the fish-free period, which means that the result of a liming treatment is affected by the top predators in the system.

Finally, it should be pointed out that the liming of the acidified Lake Gårdsjön is not representative of Swedish liming projects, since the lake was kept free of fish for several years and the original species of fish have not been re-introduced. The lack of effective planktivorous fish may, for instance, be a contributing factor in the low production in the lake during the latter years of the project. On the other hand, the Lake Gårdsjön studies have highlighted a number of important mechanisms.

7. References

Abrahamsson, I. 1993. Impact of overflows on acid-base chemistry in limed lakes. - Vatten 49: 24-33.

Andersson, B.I., Carlsson, U., Hultberg, H. and Nyström, U. 1992. Studies of liming effects in acidic forest lakes within the Gårdsjö area, southwestern Sweden. - The CEC Int. Symp., Experimental manipulations of biota and biogeochemical cycling in ecosystems, Copenhagen 18-20 May, 1992.

Andersson, F. and Olsson, B. (Eds.) 1985. Lake Gårdsjön, an acid forest lake and its catchment. - Ecol. Bull. (Stockholm) 37.

Brett, M.T. 1989. Zooplankton communities and acidification processes (a review). - Water Air Soil Poll. 44: 387-414.

Broberg, O. 1987. Nutrient responses to the liming of Lake Gårdsjön. - Hydrobiologia 150: 11-24.

Broberg, O. 1988a. Liming of Lake Gårdsjön - an acidified lake in SW Sweden; effects on organic carbon, nitrogen and phosphorus. - In: Dickson, W. (Ed.) Liming of Lake

Gårdsjön, an acidified lake in SW Sweden. - Swedish Environm. Prot. Agency 3426, pp. 135-205.
Broberg, O. 1988b. Delayed nutrient responses to the liming of Lake Gårdsjön, Sweden. - Ambio 17: 23-27.
Dickson, W. (Ed.) 1988. Liming of Lake Gårdsjön, an acidified lake in SW Sweden. - Swedish Environm. Prot. Agency 3426.
Drenner, R.W. and McComas, S.R. 1980. The role of zooplankton escape ability and fish size selectivity in the selective feeding and impact of planktivorous fish. - In: Kerfoot, W.C. (Ed.) Evolution and ecolgy of zooplankton communities. Univ. Press, New Hampshire, Hanover, N.H., pp. 587-593.
Eriksson, M.O.G., Henriksson, L., Nilsson, B.-I., Nyman, H.G., Oscarson, H.G., Stenson, J.A.E. and Larsson, P. 1980. Predator-prey relations important for the biotic changes in acidified lakes. - Ambio 9: 248-249.
Gahnström, G. 1988. The effects of liming on sediment oxygen uptake in Lake Gårdsjön, Sweden. - In: Dickson, W. (Ed.) Liming of Lake Gårdsjön, an acidified lake in SW Sweden. Swedish Environm. Prot. Agency 3426, pp. 207-244.
Grahn, O. and Sangfors, O. 1988. A comparative study of macrophytes in Lake Gårdsjön, during acid and limed conditions. - In: Dickson, W. (Ed.) Liming of Lake Gårdsjön, an acidified lake in SW Sweden. Swedish Environm. Prot. Agency 3426, pp. 281-308.
Grennfelt, P., Larsson, S., Leyton, P. and Olsson, B. 1985. Atmosphaeric deposition in the Lake Gårdsjön area, SW Sweden. - Ecol. Bull. (Stockholm) 37: 101-108.
Hasselrot, B., Alenäs, I., Andersson, I. and Hultberg, H. 1987. Response of limed lakes to episodic acid events in southwestern Sweden. - Water Air Soil Poll. 32: 341-362.
Henrikson, L. 1988. Effects on water quality and benthos of acid inflow into the limed Lake Gårdsjön. In: Dickson, W. (Ed.) Liming of Lake Gårdsjön, an acidified lake in SW Sweden. Swedish Environm. Prot. Agency 3426, pp. 309-327.
Henrikson, L. and Oscarson, H.G. 1984. Lime influence on macro-invertebrate zooplankton predators. - Rep. Inst. Freshw. Res., Drottningholm 61: 93-103.
Henrikson, L., Oscarson, H.G. and Stenson, J.A.E. 1984. Development of the crustacean zooplankton community after liming of the fishless Lake Gårdsjön. - Rep. Inst. Freshw. Res., Drottningholm 61: 104-114.
Hörnström, E., Ekström, C. and Durani, M.O. 1984. Effects of pH and different levels of aluminium on lake plankton in the Swedish west coast area. - Rep. Inst. Freshw. Res., Drottningholm, 61: 115-127.
Hultberg, H. 1985. Changes in fish populations and water chemistry in Lake Gårdsjön and neighbouring lakes during the last century. - Ecol. Bull. (Stockholm) 37: 64-72.
Hultberg, H. and Grennfelt, P. 1986. Gårdsjön Project: lake acidification, chemistry in catchment runoff, lake liming and microcatchment manipulations. - Water Air Soil Poll. 30: 31-46.
Hultberg, H. and Nyström, U. 1988. The role of hydrology in treatment duration and reacidification in the limed Lake Gårdsjön. - In: Dickson, W. (Ed.) Liming of Lake Gårdsjön, an acidified lake in SW Sweden. Swedish Environm. Prot. Agency 3426, pp. 95-134.
Johansson, S. and Nilsson, T. 1988. The hydrological balance of an acidified forested catchment area 1979–1986. - In: Dickson, W. (Ed.) Liming of Lake Gårdsjön, an acidified lake in SW Sweden. Swedish Environm. Prot. Agency 3426, pp. 9-36.
Johansson, S., Eriksson, E. and Wolf, A. 1988. Lake Gårdsjön - the chemistry of the groundwater in an acidified forested catchment area 1979-1986. - In: Dickson, W. (Ed.) Liming of Lake Gårdsjön, an acidified lake in SW Sweden. Swedish Environm. Prot. Agency 3426, pp. 37-93.
Larsson, L.S. 1988. Liming effects on phytoplankton in Lake Gårdsjön 1982-1985. - In: Dickson, W. (Ed.) Liming of Lake Gårdsjön, an acidified lake in SW Sweden. Swedish

Environm. Prot. Agency 3426, pp. 245-280.
Lazarek, S. 1982. Structure and productivity of epiphytic algal communities on *Lobelia dortmanna* L. in acidified and limed lakes. - Water Air Soil Poll. 18: 333-342.
Lazarek, S. 1986. Responses of the *Lobelia* epiphyte complex to liming. - Aquat. Bot. 25: 73-81.
Lövblad, G., Amann, M., Andersen, B., Hovmand, M., Joffre, S. and Pedersen, U. 1992. Depostion of sulphur and nitrogen in the Nordic countries: present and future. - Ambio: 339-347.
Nyman, H.G. 1990. Some biological mechanisms reducing the availability of phosphorus in acid lakes. - In: Manson, B.J. (Ed.) The surface water acidification programme. Univ. Press, Cambridge, pp. 401-408.
Nyman, H.G., Oscarson, H.G. and Stenson, J.A.E. 1985. Impact of invertebrate predators on the zooplankton composition in acid forest lakes. - Ecol. Bull. (Stockholm) 37: 239-243.
Renberg, I. and Wallin, J.-E. 1985. The history of the acidification of Lake Gårdsjön as deduced from diatoms and *Sphagnum* leaves in the sediment. - Ecol. Bull. (Stockholm) 37: 47-52.
Sangfors, O. 1991. Inventering av makrofyter i Gårdsjön, september 1991. [Inventory of macrophytes in Lake Gårdsjön, September 1991] - Miljöforskargruppen, Kil, Rep. F91/085:2.
Stenson, J.A.E. 1990. Creating conditions for changes in prey community structure by *Chaoborus* spp. in a lake in Sweden. - Hydrobiologia 198: 205-214.
Stenson, J.A.E. and Svensson, J.-E. 1994. Manipulations of planktivore fauna and development of crustacean zooplankton after restoration of the acidified Lake Gårdsjön. - Arch. Hydrobiol. 131: 1-23.
Stenson, J.A.E., Bohlin, T., Henriksson, L., Nilsson, B.-I., Nyman, H.G. and Oscarson, H.G. 1978. Effects of fish removal from a small lake. - Verh. Internat. Verein. Limnol. 20: 794-801.
Stenson, J.A.E., Svensson, J.-E. and Cronberg, G. 1993. Changes and interactions in the pelagic community in acidified lakes in Sweden. - Ambio 22: 277-282.
Sterner, R.W. 1989. The role of grazers in phytoplankton succession. - In: Sommer, U. (Ed.) Plankton ecology - succession in plankton communities. Springer, Berlin Heidelberg New York, pp. 107-170.
Svensson, J.-E. and Stenson, J.A.E. 1991. Herbivoran impact on phytoplankton community structure. - Hydrobiologia 226: 71-80.

Chapter 11

Liming strategies and effects: the Lake Västra Skälsjön case study

Per Nyberg

National Board of Fisheries, Local Authority Örebro, Trädgårdsg. 13, S - 712 12 Örebro, Sweden

1. Documentation

Around the turn of the century, Lake Västra Skälsjön was mentioned in a Swedish fishing journal in conjunction with annual figures from a fish hatchery which was situated in a small adjacent water system (Gustavsson 1889, 1901). Arwidsson (1907) provided a more detailed account of releases of Arctic char *(Salvelinus alpinus)* in the lake. The composition of benthic fauna, particularly non-biting midges (Chironomidae), was studied between 1943 and 1948 by Brundin (1949). In the 1960s, a large number of lakes in the region were surveyed and various reports on Lake Västra Skälsjön were presented by Ahl (1968), Grönberg *et al* (1967, 1968) and Mossberg and Nyberg (1972). The latter (1976) repeated Brundin's sampling of benthic fauna from 1943. Water and zooplankton samples were taken in conjunction with a survey of char lakes south of River Dalälven in 1972 (Dickson *et al* 1975). Between 1976 and 1986, the lake was included in a fish sampling programme for limed lakes (Nyberg *et al* 1986) and since 1989 it has been included in the national ISEL (Integrated Studies of the Effects of Liming) programme (Appelberg and Aldén 1992). In conjunction with fish sampling, water (Nyberg unpubl.), phytoplankton (data: E. Hörnström) and zooplankton (Nyberg unpubl.) samples were also taken. Benthic fauna samples were taken on an irregular basis between 1977 and 1987 (Mossberg and Nyberg unpubl.). Otherwise, the short-term effects of liming in the lake have been reported by Eriksson *et al* (1982, 1983) along with the re-introduction of Arctic char in 1976-78 by Nyberg (1988). The effects of liming on the rooted vegetation in the lake have been described by Eriksson (1988).

Liming of Acidified Surface Waters
L. Henrikson and Y.W. Brodin (Eds.)
© Springer-Verlag Berlin Heidelberg 1995

2. Geographical orientation

Lake Västra Skälsjön is situated in the north-west of the county of Västmanland and is a seepage lake which drains into River Hedströmmen (Fig. 1). The only visible tributary to the lake consists of a small stream which dries out many times of the year. Otherwise, water enters the lake by diffuse drainage. The catchment area consists mainly of coniferous moraine with a minor element of lichen-covered rock. There has never been any agriculture in the very small catchment area. Some morphometric and hydrological data are given in Table 1.

Figure 1. Geographical location of Lake Västra Skälsjön.

Table 1. Morphometrical and hydrological data on Lake Västra Skälsjön.

Drainage area (ha)	150	Maximum depth, m	19.2
Runoff ($l/s \cdot km^2$)	12	Mean depth, m	6.8
Lake area (ha)	42	Retention time (years)	4.9
Lake volume (m^3)	$2.8 \cdot 10^3$		

3. Acidification status

The first Arctic char fry were released in the lake in 1887. Further stockings occurred in 1899 and 1900. The original fish population probably consisted of European perch *(Perca fluviatilis)*, European minnow *(Phoxinus phoxinus)* and a few brown trout *(Salmo trutta)*. The char stocks were successful and a self-sustaining population was established. In 1943, the pH was 6.3 and weak alkalinity could still be measured in 1966 (0.004 meq/l), but in October 1972 a pH of 4.8 was measured at the surface, with 5.2 at a depth of 15 m. The transparency increased from 9.3 to 14.1 m between 1943 and 1969. Vegetational studies in 1976 showed remains of a probably quite widespread growth of the moss *Scorpidium scorpioides*. This species is found in lakes with pH >6 (Iversen 1929, Sörensen 1948). In 1976, however, the bottom in deeper areas was covered with carpets of the moss *Sphagnum auriculatum*. Judging from the size of the char which were caught in 1969 (average weight: 1.2 kg, smallest fish: 0.94 kg), recruitment must have ceased in the early 1960s. A total of 3,000 one-year-old and fry of char were released experimentally between 1970 and 1972. These survived for some years, grew well and could still be found in 1974. When samples were taken in 1976, however, they had died out completely and the perch population consisted only of ten individuals, all of the same age (hatched 1971). This was probably the last year of successful recruitment. It was possible to prove that only ten remained, since they were caught after reproducing in the spring of 1976 (following a minor liming treatment in 1975) and no new generation appeared until after the 1976 fry had reached maturity. The re-introduced char and most of the perch probably died out in the spring of 1975.

4. Liming strategy

Lake Västra Skälsjön was limed using very small quantities (1-2 g/m^3) of easily soluble unslaked lime (CaO) and slaked lime ($CaOH_2$) in the autumns of 1975 and 1976, in order to reduce the risk of further acidification damage pending more extensive treatment. In order to test methods of dispersion by helicopter, the lime was spread only on the soil around the lake and along the small tributary stream in the autumn of 1977. In 1981 and 1986, lime was spread by tractor on the lake ice and from pontoons on the open water. On the last three occasions, limestone ($CaCO_3$) was used, with a grain size of 0-0.5 mm. In 1977, 28 tonnes were spread on land around the lake and 8 tonnes along the tributary stream. The doses in 1981 and 1986 were 13.6 and 22.8 g/m^3 respectively.

5. Effects of liming

5.1. Water chemistry

The small-scale treatments in 1975 and 1976 resulted in a pH of around 5.5. Not even the large-scale land liming treatment in 1977 was able to prevent the pH from falling to below 5 during the spring seasons. Since the first lake treatment in 1981, however, the pH has not fallen below 6 at any time (Fig. 2).

Table 2 shows a comparison of the mean values for certain physical and chemical parameters in 1976 and the 1977-87 period. For iron, manganese and aluminium there is a shortage of measurement data from 1976. After liming, the pH, conductivity and water colour increased significantly (ANOVA, $p<0.001$, $p<0.01$ and $p<0.05$ respectively). Table 2 further shows that the water has been very clear and poor in nutrients since treatment; metal concentrations are low and the water is weakly buffered.

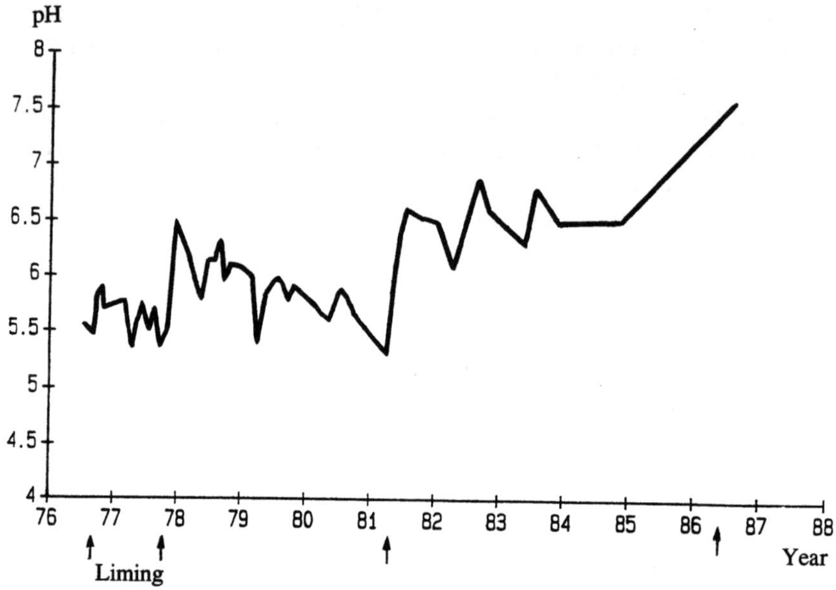

Figure 2. pH in Lake Västra Skälsjön after liming 1976-1987.

Table 2. Characteristic physical-chemical data on Lake Västra Skälsjön before (1976) and after (1977-1987) liming. Volume-weighted means of samples at 0.5, 7 and 15 m water depth respectively.

	1976	1977-1987		1976	1977-1987
pH	5.4	6.0	Tot-N (mg/l)	0.13	0.29
Alkalinity (meq/l)	0	0.03	Tot-P (mg/l)	0.005	0.005
Conductivity (mS/m)	2.3	3.1	Fe (mg/l)	-	0.046
Colour (mg Pt/l)	4	6	Mn (mg/l)	-	0.031
Ca + Mg (meq/l)	0.16	0.18	Al (mg/l)	-	0.026

5.2. Flora

Vegetation surveys were performed in 1976, 1980 and 1985. *Sphagnum* disappeared more or less completely from the limed areas, while the water lobelia *(Lobelia dortmanna)* and quillwort *(Isoetes lacustris)* increased in areas where *Sphagnum* had previously dominated. Growths of the mosses *Drepanocladus trichophyllus* and *Fontinalis antipyretica* showed signs of light deficiency in deeper water in 1985. The latter disappeared completely in deeper water after liming as a result of the reduction in transparency and the impaired light climate. At a depth of 8-10 m, the species was still found in the same abundance as before.

Phytoplankton samples were generally taken for analysis during the month of August. In 1972 and 1976-78, a total of only 12-19 phytoplankton taxa were found. The acidofilic dinoflagellates *Gymnodinium uberrimum* and *Peridinium inconspicuum* dominated. It was not until August 1979 that any far-reaching change had taken place and a number of green algae (Chlorophyceae) such as *Cosmarium*, *Elakatotrix*, *Sphaerocystis* and *Gyromitus* were found. During the 1979-85 period, 30-40 taxa occurred. Most of these were golden algae (Chrysophyceae). In 1989, after the pH had risen further as a result of the 1986 treatment, only 27 taxa were found, but in August 1990 the number of taxa had reached fifty, which can be considered a normal figure for an unacidified oligotrophic lake. The lack of nutrients is illustrated by low phytoplankton volumes (0.1-0.3 mm^3/l) and the dominance of chrysophyceans, with the exception of the first years after liming when dinoflagellates dominated.

5.3. Zooplankton and benthic fauna

Samples of zooplankton in 1966 and 1972 consisted of qualitative dip-net samples from the bottom to the surface, while the more recent samples have been quantitative and taken along a depth profile from the surface to the bottom with Ruttner or sampling tube. The zooplankton community was already significantly affected by acidification in 1966, since only *Bosmina* sp. and *Holopedium gibberum* were found among the Cladocera (Table 3). The situation was similar in 1972, except that no adult Copepoda were found. The zooplankton community in Lake Västra Skälsjön remained extremely depleted in terms of species and individuals long after liming treatments began. In 1990 the density of adult planktonic crustaceans amounted to about 10 ind/l. No major changes in species have taken place, but the cladoceran *Daphnia longispina* is found more often since treatment than it was before. In addition to the species shown in Table 3, the large predatory cladocerans *Bythotrephes longimanus* and *Leptodora kindti* have been found in the lake throughout the study. These species increased temporarily after the first liming treatment, but then decreased again as a result of increased predation from fish. They became so sparse that they could not be caught by the sampling method in use. As a result of the sparse fish population in 1976, the water-boatman *Glaenocorisa propinqua* (Corixidae) was also found in the open water and was caught in qualitative dip-net samples. This species is an attractive prey for fish, as evident from perch preferences, and is quickly reduced by predation.

Twenty-six taxa of benthic fauna were found in the lake in April 1977. This figure had risen to 34 when corresponding samples were taken in 1987. There is little point in comparing these results with the 1943 findings, since Brundin (1949) took no samples in the littoral zones. The

Table 3. Records of species (+) and abundance (individuals/l) of zooplankton before [Grönberg et al 1968 (the year 1966), Dickson et al 1975 (the year 1972)] and a short time (1977-1982) and long time (1983-1987) after liming of Lake Västra Skälsjön. Volume-weighted means for the period June-August 1976-1987.

Species	Before liming		After liming		
	1966	1972	1976	1976-1982	1983-1987
Holopedium gibberum	+	+	<0.1	0.9	0.7
Daphnia longispina			<0.1	0.2	0.8
Ceriodaphnia quadrangula			0	2.2	0
Bosmina spp.	+	+	6.5	1.3	1.0
Heterocope appendiculata			0	0.1	<0.1
Eudiaptomus gracilis	+		0	0.6	1.3
Cyclops spp.	+		5.3	0.7	<0.1
Cyclops nauplii		+	2.5	2.7	3.4
		Total	14.3	8.7	7.2

lake was divided into three depth zones for the sake of evaluation: littoral (1-2 m), sublittoral (4-6 m) and profundal (8-18 m); the time scale was divided into four periods: (1) pre-acidification (1943); (2) pre-liming (1973-78); (3) short time after liming (1979) and; (4) long time after liming (1985-87). Only comparisons between periods 2 and 4 in the three depth zones will be discussed here, along with differences and similarities between the four periods for depths from 4 m, i.e. where comparison with the 1943 findings is actually possible. The differences in abundance of individual species were tested in the first case using the Mann-Whitney U-test; in the second case a cluster analysis was performed, taking into account the occurrence and density of each individual species. The only significant change in the littoral zone was that the chironomid *Microtendipes* decreased. Several other species also decreased, albeit not significantly. In the sublittoral zone, caddis flies showed a slight significant increase as a group, as did the chironomid *Pagastiella orophila*. The pea mussel *Pisidium* (Bivalvia), which had not been found previously, also appeared. In this depth zone *Microtendipes* also decreased significantly. In the profundal zone, five genera of chironomids decreased significantly, while others showed a non-significant increase (Table 4). The overall mean density of chironomids living in the sediment in the profundal zone decreased from 1,883 ind/m^2 to 1,246 ind/m^2 between periods 2 and 4. This was probably attributable to decreased sedimentation of food, probably as a result of increased mineralization in the open water.

The cluster analysis shows that the benthic communities during the acidification phase and shortly after liming (periods 2 and 3) were very similar. During period 4, however, the community was very similar to that which was found in 1943 (period 1), i.e. prior to acidification (see Fig. 4 in Degerman *et al*, Chap. 8).

It is also worth mentioning that the mayflies (Ephemeroptera) *Leptophlebia vespertina* and *Heptagenia fuscogrisea* (both tolerant to acidification) and *Cloeon praetextum* and *Ephemera vulgata* (both susceptible to acidification) were found in the lake after liming. No snails (Gastropoda) appear to have succeeded in colonizing (Lingdell unpubl. data).

Table 4. *Occurrence of benthic taxa and changes in abundance in Lake Västra Skälsjön before (period 2, 1973-1978) and a long time after liming (period 4, 1985-1987). Increase +, decrease -, unchanged 0. Statistically significance levels; * = p<0.05, ** = p<0.01, *** = p<0.001.*

Species	Littoral	Sublittoral	Profundal
Oligochaeta	+	+	-
Asellus aquaticus	0	+	
Ephemeroptera	-		
Sialis lutaria	+	+	+
Trichoptera	0	+*	-
Pisidium sp.	+	+	+
Chironomidae			
Procladius spp.	+	-	+
Ablabesmyia spp.	-	+	-**
Heterotanytarsus apicalis	+	+	-
Heterotrissocladius marcidus	-	-	-
Psectrocladius spp.	-	+	-***
Parakiefferiella spp.	0	-	
Zalutschia sp.	-		
Dicrotendipes spp.	-	-	-***
Microtendipes spp.	-**	-***	-
Pagastiella orophila	+	+*	+
Sergentia spp.		-	-***
Stictochironomus rosenschoeldi	+	+	
Tanytarsus spp.	0	-	-***

5.4. Fish

After the first liming in 1976, char were re-introduced in the lake through release of one-summer-old fish. The roe originated from the downstream Lake Östra Skälsjön. In the springs of 1977 and 1978, fry were also stocked. Young brown trout were stocked in the outlet stream in 1978 and a few larger trout were caught in the lake during the ensuing years.

The char stock was established after liming and the density increased successively (linear regression; number/net $r^2=0.56$, $p=0.0012$, weight/net $r^2=0.26$, $p=0.04$) (Fig. 3). During the first years, the catch consisted solely of stocked fish. They probably existed in such abundance as to prevent the recruitment of younger individuals, as a result of both competition and predation. It was not until 1984 and 1985 that naturally recruited individuals appeared in the gillnets. According to the catch per unit effort (CPUE), abundance of char in the lake is now among the highest in southern and central Sweden.

European perch increased rapidly after liming, but the population then decreased over a period of years as the char stock increased (Fig. 3). This was probably attributable to the overall impact of predation on young perch by char, brown trout and older perch. The density of the perch population has since fluctuated, without any significant trend, although the weight of CPUE has declined from 0.3-0.5 kg in 1977-80 to 0.1-0.2 kg 1985-91. This normalization is probably an effect of predation and, to a lesser extent, of competition from the char population.

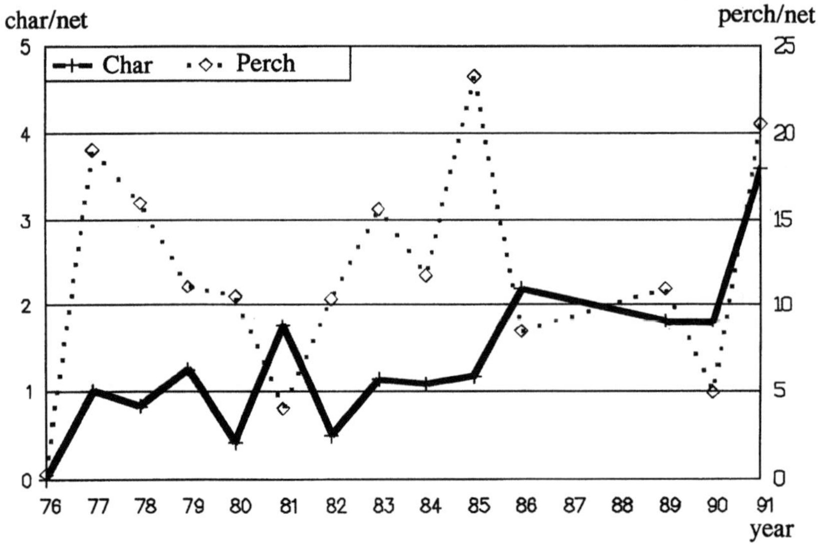

Figure 3. Catch per unit effort (CPUE) of Arctic char and European perch in Lake Västra Skälsjön after liming.

Since the char population was completely eradicated before liming, there are no data to allow comparisons of individual growth before and after liming. During the first years after liming, however, growth was exceptionally good compared with char in 25 other lakes (ANOVA for fish aged 2-6 years, $p<0.001$) (Fig. 4). At the same time, the condition of the char in

Lake Västra Skälsjön was significantly better, i.e. they were heavier for a given length than char in the 25 comparison lakes (ANOVA with age and length as covariate, $p < 0.001$), despite the fact that their July diet after liming (1983) was representative of the diet among the same species in other lakes. About half of their diet consisted of the large cladocerans *Bythotrephes longimanus* and *Daphnia longispina*. The remainder consisted in equal parts of the benthic crustacean *Eurycercus lamellatus* (Cladocera) and the water louse *Asellus aquaticus* (Isopoda) and insect larvae.

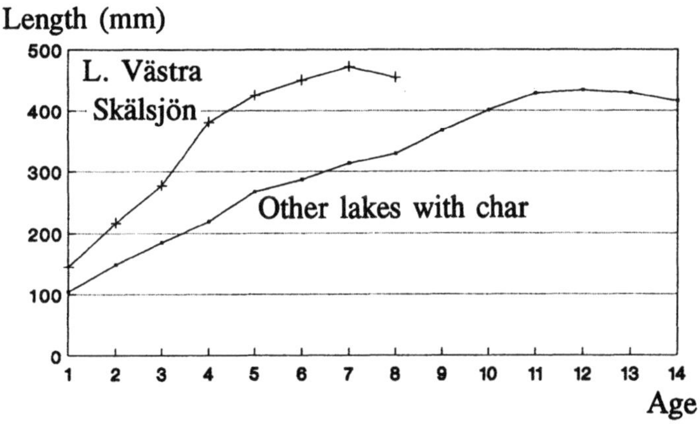

Figure 4. Growth of char in Lake Västra Skälsjön after liming (1983) in comparison with char in 25 other lakes in southern and central Sweden.

Between 1969 and 1970, growth of individual perch was typical of populations in oligotrophic forest lakes (Fig. 5). After practically the entire population had been eradicated, the survivors grew extremely well in 1976. Growth decreased soon after liming, however, returning to almost normal some years later (Fig. 5). The choice of food among large perch (> 15 cm) before liming in 1966-73 consisted of zooplankton, *A. aquaticus* and benthic insect larvae and, in 1966, some fish (Fig. 6). The choice of food among the perch which were caught in 1976 was special in that the small population could live almost exlusively on corixids. After liming and increased fish predation, corixids disappeared completely from the perch diet, and *A. aquaticus*, zooplankton and fish dominated. *A. aquaticus* appears to have been particularly important when the perch population was sparse and during the immediate post-liming years before the grazing pressure became too great. This basically constitutes a return to the pre-acidification diet, i.e. an increase in importance of benthic insect larvae and fish. Zooplankton are of lesser importance for large perch.

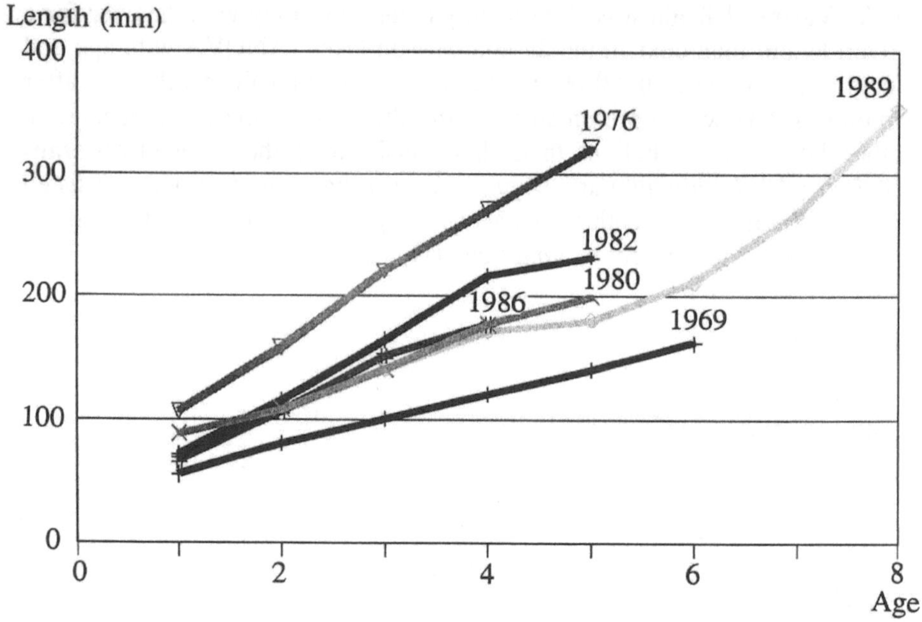

Figure 5. Growth of European perch in Lake Västra Skälsjön before (1969) and after liming (1976, 1980, 1982, 1986 and 1989).

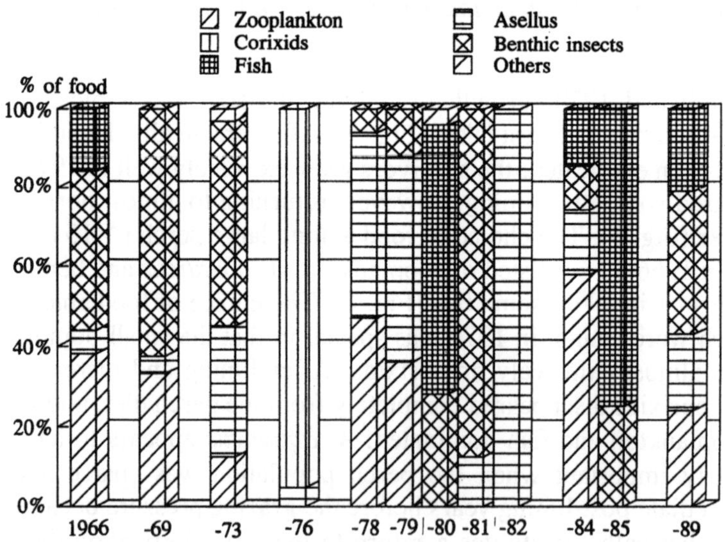

Figure 6. *Choice of food (expressed as mean volume percentage) among large (> 15cm) European perch in Lake Västra Skälsjön during July-August before (1966, 1969 and 1973) and after liming (1976 and afterwards).*

6. Conclusions

The first major treatment with limestone powder (36 tonnes) was mainly applied on the land around the lake in the autumn of 1977, and had very little effect. Dispersal of insignificantly more lime (38 tonnes) directly in the lake in the winter of 1981 yielded very good results, however. The *Sphagnum* carpets disappeared and *L. dortmanna* and *I. lacustris*, which are typical species for oligotrophic lakes, increased the limits of their range. The phytoplankton communities were dominated by chrysophyceans after the treatment, reaching 50 taxa in 1990, which is a normal figure for an unacidified lake of this sort. The zooplankton community was sparse in terms of individuals throughout the period of the study, but the cladocerans became more common after liming. Comparison of the composition of benthic fauna shows major similarities 1943-1948 (prior to acidification) and 1985-1987 (post-post-liming). The re-introduction of char resulted in a good reproducing population. Since liming, both individual growth and the choice of food in the perch population have been what one might expect in this type of lake. The results from Lake Västra Skälsjön provide one of the best examples of how normalization can take place at all trophic levels some time after liming.

7. References

Ahl, T. 1968. Limnologiska studier i Malingsbo-Klotenregionen. II. Jonsammansättning, närsalter och organisk substans. [Limnological studies in the Malningsbo-Kloten district. II. Compositions of ions, nutrients and organic matter.] - Dep. Limnol., Univ. Uppsala, Mälarundersökningen 19.

Appelberg, M. and Aldén, U. 1992. Integrerad uppföljning av kalkningens effekter på sjöar och vattendrag - en treårsrapport. [Integrated monitoring in limed lakes and streams - a three-year study.]. - Inform. Inst. Freshw. Res., Drottningholm (4): 1-60.

Arwidsson, A. 1907. En rödinginplantering i Västmanland. [Stocking of Arctic char in the county of Västmanland.] - Svensk Fisk. Tidskr. 16: 167-170.

Brundin, L. 1949. Chironomiden und andere Bodentiere der Südschwedischen Urgebirgseen. - Rep. Inst. Freshwat. Res., Drottningholm 32.

Dickson, W., Hörnström, E., Ekström, C. and Almer, B. 1975. Rödingsjöar söder om Dalälven. [English summary: Char lakes south of River Dalälven.] - Inform. Inst. Freshw. Res., Drottningholm (7).

Eriksson, F. 1988. Makrofytvegetation i kalkade sjöar. - [Macrophyte vegetation in limed lakes.] - Inform. Inst. Freshw. Res., Drottningholm (9).

Eriksson, F., Hörnström, E., Mossberg, P. and Nyberg, P. 1982. Ekologiska effekter av kalkning i försurade sjöar och vattendrag. [English summary: Ecological effects of lime treatment of acidified lakes and watercourses.] - Inform. Inst. Freshw. Res., Drottningholm (6).

Eriksson, F., Hörnström, E., Mossberg, P. and Nyberg, P. 1983. Ecological effects of lime treatment of acidified lakes and rivers in Sweden. - Hydrobiologia 101: 145-164.

Grönberg, B., Ramberg, L. and Winbladh, E. 1967. Limnologiska studier i Malingsbo-Kloten regionen. I. Fysikalisk-kemiska faktorer. [Limnological studies on the Malingsbo-Kloten district. I. Physical-chemical parameters.] - Dep. Limnol., Univ. Uppsala, Mälarundersökningen 18.

Grönberg, B., Schuisky, G. and Skommar, A.-B. 1968. Limnologiska studier i Malingsbo--Klotenregionen. I. Biologiska undersökningar. [Limnological studies on the Malningsbo-Kloten district. I. Biological studies.] - Dep. Limnol., Univ. Uppsala, Mälarundersökningen 20.

Gustafsson, C.G. 1899. Fiskodlingsföretag i Västmanlands län 1898–1899. [Fish breeding in Västmanland county 1898-1899.] - Svensk Fisk. Tidskr. 8: 184-187.

Gustafsson, C.G. 1901. Fiskodlingen i Västmanlands län 1900-1901 [Fish breeding in Västmanland county 1900-1901.]. - Svensk Fisk. Tidskr. 10: 173-178.

Iversen, J. 1929. Studien über die pH-Verhältnisse dänischer Gewässer und ihren Einfluss auf die Hydrophyten-Vegetation. - Bot. Tidskr. 40: 277-326.

Mossberg, P. and Nyberg, P. 1972. Limnologiska studier i Malingsbo-Klotenregionen. VI. Abborrens tillväxt och näringsval i fyra tjärnar. [Limnological studies on the Malingsbo-Kloten district. VI. The growth and food choice of perch in four tarns.] - Dep. Limnol., Univ. Uppsala, Mälarundersökningen 51.

Mossberg, P. and Nyberg, P. 1976. Försurningseffekter på bottendjur och fisk i Västra Skälsjön. [English summary: Effects of acidification on bottom fauna and fish in Lake Västra Skälsjön.] - Inform. Inst. Freshw. Res., Drottningholm (9).

Nyberg, P. 1988. Reclamation of acidified Arctic char (*Salvelinus alpinus* L.) lakes in Sweden by means of liming. - Verh. Internat. Verein. Limnol. 23: 1732-1742.

Nyberg, P., Appelberg, M. and Degerman, E. 1986. Effects of liming on crayfish and fish in Sweden. - Water Air Soil Poll. 31: 669-687

Sörensen, H. 1948. Studies on the ecology of Danish water- and bog-mosses. - Dansk Bot. Ark. 12: 1-47.

Chapter 12

Liming strategies and effects: the Lake Stora Härsjön case study

Magnus Appelberg

Swedish National Board of Fisheries Institute of Freshwater Research, Fish Monitoring Group, S - 178 93 Drottningholm, Sweden

1. Documentation

The first data available on Lake Stora Härsjön came from the 1930s (Institute of Freshwater Research, Drottningholm) and show that the lake was not affected by acidification at the time. The real documentation of water chemistry and biological conditions in Lake Stora Härsjön began in 1971, when the lake was already severely acidified. The lake was limed in the autumn and winter of 1977/78 and developments since then have been monitored in the form of various study projects. The results, which encompass water chemistry, phytoplankton and zooplankton, benthic fauna and fish, have been presented in a number of reports (Almer 1972, Dickson *et al* 1973, Andersson 1977, Alenäs 1979, 1986, Almer and Hansson 1980, Alenäs *et al* 1982, Eriksson *et al* 1983, Hörnström and Ekström 1983, 1986, Hörnström *et al* 1984, Appelberg *et al* 1990, Enderlein and Appelberg 1992, Appelberg and Aldén 1992). Lake Stora Härsjön has been part of the ISEL (Integrated Studies on the Effects of Liming) programme since 1989 and reports are submitted annually (Appelberg and Aldén 1992).

2. Geographical orientation

Lake Stora Härsjön is situated to the north-west of Göteborg, 90 m above sea level, in the municipalities of Lerum and Härryda (Fig. 1). The lake is part of the River Göta Älv water system and drains into River Mölndalsån via River Tväråan. Upstream of Lake Stora Härsjön there are five smaller lakes, which were all limed at the same time as Lake Stora Härsjön. The lake has an area of 258 ha, a maximum depth of 47 m and a mean depth

Liming of Acidified Surface Waters
L. Henrikson and Y.W. Brodin (Eds.)
© Springer-Verlag Berlin Heidelberg 1995

of 15.7 m. The calculated turnover time is about 3.5 years. The drainage basin has an area of 26 km^2 and is mainly covered with coniferous forest.

Figure 1. Geographical location of Lake Stora Härsjön.

3. Acidification status

Information from August 1935 states that the lake at that time had a pH of 6.7 (calometric measurement) and a Secchi depth (transparency) of about 5 m (Institute of Freshwater Research, Drottningholm). During the 1971-76 period, and up to liming in 1977/78, the pH varied between 4.9 and 5.3 (Fig. 2A). The low content of humic substances and particulate material led to a characteristic blue-green colour and a Secchi depth of 8-14 m (Fig. 2B).

The phytoplankton composition in the lake was somewhat different from that normally found in similarly acidified lakes. In spring samples prior to liming, *Asterionella formosa* (Bacillariophyceae) was abundant, while summer samples indicated sparse occurrences of *Scenedesmus* and *Staurodesmus* (Chlorophyceae, green algae). Characteristic summer forms included *Dinobryon pediforme*, *D. creulatum* and *D. sociale* var. *americanum* (Dinophyceae, dinoflagellates). The number of taxa prior to liming (i.e. before autumn 1977) was 7-21 (Fig. 3).

Prior to liming, a low number of zooplankton taxa was noted: 3-4 taxa of Rotatoria, 3-5 of Cladocera and 2 of Copepoda. The rotifer species included *Conochilus unicornis*, *Kellicottia longispina* and *Polyarthra vulgaris*, as well as *P. remata* (Fig. 4). Cladocerans, which were frequently observed, included *Bosmina longispina*, *Bythotrephes longimanus*, *Holopedium gibberum* and *Leptodora kindti*. *Diaphanosoma brachyurum* was also found. *Eudiaptomus gracilis* was the only copepod to be found in the samples. When zooplankton samples were taken in 1976, however, *Cyclops* sp. were also found (Alenäs 1986). Among the mayflies (Ephemeroptera), which are often used as an indicator of acidification, *Leptophlebia* was the only genus found the year before the liming treatment.

In 1935, the lake supported northern pike *(Esox lucius)*, European perch *(Perca fluviatilis)*, roach *(Rutilus rutilus)* and eel *(Anguilla anguilla)* (Institute of Freshwater Research, Drottningholm). Vendace *(Coregonus albula)* were introduced in the lake in the 1940s, followed by Arctic char

A

B

C

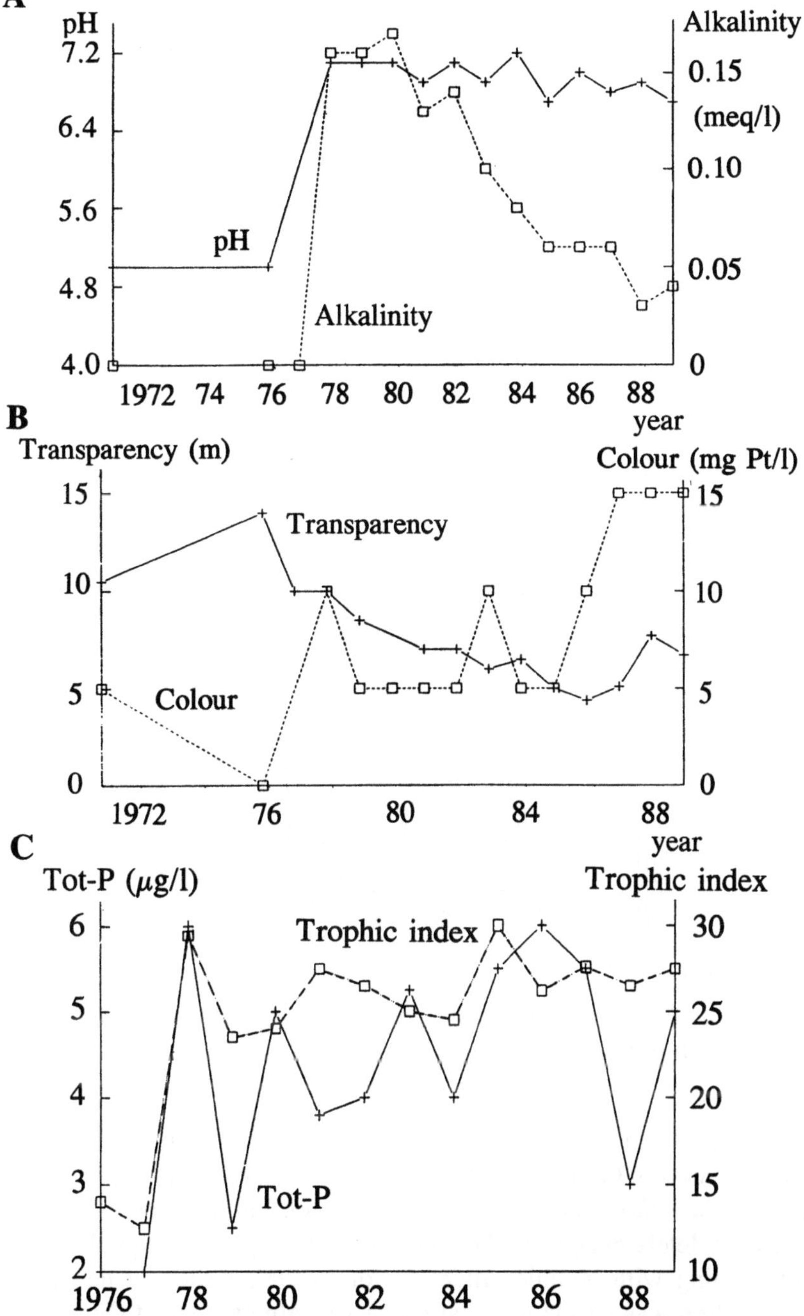

Figure 2. Development of chemical parameters and trophic index based on the composition of phytoplankton in Lake Stora Härsjön (cf. text) before and after liming (i.e. after 1977/78). A) pH and alkalinity, B) transparency and water colour, C) total phosphorus and trophic index (Hörnström 1981).

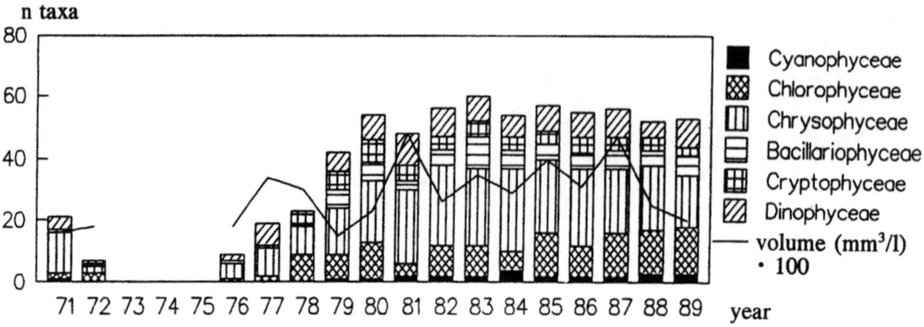

Figure 3. Groups and volume of phytoplankton in Lake Stora Härsjön before and after liming.

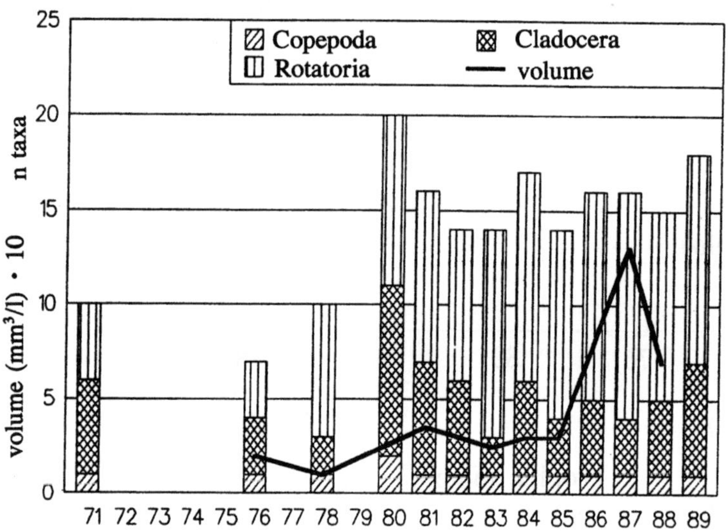

Figure 4. Taxa and relative volume of zooplankton in Lake Stora Härsjön before and after liming.

(Salvelinus alpinus) and whitefish *(Coregonus* sp.) in the 1960s. The roach population decreased in the 1950s, while the brown trout *(Salmo trutta)* and European minnow *(Phoxinus phoxinus)* in the runoff stream decreased in number and completely disappeared by the early 1970s (Almer unpubl. data 1971). Test fishing with benthic gillnets prior to liming showed that the vendace population was sparse and that roach was almost extinct (Fig. 5).

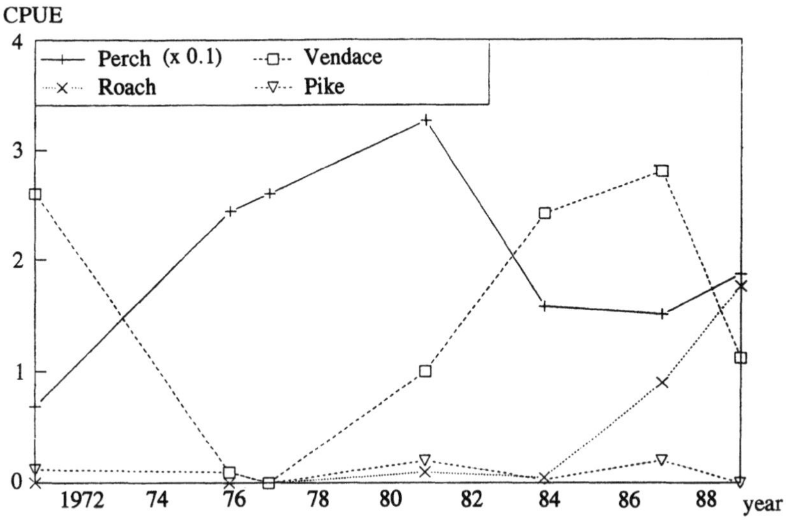

Figure 5. Catch per unit effort (CPUE) of fish in benthic gillnets in Lake Stora Härsjön before and after liming (1977/1978).

4. Liming strategy

The lake was limed in the autumn of 1977/spring of 1978 with 800 tonnes of ground limestone ($CaCO_3$) with a grain size of 0-0.5 mm. The calculated dose was 18 g/m^3 and duration was estimated to be about ten years. The relatively low dose was calculated in the light of the fact that five waters upstream were also being limed. The lake was re-limed in 1991.

5. The effects of liming

After liming in 1977/78, the alkalinity rose from 0 to approximately 0.16 meq/l and high values were noted over a number of years. By 1988/89, however, the summer alkalinity had decreased to below 0.05 meq/l, while the pH was still relatively high (Fig. 2A). In August 1978, a brief but strong increase in aluminium concentration was observed, with a subsequent decrease to below the 1976 levels. The concentration was still high only during spring the years after liming, with a peak of 155 μg/l in 1984. Secchi depth remained high during the initial years after liming, but decreased thereafter by about 50% (Fig. 2B).

Total phosphorus in Lake Stora Härsjön (summer values) rose during the first ten years after liming (linear regression $r^2=0.36$, $p<0.05$), while the Secchi depth decreased and the water colour increased (Fig. 2C). However, the low phosphorus concentrations indicate that these figures are unreliable. Apart from an immediate peak the same year as the treatment was carried out, the phosphorus concentration rose from about 2 μg/l before liming (1971, 1976) to about 6 μg/l in 1986. The somewhat higher levels are probably a result of the increased biomass in the higher trophic levels (zooplankton, benthic fauna and fish) which helped to retain a higher level of phosphorus in suspended form. The changes in the fish community have probably also affected the lower trophic levels and may have been crucial in determining the total phosphorus concentrations after liming.

The concentration of calcium + magnesium increased after liming, from 0.26 to 0.44 meq/l, while sodium and potassium concentrations varied between 0.24 and 0.30, and 0.02 and 0.03 meq/l respectively. The sulphate levels declined from about 0.30 meq/l between 1976 and 1981, to about 0.20 meq/l in 1989.

5.1. Phytoplankton

During the first summer season after liming (1978) the composition of phytoplankton species changed. Acidophilic species such as *Peridinium inconspicuum* became mixed with forms which are common in more eutrophic waters, such as *Ankistrodesmus* and *Closterium*. The total number of phytoplankton species was still low, however. In 1979, the acidophilic species of diatoms *(Cyclotella comensis* and *C. stelligera)* and chrysomonads were replaced. From 1980, 50-60 taxa were found at each sampling (Fig. 3). The volume of plankton was high in the spring of 1980 at 1.4 mm^3/l and 65 taxa were identified. The number of forms increased constantly, and species such as *Cyclotella comta*, *Rhizosolenia longiseta*, *Chrysosphaerella longispina* and *Gymnodinium helveticum* were observed. During the 1981-84 period, species of blue-green algae (Cyanophyceae) were also found; first *Anabaena flos-aquae* and then, in 1984, *Gomphosphaeria lacustris*. After liming, the phytoplankton volumes varied between 0.15 and 0.48 mm^3/l. The highest summer levels were observed in 1981 and 1987 (Fig. 3).

The composition of phytoplankton may be used to describe lake productivity by calculating a trophic index (Hörnström 1981). In Lake Stora Härsjön the trophic index rose sharply after liming from 12-15 to 23-30 (ANOVA, $p<0.05$, Fig. 2C). Maximum values were observed in 1978, 1981 and 1985, while the minimum was noted in 1979. The trophic index showed a positive correlation to the phytoplankton volume after liming (linear regression, $r^2=0.28$, $p<0.05$).

Although liming resulted in a sharp increase in pH and a reduced concentration of aluminium, the diversity of phytoplankton species reached its highest figure several years after liming. This increase occurred together with an increased concentration of phosphorus and humic substances. The fact that the phytoplankton volume increased as the transparency decreased, the colour of the water - i.e. the amount of organic carbon - and the total phosphorus level increased indicate that phytoplankton may have been limited by phosphorus and, indirectly, also by the organic carbon prior to liming.

5.2. Zooplankton

Rotatoria was the dominating zooplankton group during the year following the initial liming in 1978, while cladocerans and copepods were almost completely absent. Between 1980 and 1989 the number of species increased from 7-10 to 14-19 (Fig. 4). The number of cladoceran species reached a peak as early as 1980, while the number of rotifer species peaked in 1983.

The large cladoceran *B. longimanus* was not noted in the zooplankton samples after liming, although it was found in fish stomachs. *D. brachyurum*, *H. gibberum* and *E. gracilis*, which were observed prior to liming, decreased sharply afterwards. *H. gibberum* could not be found in 1982, but was subsequently discovered again in samples from 1988 onwards. New species which appeared after liming included the rotifers *Ploesoma truncatum*, *P. hudsoni*, *Gastropus stylifer*, *Collotheca* sp., *Synchaeta* sp. and the cladocerans *Ceriodaphnia quadrangula*, *Daphnia cristata*, *D. hyalina* and *D. galatea* (in other cases these are brought together under the name of *D. longispina)* and *Polyphemus pediculus*. *Cyclops* sp., which could not be found in the pre-liming samples, were found in all post-liming samples. The total volume of zooplankton rose significantly after liming (linear regression, $p<0.05$, $n=7$) and peaked in 1987.

The change in phytoplankton species composition after liming most probably had a major impact on the food available to the zooplankton. The available food particles probably covered a broader range of sizes, particularly in the range below 5μm, where there was basically neither plankton, bacteria nor detritus prior to liming. All the *Daphnia* species, *D. brachyurum*, and several rotifer species filter small particles and should therefore have benefited from the development of phytoplankton.

The post-liming changes in the zooplankton community in Lake Stora Härsjön suggest that several factors have interacted: improved water quality allowed species susceptible to acidification to increase and/or colonize, while the improved food supply in the form of smaller particles benefited many zooplankton forms, and the increased predation from fish reduced larger zooplankton forms to the benefit of smaller ones (Stenson *et al* 1978, Appelberg, Chap. 9). On the basis of the qualitative studies which have been performed, it is not possible to say which of these factors has been of the greatest importance. It can, however, be observed that all four factors have probably contributed to the structure of the zooplankton community.

5.3. Benthic fauna

After liming, mayflies of *Leptophlebia* were observed, along with *Heptagenia fuscogrisea*. A greater density of the cladoceran *Eurycercus lamellatus* was also observed in littoral samples two years after treatment, but this was followed by decreses in 1981 and 1984. The water louse *Asellus aquaticus* increased from a pre-1977 density (the year before liming) of 0.2 ind/m^2 (SD=$\pm$0.13) in the sublittoral zone (3-10 m deep) to 0.7 $\pm$0.11 in 1979, 0.9 $\pm$ 1.8 in 1981 and 1.5 $\pm$2.0 ind/m^2 in 1984 (Kruskal-Wallis, one-way ANOVA, $p<0.05$, data from Alenäs 1986). After liming, in 1979, 1981 and 1984, more acid-sensitive species of the mayfly families Ephemeridae, Caenidae and Baetidae were found. In 1977, caddis flies (Trichoptera) in the Polycentropidae family dominated, while species in the Molannidae and Leptoceridae families were more common after liming. This also applied to the alder fly *(Sialis* sp.). Alenäs (1986) also reported increased numbers of both phantom midge larvae *(Chaoborus* sp.) and Ostracoda in the profundal zone.

A. aquaticus represented 51% of the individuals in the littoral benthic fauna samples from August 1989. Chironomidae were abundant, and *Leptophlebia* and *Eurycercus* were also observed. Of the other 42 taxa observed, the dragon fly *Coenagrion* sp. (Odonata), *Sida crystallina*, oligochaetes and several mayfly species were fairly abundant.

5.4. Fish

After liming, the number of vendace caught in benthic gillnets increased. The number of perch also continued to increase until 1981, after which a decrease was observed. Few roach were caught at the test fishing in 1981, while the catch per unit effort (CPUE) was 0.5 (SD$\pm$0.03) in 1984, 0.6 ($\pm$1.9) in 1987 and 1.76 ($\pm$1.61) in 1989 (Fig. 5). Catches of fish in the pelagic zone differed considerably from those in the littoral and benthic zones. In August 1989, CPUE of vendace in pelagic gillnets was nearly eight times higher than in benthic gillnets (in relation to net area).

Vendace increased considerably in the pelagic gillnets during the 1984-89 period, from 0.6 ind/net in 1984 to 9.2 in 1987 and 35 in 1989. A corresponding increase in vendace was not seen in the benthic gillnets, however, where the CPUE decreased from 2.9 ind/net in 1984 to 2.7 in 1987 and 1.1 in 1989. Among perch and roach, only one individual of each species was caught in the pelagic gillnets in 1989. The depth distribution of the fish (in benthic gillnets) during the same period showed that perch dominated at depths above 12 m in all three years, while vendace were found at greater depths. Test fishing in 1984 found vendace in the benthic gillnets at all depths, with the largest catches in the 12-20 m zone. In 1989, however, vendace only appeared in the deeper zones with a peak at 20-35

m. The depth distribution differed significantly from that of 1984 (Chi-square, $p<0.05$).

The mean length of perch at different year classes decreased during the period (ANOVA, $p<0.0001$ for covariate (age) and $p<0.01$ for effect (year), Fig. 6). Perch caught in test fishing 1971 exhibited a length at a given age above the average for limed waters, but it was closer to the average in 1989 (Appelberg unpubl. data).

In a stable community, a planktivore such as vendace is usually a superior competitor to species which have a more mixed diet (Svärdson 1976). One might therefore expect that vendace would dominate over perch in Lake Stora Härsjön as well, and this seems to be the case. The fact that perch were limited by competition was indicated by catches in benthic gillnets, which increased as the vendace catches declined during the pre-liming period. The opposite was seen after liming, when catches of vendace and roach in benthic gillnets increased.

The difference between catches in the pelagic and in benthic gillnets and the depth distribution of the catches in the benthic gillnets clearly demonstrates the vertical and horizontal habitat segregation between more littoral species, perch and roach, and the pelagic vendace. Comparison of food choice among planktivorous perch and vendace in 1989 also supports this: perch diet consisted mainly of littoral/profundal benthic fauna, while

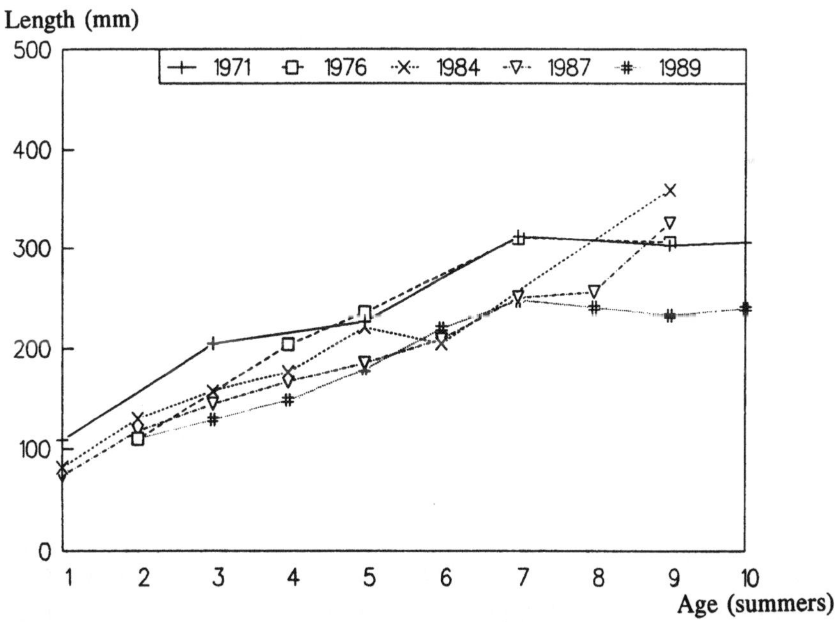

Figure 6. *The relationship between perch length and age in Lake Stora Härsjön before and after liming.*

vendace diet consisted exclusively of pelagic forms. Analyses of perch diet in 1977 suggest that this species lived more in the pelagic zone on large zooplankton before liming, since the catches of vendace in the benthic gillnet were low. After liming, the increased occurrence of littoral zooplankton species such as *S. crystallina* and *E. lamellatus* in perch diet shows that they had switched their habitat to the littoral zone.

5.5. Comparison of different trophic levels

The change in lake productivity is reflected in several variables. Secchi depth showed a negative correlation to the total phosphorus concentration (r^2=0.46, $p<0.01$, Fig. 7A). The phosphorus concentration rose in conjunction with liming in 1977/78, however, with no corresponding decrease in Secchi depth, which indicates that phosphorus was added with the lime. The volume of phytoplankton and the trophic index showed a positive correlation to the total phosphorus concentration (r^2=0.30, $p<0.05$). The total phosphorus concentration in the lake was also correlated to the overall CPUE of the two planktivorous species, vendace and roach, after liming (r^2=0.92, $p<0.001$, n=6, Fig. 7B).

The increased catches of roach in the benthic gillnet during the late 1980s and the increased catches of vendace in the pelagic gillnets coincided with the increased volumes of zooplankton which were observed during the same period. The catches of the different fish species in the benthic gillnets were also correlated to some extent. A slight negative correlation between catches of perch and the total catch of vendace and roach in the benthic gillnet was observed (r^2=0.38, p=0.08).

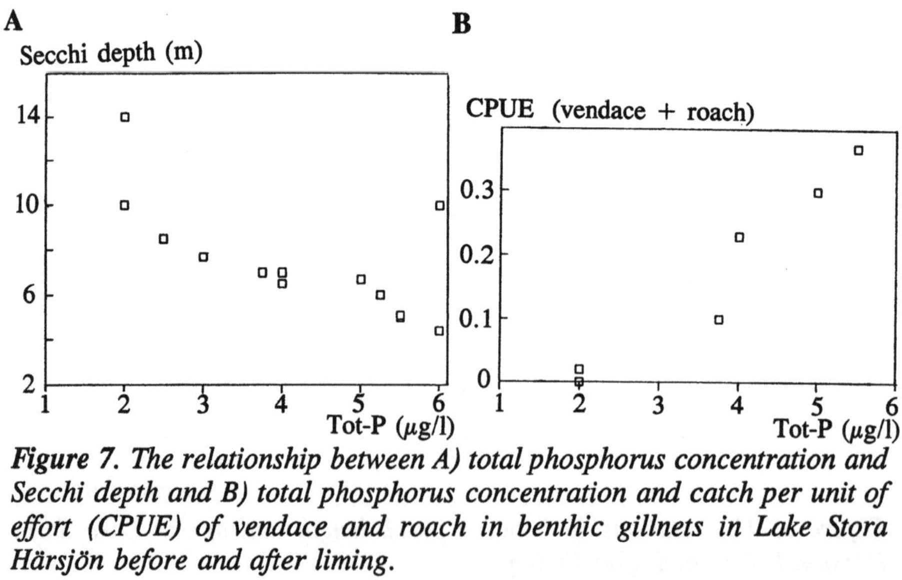

Figure 7. The relationship between A) total phosphorus concentration and Secchi depth and B) total phosphorus concentration and catch per unit of effort (CPUE) of vendace and roach in benthic gillnets in Lake Stora Härsjön before and after liming.

The relationship between fish and lower fauna is illustrated by the choice of food among perch and vendace. Before liming, the stomach content of perch was inconsistent with the samples of pelagic zooplankton. Despite the dominance of the cladocerans *B. longispina* and *E. gracilis* in the lake in 1977, the diet of perch of all sizes (10-35 cm, n=16) consisted mainly of large zooplankton such as *B. longimanus*. The year after liming, littoral faunas such as Sididae were seen to dominate in perch below 15 cm, while those over 20 cm mainly fed on vendace and *A. aquaticus*. The increase in *A. aquaticus* and vendace in the diet of the larger perch coincided with the increased occurrence of *A. aquaticus* in benthic fauna samples and the increased catches of vendace in the benthic gillnets.

The choice of food among perch three years after liming (1981) was more varied. Among perch of less than 15 cm, *E. lamellatus*, *P. pediculus* and *Bosmina* sp. were seen to dominate. Vendace constituted the main food for larger perch, but *D. brachyurum* was also found in the stomachs. Both *B. longispina* and *P. pediculus* dominated in plankton samples at the same time, while catches of vendace in the benthic gillnets averaged more than 3 ind/net. In August 1989, the occurrence of the littoral cladocerans coincided with the diet of the zooplanktivorous perch. Both *S. crystallina* and *E. lamellatus* were represented in the stomach of perch of less than 10 cm in length. In the stomachs of the larger perch, *P. pediculus* and *L. kindtii* dominated, although the former species was found neither in the zooplankton samples nor in the littoral samples. The occurrence of benthic fauna agreed to some extent with the choice of food among perch of 15-25 cm. *A. aquaticus* and chironomids, which were found in the greatest numbers among the littoral faunas, appeared to a large extent in the perch diet as well. Among piscivorous perch, the main diet consisted of vendace, which also dominated in the pelagic zone.

6. Conclusions

The development of aquatic communities in Lake Stora Härsjön was influenced by both chemical processes and post-liming predation from fish. Several of the relationships show that biotic interactions, particularly in terms of the post-liming change in fish fauna, have played an important part. This applies particularly to production in the pelagic zone. The pre-liming community was probably influenced mainly by the lack of phosphorus and organic carbon, which meant that the total biomass was low, while the toxic effects of low pH and high aluminium concentrations affected the composition of species within different groups of organisms. After liming, which resulted in increased biomass at all trophic levels, both abiotic and biotic processes played an important part. Turnover of nutrients (phosphorus and organic carbon) in the lake should have increased as a result of the rise in pH, whereas the toxicity decreased. This may, in turn, have been reflected higher up in the ecosystem. It is probable that the controlling biotic factors may have had an increasing influence on the development of the system. It may be assumed that, particularly towards the end of the period, the interactions between and within the different trophic levels played an important part.

Monitoring of the liming effort in Lake Stora Härsjön has shown that liming may produce long-term changes in the community as an indirect result of the initial treatment. Such effects may still be observed after 16 years. Studies have also shown that changes at one trophic level may have a significant effect on other levels in the system, both higher and lower. In order to understand the effects of liming, therefore, it is essential to take as many parts of the ecosystem as possible into consideration and to evaluate the effects in terms of changes to the system as a whole.

7. References

Alenäs, I. 1979. Rapport över biologiska och fysikalisk-kemiska provtagningar för kalkningsprojektet Härskogen 1977-78. [Report on biological and physical-chemical sampling of the Härskogen liming project 1977-78.] - Swedish Environm. Res. Inst., unpubl. rep.

Alenäs, I. 1986. Kalkningsprojektet Härskogen 1976-1986. [The Härskogen liming project 1976-1986.] - Swedish Environm. Res. Inst., B 846.

Alenäs, I., Hultberg, H. and Andersson, I. 1982. Sammanställning av för- och uppföljande undersökningar i Härskogenområdets sjöar. [Compilation of studies before and after liming in the lakes of the Härskogen district.] - Swedish Environm. Res. Inst., unpubl. rep.

Almer, B. 1972. Försurningens inverkan på fiskbestånd i västkustsjöar. [English summary: The effect of acidification on fish stocks in lakes on the west coast of Sweden.] - Inform. Inst. Freshw. Res., Drottningholm (12).

Almer, B. and Hansson, M. 1980. Försurningseffekter i västkustsjöar. [English summary: Effects of acidification in west coast lakes.] - Inform. Inst. Freshw. Res., Drottningholm (5).

Andersson, I. 1977. Fysikalisk-kemiska och biologiska undersökningar i sjöar inom Härskogens naturområde. [Physical-chemical and biological studies on lakes in the Härskogen Reserve.] - Swedish Environm. Res. Inst., unpubl. rep.

Appelberg, M. and Aldén, U. 1992. Integrerad uppföljning av kalkningens effekter på sjöar och vattendrag - en treårsrapport. [Integrated monitoring in limed lakes and streams - a three-year study.] - Inform. Inst. Freshw. Res., Drottningholm (4): 1-60.

Appelberg, M., Ekström, C. and Hörnström, E. 1990. Stora Härsjön - ett exempel på integrerad uppföljning av kalkningens effekter. [English summary: Lake Stora Härsjön - a case study for integrated monitoring of the effects of liming.] - Inform. Inst. Freshw. Res., Drottningholm (1).

Dickson, W., Ekström, C., Hörnström, E. and Miller, U. 1973. Försurningens inverkan på västkustsjöar. [The impact of acidification on Swedish west coast lakes.] - Inform. Inst. Freshw. Res., Drottningholm (4).

Enderlein, O. and Appelberg, M. 1992. Sampling stocks of *Coregonus lavaretus* and *C. albula* with pelagic gillnets, trawls and hydroacoustic gear. - Pol. Arch. Hydrobiol. 39:

771-777.

Eriksson, F., Hörnström, E., Mossberg, P. and Nyberg, P. 1983. Ecological effects of lime treatment of acidified lakes and rivers in Sweden. - Hydrobiologia 101: 145-164.

Hörnström, E. 1981. Trophic characterization of lakes by means of qualitative phytoplankton analysis. - Limnologica 13: 249-261.

Hörnström, E. and Ekström, C. 1983. pH-, närings- och aluminiumeffekter på plankton i västkustsjöar. [Effects of pH, nutrients and aluminium on plankton in lakes on the west coast.] - Swedish Environm. Prot. Agency 1704.

Hörnström, E. and Ekström, C. 1986. Acidification and liming effects on phyto- and zooplankton in some Swedish west coast lakes. - Swedish Environm. Prot. Agency 1864.

Hörnström, E., Ekström, C. and Duraini, M.E. 1984. Effects of pH and different levels of aluminium on lake plankton in the Swedish west coast area. - Rep. Inst. Freshw. Res., Drottningholm 61: 115-127.

Stenson, J.A.E., Bohlin, T., Henrikson, L., Nilsson, B.-I., Nyman, H.G., Oscarson, H.G. and Larsson, P. 1978. Effects of fish removal from a small lake. - Verh. Internat. Verein. Limnol. 20: 794-801.

Svärdson, G. 1976. Interspecific population dominance in fish communitites of Scandinavian lakes. - Rep. Inst. Freshw. Res., Drottningholm 55: 145-171.

Chapter 13

Liming strategies and effects: the Lake Gyslättasjön case study

Magnus Appelberg

Swedish National Board of Fisheries Institute of Freshwater Research, Fish Monitoring Group, S - 178 93 Drottningholm, Sweden

1. Documentation

Proximity to the Aneboda field station has meant that information on Lake Gyslättasjön has been documented for the past half century. The first data orginated in the early 1930s (Lang 1931, Lönnerblad 1931a,b). Since then, Malmer (1961) and Andersson (1971) have documented the limnology of the lake. More recent studies have focused on the effects of acidification, liming and metals. Lake Gyslättasjön is included in two national projects: "Liming-mercury" (Andersson *et al* 1982, 1989, Degerman and Nyberg 1987, Borg *et al* 1989) and, since 1989, the ISEL (Intergrated Studies on the Effects of Liming) programme, with regular monitoring of the post-liming ecosystem (Andersson *et al* 1993, Appelberg and Aldén 1992, Ekström 1992, Hörnström 1992, Lingdell and Engblom 1992).

2. Geographical orientation

Lake Gyslättasjön belongs to the River Lagan water system at 236 m above sea level (Fig. 1). The lake is characteristically dystrophic with a total area of 33 ha and a maximum depth of 9.0 m. The average depth is 2.3 m. The vegetation in the lake is dominated by reeds *(Phragmites australis)*, which are abundant in places. Theoretical turnover time has been estimated at 1.3 years. The lake is affected by the high proportion (15%) of bogs and fens in the 2.4 km^2 catchment area, which means that there is a high humus content and therefore also a high colour value of the lake.

Liming of Acidified Surface Waters
L. Henrikson and Y.W. Brodin (Eds.)
© Springer-Verlag Berlin Heidelberg 1995

3. Acidification status

pH measurements during the early 1930s showed that the lake was affected by humic acids. The pH varied between 5.8 (bottom samples) and 6.3 (surface samples) (Lönnerblad 1931b). In the early 1980s, the lake had become severely acidified. Alkalinity was zero and the annual mean pH during the 1981-84 period was 4.8. A relatively high water colour (annual mean: 59 mg Pt/l) and high TOC concentrations (5.4 mg/l) show that the lake was greatly affected by the surrounding forest, bogs and fens. The phosphorus concentrations were also high (10 μg/l), mainly as a result of the high humic content.

The year before liming, the phytoplankton volume varied between 0.15 and 2.4 mm^3/l (Table 1). There were few species in the plankton community, and *Gonyostyomum semens* dominated almost entirely. Studies of the zooplankton community before liming (1982-84)

Figure 1. The geographic location of Lake Gyslättasjön.

Table 1. Phytoplankton volume and number of taxa in Lake Gyslättasjön before and after first liming (1985). Data from Hörnström (1992).

Date	Volume, mm^3/l	n taxa
21.09.1981	0.30	<10
27.08.1982	2.40	20
16.08.1983	0.15	28
27.09.1984	0.57	28
17.08.1989	4.63	30
01.08.1990	1.83	43
14.08.1991	0.43	47

showed that the number of species was diverse despite the low pH (Fig. 2). During these years, 14 species of rotifers (Rotatoria) were observed, together with 5 species of Cladocera and 2 of Copepoda. The phantom midge larvae *Chaoborus flavicans* (Chaoboridae) was also observed.

No information is available on the composition of the fish community during the 1930s, but considering the situation and physical factors, it is reasonable to believe that the lake supported several species of cyprinids, such as roach *(Rutilus rutilus)*. This is also supported by information from owners of fishing rights, who have stated that there was an indigenous population of roach in the lake, while bream *(Abramis brama)* was probably introduced between 1940 and 1950. During the 1960s, the roach disappeared, while large individual bream (about 5 kg) were still being caught in the early 1970s (G. Andersson pers. comm.). The lake has also

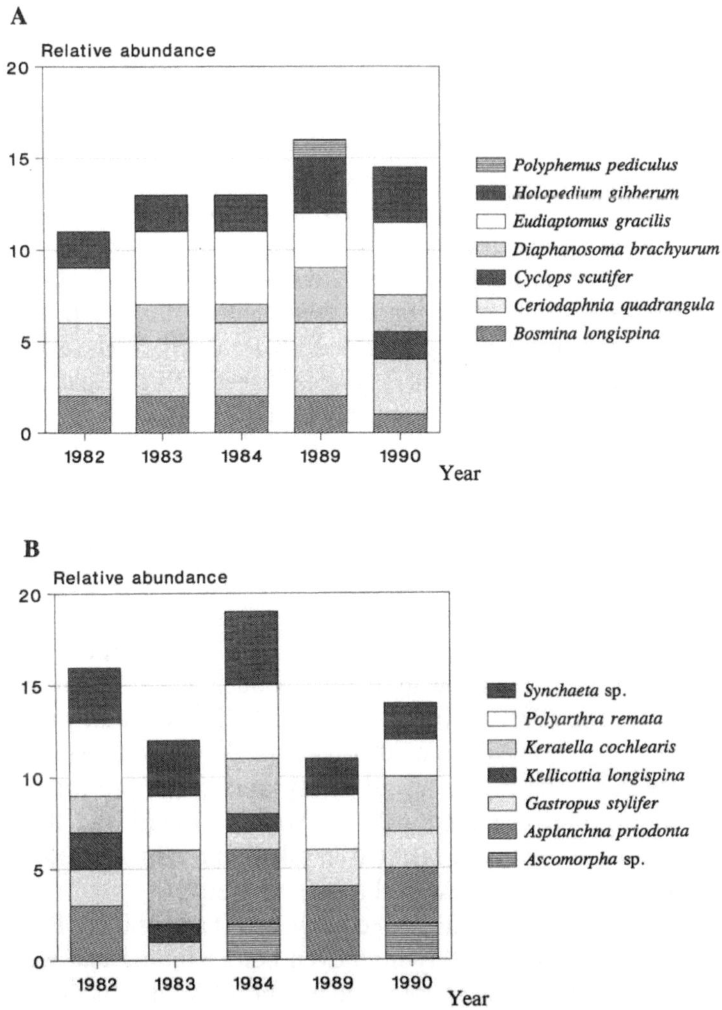

Figure 2. *Relative abundance of zooplankton in Lake Gyslättasjön before and after first liming (1985), based on semi-quantitative sampling. A) Cladocerans and copepods, B) rotifers.*

supported eel *(Anguilla anguilla)*. In test fishing with multi-mesh gillnets in 1983 only European perch *(Perca fluviatilis)* and northern pike *(Esox lucius)* were caught. The perch population, which was abundant at the time, showed no signs of acidification impact.

According to available data the lake was severely damaged by acidification by the early 1980s. The phytoplankton community was dominated by *G. semens*, while the zooplankton community was relatively intact. Acidification had eradicated cyprinids from the fish community.

4. Liming strategy

The initial liming treatment was carried out on 12 March 1985 using 32 tonnes of crushed limestone ($CaCO_3$) (CaO=46.1%) with a grain size of 0-0.25 mm, which was spread by tractor on the ice on the lake. One further treatment was performed on 7 March 1987, with 12 tonnes of crushed limestone being spread on the ice by hand. A more extensive treatment (132.5 tonnes) was performed on 27 October 1988 from the air, covering the discharge areas around Lake Gyslättasjön and the nearby Lake Förhultasjön. At the same time, 45.5 tonnes were spread directly in the lake. The most recent treatment was performed on 10 April 1991, using 1.4 tonnes in the discharge area and 14.3 tonnes directly in the lake. A limestone grain size of 0-0.2 mm was used throughout.

5. Effects of liming

5.1. Water chemistry

After the initial treatment in 1985, pH increased from around 4.8 to over 6.0 during the summer (Fig. 3). This was followed by a decrease in 1986, however, and after a further treatment in 1987 to values which remained fairly stable above 6.0. The 1988 treatment had a beneficial effect on water chemistry and more frequent measurements during 1990 showed that the pH never fell below 6.1. There was no alkalinity during the pre-liming period, but after liming it rose temporarily to over 0.1 meq/l and then decreased again. The 1988 treatment resulted in alkalinity values of up to 0.42 meq/l. During 1990, the figure was above 0.15 meq/l.

After the 1985 treatment, the colour varied between 80 and 160 mg Pt/l. Phosphorus concentration was relatively high even before liming and was not affected by the treatment to any appreciable extent. Oxygen concentrations in 1989 and 1990 were low in the water immediately above the lake bottom during the stagnation period, probably as a result of high levels of organic substances. There is no information on Secchi depth for the period prior to liming, but it varied between 2 and 3.5 m in 1989 and 1990.

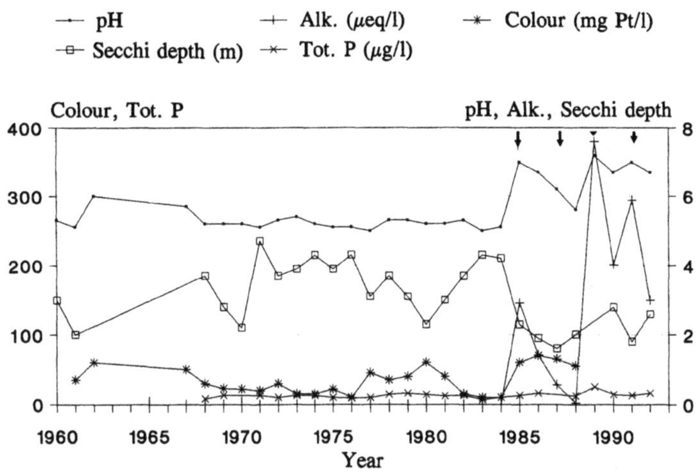

Figure 3. pH, alkalinity, total phosphorus, Secchi depth and water colour in Lake Gyslättasjön. Lime treatments indicated by arrows.

5.2. Phytoplankton

The mean volume of phytoplankton in 1990, the sixth growth season after the initial liming treatment, was 0.63 mm^3/l (0.43 mm^3/l in August). Earlier, however, high values were observed, 4.5 mm^3/l in August 1989, 1.75 in April 1990 and 1.83 in August 1990 (Table 1).

The species composition was mixed and even included Euglenophyceae in the form of *Trachelomonas* spp. *Gonyostomum semens* was particularly common and dominated in terms of volume in 1989. In 1990, Chrysophyceae dominated at about 38 % (usually *Uroglena americana*). Certain phytoplankton genera only appeared after liming, e.g. *Scenedesmus*, *Quadrigula* (Chlorophyceae), *Asterionella*, *Cyclotella* and *Synedra* (Bacillariophyceae). The acidophilic *Dinobryon pediforme* (Chrysophyceae) was succeeded by *D. divergens*. The dominant species varied with time: *G. semen* dominated in 1989, while several species dominated in 1990. The most common of these were *Uroglena*, *Anobaena flos-aquae* and *G. semen*, but *Peridinum cinctum* (Dinophyceae) and several chrysomonads and cryptomonads were also found in abundance.

A total of 30 taxa were observed in August 1989 and 43 in 1990. The main species, in addition to those already mentioned, were *Chryptomonas* spp., *Rhodomonas minuta* (Cryptophyceae), *Dinobryon sociale*, *Mallomonas* spp., and *Synura* sp. (Chrysophyceae) and *Monoraphidium dybowskii* (Chlorophyceae). Some of these species (e.g. *Synura* and *Gonyostomum)* may be considered demanding with regard to nutrient requirements and are indicators of humus.

5.3. Zooplankton

In the 1989-90 studies, 5-6 years after the initial liming treatments, 13 species of rotifers were found, along with 7 cladoceran species, 3 copepod species and *Chaoborus flavicans*. The number of zooplankton taxa had thus increased somewhat compared with the pre-liming situation. The number of cladoceran species showed a particular increase (Fig. 2).

The most common rotifers were *Asplanchna priodonta*, *Polyarthra remata* and *Synchaeta* sp. *Filina longiseta* (occurred during August 1989 and May-June 1990) and *Polyarthra euryptera* (May and June 1990) are seen as indicators of eutrophication. These species were not observed prior to liming. *Ceriodaphnia quadrangula*, *Bosmina longispina* and *Diaphanosoma brachyurum* were the most common cladocerans. Among the copepods, *Eudiaptomus gracilis* was the most abundant.

Certain rotifer species would seem to have suffered from the treatment. *Conochilus unicornis* occurred in samples from 1982-84, but not in 1989 or 1990. *Kelicottia longispina* was found abundantly in 1982-83, but was more sparse in 1984 and 1990 on one occasion. This is quite opposed to the usual case in limed lakes. *Keratella serrulata* also disappeared, which is to be expected since it prefers more acidic environments.

5.4. The littoral benthic fauna

Information on the benthic fauna in the littoral zone is only available for the 1989-91 period, i.e. 5-7 years after the initial liming. The benthic fauna was depleted in terms of species and was characterized in both 1989 and 1991 by the water louse *Asellus aquaticus* (Isopoda) and the non-biting midge larvae (Chironomidae). Colonization by the mayfly (Ephemeroptera) *Caenis* sp. in 1990 meant that the composition of benthic fauna came to resemble that of neutral reference lakes more closely. Despite a certain decline in the total number of individuals, the development of the fauna suggests that the treatments had a cumulative effect. In 1992, however, typical species (for the biotope) were still not found, including the snails (Gastropoda) *Lymnaea peregra* and *Gyraulus* sp. and the mayfly *Ephemera vulgata*.

5.5. Fish

Test fishing with multi-mesh gillnets has been performed in the lake on seven occasions - 1983, 1987 and 1989-93 - using similar methods (Fig. 4). European perch was the dominant species in the lake throughout the period, although northern pike was also found. Roach and bream were not caught either in 1983 (prior to the initial liming) or in 1987. They were reintroduced in 1988 and 1989, which resulted in a rapidly growing roach population and a weak, but growing, bream population. In conjunction with test fishing in 1992, some introduced roach were obtained, as well as a considerable number of roach which were two summers old (hatched 1991). Bream which were four summers old were caught in the pelagic nets. The total biomass was 0.80 kg (15.9 ind)/benthic gillnet and 1.77 kg (52.5 ind)/pelagic net in 1991, which is close to the average for similar lakes (Andersson *et al* 1993). Test fishing in 1993 showed that the roach expansion had ceased, while the bream were increasing steadily in number.

The diet of the fish in Lake Gyslättasjön was studied five years after the initial liming treatment. The perch had eaten relatively few benthic fauna, mainly *A. aquaticus*. The smallest size class of perch had mostly fed on chironomids. No roach smaller than 150 mm were caught for diet analysis. More than half the diet for the largest size class of roach consisted of large benthic fauna. The main part of the diet consisted of caddis flies and mussels. The smaller roach had also fed on cladocerans in addition to benthic fauna, mainly in the form of *Daphnia* sp. and *Bosmina* sp., which confirms that they used the pelagic zone to a large extent.

The growth of perch during their first year decreased from about 75 mm in 1980 to the lowest value of 63 mm in 1987. In 1988, the growth averaged just over 70 mm, which is close to the average of perch in similar lakes in Sweden. The growth of four-year-old perch decreased from 21 mm in 1982 to 14 mm in 1989, which was lower than in comparable lakes. The growth of 8-year-old perch was also low during the years after liming,

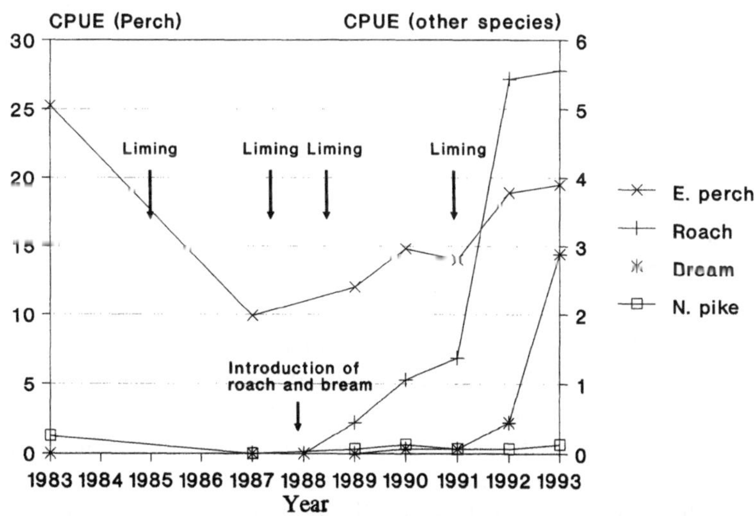

Figure 4. Relative abundance, expressed as CPUE (Catch Per Unit Effort), of different fish species in Lake Gyslättasjön before and after liming.

1986-88 (about 10 mm) but increased to 21 mm in 1989. This increase was probably an effect of high recruitment among re-introduced roach during 1988, providing a greater food supply for piscivorous perch.

Roach were re-introduced into Lake Gyslättasjön in 1988 and fry which hatched in the lake were caught in test fishing for the first time in 1990. Growth analyses showed that the roach which had been introduced into Lake Gyslättasjön grew much better than in their original lake. Growth varied considerably, but was higher on average than in other limed lakes during 1988 and 1989.

6. Conclusions

Lake Gyslättasjön provides an example of a lake which was primarily subjected to humic acids but which became progressively affected by acidification from increasing acid atmospheric deposition. The major element of humic substances has probably had an ameliorating effect on the damage caused by acidification to several groups of organisms. The phytoplankton community had a low species diversity while the zooplankton community appears to have been affected only moderately by the pre-liming acidification. Although roach had disappeared from the lake, the perch population showed no signs of acidification injuries prior to acidification, although the pH fell to low levels at times.

Liming efforts in Lake Gyslättasjön affected the development in the lake to a relatively moderate degree. This applies particularly to the treatments in 1985 and 1987. The most recent treatments, in 1988 and 1991, were more effective and more comprehensive, however, and resulted in far-reaching changes. It is difficult to distinguish the direct effects of liming from those caused by the introduction of roach and bream, however. Both the phytoplankton and zooplankton compositions indicate that the lake has a comparatively high trophic status. Several species which indicate eutrophication were observed in the lake after liming. Species richness among phytoplankton and zooplankton increased after liming. This also applied to fish communities, as a result of restocking. Samples taken in the early 1990s, more than six years after the initial liming treatment, showed that the fauna in the lake is in a phase of expansive development. Benthic faunas, which are susceptible to acidification, are recolonizing the lake, while the fish faunas are undergoing a shift from a community consisting of perch alone to a more typical cyprinid community. It might be expected that the roach and bream populations will stabilize at a relatively moderate level. This will also have a major impact on the perch population, and will benefit piscivorous perch, as well as pike. It is also likely that the expanding roach population will affect the composition of zooplankton, which will in turn affect the phytoplankton.

Nine years after the first liming, it is still too early to predict what the flora and fauna in the lake will be like once the ecosystem has stabilized. Developments suggest, however, that the species composition will be similar to that normally found in neutral, humic forest lakes in the same region.

7. References

Andersson, G. 1971. Vattenkemi och långtidsmässiga förändringar i Anebodaområdets sjöar. [Water chemistry and long-term changes in the lakes of the Aneboda district.] - PhD Thesis, Dep. Limnol., Univ. Lund.

Andersson, H., Aldén, U., Appelberg, M., Bergquist, B. and Ångström, C. 1993. Årsrapport 1992. Resultat från kalkprojektets provfisken. [Annual report 1992. Results of test fishing within the liming project.] - Inform. Inst. Freshw. Res., Drottningholm PM 1:1993.

Andersson, P., Borg, H., Johansson, K. and Kulinski, I. 1982. Effects of acidification on the sedimentation of metals in forest lakes. - Swedish Environm. Prot. Agency, NE-project 25.

Andersson, P., Borg, H., Olsson, B., Nilsson, Å. and Håkansson, L. 1989. Bakgrundstillstånd och genomförda åtgärder i PU-laboratoriets sjöar. Projekt Kalkning-kvicksilver-cesium. [Background situation and measures carried out in the lakes of the PU laboratory. The Project Liming-mercury-caesium.] - Swedish Environm. Prot. Agency 3608.

Appelberg, M. and Aldén, U. 1992. Integrerad uppföljning av kalkningens effekter på sjöar och vattendrag - en treårsrapport. [Integrated monitoring in limed lakes and streams - a three-year study.] - Inform. Inst. Freshw. Res., Drottningholm (4): 1-60.

Borg, H., Andersson, P. and Johansson, K. 1989. Influence of acidification on metal fluxes in Swedish forest lakes. - Sci. Tot. Env. 87/88: 241-253.

Degerman, E. and Nyberg, P. 1987. Fiskfaunans sammansättning och täthet i försurade och kalkade sjöar - en arbetsrapport. [English summary: The composition and abundance of the fish fauna in acidified and limed lakes in Sweden.] - Inform. Inst. Freshw. Res., Drottningholm (7).

Ekström, C. 1992. Underlag till utvärdering av den integrerade kalkningseffektuppföljningen (IKEU). [Basic data for analysis of the Integrated Studies on the Effects of Liming (ISEL).] - Swedish Environm. Prot. Agency, unpubl. rep.

Hörnström, E. 1992. Underlag till utvärdering av den integrerade kalkningseffektuppföljningen (IKEU). [Basic data for analysis of the Integrated Studies on the Effects of Liming (ISEL).] - Swedish Environm. Prot. Agency, unpubl. rep.

Lang, K. 1931. Faunisch-Ökologische Untersuchungen in einigen seichten oligotrophen bzw. dystrophen Seen in Südschweden - mit besonderer Berücksichtung der Profundalfauna. - Lunds Univ., Årsskrift, NF avd. 2, bd 27, No. 8.

Lingdell, P.-E. and Engblom, E. 1992. Underlag till utvärdering av den integrerade kalkningseffektuppföljningen (IKEU). [Basic data for analysis of the Integrated Studies on the Effects of Liming (ISEL).] - Limnodata Consultant Co., Gunnilbo.

Lönnerblad, G. 1931a. Zur kenntnis der Chemie einiger Humusseen. - Arch. Hydrobiol. 22.

Lönnerblad, G. 1931b. Über den Sauerstoffhaushalt der dystrophen Seen. - Lunds Univ. Årsskrift, NF avd. 2 bd 27, No. 14.

Malmer, N. 1961. Some ecological studies on lakes and brooks in south Sweden. - Bot. Notiser 114.

Chapter 14

Liming strategies and effects: the River Högvadsån case study

Ingemar Alenäs[1], Erik Degerman[2] and Lennart Henrikson[3]

[1] Municipality of Falkenberg, S - 311 80 Falkenberg, Sweden
[2] National Board of Fisheries, Local Authority Örebro, Trädgårdsg. 13, S - 712 12 Örebro, Sweden
[3] Municipality of Mark, S-511 80 Kinna, Sweden

1. Documentation

The main stream of River Ätran was developed for power generation in the 1920s. River Högvadsån, a relatively unspoiled tributary, thereby became very important in the reproduction of the genetically unique population of Atlantic salmon *(Salmo salar)*. River Ätran and River Högvadsån are also among the finest waters in western Sweden for recreational fishing (Lindgren 1992).

The salmon population in River Högvadsån has been documented at the Nydala Mill salmon trap in the form of counts of salmon ascending for spawning since 1954 and migratory smolt since 1959 (e.g. Edman 1959, 1983, 1986). The traps do not catch all the fish, so they only provide a relative estimate of the size of the population. Supplementary studies using electro-fishing have been performed over a considerable period of time (Sjöstrand 1988, 1989, Degerman *et al* 1990, National Board of Fisheries 1991, 1992a,b, 1993, Alenäs 1992 a,b, Dellerfors and Faremo 1993). Since 1991, the gill parasite *Gyrodactylus salaris* has been studied on salmon fry (Alenäs 1992a,b, Malmberg 1993, Malmberg and Malmberg 1993). Attention has been drawn to the importance of the conditions under which the salmon grow in the sea in conjunction with occasional smolt deaths caused by algal blooms (Fleischer *et al* 1993) and poor results in terms of returning marked smolt (Karlsson 1994). Crayfish *(Astacus astacus)* has been studied (Appelberg and Odelström 1990) as have the benthic fauna in running water (Engblom and Lingdell 1985, Medin and Oscarson 1989, Medin 1991a,b).

Lake Högsjön and the reference lake, Lake Tjärnesjön, have been moni-

Liming of Acidified Surface Waters
L. Henrikson and Y.W. Brodin (Eds.)
© Springer-Verlag Berlin Heidelberg 1995

tored in the form of special studies of fish, benthic fauna, plankton and glucose breakdown (Johansson 1979, Gahnström *et al* 1980, Chap. 6, Andersson 1985). Sample fishing using survey gillnets has been carried out in several lakes in the system (Edman 1981, Edman *et al* 1988, Halland County Council, unpubl. data). The following only deals with running waters, however. The lakes are not discussed.

Liming in River Högvadsån began on the basis of a 100% state subsidy in 1978 under the auspices of the municipality of Falkenberg. The liming project was the biggest in terms of running waters in Sweden. The project and the findings have been reported in a number of publications (see above, and Bengtsson *et al* 1977, Edman *et al* 1988, Fleischer 1981a,b, 1984, Haux 1981, Svensson 1981, 1983, Henrikson and Oscarson 1983, Sverdrup and Warfvinge 1988, Fritz and Stibe 1991, Carlsson 1992). River Högvadsån is also included in the national ISEL programme (Integrated Studies on the Effects of Liming) (Appelberg and Aldén 1992).

2. Geographical orientation

River Högvadsån belongs to the River Ätran water system, which enters the Kattegatt on the Swedish west coast at the city of Falkenberg (Fig. 1). River Högvadsån enters River Ätran about 30 km upstream of Falkenberg. The catchment area largely consists of coniferous forest, bogs and fens with a low buffer capacity (Table 1). There are relatively few lakes in the area, which means that there is considerable variation in discharge flow (0.6-120 m³/s). Ditching of wetlands in the forests probably contributes to the fluctuations in flow and water chemistry.

***Figure 1.** Geographical location of River Högvadsån.*

***Table 1.** Morphometrical and hydrological data on the River Högvadsån.*

Drainage area (ha)	47,600
Forests and bogs (%)	85
Arable and pasture land (%)	9
Lakes (%)	6
Mean water discharge (m^3/l)	8.5
Precipitation (mm/year)	800-1,200

The production of smolt in River Högvadsån has been estimated at about 40,000 per year. The area which is suitable for them to grow totals about 15 ha (Appelberg *et al* 1989). According to this study, the river accounted for about one-sixth of the total salmon production on the Swedish west coast. Sea trout and stationary brown trout *(Salmo trutta)* are also found in the River Högvadsån water system. Freshwater pearl mussel *(Margaritifera margaritifera)* (Eriksson *et al* 1985) and other rare invertebrates (Medin and Oscarson 1989, Engblom and Lingdell 1985) are also found in the river, along with birds such as dipper *(Cinclus cinclus)* which are otherwise rare in the region. River Högvadsån is of national interest in terms of nature conservation and leisure.

3. Acidification status

The sources of River Högvadsån are situated in areas which are subjected to the highest acid load in Sweden, with a sulphur deposition of 20-25 kg/ha and year, which is considerably higher than the critical load of around 3-5 kg (Nilsson and Grennfelt 1988).

In the 1960s, the salmon population in River Högvadsån was relatively unaffected by acidification, but the impact became increasingly apparent during the late 1970s, with an imminent risk of eradication. The water chemistry in 1972-76 was registered via on-going pH measurements at Nydala Mill, showing the pH to be below 6.0 for long periods and even below 5.0 on a number of occasions in the spring and autumn.

4. Liming strategy

The prime goal of liming treatment is to create acceptable conditions in terms of water chemistry for the salmon in the main stream of River Högvadsån. During the trial period, 1978-81, the strategy consisted of liming via two dosers placed by the main stream (Älvsered) and a tributary, River Skärhultaån (Kärnebygd), and of liming of lakes, arable land, shoreline areas and bogs and fens in the catchment area. Lake Tjärnesjön, which is severely acidified, was reserved as a reference lake.

Limestone ($CaCO_3$) was used. During the initial period of the project, the grain size 0-0.5 mm was used. This was subsequently replaced by a finer powder of 0-0.2 mm. On agricultural land, 0-0.3 mm was used, along with crushed limestone (0-25 mm) in rivers.

Volumes of lime distributed in the various areas are shown in Table 2. The strategy for the second liming period (1982-86) was subsequently revised. The effect of liming on agricultural land, forest soil and bogs and fens was very slight - only 0.8-1.2% of the lime added dissolved into the water system - as was the effect of crushed limestone and dispersal via bulk truck directly into the rivers (cf. Wilander *et al*, Chap. 5). Therefore more extensive dispersal from boats in lakes and via the Älvsered doser was carried out.

The strategy was revised once again for the third liming period, 1987-91, and the amount of lime was increased by 37%. Dispersal from bulk trucks directly into the littoral zones was replaced by lake liming from boats. No liming was performed in River Fageredsån between 1984 and 1988, but in 1989-91, wetlands were limed by helicopter and the larger lakes by boat.

The strategy for the 1992-96 period involved continued lake and wetland liming, supplemented by new wetland liming in the vicinity of the main stream and certain tributaries. As a result, it may be possible to reduce the consumption of lime in the dosers. The strategy for the current period means that the total volume of lime will be increased.

The liming efforts have also been supplemented with fishery conservation measures such as the building of a fish-way at Lia and compensatory re-stocking with salmon fry during the 1970s and 1980s.

Table 2. *Percentage limestone spread and planned to be spread by different methods in the drainage area of River Högvadsån during different periods.*

	1978-81	82-86	87-91	92-96
Truck; lakes	32	12	1	-
Boat; lakes	15	44	47	45
Truck; streams	5	7	-	-
Tractor; coarse limestone in streams	4	-	-	-
Doser; streams	12	37	29	24
Tractor; wetlands and shoreline areas	19	-	-	-
Helicopter; wetlands	-	-	23	31
Tractor; agricultural land	13	-	-	-
Total amount of limestone (tonnes)	10,446	12,001	16,490	19,970

5. Effects of liming

5.1. Water chemistry

Experience gained from the initial project period, 1978-81, showed that the best effect on water chemistry was obtained in conjunction with lake liming and dispersal using dosers. In Lake Högsjön, which was limed by boat, 46.4% of the lime was dissolved. For the dosers, the amount of lime dissolved amounted to 54 and 43.6% respectively. The dosers had considerable initial problems, which affected the continuity of the treatment. Wilander *et al* (Chap. 5) showed, however, that the reliability of the doser at Älvsered increased, i.e. measurements of pH and alkalinity showed values exceeding 6 and 0.1 meq/l respectively for 66% of the time between 1975 and 1987, and for 100% of the time between 1983 and 1987. Of the total amount of lime distributed, 1,945 tonnes, or 18.6%, were estimated to have dissolved during the initial period. Measurements of water chemistry in the reproduction areas for salmon upstream of Nydala Mill during the initial period (1978-81) showed that the objective - pH above 6.0 and alkalinity above 0.05 meq/l - was often not achieved (Fig. 2). Only on one occasion was there no alkalinity.

During the second period (1982-86), the results in terms of water chemistry were somewhat better, although the values measured were still below target. A further improvement was observed during the third period (1987-91). Despite the liming, however, there remains a risk of acid surges, as is apparent from a comparison between measurements of water chemistry and 24-h flow in 1988 (Fig. 3). It is, however, difficult to find evidence of these acid surges in routine measurements. Such surges may have an undesirable biological impact as a result of increased metal concentrations, for instance (Degerman *et al*, Chap. 8).

Synoptic aluminium sampling in 1980 and 1984 showed concentrations of inorganic aluminium of over 0.2 mg/l at several stations. Such concentrations may be harmful to salmon fry. Lower concentrations (0.021-0.084 mg/l) were measured on nine occasions between 1987 and 1991. On these occasions, however, the minimum pH was 6.5, which means that the concentration of inorganic aluminium was probably much higher when the pH fell to 5.5-6.0 in conjunction with a high discharge flow.

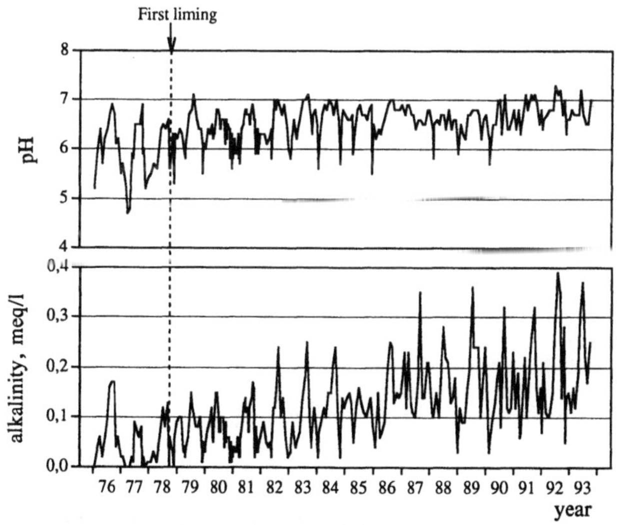

***Figure 2.** pH and alkalinity at Nydala Mill in River Högvadsån, 1978-1988.*

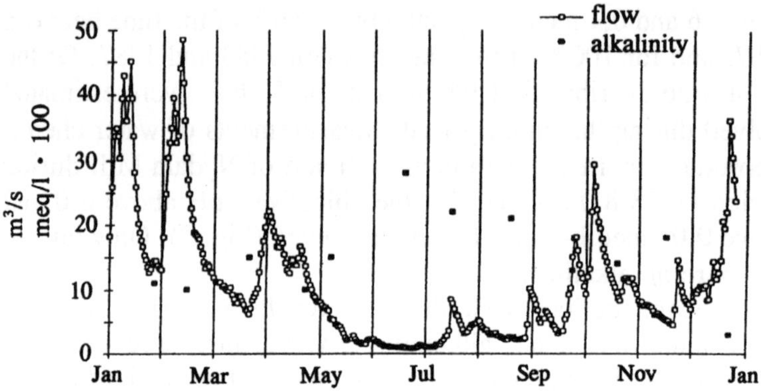

Figure 3. Alkalinity and 24-h mean water flow at Nydala Mill in River Högvadsån during 1988.

5.2. Salmon stock

Catches of ascending salmon at Nydala Mill between 1954 and 1977 - i.e. prior to liming - amounted to a few hundred a year and exceeded 500 on only two occasions (1958 and 1970). Toxic metals in industrial effluent deterred upstream migration for some years during the late 1960s. Catches of ascending salmon increased significantly after liming (ANOVA, $p=0.03$, comparison between the 1955-76 and 1977-90 periods) (Fig. 4). The number of salmon ascending the river to spawn increased until 1987, when over 1,600 were counted at Nydala. Catches of salmon by anglers have also

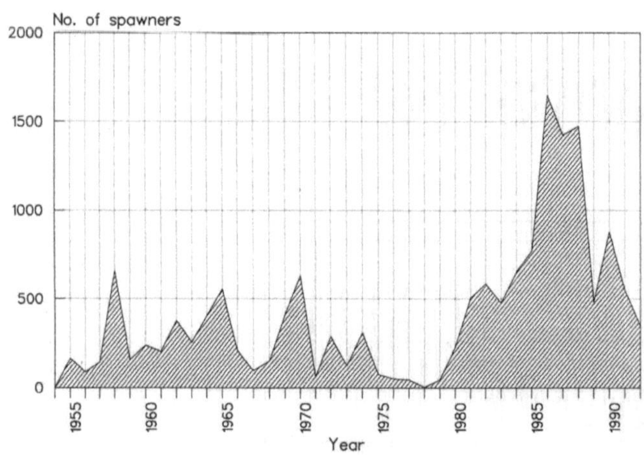

Figure 4. Numbers of spawning salmon at Nydala Mill in River Högvadsån, 1955-1992. First liming occurred in 1978.

improved since liming commenced.

Catches of migrating smolt at Nydala Mill fluctuated widely from year to year during the pre-liming period (1959-77) with figures of between a few hundred and about 10,000. These variations are probably attributable to the fact that the effectiveness of the smolt trap varies depending on the water level. After liming, catches increased and reached about 5,000 (1984-85). There was no significant difference between numbers of smolt migrating downstream before or after liming. During the 1990s, a decline was observed, and the figures fell to relatively low levels. The results of electro-fishing in recent years show good covariation with data from the salmon trap and significant (linear regression, $p<0.02$) increases in salmon fry density since the start of liming (Fig. 5). Since the density of salmon fry and the number of salmon migrating upstream have increased, it is quite possible that the number of smolt leaving the river has also increased, although the ineffectiveness of the smolt trap makes this impossible to prove.

Other species, such as European minnow *(Phoxinus phoxinus)*, benefited from the increased alkalinity (linear regression, $p<0.01$). It should be noted that the above-mentioned increase in the fish population has not been observed in unlimed watercourses on the Swedish west coast, which means that climatic effects can be excluded (Degerman *et al* 1985, 1990).

During the 1990s, occasional decreases in the numbers of fry, smolt and spawning fish have been observed despite increased liming efforts and improved pH and alkalinity. The reason for this decline is unclear. One serious problem may be that ongoing acidification is causing increased

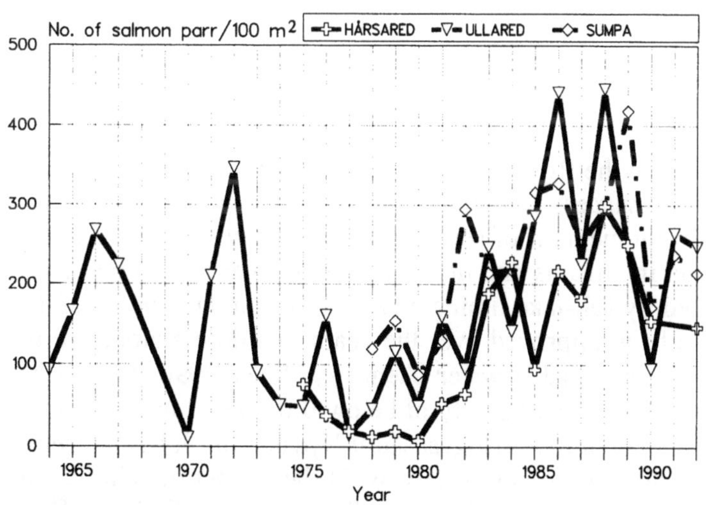

Figure 5. *Abundance of salmon parr at three sites in River Högvadsån, 1965-1992. First liming occurred in 1978.*

leaching of toxic metals to the watercourses, which may be particularly vulnerable during dry periods. The summer of 1992 and the early summer of 1993 were extremely dry. In combination with water regulation and water extraction, increased predation and metal oxidation, this may have done much to counteract the positive effects of liming. The densities of salmon fry in nearby River Fylleån were even lower in 1993, which suggests that the effects apply to the west coast salmon populations as a whole. Poorer conditions for growth and algal blooms in the sea have also played a part in the decline of recent years. It will be up to future studies to show whether there is any link between infection with the gill parasite *Gyrodactylus salaris* (studies indicate that River Högvadsån has a higher frequency of infection than any other rivers in Sweden) and declining numbers of salmon fry.

In 1987 a number of factors combined to exert a positive influence. The effect in 1992 was quite the opposite. There may thus be considerable fluctuation in upstream migration of salmon or the densities of salmon fry from year to year. In the short term, this may overshadow the effects of liming treatments. In River Högvadsån, however, the single most positive factor affecting the salmon population has been the liming of the river itself.

Sea trout did not increase to the same extent as the salmon after liming, which can probably be explained by the increased interspecies competition from the larger salmon population (Degerman and Appelberg 1992) and that the liming strategy was, from the start, aimed at improving conditions for the salmon and therefore did not include smaller streams and watercourses.

The crayfish population increased as soon as pH improved (Fig. 6). In the spring of 1992, however, dead crayfish were found and electro-fishing thereafter yielded nothing. There are fears that River Högvadsån is infected with parasitic mould, but analyses have been unable to confirm this.

6. Conclusions

The liming strategy in the River Högvadsån water system has been revised several times, with successive improvements of the water quality as a result. The inefficient methods of spreading lime on forest and agricultural land, crushed lime in streams and directly into the lake using bulk trucks have been replaced by full-lake liming and wetland liming. The reliability of dosers has also been improved over the years, but dosers alone are not sufficient to achieve acceptable water chemistry. Since 1990, the pH has not fallen below 6.0 and no alkalinity of below 0.05 meq/l has been measured in the main stream of River Högvadsån at Nydala Mill. It has, however, proven to be very difficult to completely avoid the impact of acidification in River Högvadsån, as a result of the major fluctuations in discharge flow. One might suspect that limited acid surges still occur, although this has not been proved despite detailed measurements of water chemistry.

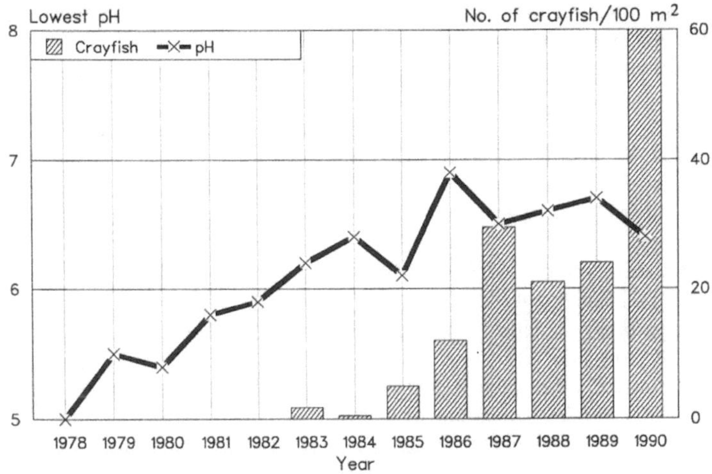

Figure 6. Minimum pH and abundance of crayfish at Ullared in Högvadsån, 1978-1990. First liming occurred in 1978.

Liming has been successful in improving the water chemistry for salmon. During the first ten years after the initial treatments, the salmon population showed gradual recovery, although there has been a decline in recent years. It has not been possible to relate this decline to shortcomings in liming, however, since it is more evident in the main stream, with its better water quality i.e. buffer capacity, than in the tributaries. The decline may well be caused by a combination of other natural or anthropogenic factors.

Finally, it is worth noting that liming of the River Högvadsån water system has been profitable in socio-economic terms, which is quite clear when the liming costs are seen in relation to the recreational value of the salmon and salmon fishing (Bengtsson and Bogelius, Chap. 18).

7. References

Alenäs, I. 1992a. Sammanställning av elfisken Falkenbergs kommun 1991. [Compilation of electro-fishing in the municipality of Falkenberg during 1991.] - Municipality of Falkenberg, Rep. 1992:1.

Alenäs, I. 1992b. Sammanställning av elfisken Falkenbergs kommun 1992. [Compilation of electro-fishing in the municipality of Falkenberg during 1992.] - Municipality of Falkenberg, Rep. 1992:3.

Andersson, G. 1985. Decomposition of alder leaves in acid lake waters. - Ecol. Bull. (Stockholm) 37: 293-299.

Appelberg, M. and Aldén, U. 1992. Integrerad uppföljning av kalkningens effekter på sjöar och vattendrag - en treårsrapport. [Integrated monitoring in limed lakes and streams - a three-year study.] - Inform. Inst. Freshw. Res., Drottningholm (4): 1-60.

Appelberg, M. and Odelström, T. 1990. Kräftor i sura och kalkade vatten. [English summary: Effects of acidification and liming on the freshwater crayfish *Astacus astacus*.] - Inform. Inst. Freshw. Res., Drottningholm (4): 1-25.

Appelberg, M., Degerman, E., Johlander, A. and Karlsson, L. 1989. Liming increases the catches of Atlantic salmon on the west coast of Sweden. - Nordic J. Freshw. Res. 65: 44-53.

Bengtsson, B., Dickson, W. and Fleischer, S. 1977. Högvadsån. Försurning, kalkning och uppföljning. [River Högvadsån. Acidification, liming and monitoring.] - Reg. Fish. Office of the Nat. Board Fish., Swedish Envvironm. Prot. Agency and Halland County Council, unpubl. rep.

Carlsson, U. 1992. Kalkningsprojekt Högvadsån 1987-1991. [The River Högvadsån liming project 1987-1991.] - Municipality of Falkenberg, Rep. 1992:2.

Degerman, E. and Appelberg, M. 1992. The response of stream-dwelling fish to liming. - Environm. Poll. 78: 149-155.

Degerman, E., Fogelgren, J.E., Tengelin, B. and Thörnelöf, E. 1985. Förekomst och täthet av lax, öring och ål i mindre vattendrag på svenska västkusten. [English summary: Occurrence of salmonid parr and eel in relation to water quality on the west coast of Sweden.] - Inform. Inst. Freshw. Res., Drottningholm (1).

Degerman, E., Sjölander, E., Johlander, A., Sjöstrand, P., Höglind, K., Thorsson, L. and Carlstrand, H. 1990. Kalkning för att motverka försurningpåverkan på fisk i rinnande vatten. [English summary: Liming of acidified running waters to improve conditions for fish.] - Inform. Inst. Freshw. Res., Drottningholm (4): 27-214.

Dellerfors, C. and Fahlemo, U. 1993. Fiskeribiologisk undersökning inom Högvadsåns kalkningsprojekt. [Fishery-biological study within the River Högvadsån liming project.] - Municipality of Falkenberg 1993:2.

Edman, G. 1959. Redogörelse för märkning av uppvandrande laxöring vid Herting i Ätran åren 1945-50 samt återfångster därav. [Report on marking of migrating trout at Herting in River Ätran 1945-50, and recaptures.] - Svensk Fiskeri Tidskrift 1959: 98-103.

Edman, G. 1981, Resultat av provfisken inom Högvadsåprojektet. [Report on test fishing within the River Högvadsån liming project.] - In: Johansson, B. and Edman, G. (Eds.) The River Högvadsån liming project 1978-1981. Municipality of Falkenberg, Rep. 1981, Suppl. 4.

Edman, G. 1983. Redogörelse för fiskeriundersökningarna inom Ätrans vattensystem i anslutning till kalkningsprojekt inom Högvadsåns avrinningsområde under tiden 1978-1981. [Report on fishery studies in the water system of River Ätran in conjunction with the River Högvadsån liming project during 1978-1981.]. - In: Peterson, A., Johansson, B. and Edman, G. (Eds.) The River Högvadsån liming project 1978-1981. Municipality of Falkenberg, Rep. 1983.

Edman, G. 1985. Delrapport för Högvadsån kalkningsprojekt 1982-1986. Delrapporten omfattar 1982-1984. [Report on the River Högvadsån liming project during 1982-1986. The report includes 1982-1984.] - Municipality of Falkenberg, Rep. 1985.

Edman, G. 1986. Redogörelse för fiskeribiologiska undersökningar i Högvadsån 1985. [Report on fishery-biological studies in River Högvadsån during 1985.] - Municipality of Falkenberg, Rep. 1986.

Edman, G. and Fleischer, S. 1980. The River Högvadsån liming project - a presentation. - In: Drabløs, D. and Tollan, A. (Eds.) Proc. Int. Conf. Ecol. impact acid precip., Norway, SNSF project., Oslo-Ås, pp. 300-301.

Edman, G., Fleischer, S., Fritz, Ö. and Stibe, L. 1988. Högvadsån 1978-1986. Försurad - Kalkad - Pånyttfödd. [English summary: River Högvadsån. Acicified - limed - regenerated.] - Halland County Council, Rep. 1988:4.

Engblom, E. and Lingdell, P.-E. 1985. Hur påverkar kalkdoserare bottenfaunan? [How do lime dosing devices affect the benthic fauna?].- Swedish Environm. Prot. Agency 1994.

Eriksson, M.O.G., Henrikson, L. and Oscarson, H.G. 1985. Flodpärlmusslan i Hallands län 1984. [The freshwater pearl mussel in the county of Halland, 1984.] - Halland County Council, Rep. 1985:13.

Fleischer, S. 1981a. Resultat från fysikalisk-kemiska undersökningar inom Högvadsåprojektet. [Results of physical-chemical studies within the River Högvadsån project.] - In: Johansson, B. and Edman, G. (Eds.) Report on the River Högvadsån liming project 1978-1980. Municipality of Falkenberg, Rep. 1981, Suppl. 2.

Fleischer, S. 1981b. Organismundersökningar inom Högvadsåprojektet. [Studies on organisms within the River Högvadsån project.] - In: Johansson, B. and Edman, G. (Eds.) Report on the River Högvadsån liming project 1978-1980. Municipality of Falkenberg, Rep. 1981, Suppl. 3.

Fleischer, S. 1984. Högvadsån Kalkningsprojekt 1978-1981, III. Vattenkemiska undersökningar. [The River Högvadsån liming projekt 1978-1981. III. Water chemistry studies.] - Municipality of Falkenberg, Rep. 1984.

Fleischer, S., Andersson, G., Brodin, Y.-W., Dickson, W., Herrmann, J. and Muniz, I.P. 1993. Acid water research in Sweden - knowledge for tomorrow? - Ambio 22: 258-263.

Fritz, Ö. and Stibe, L. 1991. Aluminium i Högvadsån, en studie för val av kommande kalkningsstrategier. [Aluminium in River Högvadsån, a study before deciding on future liming strategies.] - Halland County Council, PM 1991-01-17.

Gahnström, G., Andersson, G. and Fleischer, S. 1980. Decomposition and exchange processes in acidified lake sediment. - In: Drabløs, D. and Tollan, A. (Eds.) Proc. Int. Conf. Ecol. impact acid precip., Norway, SNSF project, Oslo-Ås, pp. 306-307.

Haux, C. 1981. Utvärdering av preliminärstudie av jonförändringar i laxyngel från Högvadsån och Stensån 1978-1980. [Evaluation of a preliminary study on ion changes in salmon parr from River Högvadsån and River Stensån, 1978-1980.] - In: Johansson, B. and Edman, G. (Eds.) Report on the River Högvadsån liming project 1978-1980. Municipality of Falkenberg, Rep. 1981, Suppl. 7.

Henrikson, L. and Oscarson, H.G. 1983. Högvadsån - Laxungarna och deras födoval. [River Högvadsån - the salmon parr and their food choice.] - In: Peterson, A., Johansson, B. and Edman, G. (Eds.) The River Högvadsån liming project, 1978-1981. Municipality of Falkenberg, Rep. 1983.

Johansson, P.-A. 1979. Nedbrytning av al-löv, makrobentossammansättning samt transport av organiskt material i tre halländska bäckar. [Breakdown of alder leaves, composition of macrobenthos and transportation of organic matter in three streams in the county of Halland.] - Application project, Dep. Limnol., Univ. Lund.

Karlsson, L. 1994. Beståndssituationen i Sverige. [The situation for Swedish populations.] - Salmon symposium, May 1994, Falkenberg, Sweden. Swedish Com. Preserv. Atlantic Salmon. Abstract.

Lindgren, B. 1992. Där laxen vandrar. [Where the salmon migrates.] - Settern Publ. Comp., Örkelljunga, Sweden.

Malmberg, G. 1993. Undersökningar angående *Gyrodactylus* på laxfisk i fria vatten och odlingar under åren 1951-72 och 1986-maj 1991. [Studies of *Gyrodactylus* on salmonids in open water and farms during 1951-72 and 1986-May 1991.] - Inform. Inst. Freshw. Res., Drottningholm (2):1-30.

Malmberg, G. and Malmberg, M. 1993. Species of *Gyrodactylus* (Platyhelminthes, Monogena) on salmonids in Sweden. - Fish. Res. 17: 59-68.

Medin, M. 1991a. Bottenfaunan vid tre lokaler i Högvadsån våren 1991, en biologisk bedömning av miljön. [The bottom fauna at three localities in the River Högvadsån during spring 1991, a biological assessment of the environment.] - Medins Sjö- och Åbiologi Consultant Co., Mölnlycke.

Medin, M. 1991b. Bottenfaunan på tre lokaler i Fageredsån och Skrockån våren 1991. [The bottom fauna at three localities in the streams Fageredsån and Skrockån during spring 1991.] - Medins Sjö- och Åbiologi Consultant Co., Mölnlycke.

Medin, M. and Oscarson, H. 1989. Bottenfaunan i Högvadsåns vattensystem 1988. [The bottom fauna in the water system of River Högvadsån during 1988.] - Halland County Council, Rep. 1989:5.

NBF (National Board of Fisheries) 1991. Uppföljande fiskeriundersökningar 1990 inom Högvadsåns kalkningsprojekt. [Fishery monitoring studies during 1990 within the River Högvadsån liming project.] - Nat. Board Fish., Jönköping Reg. Fish. Off., Rep.

NBF (National Board of Fisheries) 1992a. Uppföljande fiskeriundersökningar 1987-91 inom Högvadsåns kalkningsprojekt. [Fishery monitoring studies during 1987-91 within the River Högvadsån liming project.] - Nat. Board Fish., Jönköping Reg. Fish. Off., Rep.

NBF (National Board of Fisheries) 1992b. Sammanställning av uppföljande fiskeriundersökningar 1991 inom Högvadsåns kalkningsprojekt. [Compilation of fishery monitoring studies during 1991 within the River Högvadsån liming project.] - Nat. Board Fish., Jönköping Reg. Fish. Off., Rep.

NBF (National Board of Fisheries) 1993. Uppföljande elfiskeundersökningar 1992 inom Högvadsåns kalkningsprojekt. [Electro-fishing monitoring studies during 1987-91 within the River Högvadsån liming project.] - Nat. Board Fish., Jönköping Reg. Fish. Off., Rep.

Nilsson, J. and Grennfelt, P. (Eds.) 1988. Critical loads for sulphur and nitrogen. - Nordic Counc. Minist. Miljørap. 1988:15.

Sjöstrand, P. 1988. Sammanställning av fiskeriundersökningar 1987 inom Högvadsåns kalkningsprojekt. [Compilation of fishery monitoring studies during 1987 within the River Högvadsån liming project.] - Nat. Board Fish., Gothenburg Reg. Fish. Off., Rep.

Sjöstrand, P. 1989. Sammanställning av uppföljande fiskeriundersökningar 1988 inom Högvadsåns kalkningsprojekt. [Compilation of fishery monitoring studies during 1987 within the River Högvadsån liming project.] - Nat. Board Fish., Gothenburg Reg. Fish. Off., Rep.

Svensson, B.-A. 1981. Praktiska problem och erfarenheter vid Högvadsåprojektets genomförande. [Practical problems and experience in carrying out the River Högvadsån liming project.] - In: Johansson, B. and Edman, G. (Eds.) Report on the River Högvadsån liming project 1978-1980. Municipality of Falkenberg, Rep. 1981, Suppl. 1.

Svensson, B.-A. 1983. Kalkningsprojekt Högvadsån. Praktiskt genomförande och erfarenheter 1978-1981. [The River Högvadsån liming project. Practical performance and experiences 1978-1981.] - In: Peterson, A., Johansson, B. and Edman, G. (Eds.) The River Högvadsån liming project 1978-1981. Municipality of Falkenberg, Rep. 1983.

Sverdrup, H. and Warfvinge, P. 1988. Lake liming in different types of acid lakes using various types of calcite powders and methods. - Water Air Soil Poll. 41: 189-222.

Chapter 15

Liming strategies and effects: the Prästvallsbäcken Stream case study

Paul Andersson

Institute of Applied Environmental Research, Stockholm University, S - 171 85, Solna, Sweden

1. Documentation

Limnological measurements have been carried out in Prästvallsbäcken Stream and the Lake Källsjön catchment area since the mid-1970s. Since 1983, the catchment area has been included in various programmes for the study of liming effects on chemical and biological conditions. The hydrological conditions have been described in Lakes Lorttjärn and Källsjön (Andersson and Kulinski 1985) and studies have been published on phytoplankton, zooplankton, water chemistry (Hörnström *et al* 1992), benthic fauna (Lindgren 1994), fish biology and mercury in fish (Andersson *et al* 1987, Andersson 1989, Andersson 1991, Appelberg and Aldén 1992).

In regard to Prästvallsbäcken Stream, briefer reports have been published on water chemistry and benthic fauna (Engblom and Lingdell 1983) and on the condition and development of the fish population (Andersson and Andersson 1984, Appelberg and Aldén 1992). Prästvallsbäcken Stream is also part of the national ISEL programme (Integrated Studies on the Effects of Liming) (Appelberg and Aldén 1992).

2. Geographical orientation

Prästvallsbäcken Stream is situated in the River Nianån water system in central Sweden (Fig. 1). Its highest tributary starts 350 m above sea level and reaches a level of 272 m above sea level about 7 km later when it reaches Lake Lorttjärn. In its last steep km down to Lake Källsjön, with natural migration barriers to fish other than brown trout *(Salmo trutta)*, it

Liming of Acidified Surface Waters
L. Henrikson and Y.W. Brodin (Eds.)
© Springer-Verlag Berlin Heidelberg 1995

falls to 232 m above sea level. At the sampling site just downstream of Lake Lorttjärn the average flow is 80 l/s, fluctuating from less than 5 to 2,000 l/s. Two-thirds of the bedrock in the drainage basin consists of acidic types. The remaining third consists equally of intermediate and basal types, while there is a lack of more calcium-rich types. The soil consists mainly of moraine. Bogs and fens constitute approximately one quarter of the drainage basin, which is otherwise covered in forest which consists mainly of spruce and pine less than 40 years old. Table 1 provides detailed information about the drainage basin.

Figure 1. Geographical location of Prästvallsbäcken Stream.

Table 1. Morphometrical and hydrological data from two sampling sites in the Prästvallsbäcken Stream. Area cover percentage for forests, wetlands and lakes are shown.

Site	Finnbaracken	Militärvägen
m above sea level	275	270
Drainage area (ha)	611	850
Runoff (l/s)	60	80
Forest (%)	73	76
Wetlands (%)	26	23
Lakes (%)	<1	<1

3. Acidification status

Several lakes and running waters in the surroundings of the Prästvallsbäcken Stream have taken on a much browner colour over the past 30-40 years. This has coincided with the clear-cutting and ditching of large areas of the drainage basins. Lake Källsjön is now brown, its name referring to a clear-water lake, however. Deterioration of the water quality, particularly during the spring seasons, became apparent in the 1970s, when northern pike *(Esox lucius)* caught in nets in the lake during the spawning season were often dead.

In Prästvallsbäcken Stream, the density of brown trout has gradually declined since the early 1960s. Fry were still found in the lower reaches of the stream in 1978, and in 1984 also in a section close to a groundwater spring in the upper reaches. Scarcity of older fish showed that recruitment had been poor for several years. High density of older fish at the outlet from Lake Källsjön in 1984 also suggests that the trout migrated downstream in order to avoid the poorer water quality upstream of the lake.

The first measurements from the stream during the spring thaw in 1976 and in August 1977 indicated acidic conditions. The acidic water was associated with high runoff with low pH and high aluminium concentrations. At Finnbaracken the pH declined by more than one unit with only a moderate increase in runoff. The difference in acidity between the highest and the lowest discharge flow was almost three pH units (Fig. 2). The water quality was better when the discharge was lower.

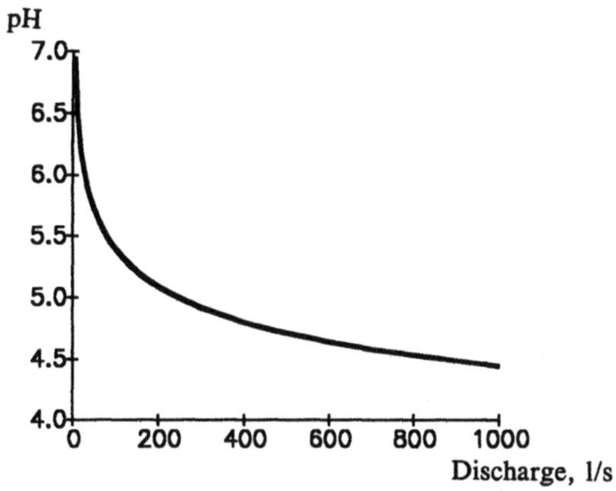

***Figure** 2. The relationship between pH and runoff prior to liming at the Finnbaracken site in the Prästvallsbäcken Stream.*

4. Liming strategy

The aim of liming was to avoid low pH at high water runoff in Prästvallsbäcken Stream and to improve the buffer capacity of Lake Källsjön. The project began with liming by helicopter during the autumn of 1984. The treatment has since been repeated every second autumn (October--November), usually after the first snow has settled. On each occasion, about half of the lime was spread on defined wet areas and the other half in a zone along the stream in areas where the shoreline was damp. The breadth of the zones on the first occasion was 2x10 m (a total of 15 ha), 2x20 m (22 ha) on the second and 2x40 m (36 ha) on the two most recent occasions. Limestone ($CaCO_3$) of bedrock or sedimentary type was used

(CaO content 53-54%) with a grain size of 0-0.02 mm. A total of 425-450 tonnes was used in the first three treatments, while the most recent treatment involved only 212 tonnes. This corresponds to dosages of 31.5, 20.5, 12.5 and 6.4 t/ha repectively. At the Finnbaracken sampling site the treatments corresponded to entire drainage area doses of 0.7 t/ha for the first three occasions and half of it the last time.

5. Effects of liming

Liming resulted in an increase in pH and reduced fluctuations in pH with the water flow (Fig. 3A). The highest values were consequently observed in the vicinity of the limed area and declined downstream as acidic water flowed in. Like the effect of each re-liming treatment, this is most clearly seen as an average annual pH (Fig. 3B). Such apparent effects have not been noted for chemical parameters indirectly dependent on the lime dose. There has been little or no change in the water colour. Among the potentially toxic metals such as aluminium and iron, the levels were 30-40% lower during the first years of liming than in 1983-84 (Fig. 4). Still lower levels were measured in 1989 when the runoff was extraordinary low. The same pattern is also found in manganese and zinc. The lower levels may be attributable to fixation of the metals in the limed areas around the stream, where they may constitute a dangerous store from which they can be leached out if the acidity rises again - perhaps temporarily in higher concentrations than if the stream had never been limed.

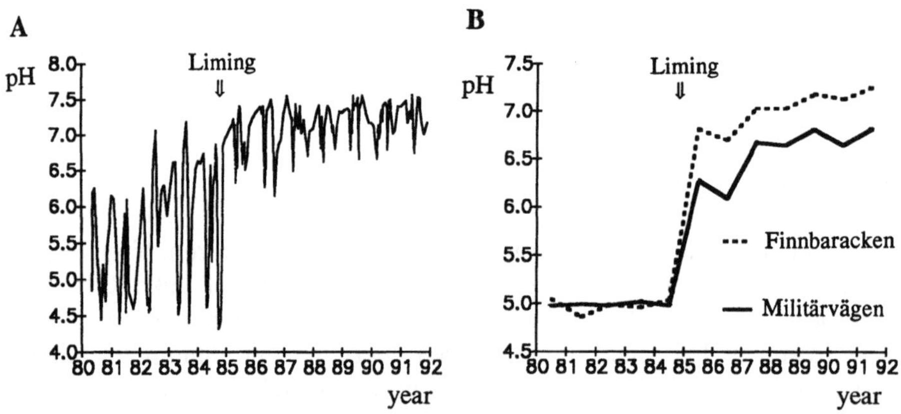

Figure 3. pH in the Prästvallsbäcken Stream before and after liming which commenced in autumn 1984 and then every second year after that. A) pH changes at the Finnbaracken site and B) yearly mean pH at the Finnbaracken site and Militärvägen site further downstream.

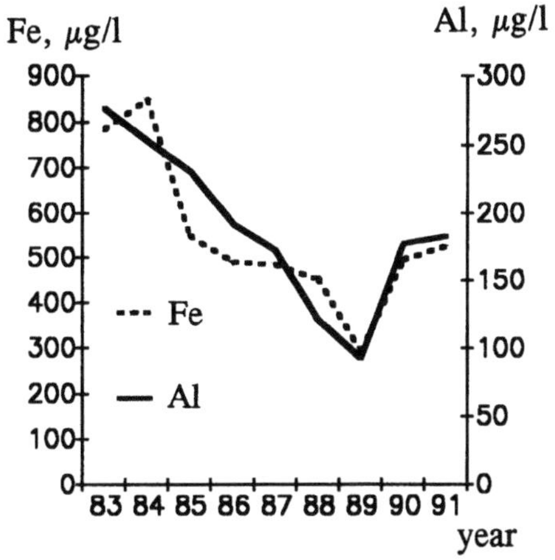

Figure 4. Concentrations of aluminium and iron at the Finnbaracken site in Prästvallsbäcken Stream before and after liming commenced (1984).

The benthic fauna has successively improved, with successive colonization by species which demand good water quality. Before liming, the fauna was characterized by tolerant species such as the mayflies *Leptophlebia vespertina* and *L. marginata* (Ephemeroptera). Afterwards, the total number of species increased, with colonization by mayflies such as *Metretopus* sp., which are susceptible to acidification, and a number of *Baetis* species. The latter represent a very important food for growing brown trout. Before liming, no individuals of *Baetis* were observed, neither in summer nor winter or autumn. The number of individuals has varied widely since treatment, indicating unstable water chemistry (Fig. 5).

Initially, the main species to colonize consisted of those which survive as eggs during the winter and grow into adults during the summer. In 1991 and 1992, the proportions of *Baetis rhodani* and *B. niger* increased. These exist as larvae during the winter. *B. rhodani* can be found as larvae of varying sizes throughout the year, which means that this species, unlike the summer species, can serve as food for the brown trout all year-round. The fluctuations in the numbers of *Baetis* species and individuals are probably attributable to their not having stabilized after liming and to the site periodically being affected by acidic water from unlimed ditches.

Higher and more stable pH values were reflected in better recruitment among the brown trout. In 1984, four fries were caught at a site in the upper part of Prästvallsbäcken, all of them close to the groundwater spring. In 1986 the number had increased to 100 and the fries were dispersed along the entire test fishing stretch. For the stream as a whole, there was a density increase of between three- and fivefold (Fig. 6A). By autumn of 1992, almost the entire stream system had been colonized by trout of various ages (Fig. 6B). The size composition has varied between years and fishing sites. As a result of competition and predation, the abundance of fries was often lower at high abundance of older fish.

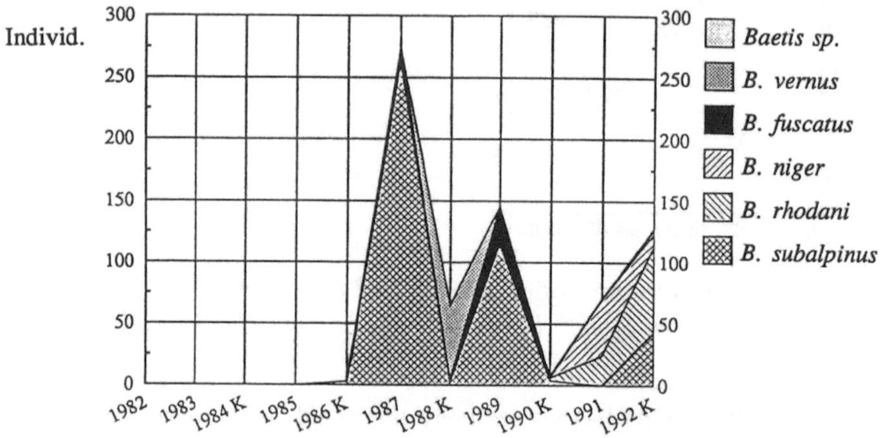

Figure 5. Baetis mayfly larvae caught (30 samples per occasion) at a site 700 m upstream of the Finnbaracken site in the Prästvallsbäcken Stream before and after liming (K = years when liming occurred).

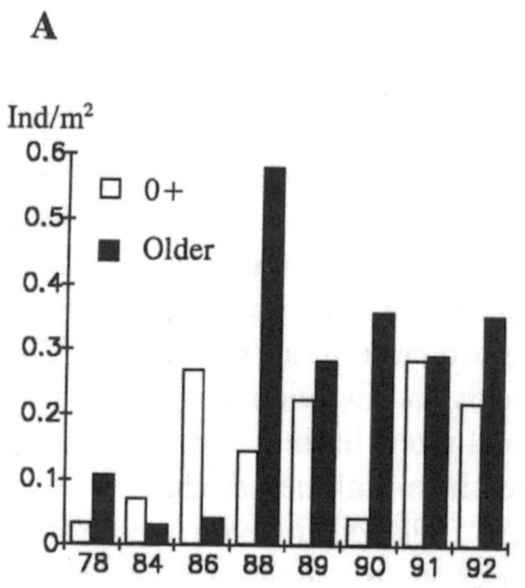

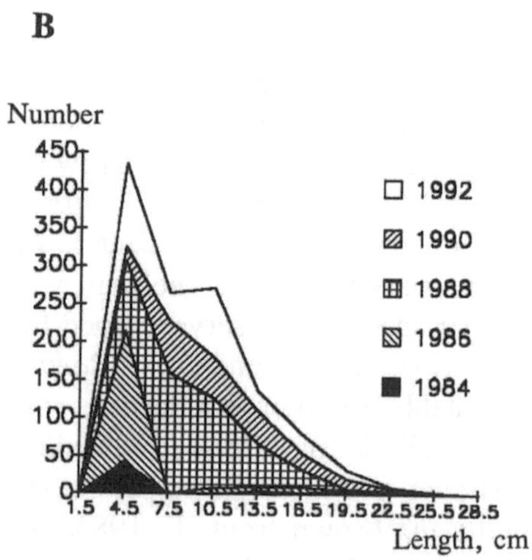

***Figure 6.** Brown trout in the Prästvallsbäcken Stream. A) Mean abundance of one-summer-old (0+) and older brown trout at two sites (1978) and four sites (other years) before (1978 and 1984) and after (1986-1992) liming. B) Size distribution among brown trout 1984-1992.*

6. Conclusions

The studies in Prästvallsbäcken Stream have shown that:
- higher and more stable pH values were obtained almost immediately downstream of the limed area,
- concentrations of aluminium and iron have declined,
- the number of benthic fauna species increased, while the number of individuals varied from isolated occurrences to levels which may be considered normal for a watercourse of this type,
- brown trout reproduction improved immediately, which was apparent in the form of an increased number of fry and a subsequent colonization of previously empty stretches of the stream by adult fish.

7. References

Andersson, B. and Andersson, P. 1984. The distribution of trout (*Salmo trutta* L.) in relation to pH - an inventory of small streams in Delsbo, central Sweden. - Rep. Inst. Freshw. Res., Drottningholm 61: 28-33.

Andersson, P. and Kulinski, I.P. 1985. Sjöar i Delsboområdet - Hydrologiska basuppgifter och lodkartor. [Lakes in the Delsbo district - hydrological data and depth charts.] - Swedish Environm. Prot. Agency 2007.

Andersson, P., Borg, H., Holmgren, K. and Håkansson, L. 1987. Typsjöar och tillrinnings-områden i projektet Kalkning - kvicksilver vid naturvårdsverkets PU-lab. [Type lakes and drainage areas of the liming-mercury project at PU laboratory, Swedish Environm. Prot. Agency.] - Swedish Environm. Prot. Agency 3396.

Andersson, P., Borg, H., Olsson, B., Nilsson, Å. and Håkansson, L. 1989. Bakgrundstillstånd och genomförda åtgärder i PU-laboratoriets sjöar. [Background status and performed measures in lakes of the PU laboratory.] - Swedish Environm. Prot. Agency 3608.

Andersson, P., Borg, H. and Kärrhage, P. 1991. Kalkningseffekter på kvicksilver i fisk. Läget efter genomförda åtgärder i 19 typsjöar. [Effects of liming on mercury in fish. The status after measures in 19 type lakes.] - Swedish Environm. Prot. Agency 3883.

Appelberg, M. and Aldén, U. 1992. Integrerad uppföljning av kalkningens effekter på sjöar och vattendrag - en treårsrapport. [Integrated monitoring in limed lakes and streams - a three-year study.] - Inform. Inst. Freshw. Res., Drottningholm (4): 1-60.

Engblom, E. and Lingdell, P.-E. 1983. Bottenfaunans användbarhet som pH-indikator. [The use of bottom fauna as indicator of pH.] - Swedish Environm. Prot. Agency 1741.

Håkanson, L., Andersson, P., Andersson, T., Bengtssson, Å., Grahn, P., Johansson, J.-Å., Jönsson, C.-P., Kvarnäs, H., Lindgren, G. and Nilsson, Å. 1990. Åtgärder mot höga kvicksilverhalter i insjöfisk. Slutrapport för kvicksilverdelen av projektet: Kalkning-Kvicksilver-Cesium. [Measures to reduce mercury in lake fish. Final report from the project Liming-Mercury-Caesium.] - Swedish Environm. Prot. Agency 3818.

Hörnström, E., Ekström, C. and Andersson, P. 1992. Tio mellansvenska sjöar. Kalkningseffekter på plankton och vattenkemi. [English summary: Ten lakes in central Sweden, liming effects on plankton and water chemistry.] - Swedish Environm. Prot. Agency 4048.

Lindgren, G. 1994. Bottenfaunan i kalkade sjöar i Delsboområdet 1983–91. [The bottom fauna of limed lakes in the Delsbo district.] - Gävleborg County Council, unpubl. rep.

Chapter 16

Liming effects on mercury concentrations in fish

Markus Meili

Uppsala University, Institute of Earth Sciences, Norbyvägen 18 B, S - 752 36 Uppsala, Sweden

Summary

A variety of different liming operations were conducted in 46 Swedish lakes with a mean summer pH between 5.0 and 7.0. This report deals with the effect on mercury concentrations in pike *(Esox lucius)*. Before treatment, mercury concentrations in 1-kg pike (HgPike) varied between 0.3 and 2.0 μg/g (wet weight). The concentrations were highest in acidic lakes, but lowest in the lakes of the southernmost county, where atmospheric deposition both of Hg and acid is highest. After 1-2 reference years, half of the lakes were subjected to lake liming and half to watershed liming, using varying dosage and dispersal strategies. Water chemistry and pike Hg concentrations were monitored during the following 5-6 years. Substantial changes in HgPike were observed during this period. Both increases and decreases occurred (-40 to +60%), as well as considerable interannual variations of a systematic or apparently random type (about ±20%). While no overall change in HgPike could be observed during the first 1-2 years after liming, concentrations often declined during the following years, but only in lakes where liming had a substantial impact on water chemistry. After effective lake liming, HgPike decreased by about 20% on average, whereas after equally effective watershed liming the average decrease was only half as much, despite a substantial rise in pH. HgPike increased in several lakes, dramatically in one. A general causal link to simultaneous changes in lake water pH could not be established. The re-equilibration time of HgPike after treatment was about four years. The costs of liming per kg fish yield were substantial.

Liming of Acidified Surface Waters
L. Henrikson and Y.W. Brodin (Eds.)
© Springer-Verlag Berlin Heidelberg 1995

1. Introduction

High concentrations of mercury (Hg) occur in many lakes which have never been affected by any known direct discharge of Hg. In about 10,000 Swedish lakes, Hg concentrations in 1-kg pike *(Esox lucius)* exceed health administration guidelines of 1 μg/g wet weight (Håkanson *et al* 1988, 1990a). The concentrations are often highest in acidic soft-water lakes. Many survey data (Mannio *et al* 1986, Håkanson 1991a, Lathrop *et al* 1991, Meili 1994) show an approximate doubling of animal Hg per unit of lower surface water pH.

Lake liming has been used for many years to counteract acidification in Swedish lakes (Bernes 1986, 1991, Henrikson and Brodin, Chap. 1). It was suggested at an early stage that liming might also reduce Hg concentrations in fish (Andersson and Kärrhage 1984). In 1985, a large project was initiated with the aim of reducing Hg concentrations in fish by chemical and biological manipulation of lake waters. This project included 95 lakes which where subjected to various treatments, including experimental liming of acid-sensitive lakes and watersheds (Håkanson 1986, 1991b, Håkanson *et al* 1990b).

Most of the data within this project were collected during the 1986-1989 period. It has been reported that a decrease in Hg concentrations in small perch *(Perca fluviatilis)* occurred during the first two years after liming, but no systematic change was found in pike, the most common predatory fish in the type of lake studied (Håkanson *et al* 1990b, Håkanson 1991b). Since this first period was considered to be too short to allow observation of delayed responses to liming, the study was continued between 1990 and 1992 in fewer lakes according to a reduced program. This report, which is the first to include data from the second period, focuses on the Hg concentrations in pike in 46 lakes where the primary treatment consisted of liming.

2. Methods

2.1. Study lakes

The 46 study lakes are situated in four counties in central and southern Sweden (Fig. 1). The watersheds consist mainly of shallow podzolic soils on igneous bedrock, covered by coniferous forest (Bengtsson *et al* 1987, Johansson *et al* 1987, Grahn *et al* 1987, Jönsson *et al* 1987). The lakes are mostly small and oligotrophic forest lakes (surface area 0.03 to 5.2 km^2; total P typically 5 to 20 μg/l). Before treatment the lakes contained soft water of varying acid and humic status, as shown by the medians and ranges of lake mean values at a depth of 2 m during summer 1986 (n = 2-6 at a temperature >12°C): pH = 6.3 (5.0-6.9); ANC (alkalinity) = 31 (0-

163) μeq/l; conductivity = 2.9 (1.5-10.5) mS/m; colour = 77 (20-475) mg Pt/l; estimated TOC ≈ 3-50 mg/l (recalculated data from Bengtsson *et al* 1988, Johansson *et al* 1988, Grahn *et al* 1988, Jönsson *et al* 1988).

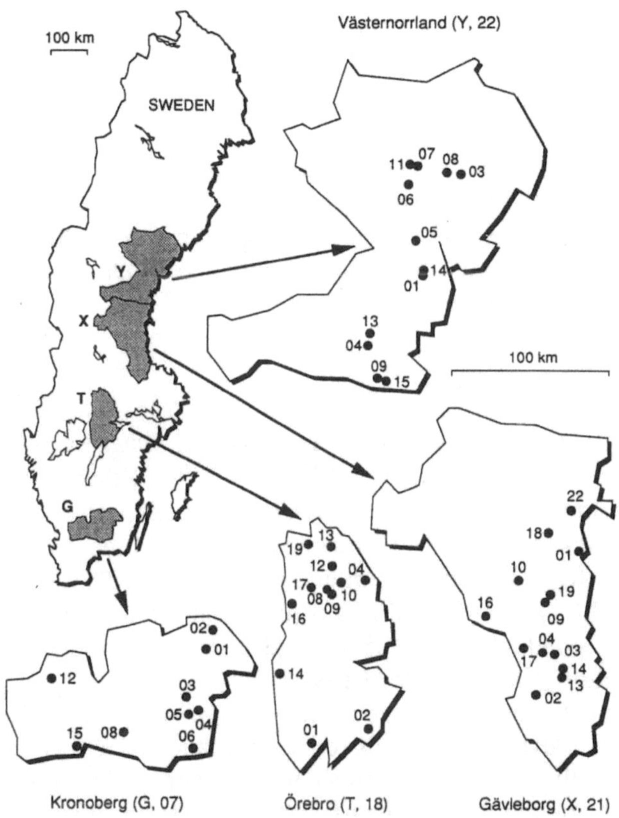

Figure 1. Location of the 46 study lakes in four Swedish counties.

2.2. Treatments

Liming was performed directly in the water in 23 lakes. Eight of these received an excess dose (about threefold) in order to obtain a pH >7. Extensive fishing was conducted simultaneously in five lakes, and in six lakes potash was added to counteract the bioaccumulation of ^{137}Cs from the Chernobyl fallout.

Indirect liming by dispersal on surrounding land was applied in 23 lakes, of which four were subjected to a whole-scale watershed liming, and 19 to selective liming of wetlands in discharge areas. In seven of these lakes, potash was also added to counteract the bioaccumulation of ^{137}Cs from the Chernobyl fallout.

All liming was conducted using powder of primary and sedimentary limestone ($CaCO_3$), except in the case of whole-scale watershed liming where dolomite ($CaMg(CO_3)_2$) was used (Håkanson 1986, 1991b, Bengtsson *et al* 1988, Johansson *et al* 1988, Grahn *et al* 1988, Jönsson *et al* 1988). Dosage per lake volume was calculated according to current liming guidelines and was typically 10 to 200 g/m^3 in the case of direct lake liming, while it corresponded to 7 to 2,000 g/m^3 in the case of watershed liming.

The initial liming of the lakes or watersheds was performed between February 1987 and February 1988, usually during the ice-free season of 1987, more rarely on ice or snow during the preceding or following winters. Lake liming was repeated during the following years as required in order to maintain ANC (alkalinity).

2.3. Sampling and analyses

Between 1986 and 1989, fish were collected annually and water samples monthly. The study was continued between 1990 and 1992, although the sampling schedule and the number of lakes studied were reduced.

Water samples were taken with a Ruttner sampling bottle in the deepest or central part of a lake from a depth of 2 m and stored cool in PVC bottles. Samples were usually analysed within 24 h for pH at 20°C, ANC (to pH 5.4, $\approx$ alkalinity), conductivity at 25°C, and water colour using a Hellige colour comparator (Håkanson *et al* 1990c).

Pike were mostly caught during the spawning season in spring using traps of various types. Several measures of length and weight were recorded. Fish of typically 0.5 to 2 kg fresh weight were selected for analysis. One lake had to be omitted from part of this evaluation since only four pike were caught during the whole study period (Grahn 1990). Hg in pike was analysed in wet dorsal muscle samples from several individuals (n = 5-20, usually 10-15) using FIA-CVAAS after digestion in HNO_3 + H_2SO_4 in sealed vials at 250°C (Uhrberg 1982, Håkanson *et al* 1990c).

2.4. Calculations

New averages for each lake and year were calculated from raw data. Hg concentrations in pike were standardized to a somatic body weight of 1 kg, based on a non-linear standard relationship with body weight. From standardized individual concentrations in muscle tissue, geometric annual mean values (henceforth referred to as HgPike) were calculated for each lake. In order to stabilize comparisons of naturally varying or incomplete data, values representing periods of several years were calculated by linear interpolation of annual means.

The effect of water chemistry on the mercury uptake in lacustrine food webs is likely to be most pronounced during the summer, since fish in boreal lakes take up virtually all mercury from the food (e.g. Meili 1991a), most of which is produced during a rather short growing season. For this reason, the Hg concentrations in fish were compared to a restricted subset of water data that were collected during periods when the epilimnetic water temperature exceeded 12°C in most lakes. Although these well-defined periods differ between regions, lakes and years, they represent the growing season at a given site better than pre-defined calendar periods. Water chemistry data from these periods were converted into annual summer mean values based on 2 to 8 measurements in each lake.

Trends in HgPike after treatment were tested for significance using the two-sided one-sample t-test on relative deviations from pre-treatment values.

3. Results and discussion

3.1. Pre-treatment situation

Mercury concentrations in standard pike (HgPike = geometric mean of spring concentrations at a somatic body weight of 1 kg) varied between 0.3 and 2.0 μg/g (wet weight) in the lakes before liming (Fig. 2). Natural background values in such lakes are estimated to be about 0.1 to 0.5 μg/g (based on Verta *et al* 1986). Most values were far above governmental guidelines, although only few of the lakes had been exposed to local emissions of Hg to tributaries or to the atmosphere. Only about one third of the lakes showed values <1 μg/g, and less than 5% of the lakes showed values <0.5 μg/g. In large pike (around 3-5 kg), Hg concentrations are usually about twice the standardized HgPike values (e.g. Meili 1991a).

In all counties, HgPike increased with lake acidity (Fig. 2). Mean summer pH values in 1986 before liming were between 5.5 and 7.0 in most lakes; three lakes were more acidic, with pH down to 5.0. HgPike typically varied around 1 μg/g in the circumneutral lakes and around 1.5 μg/g in the acidic lakes, when comparing HgPike during the reference period (mean of 1987 and 1988) with summer water pH from 1986 (a procedure which is justified by the time lag of about two years in the response of HgPike to treatments; see below). At the same time, however, Hg concentrations were lowest in the lakes of the southernmost county, where atmospheric deposition of both Hg and acid is highest (cf. Meili 1991b, Bernes 1991).

If there were to be a direct causal relationship between HgPike and pH, neutralization of lake water with lime could be expected to result in lower

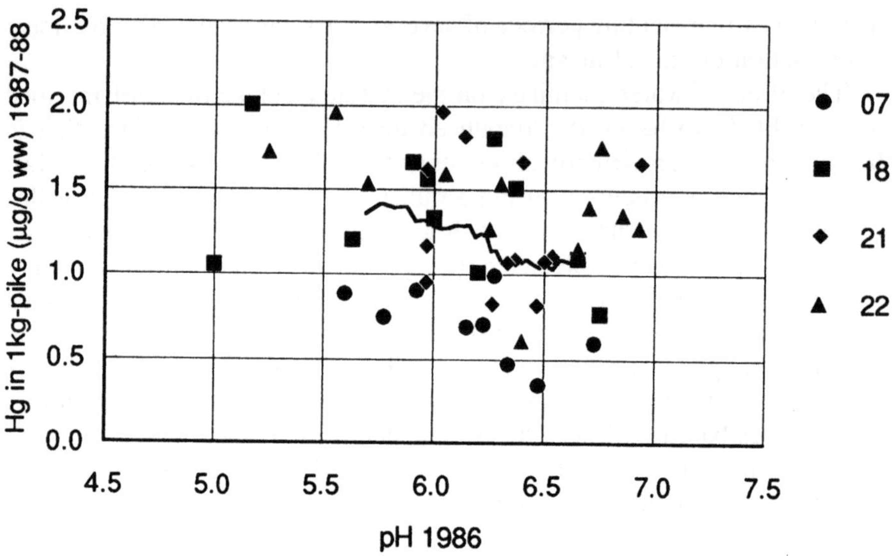

Figure 2. Variation and pre-liming relationship of mercury concentrations in standard pike (Esox lucius) and mean summer pH (at a depth of 2 m during periods with water temperatures above 12°C) in 45 forest lakes located in four Swedish counties (number codes). Standard Hg values in pike (HgPike) represent annual geometric means of spring concentrations in muscle tissue, standardized non-linearly to a somatic body weight of 1 kg. The line shows the running Hg mean (n=15) along the pH axis.

Hg concentrations in fish, and this decline would be a function of the pH change. An overall running mean of HgPike versus summer pH (Fig. 2) suggests that after each unit of increase in water pH, HgPike may eventually decrease by about 30%, or even by half according to the relationship found in other survey studies (Mannio *et al* 1986, Håkanson 1991a, Lathrop *et al* 1991).

On the other hand, lake water acidity accounted at best for a minor fraction of the observed variation in HgPike (r^2 = 6% in 45 lakes, pH 5.0-7.0; and r^2 = 2% in 42 lakes, pH 5.5-7.0). It was also evident that HgPike may vary more than threefold even among neighbouring lakes of similar type (no large or eutrophic lakes are considered here) and acidity (Fig. 2).

3.2. Liming

The effect of the liming operations on the ionic status of the lake waters varied widely, ranging from absent to dramatic. From 1986 to 1988-89 (mean), the mean summer pH increased by up to 1.5 units (typically about 0.7 units), and ANC by up to 0.6 (mostly 0.03 to 0.3) meq/l. The values within lakes were very similar during the first three years following liming, suggesting that a rather constant chemical status was maintained. In part, this was achieved by repeated dispersal of lime where required, although rarely after watershed liming which has a long durability. In the following three years, the chemical status of most lakes still differed clearly from the original status. Any changes during this period toward the pre-liming status are unlikely to influence the results reported here in any case, since HgPike started responding to changes in water chemistry only after an initial delay of about two years (see below).

Unexpectedly, both the mean and the range of chemical effects of watershed liming were similar to those of direct lake liming. While, in many cases, watershed liming significantly influenced lake water chemistry without the expected delay, lake liming failed in some cases to provide the projected response.

The average cost per lake of the liming operations was in the region of US$ 100,000 (30,000-300,000) in the case of whole-scale watershed liming, US$ 50,000 (20,000-250,000) in the case of selective wetland liming, and US$ 10,000 (1,000-100,000) in the case of direct lake liming (or about three times as much when applying excess dosage). The corresponding amounts per ha lake surface were around US$ 2,500 (1,000-7,000), US$ 1,000 (500-5,000), and US$ 150 (20-500).

3.3. Development of Hg concentrations in pike after liming

Substantial changes in HgPike were observed in many lakes during the study period (five to six years after liming). Both increases and decreases occurred, as well as considerable interannual variations of a systematic and apparently random type. To facilitate comparison of Hg concentrations both within and between lakes, mean values were calculated from periods of one to three years: one reference period and two post-treatment periods (cf. Fig. 3). Hg concentrations in standard pike are available for 45 lakes from the first post-treatment period (1989, 1 to 2 years after liming), and for 41 lakes from the second period (mean values from 1990 to 1992, i.e. 2 to 5 years after liming, normalized to 1991, i.e. 3 to 4 years after liming).

Little overall change in HgPike was observed during the first period of 1 to 2 years after liming (cf. Håkanson *et al* 1990b, Håkanson 1991b). Statistical evaluation revealed a slight increase (+3.8%, $p = 0.07$, $n = 45$). This was valid for both watershed liming (+4.6%, $p = 0.19$, $n = 23$) and lake liming (+2.9%, $p = 0.19$, $n = 22$). This increase may partly be explained by the wet periods which preceded the treatments, resulting in increased levels of mercury in lake waters and among zooplankton during 1986 (Meili 1991a), and possibly in a delayed weak response of HgPike during the first years of the study. Since the changes in HgPike were not

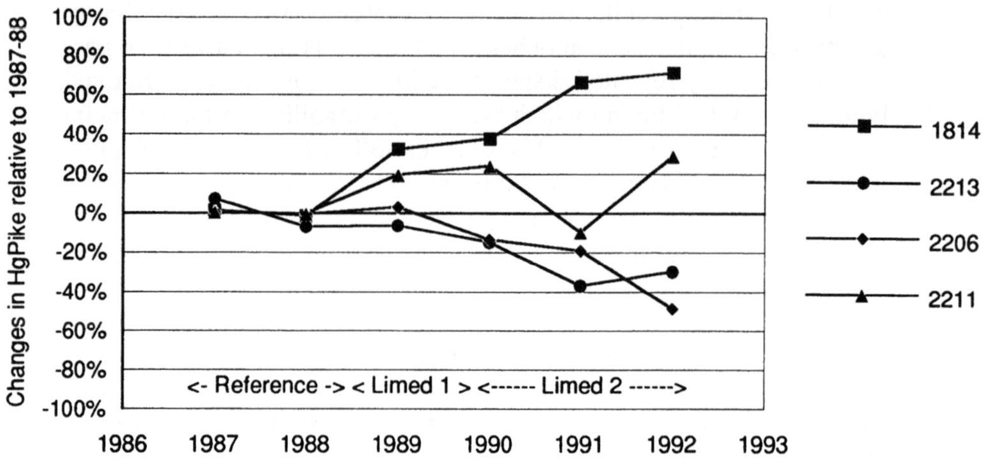

Figure 3. Variations in the development of Hg concentrations in standard pike in four Swedish lakes after dispersal of lime powder on surrounding land. Values are given as the change relative to pre-liming conditions (reference = mean of spring 1987 and spring 1988). All these lakes were limed after the growing season of 1987 (Sep. to Nov.), shortly before ice accretion. The first two lakes were treated by whole-scale watershed liming with dolomite, the last two by selective wetland liming with primary limestone.

related to changes in pH or the type of treatment, it is less likely that the rather early overall increase was a liming effect.

In individual lakes, the changes in HgPike during 1987-1989 were usually within ±20% (Fig. 4). This may be the range of natural interannual variability, controlled either by climatic variability or by in-lake processes such as fluctuations in mercury transformations, fish biomass or growth rates. Consequently, it is difficult to quantify the effect of liming in a single lake without HgPike data from several pre- and post-treatment years. Similar temporal changes are known from untreated lakes not included in this study (Meili unpubl. data).

During the second period (2 to 5 years after liming), an overall decrease in HgPike was observed. Although the changes in HgPike after 3 to 4 years varied from about -40 to +60%, the overall decline was significant ($p = 0.013$, $n = 41$). The decline was most pronounced in lakes where liming was successful and yielded substantial chemical changes in pH, ANC and conductivity. The overall decrease in HgPike depended on the treatment, however. After effective lake liming (here defined by a pH increase >0.5), HgPike decreased on average 20% ($p < 0.0001$, $n = 13$). After watershed liming the overall change was only 9.5% ($p = 0.046$, $n = 14$, lake 1814 excluded), despite an equally dramatic rise in pH.

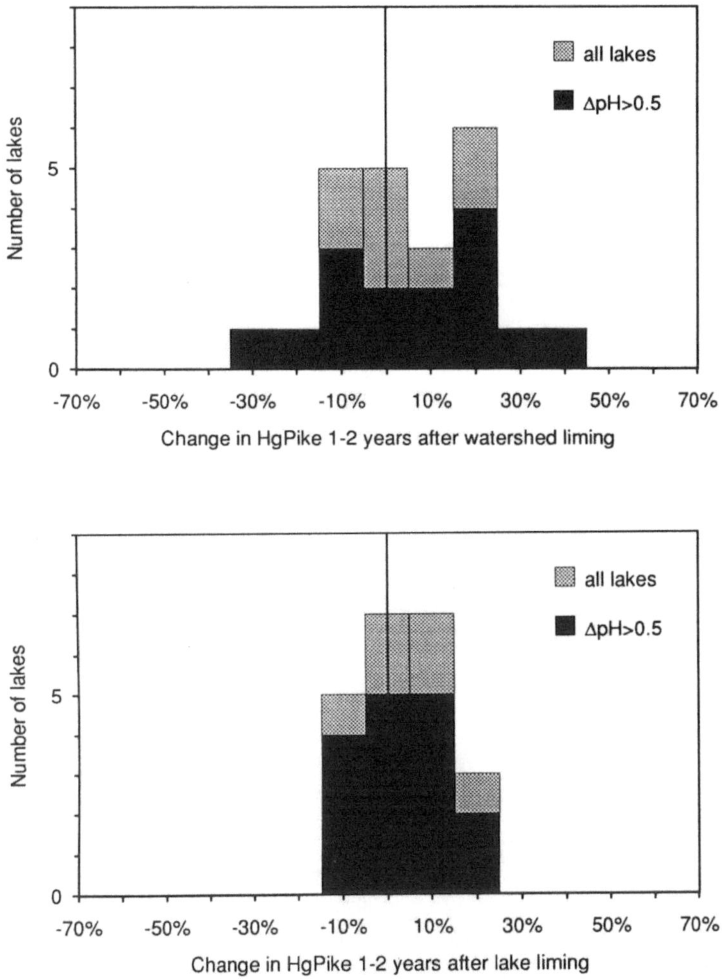

Figure 4. Relative changes in the Hg concentrations of standard pike 1 to 2 years after application of lime to the watershed (upper) or the lake water (lower) in 45 Swedish forest lakes, divided according to the resultant change in mean lake water pH during the summer, two years after the initial treatment (ΔpH < > 0.5).

However, the variations were again substantial, roughly ±20% around the mean after lake liming and ±30% after watershed liming (Fig. 5). These numbers indicate that the effects of watershed liming on HgPike are not only less pronounced, but also less predictable than the effects of direct lake liming.

It should be pointed out that an apparently ineffective treatment, especially when applied to the watershed, is not necessarily harmless - the con-

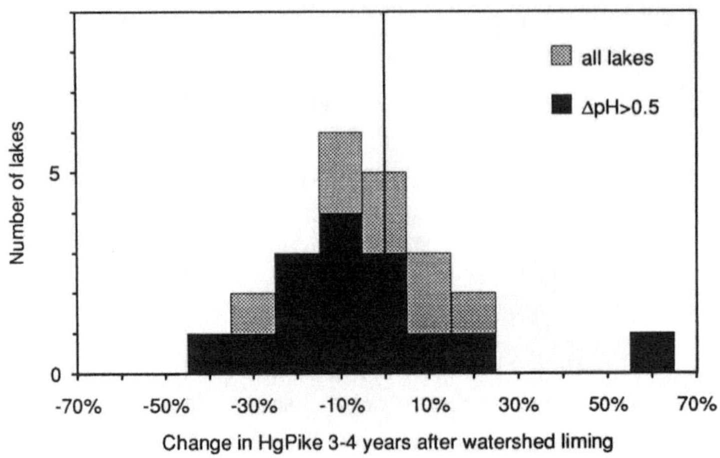

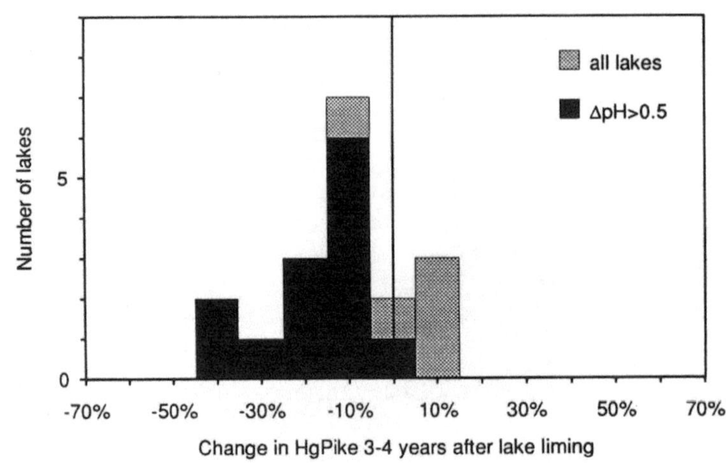

Figure 5. *Relative changes in the Hg concentrations of standard pike 3 to 4 years after application of lime to the watershed (upper) or the lake water (lower) in 41 Swedish forest lakes, divided according to the resultant change in mean lake water pH during the summer, two years after the initial treatment (ΔpH < > 0.5).*

sequences may be disastrous. One example is provided by a brown-water lake (1814 in Fig. 3), where complete watershed liming resulted in a steady increase in HgPike by over 70% during the first five years after liming. The pH increased even in this lake by 0.75 units after liming, but the increase occurred at the lower end of the pH scale, starting at 5.0 and ending at 5.8, i.e. below the starting values of many other lakes in this study (Fig. 2).

At the end of the study period, HgPike in many lakes appeared to have passed most of the shift to a new equilibrium. It would therefore seem unlikely that there will be any further substantial changes not covered by the study period. This is supported by data from concurrently limed lakes which were also treated with selenium and phosphorus, suggesting that HgPike stabilizes about four to five years after treatment. It should be noted that this re-equilibration time is not a physiological parameter but an ecological one, not only because the bioavailability of Hg may change slowly after liming, but also because the response of biotic Hg concentrations is naturally retarded along food chains, and finally because HgPike refers to a different cohort within a growing pike population each year. The initial time lag of HgPike to liming was, not surprisingly, about two years. This justifies the use of the mean HgPike from spring 1987 and spring 1988 as a reference value, even where liming was conducted during the growing season of 1987.

3.4. Potential mechanisms

In order to judge whether or not liming was the main cause of the overall decrease in Hg concentrations in pike, a comparison was drawn with the impact of liming on water chemistry, i.e. to the actual effect of the treatment on mean summer pH, ANC, conductivity, Ca concentrations, etc. Both watershed and lake liming showed a wide range in chemical effect. In the case of pH, values increased by up to 1.5 units. In one lake, however, both pH and ANC actually decreased after liming. This proved to be because the lake was unofficially limed in the spring of 1985 (Jönsson *et al* 1987), i.e. one year before the reference period, and this was followed by an inadequate whole-scale liming of the wetland-poor watershed.

If changes in fish Hg were not due to liming, it would be unlikely that any correlation with simultaneous changes in acidity variables would be found (even at overall changes in HgPike). However, 3 to 4 years after liming, a relationship was evident. Whereas HgPike generally decreased in lakes where the pH increased by more than 0.5 units after liming, no significant overall changes were observed in lakes where pH was changed by less than 0.5 (Fig. 5). The same applied in lakes where ANC was changed by less than 50 μeq/l. These lakes may be regarded as unforeseen reference lakes. In another set of lakes which was studied concurrently, an absence of any change in the pH value was accompanied by a generally constant HgPike. These lakes were not treated with any chemicals, but subjected only to extensive fishing. Similarly, the results in limed lakes did not seem to be influenced by simultaneous fish removal.

The decrease in HgPike was often less than expected from the change in pH (Fig. 2), especially in the case of watershed liming. Many other survey data (Mannio *et al* 1986, Håkanson 1991a, Lathrop *et al* 1991,

Meili 1994) show an approximate doubling of animal Hg per unit pH. However, experimental acidification or liming of lakes rarely has such dramatic effects (see also Andersson and Kärrhage 1984, Andersson *et al* 1989, Wiener *et al* 1990, Håkanson 1991). It is likewise puzzling that the change in HgPike was the same after the pH was raised by either 0.5 or 1.5 units, although this corresponds to a tenfold difference in acidity change. Experimental and empirical studies of the influence of pH on the Hg cycle actually provide contradictory results (Richman *et al* 1988, Winfrey and Rudd 1990). This implies that neither the direct (e.g. chemical) nor indirect (e.g. biological) effects of changes in the pH of the lake water can account for its strong empirical relationship with fish Hg in temperate lakes. Statistically significant correlations of acidity variables with biotic Hg concentrations may result from collinearity with both the concentration and the biodilution of Hg (Richman *et al* 1988, Verta 1990, Håkanson *et al* 1990), given the correlations between lake water pH, precipitation pH and Hg loading to lakes and their catchment (e.g. Håkanson *et al* 1990c, Meili 1994), between weathering rate, buffering capacity and nutrient concentration, and between water colour and pH resulting from the natural acidity of humic substances (e.g. Mannio *et al* 1986).

It can be concluded from these findings that the effect of liming is not necessarily an effect of changes in lake water pH (which here was used to represent acidity variables in general). Liming effects may also be related to the simultaneous increase in concentration of ions or particles (Andersson *et al* 1991, Meili *et al* 1989). This would support a non-causality of the frequently reported relationship between acidity and animal Hg. In this case, the comparative success of lake liming remains to be explained, as do the weak effects of watershed liming, even after apparently satisfactory neutralization of the lake water. The slow reaction of soils to liming may substantially delay the effects on surface waters, but may eventually be compensated by a longer durability of watershed liming effects. At any rate, the wider variability of changes in HgPike after land liming which affects both watershed and recipient, implies that soil processes strongly influence biotic Hg concentrations in drainage lakes.

Liming effects on fish Hg concentrations may also vary with the initial status of the lake. It has been suggested that Hg concentrations may increase initially in very acidic systems before they decrease, and that increases in Hg concentrations after liming may result from ion exchange processes (Andersson *et al* 1989, Andersson *et al* 1991). Humic lakes may react differently from clear-water lakes, but this difference may again depend on the initial acidity.

3.5. Cost-benefit analysis

The cost of reducing HgPike by liming can be illustrated by comparing average fish yields to the cost and durability of the treatment in the case of the most inexpensive type of liming, for instance, i.e. lake liming at normal dosage. The total yield of predatory fish of a size suitable for human consumption is unlikely to exceed 10 kg/ha per treatment, a value based on an average biomass of around 10 kg/ha, an annual production per biomass of 50%, an intensive fishing activity yielding 50% of the production, and a treatment durability of four years. Since the cost per ha for this treatment varies around US$ 150 (see above), the cost for each consumed kg of fish containing 0 to 40% less mercury amounts to at least US$ 15, and thus exceeds the commercial value of the fish by ten or even a hundred times.

The corresponding cost of the other treatments described here may be ten or more times higher, although durability may be longer instead in these cases. Fishing in forest lakes is likely to be far less intense than considered above, moreover, and the costs per yield may accordingly increase several times. However, it should also be pointed out that, in these calculations, the cost comparisons are seen exclusively in the light of the mercury issue, although liming may have other beneficial effects on surface waters.

4. Conclusions

Liming of lakes, especially direct liming of lake waters, usually appears to reduce mercury concentrations in predatory fish. However, substantial increases may also occur, e.g. possibly after inadequate treatment of humic lake watersheds. In Swedish lakes, the total ecological reaction time of Hg concentrations in 1-kg pike to water treatment is at least four years.

A cost-effective operation which would promise a decline in Hg concentrations in pike by 0 to 40% after an initial delay of several years would be to add lime to the water of a lake with low salinity (conductivity), low ANC (alkalinity <100 μeq/l) and low pH (<6.5), to achieve a new status where ANC >100 μeq/l and pH >6.5 by simultaneously increasing the salinity by at least twofold, ANC by at least threefold and pH by one unit. However, the effect of liming on mercury concentrations in fish varies widely, and surprises cannot be excluded as long as the mechanisms are still obscure.

Given a re-equilibration time for HgPike of about four years after water treatment, further substantial changes seem unlikely to occur, but the long-term (5-20 years) effects of liming on fish mercury are not yet understood. The apparent initial failure of many watershed treatments, for example, may partly be caused by a longer reaction time and eventually be compensated by a longer durability. The durability of lake liming effects on fish mercury after return to pre-treatment water chemistry is also an aspect which requires study.

5. Acknowledgements

This large-scale project could not have been accomplished without Lars Håkanson's intrepid initiation and efficient management, the enormous year-after-year and year-around efforts in liming, sampling and analysis by the numerous staff of four county administrations, the reliable analyses of a.o. mercury in fish by Roland Uhrberg, the data compilation performed by Tord Andersson, and generous funding by the Swedish Environmental Protection Agency. I should like to thank Y.-W. Brodin and Alain Tremblay for their dedicated and detailed review of an earlier draft. Valuable comments were also provided by Lennart Henrikson, Paul Andersson, Kjell Johansson, Lars Håkanson, Carl-Philip Jönsson and Åke Bengtsson.

6. References

Andersson, P. and Kärrhage, P. 1984. Effekter av kalkning på kvicksilverhalten i fisk från Åvaåns vattensystem. [Effects of liming on the mercury content of fish in the watershed of River Åva.] - Swedish Environm. Prot. Agency 1771.

Andersson, P., Grahn, P., Hörnström, E., Nyberg, P. and Dahlquist, K. 1989. Kalkning och gödsling i Rammsjön och Ämten, Örebro län 1980-1987 - ett försök att minska kvicksilverhalten i fisk. [Liming and fertilization of Lake Rammsjön and Lake Ämten, county of Örebro, 1980-1987: an attempt to decrease mercury levels in fish.] - Swedish Environm. Prot. Agency 3584.

Andersson, P., Borg, H. and Kärrhage, P. 1991. Kalkningseffekter på kvicksilver i fisk; Läget efter genomförda åtgärder i 19 typsjöar. [English summary: Effects of liming on mercury in fish; the status after measures in 19 study lakes.] - Swedish Environm. Prot. Agency 3883.

Bengtsson, Å., Viklund, T., Häggbom, M., Andersson, T. and Håkanson, L. 1987. Typsjöar och tillrinningsområden i Västernorrlands län. Projekt Kalkning-Kvicksilver-Cesium. [Study lakes and catchments in the county of Västernorrland. Project Liming-Mercury-Caesium.] - Swedish Environm. Prot. Agency 3402.

Bengtsson, Å., Göthe, L., Andersson, T. and Håkanson, L. 1988. Läget före åtgärder i Västernorrlands län. Projekt Kalkning-Kvicksilver-Cesium. [The situation before treatment in the county of Västernorrland. Project Liming-Mercury-Caesium.] - Swedish Environm. Prot. Agency 3564.

Bernes, C. 1986. Monitor 1986 - Sura och försurade vatten. [Acidic and acidified waters.] - Swedish Environm. Prot. Agency Informs.

Bernes, C. 1991. Monitor 12 - Acidification and liming of Swedish waters. - Swedish Environm. Prot. Agency Informs.

Grahn, P. 1990. Projekt Kalkning-Kvicksilver-Cesium - en sammanfattning av projektet i Örebro län. [English summary: Project Liming-Mercury-Caesium - a summary of the project in the county of Örebro, Sweden.] - Vatten 46: 211-223.

Grahn, P., Thorssell, S., Nilsson, Å. and Håkanson, L. 1987. Typsjöar och tillrinningsområden i Örebro län. Projekt Kalkning-Kvicksilver-Cesium. [Study lakes and catchments in the county of Örebro. Project Liming-Mercury-Caesium.] - Swedish Environm. Prot. Agency 3384.

Grahn, P., Thorssell, S., Nilsson, Å. and Håkanson, L. 1988. Läget före åtgärder i Örebro län. Projekt Kalkning-Kvicksilver-Cesium. [The situation before treatment in the county of Örebro. Project Liming-Mercury-Caesium.] - Swedish Environm. Prot. Agency 3530.

Håkanson, L. 1986. Arbetsplan 1985-1989. Projekt Kalkning-Kvicksilver. [Working plan 1985-1989. Project Liming-Mercury.] - Swedish Environm. Prot. Agency 3099.

Håkanson, L. 1991a. Mercury in fish - geographical and temporal perspectives. - Water Air Soil Poll. 55: 159-177.

Håkanson, L. 1991b. Measures to reduce mercury in fish. - Water Air Soil Poll. 55: 193-216.

Håkanson, L., Nilsson, Å. and Andersson, T. 1988. Mercury in fish in Swedish lakes. - Environ. Poll. 49: 145-162.

Håkanson, L., Nilsson, Å. and Andersson, T. 1990a. Mercury in fish in Swedish lakes - linkages to domestic and European sources of emission. - Water Air Soil Poll. 50: 171-191.

Håkanson, L., Andersson, P., Andersson, T., Bengtsson, Å., Grahn, P., Johansson, J.-Å., Jönsson, C.P., Kvarnäs, H., Lindgren, G. and Nilsson, Å. 1990b. Åtgärder mot höga kvicksilverhalter i insjöfisk. Slutrapport för kvicksilverdelen av projektet Kalkning-Kvicksilver-Cesium. [English summary: Measures to reduce mercury in lake fish. Final report from the project Liming-Mercury-Caesium.] - Swedish Environm. Prot. Agency 3818.

Håkanson, L., Borg, H. and Uhrberg, R. 1990c. Reliability of analyses of Hg, Fe, Ca, K, P, alkalinity, conductivity, hardness and colour from lakes. - Int. Rev. ges. Hydrobiol. 75: 79-94.

Johansson, J.-Å., Lindgren, G., Andersson, I., Nilsson, Å. and Håkanson, L. 1987. Typsjöar och tillrinningsområden i Gävleborgs län. Projekt Kalkning-Kvicksilver-Cesium. [Study lakes and catchments in the county of Gävleborg. Project Liming-Mercury-Caesium.] - Swedish Environm. Prot. Agency 3383.

Johansson, J.-Å., Lindgren, G., Sveden, I., Nilsson, Å. and Håkanson, L. 1988. Läget före åtgärder i Gävleborgs län. Projekt Kalkning-Kvicksilver-Cesium. [The situation before treatment in the county of Gävleborg. Project Liming-Mercury-Caesium.] - Swedish Environm. Prot. Agency 3553.

Jönsson, C.P., Karlsson, L., Mårtenson, B., Andersson, T. and Håkanson, L. 1987. Typsjöar och tillrinningsområden i Kronobergs län. Projekt Kalkning-Kvicksilver-Cesium. [Study lakes and catchments in the county of Kronoberg. Project Liming-Mercury-Caesium.] - Swedish Environm. Prot. Agency 3360.

Jönsson, C.P., Karlsson, L., Andersson, T. and Håkanson, L. 1988. Läget före åtgärder i Kronobergs län. Projekt Kalkning-Kvicksilver-Cesium. [The situation before treatment in the county of Kronoberg. Project Liming-Mercury-Caesium.] - Swedish Environm. Prot. Agency 3540.

Lathrop, R.C., Rasmussen, P.W. and Knauer, D.R. 1991. Mercury concentrations in walleyes from Wisconsin (USA) lakes. - Water Air Soil Poll. 56: 295-307.

Mannio, J., Verta, M., Kortelainen, P. and Rekolainen, S. 1986. The effect of water quality on the mercury concentration of northern pike *(Esox lucius* L.) in Finnish forest lakes and reservoirs. - National Board of Waters, Helsinki, Water Res. Inst. 65: 32-43.

Meili, M. 1991a. Mercury in boreal forest lake ecosystems. - Acta Univ. Upsaliensis 336.

Meili, M. 1991b. Fluxes, pools and turnover of mercury in Swedish forest lakes. - Water Air Soil Poll. 56: 719-727.

Meili, M. 1994. Aqueous and biotic mercury concentrations in boreal lakes: model predictions and observations. - In: Huckabee, J. and Watras, C.J. (Eds.) Mercury as a global pollutant: toward integration and synthesis. Lewis Publ. Inc., Chelsea, Michigan. (in press)

Meili, M., Rudebeck, A., Brewer, A. and Howard, J. 1989. Cs-137 in Swedish forest lake sediments, 2 and 3 years after Chernobyl. - In: Feldt, W. (Ed.) The radioecology of natural and artificial radionuclides. Verlag TÜ V Rheinland GmbH, Köln, Germany, Publ. Ser. Prog. Radiat. Prot. 22: 306-311.

Richman, L.A., Wren, C.D. and Stokes, P.M. 1988. Facts and fallacies concerning mercury uptake by fish in acid-stressed lakes. - Water Air Soil Poll. 37: 465-473.

Uhrberg, R. 1982. Acid digestion bomb for biological samples. - Anal. Chem. 54: 1906-1908.

Uhrberg, R., Borg, H. and Håkanson, L. 1989. Kvalitetskontroll av analyser inom projektet Kalkning-Kvicksilver-Cesium. [English summary: Quality control of analyses in the project Liming-Mercury-Caesium.] - Swedish Environm. Prot. Agency 3572.

Verta, M. 1990. Mercury in Finnish forest lakes and reservoirs: anthropogenic contribution to the load and accumulation in fish. - National Board of Waters and the Environment, Helsinki, Water Environm. Res. Inst. 6.

Verta, M., Rekolainen, S., Mannio, J. and Surma-Aho, K. 1986. The origin and level of mercury in Finnish forest lakes. - National Board of Waters, Helsinki, Water Res. Inst. 65: 21-31.

Wiener, J.G., Fitzgerald, W.F., Watras, C.J. and Rada, R.G. 1990. Partitioning and bioavailability of mercury in an experimentally acidified Wisconsin lake. - Environ. Toxicol. Chem. 9: 909-918.

Winfrey, M.R. and Rudd, J.M.W. 1990. Environmental factors affecting the formation of methylmercury in low pH lakes. - Environ. Toxicol. Chem. 9: 853-869.

Chapter 17

Supplementary measures to aquatic liming

Björn C. Bergquist

National Board of Fisheries, Institute of Freshwater Research, S - 178 93 Drottningholm, Sweden

Summary

In many limed waters, particularly in waters which have been severely acidified prior to treatment, it may often take more than ten years before locally extinct fish and benthic fauna can recolonize. In certain cases recolonization may also be impossible. The slow recovery is primarily attributable to colonization difficulties, but also to the problems involved in re-establishing populations. Natural recolonization is often impeded by the existence of migration obstacles or to the limited colonization capability of individual species. Establishment difficulties are mainly attributable to changes in the predation and competition situation, during both the acidification phase and after liming.

In order to achieve re-establishment of locally extinct or depleted species in limed waters, therefore, supplementary measures are often required in the form of restocking, habitat restoration and elimination of migration obstacles. Measures of this sort may be called for when it is quite clear that re-establishment will not occur under its own impetus or when earlier occurring numbers of species and individuals cannot be achieved within a reasonable time span (10-15 years) after the initial treatment. The ultimate objective is to restore and sustain the biodiversity of the limed water at a natural level.

Post-liming biological restoration may be divided into two categories: a) measures which facilitate natural recolonization and re-establishment of locally extinct species; b) reintroduction of locally extinct species through restocking. Measures which facilitate both colonization and establishment include removal of migration obstacles, habitat improvement and restoration of the balance of species. As in the case of liming, state subsidies are available for implementation of such measures and have been paid out since 1986.

Liming of Acidified Surface Waters
L. Henrikson and Y.W. Brodin (Eds.)
© Springer-Verlag Berlin Heidelberg 1995

In running waters, biological restoration includes measures such as building fish-ways, removing old dams, habitat improvement and restocking with trout *(Salmo trutta)*, grayling *(Thymallus thymallus)* and Atlantic salmon *(Salmo salar)*. Certain benthic faunas, such as freshwater pearl mussels *(Margaritifera margaritifera)*, have also been reintroduced after liming. In lakes, restoration has mainly involved the release of noble crayfish *(Astacus astacus)* and signal crayfish *(Pacifastacus leniusculus)*, Arctic char *(Salvelinus alpinus)* and roach *(Rutilus rutilus)*. In some cases, decimation fishing of populations has been undertaken.

Measures of this sort have generally had a positive effect, resulting in re-establishment of locally extinct or depleted species. Particularly successful measures include the removal of dams, building of fish-ways and reintroduction of trout and char. It is, however, difficult to assess the overall success and cost-effectiveness of the measures, as a satisfactory documentation after restoration is missing in most cases. Results and general experience suggest, however, that measures which promote natural colonization and establishment are to be preferred to direct restocking, for several reasons. Whether restocking is performed in order to reintroduce a species or to replenish existing stocks, it is often difficult to find stocks whose genetic characteristics agree with or resemble those of the original populations. In some cases, however, restocking is essential since the locally extinct species cannot return without help.

Invertebrate benthic faunas such as mayflies, crustaceans, mussels and freshwater snails have sometimes been reintroduced into limed waters. The lack of background data and the risk of confusion between species mean, however, that active reintroduction of benthic faunas should only be considered for a few species and in very few waters.

1. Introduction

Despite the fact that liming has a number of beneficial effects on flora and fauna in acidified waters (Larsson, Chap. 7, Degerman *et al*, Chap. 8, Appelberg, Chap. 9) recolonization and re-establishment often take place very slowly, and sometimes it is not possible (Alenäs 1986, Degerman and Nyberg 1989, Bergquist 1991, Degerman and Appelberg 1992, Bergquist *et al* 1992). Although colonization may occur relatively rapidly after liming, the establishment of reproducing populations may take a long time and it is not unusual for stabilization of communities to require up to fifteen years after the initial treatment (Appelberg and Degerman 1991, Appelberg *et al* 1992). Crayfish populations find it particularly difficult to recover after acidification (Appelberg and Odelström 1990) and it is also hard for populations of other crustaceans and freshwater molluscs such as snails and mussels to re-attain their normal levels in many waters (Eriksson *et al*

1983, Bergquist *et al* 1992).

Recolonization success is largely linked with the abundance and colonization capability of the individual species, but the extent of the acidification damage is also important. If, for example, the species in question has been eradicated from an entire water system, leaving no residual population whatsoever, recolonization may be totally impossible. For fish, crayfish and some species of benthic faunas, moreover, natural recolonization is impeded by migration obstacles such as dams, culverts and various natural barriers.

Even if recolonization does occur, changes in the density of individuals during the acidification and liming phases - increased post-liming populations of European perch *(Perca fluviatilis)*, for instance - may represent an obstacle to the re-establishment of species such as noble crayfish, roach and vendace *(Coregonus albula)* (Appelberg and Odelström 1990, Appelberg *et al* 1990, 1992). In many cases, moreover, the habitat may have been damaged by human activities such as floatway clearance, ditching, waterway excavation and various water regulation measures. Success in improving water quality with liming, i.e. the degree to which the treatment has succeeded in restoring water quality in the form of pH and stability, for instance, is also crucial to colonization and establishment.

All these colonization and establishment difficulties often mean that, despite extensive liming, supplementary measures are required in order to fully restore pre-liming populations. Such biological restoration is attaining increasing importance as more insight is gained into the post-liming recovery process. Different fishery management measures, such as restocking with locally extinct species, habitat improvement and removal of migration obstacles, have been shown to play an important part in efforts to restore lakes and watercourses after acidification. The best measure in terms of preserving biodiversity is, however, to lime waters before serious damage occurs, i.e. during the earliest phase of acidification.

The objective of biological restoration measures in limed waters is to restore and sustain the biodiversity of the water system as a whole. This can be achieved by promoting natural recolonization and the establishment of locally extinct or depleted species and through direct reintroduction of locally extinct species. Biological restoration can therefore be divided into two categories: a) measures which facilitate natural recolonization and re-establishment of locally extinct species; b) reintroduction of locally extinct species through restocking. Since 1986, both types of action have been eligible for state subsidies (Henriksson and Brodin, Chap. 1). Measures which facilitate colonization and establishment of locally extinct species include removal of migration obstacles, habitat improvement and restoration of the balance of competition and predation.

To date, biological restoration measures have mainly focused on fish and crayfish. Quite apart from the fact that these are more important than flora and lower fauna in economic terms, they are also key organisms which play an important part in structuring the ecosystem. Moreover, flora and lower fauna often recover more rapidly, thanks to their shorter generation periods, and in many cases thanks to the fact that they can spread more effectively between individual waters and different water systems. They may still require restoration measures, however, but little has been done since less is known of their pre-acidification abundance and range of occurrence.

2. Measures to promote colonization and re-establish locally extinct species

Post-liming colonization and re-establishment of locally extinct species are often impeded by other aspects of anthropogenic impact on the water concerned. If a limed water is to be fully restored, it is essential that such obstacles are removed or minimized.

Running waters contain many migration obstacles such as dams. Such obstacles mean that it is often necessary to build fish-ways, e.g. salmon ladders, or to actively reintroduce locally extinct species upstream of the obstacle in order to restore the fauna in the watercourse. Various forms of habitat improvement have been necessary in many watercourses which have suffered from the impact of agriculture and forestry. In certain limed waters, measures such as population replenishment through stocking, decimation fishing, fertilization and fishing regulations have been used in order to restore the original balance of competition and predation between species (Table 1).

Table 1. *Biological restoration measures which facilitate colonization and re-establishment. Governmental subsidy: 1986-1991 (*), 1986-1991 and 1992-1993 (**), none (-). Long-term effect: good (+), limited (-).*

Measure	Grant	Result	Comments
Habitat improvement	**	+	Often combined with other measures
Removal of migration obstacle	**	+	Removal of dams; building fish paths
Stock replenishment	*	-	Difficulties in obtaining suitable stock
Decimation by fishing	*	-	Highly laborious; short duration
Reduction in fishing pressure	-	+	Catch restrictions
Fertilization	-	-	Requires careful control of dosage

2.1. Habitat improvement

Many of the acidified watercourses in Sweden have also been damaged in other ways as a result of floatway clearance, clear-felling, land preparation, ditching and waterway excavation, which often impede post-liming colonization and re-establishment of populations. In order to improve the chances for colonization and re-establishment, therefore, habitat improvement measures are performed, when needed, so as to restore natural structure and function. Important measures are restoration of floatways and conserving or planting vegetation zones along watercourses.

The main objective of habitat restoration is to provide a better environment in which salmon, trout and grayling can spawn and grow, while other objectives may include providing protection and refuges for crayfish. Trout waters have received a great deal of attention, with measures aimed at providing or improving spawning grounds (laying out of gravel for spawning), improving sites for the growth of fry (laying out of stones), modifying currents to create cover and feeding stations for older fish, digging pools and building thresholds or low dams. Habitat restoration measures are often combined with removal of migration obstacles and fish reintroduction, which makes individual assessment difficult.

Little documentation is available on the effects of habitat improvement in limed watercourses, but a number of studies of unlimed waters have shown that it has a positive effect on salmonid fish populations in the form of increased density, biomass and winter survival (Saunders and Smith 1962, Lowry 1971, Hunt 1969, Karlström 1977, Ward and Slaney 1979, Cueto 1981, Näslund 1987, 1989, Linløkken 1989). However, the same authors report that growth was not significantly improved by such measures. Assuming that the water quality is restored, the corresponding effects in limed waters are unlikely to be significantly different. If such restoration measures are correctly performed, they can provide watercourses with a more varied structure in the form of greater average depth, surface area and volume (Crispin *et al* 1993). When damaged habitats are restored by habitat improvement measures, the winter survival rate and spawning activity of the fish are often improved, but if this effect is to be sustained, it is important to establish protective zones along the watercourses. According to Crispin *et al* (1993), satisfactory results may be achieved in some cases simply by providing proper protective zones of this sort.

In general, the greatest benefits have been obtained in watercourses where the original biotope/habitat has been damaged as a result of human activity (Bergquist and Degerman unpubl.). In watercourses with large flow variations and difficult ice conditions, however, such benefits may not last. The best and most cost-effective results are probably achieved in small and medium-sized watercourses with moderate flow variations and low levels of sediment transport. In cases of high sediment transport, high flow peaks or unstable shorelines, additional measures in the drainage area are often required in order to achieve lasting restoration (Bergquist and Degerman unpubl., Frissell and Nawa 1992).

2.2. Removal of migration obstacles and building of fish-ways

When migration obstacles are removed or fish passes are built, migrating fish are assured of a free passage so that recolonization can be facilitated. If action is taken to remove old dams and remedy incorrectly built culverts, the chances of recolonization by crayfish and other benthic faunas (e.g. fresh-water shrimp, *Gammarus pulex*) which live their entire lives in the aquatic environment can also be improved. It is thus likely that elimination of migration obstacles in running waters and lake outlets is the most important biological restoration measure in limed waters. From the genetic point of view, it is preferable to stocking and reintroduction using hatchery-reared fish.

In many watercourses, dams have meant that the fish only have limited access to the stretches which previously produced salmon and sea trout smolt (Hannerz and Degerman 1984). In terms of general fishery management, therefore, the removal of such obstacles is important step in the conservation of the original wild fish stocks (Berg and Myhre 1990).

Restoration measures have mainly focused on fish which migrate to the open sea, such as salmon and sea trout, but in many limed waters, paths have also been opened for trout populations which are stationary or which migrate between lakes. As a rule, old mill dams, sluices and mini-power station dams are either demolished or by-passed with fish-ways such as salmon ladders. Passage through culverts and existing fish-ways has also been improved in a number of watercourses.

By eliminating migration obstacles, the number of sites available for spawning and growing can be increased. In a number of watercourses, increased production of young salmon and trout has been observed after liming and removal of migration obstacles. It has been estimated that a combination of these two measures has improved the production of smolt on the Swedish west coast by 49,000 to 92,000 smolts per year (Appelberg *et al* 1989). In the county of Halland, southern Sweden, new spawning sites have been opened in a dozen or so water systems and 28 fish-ways have been built, twenty of them since 1978. Upstream migration of fish has also been improved in a number of watercourses by demolishing dams. The importance of improving upstream migration so that salmon can re-establish in a limed water system is illustrated by data from River Fylleån, south-western Sweden. Since liming began in this river in 1982 and a salmon ladder was built in 1986, the production of salmon parr has increased sharply on electro-fished stretches upstream of the migration obstacles (Fig. 1). By elimination of the migration obstacle the potential for salmon smolt production has been calculated to increase by 10,000 to 15,000 individuals per year (Sjöstrand 1990, Widarsson 1992).

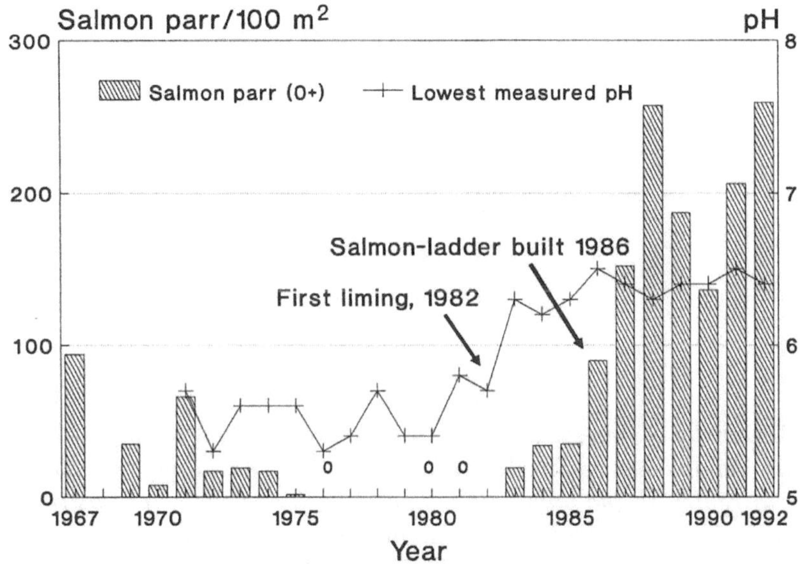

Figure 1. Density of one-summer-old salmon parr (0+) at Tolarp in River Fylleån in the county of Halland, S Sweden, before and after the start of a lime doser in 1982 and the installation of a salmon ladder in 1986 (data from B. Almer, A. Johlander and E. Degerman).

2.3. Restocking existing populations

Efforts have been made to speed up the re-establishment of weak fish and crayfish populations in limed waters by introducing extra stocks. The main species of interest here are trout (sea trout and brown trout living in lakes and stationary trout in streams), but salmon and grayling have also been target species. In some cases, stocking of trout has been conducted to save threatened stocks of freshwater pearl mussels, which are dependent upon salmon and trout as host fish for their larvae. In the stream Venabäcken, southern Sweden, 5,000 trout fry were released both 1989 and 1990 at sites where the stocks were weak. It is still too early to assess the effects on the mussel population, and perhaps also the long-term effects on the trout population. However, previous results from electro-fishing indicate that recruitment of one-summer-old trout - which is important for the survival of the mussel larvae - has not increased as much as was expected (Fig. 2).

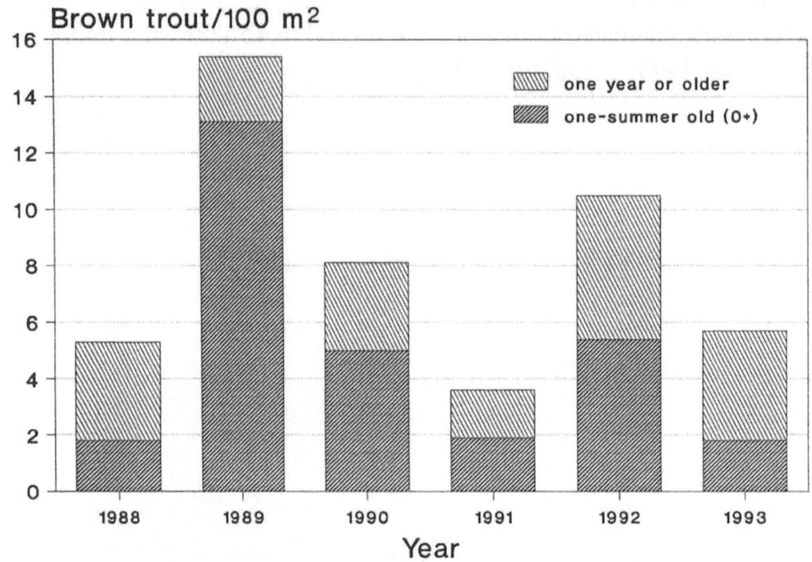

Figure 2. Density of one-summer-old trout (0+) and older trout (>0+) in the stream Venabäcken, S Sweden, 1988-1993, before and after releases of trout fry to replenish the population. Lake Vågsjön, which lies upstream, was limed in 1983 and 1988, and the wetlands in the catchment area of the stream Venabäcken were limed in 1988. Electro-fishing was performed after releases of trout in 1989 and 1990.

In running waters, the fish used for stocking often consist of one-summer-old or one-year-old individuals. For salmon and sea trout, migratory smolts have also been used. The fish released in lakes are usually more than two years old. Stocking may be performed on various occasions between June and September, depending on the age of the fish, but August and September are the main months despite the fact that spring and early summer are preferable from the point of view of survival (Cresswell *et al* 1984). This is probably mainly due to poor information and the general view that "the bigger the fish, the better the result". Another reason may be that it is more profitable for fish farms to supply fish in the autumn than in the spring.

Despite the fact that general guidelines and recommendations state that fish orginating from the existing population in the watercourses should be used, the stocks are often dominated by fish which have been hatchery-reared for many generations. The reason for this is that it is often difficult to obtain sufficient stock of the same or similar strain as the original population. In many cases, trout and salmon populations have been severely reduced, not only by acidification but also by dams, sluices and floatway regulation and this has limited the possibilities of obtaining stock with the special characteristics of the original population. However, local trout strains are genetically much better adapted than farmed stock to the specific conditions of a water and are therefore preferable in most situations.

In most cases, measures taken to replenish existing trout populations have met with limited success. No specific recapture data are available for limed waters, but for unlimed waters it has been shown that stocking with one-summer-old trout usually results in a 5-30% survival rate during the first year. However, only 3-7% of the fish live long enough to reach a catchable size (2-3 years of age) (Berg and Jørgensen 1991, Näslund 1990, 1992). Survival rates and recapture depend on competition and density, which means that the higher the density of the existing stock, the lower the survival and recapture rates (L'Abbee-Lund 1986, Näslund 1992). This also means that the recapture rate depends on the relationship between the number of fish released and the production capability of the water. Hatchery-reared trout strains generally involve lower survival and recapture rates than the offspring of wild fish (O'Grady 1984, L'Abbee-Lund 1986). In general, the shorter the time the fish has been reared, the better the result. Recapture rates also depend largely on the type of hatchery-reared strain involved (Gönzi *et al* 1986, Hulbert 1985).

One problem associated with the release of hatchery-reared fish is the genetic impact such fish may have on the original population (Krueger *et al* 1981, Hindar *et al* 1991, Waples 1991). When a population of hatchery-reared fish becomes established in a water, it may have a negative impact on the natural population through hybridization and the introduction of foreign genes. This represents a threat to the integrity of the genetic pool and character of the local population and, in the short term, may impair its ability to adapt to the local conditions and thus to survive (Taggart and Ferguson 1986, Hindar *et al* 1991, Ryman 1991). In waters where hatchery-reared fish have been released over long periods of time, trout populations often have a high percentage of genes from reared fish (Taggart and Ferguson 1986). The trout populations of Europe and North America (USA and Canada) display considerable genetic variation both within and between different water systems (Ferguson 1989). If this variation is to be sustained in limed waters, release of hatchery-reard fish should be avoided and efforts should concentrate on identifying and preserving the genetic diversity of the original stock.

In certain limed waters, salmon smolts have also been released in order to replenish the existing population. In these cases, however, the objective has in general not been to remedy acidification damage, but to compensate for the effects of water regulation. Water systems where this has been done include Göta Älv, Viskan, Ätran, Lagan and Mörrumsån, all in southern Sweden. Young sea trout have also been released in certain waters for the same reason. The effects of such releases in limed waters have not been evaluated, however. Toivonen (1975) reports that experiments with marked fish in unlimed waters in northern Sweden (Kalix, Pite and Lule Rivers) show an average recapture rate of specimens of catchable size of 10%. The majority of these fish were caught in the sea, however, and the proportion returning to the freshwater is even lower. According to Shearer (1965), only 0.3-5% of the smolts released in the freshwater return as adult fish. The proportion for fry stocking is even lower, particularly when the stock does not come from the river concerned. Althukov and Salmenkova (1990) found that only about 0.05% of the salmon fry released returned as adult fish to the "foreign" waters where they had been released. As in the case of trout, replenishment of existing populations with hatchery-reared fish should be avoided if one wishes to preserve the original wild genetic strain.

Running waters and lakes have also been stocked with (mainly) adult and sexually mature crayfish in order to replenish existing populations. As a result of the spread of the crayfish plague fungus *(Aphanomyces astaci)* among crayfish in Sweden, it has been difficult to obtain noble crayfish from nearby waters. The only options have been to use either hatchery-reared crayfish or individuals caught in completely different water systems. However, this has probably less impact on the genetic integrity of the original population than in the case of salmon and trout, since there is very little genetic variation between different populations of Swedish noble crayfish. Using electrophoresis, Agerberg (1990) was unable to demonstrate any genetic variation among four populations of noble crayfish in a geographic range from Poland to Norrbotten in northern Sweden.

The need to replenish stocks of crayfish is greater than in the case of fish, since the crayfish populations spread and become established more slowly than fish in limed waters. Repeated stocking is often required, both when reintroducing and when replenishing stocks, in order to achieve a satisfactory population density. The release of adult crayfish rather than fry is justified from the point of view of the better survival rate. Adult specimens are less vulnerable to predation from perch, for instance. In many cases, crayfish have been released in the same water on repeated occasions over a number of years. Studies performed by the Institute of Freshwater Research, Drottningholm, indicate that re-establishment is very slow despite repeated replenishment (Appelberg and Bergquist, unpubl. data).

2.4. Decimation fishing

In acidified and limed waters, reproductive damage and local extinction of species often lead to changes in the species balance and composition of fish populations. When more tolerant species such as perch and northern pike *(Esox lucius)* increase in number after liming, it may be desirable to decimate the stocks by fishing in order to modify the dominance and size structure of the populations and reduce the predation and competition pressure on species which are more susceptible to acidification, such as roach, vendace and crayfish, thus promoting a return to more normal conditions.

There are two models for decimation fishing. One is to target one or two species, while the other is to deplete stocks by removing all the fish caught. The latter method has been used to reduce mercury levels in perch and pike, for instance (Håkanson *et al* 1990, Verta 1990). This is usually done in the breeding season using fish traps and hoop nets, but also at other times of the year using nets and long-lines.

Single species are usually targeted in waters where the objective is to change the dominance and size structure of the population. This method has been used with varying success in unlimed waters to change the numbers, size structure and individual growth among perch, Arctic char and whitefish *(Coregonus* spp.) in nutrient-deficient lakes (Langeland 1986, Craig 1987, Amundsen 1989). Targeted fishing has also been employed in more nutrient-rich waters where the purpose has been to reduce the numbers of coarse fish such as roach and bream *(Abramis brama)* (Andersson 1988, Hamrin *et al* 1991).

Several studies have indicated that decimation fishing may change the composition of a fish population in terms of size, age, individual growth and new recruitment of young individuals temporarily, over a period of a few years (Svärdson 1970, Langeland 1986, Craig 1987, Sjöstrand 1991). Of these, only Sjöstrand (1991) has studied decimation fishing in a limed lake, Lake Förhultasjön (100 ha), southern Sweden. In this lake, where an average of 1.2 kg perch was removed per hectare and year over a five-year period, both new recruitment of young individuals (one-summer and one-year-old perch) and growth of one-year-old perch were improved. In another limed lake, Lake Bottentjärn (16 ha) mid-Sweden, an average of 3.7 kg perch/ha and year was removed over a three-year period from 1987-1989 (Lindgren 1992). This resulted in a change in the structure of the community in the form of a sharp reduction in fish measuring between 10 and 15 cm, while new recruitment of small fish increased.

In order to have an impact on the size of the total population and the balance of species, however, the number of fish removed must exceed production/yield. According to Toivonen (1991), it is possible to remove up to 25% of the biomass of a fish species without significantly changing the dominance situation among the populations. In nutrient-deficient forest lakes with perch as the dominant species, the biomass of fish averages around 20 kg/ha (Toivonen 1991). This means that at least 5 kg perch/ha must be removed each year if the total number of individuals in the population is to be reduced. In most limed waters where decimation fishing has been carried out, it has not been possible to reach this level. The average number of perch removed from sixteen limed waters only amounted to 3.8 kg/ha and year and varied from 1 to 8 kg (Bergquist, unpubl. data). Another problem is that there is an initial increase in production/yield when more fish are removed, which makes it even more difficult to remove sufficient fish to decimate the overall population. The conclusion is that decimation fishing as a supplementary measure to liming should be limited to smaller lakes (<100 ha) with low production levels, otherwise it cannot justify the work involved.

Targeted fishing for perch has also been employed in order to reduce predation on crayfish. The results of experimental fishing in Lake Hjärtasjön, southern Sweden, suggest, however, that the perch population must be drastically reduced if there is to be any appreciable result in terms of the abundance and growth of individuals among crayfish populations (Appelberg and Bergquist, unpubl. data). In Lake Hjärtasjön, only 2 kg/ha and year were removed, which was not sufficient to achieve the desired effect. Elsewhere, fishing in running waters has been targeted at pike in order to reduce predation on trout, but the results have not been documented.

2.5. Other measures

Other measures which have been used to supplement liming in certain cases include regulating the pressure caused by fishing, and fertilization. Such measures have not been eligible for state subsidies, however.

In lakes which are popular for leisure fishing, one obvious measure is to regulate the fish pressure in order to facilitate re-establishment of locally extinct or weakened stocks after liming. The most common regulatory methods consist of setting minimum sizes for removal of fish, limiting the number of fish which may be caught, seasonal restrictions (e.g. prohibiting fishing during the spawning season) or a complete ban on fishing over a period of years. The latter is the most suitable method in limed waters and is particularly appropriate in places where stocks have been released for reintroduction or replenishment, especially if crayfish are being reintroduced. In nutrient-deficient mountain lakes where Arctic char and trout are heavily fished, fishing restrictions have had positive effects in terms of the number of fish above 25 cm in length (Filipsson 1989).

As acidification progresses, phosphorus levels decline in certain lakes. This is thought to be attributable to reduced leaching of phosphorus from the surrounding soil (Wilander *et al*, Chap. 5). In order to restore the nutrient levels in limed water it may be necessary in certain cases to perform cautious post-liming fertilization in order to facilitate the re-establishment of roach, for instance. In other lakes, phosphorus levels may increase after liming (Wilander *et al*, Chap. 5) which means that there is no general need of fertilization. A combination of liming and fertilization has been tested in three lakes in southern Sweden involving both nitrogen and phosphorus (Andersson *et al* 1989, Olofsson *et al* 1988, Olsson *et al* 1992). In two lakes, where the purpose was primarily to reduce the mercury levels in the fish, an increased number of eutrophic species and higher species diversity were observed among phytoplankton and zooplankton. In the third lake, the fertilization trials in Lake Hecklan resulted in increased phytoplankton and zooplankton production without any noticeable changes in the species composition structure. This lake was only mildly acidified, however, and no noticeable changes were observed in the species composition of phytoplankton and zooplankton as a result of acidification.

3. Reintroduction of locally extinct species

Measures aimed at reintroducing locally extinct species in limed waters have hitherto focused on the species which are interesting in terms of sport fishing, such as salmon, trout, grayling and Arctic char, and crayfish. Less attractive species, albeit susceptible to acidification, such as roach, European minnow *(Phoxinus phoxinus)*, whitefish and vendace have been reintroduced only in relatively rare cases despite the fact that they are often key species with a structuring function in the ecosystem. In many cases, it is just as important to reintroduce these species as it is to reintroduce salmon and trout. Flora and lower fauna have received even less attention, as a result of inadequate insight into the need of such measures and the methods which should be used. Reintroduction of locally extinct species should be considered in cases where natural recolonization is not possible or is expected to take far too long time, i.e. more than 5-10 years. Table 2 shows the species which have been reintroduced in limed waters.

3.1. Salmonid fish (Atlantic salmon, trout, grayling and Arctic char)

Reintroduction of trout (both sea trout and brown trout) has been undertaken extensively in running waters and lakes. As trout often occur far upstream in a water system, they are perhaps more severely affected by acidification than many other species of fish. Besides the fact that this is a highly interesting species in terms of sport fishing, this explains why it is the target of a large proportion of the measures undertaken. Atlantic salmon and grayling have also been reintroduced in running waters, albeit to a lesser extent, while Arctic char has only been reintroduced in lakes.

Roe and fry are mainly employed for reintroduction of trout in running waters, although one-summer-old and sexually mature fish may also be used. In lakes, the fish are normally two years old or more in order to reduce the risk of predation from perch and pike, for instance. It is not clear, however, whether this has improved the survival rate. As a rule, trout strains from the same water system have been used where available. Otherwise local strains from nearby waters are used, but in many cases the only option has been to use farmed strains from elsewhere in the country.

In running waters, the reintroduction of trout by way of roe and fry has

Table 2. Species or groups of species reintroduced to limed waters in Sweden. Governmental subsidy: 1986-1991 (*), 1986-1991 and 1992-1993 (**), none (-). Long-term effect: good (+), limited (-), no evaluation performed (0).

Taxa	Grant	Result	Comments
Trout	**	+	Difficulty in obtaining suitable stock
Atlantic salmon	**	(+)	Poor remigration after stocking
Grayling	**	+	Difficulty in obtaining suitable stock
Arctic char	**	+	Difficulty in obtaining suitable stock
Whitefish	**	0	No documentation/evaluation available
Vendace	**	0	No documentation/evaluation available
European minnow	**	0	No documentation/evaluation available
Roach	**	+	Breeding stock from nearby lake
Bream	**	+	Breeding stock from nearby lake
Perch-pike	**	-	No documented reproduction
European perch	**	+	Breeding stock from nearby lake
Northern pike	**	+	Breeding stock from nearby lake
Noble crayfish	**	+	Very slow establishment/population growth
Signal crayfish	*	+	No governmental subsidies after 1991
Other crustaceans	**	0	No documentation
Freshwater pearl mussel	**	0	Evaluation in progress
Other molluscs and mussels	*	0	No documentation
Other benthic faunas (insects)	*	0	Little need of reintroduction; natural air-borne recolonization
Flora	-	0	Little need of reintroduction; natural air-borne recolonization (seed dissemination)

yielded good results in terms of re-establishment, although there is a shortage of documentation. Few studies have been performed on this sort of measure, even in unlimed waters. However, if the water quality has been restored and releases of roe and fry have been performed in the correct way and over the correct type of bottom, the survival rates should be at least as good as in unlimed waters with residual fish populations. According to studies performed in such waters, the survival rate of fry varies during the first summer from 0 to 80%, but because of a high death rate during the winter, it averages only 3-4% after the first year (Cooper 1959, Näslund 1992). Where roe, fry and one-summer-old fish have been released, the survival rate for the first year is usually equal to that of the existing population (Näslund 1992). If the fish survive the first year, the mortality declines, although it is still high. Only about 25% of the one-year-old fish are expected to survive two years (Cooper 1959).

The survival rate is higher for fish which have passed the roe and fry stage at the time of release, but mortality is still high during the first year (Näslund 1992). Releases of older fish, despite better survival rates, yield relatively low harvests of fish of catchable size (Hume and Parkinson 1988, Berg and Jørgensen 1991, Näslund 1992). Releases of roe and fry give the best results in waters which are devoid of fish. Where some fish remain, one-summer-old fish are usually used for stocking. The choice of trout strain is also of great importance in terms of survival and recapture rates. It is known from early studies that the recapture rate for wild fish is much higher than for hatchery-reared strains (Strange and Kennedy 1979, O'Grady 1984). The best result is obtained if wild fish from the same water system are used. This can be done in the form of releases of roe or fry from wild fish or by moving spawning fish upstream. If no stock is available from the same river, local strains from nearby water systems should be used.

Trout of over one year of age are the first choice for restocking limed lakes, although fry have been used in lakes which are devoid of fish. As in the case of reintroduction of roe and fry in running waters, there is a shortage of documented results. One case has been recorded from Lake Södra Boksjön, southern Sweden, where sixteen downstream-spawning trout from a nearby stream were released in 1980. Test catches subsequently showed that the trout were established in the lake (Degerman 1989). In Lakes Blanksjön and Iglafallsjön, southern Sweden, trout populations were re-established after restocking (Nyberg 1984). In Lake Blanksjön, 3,000 trout hatched the same year were released, together with 5,000 free-swimming fry in Lake Iglafallsjön. At the time this was done, there were no other fish in either lake. Turnpenny (1991) also reports successful reintroduction of trout after liming in Loch Fleet in Scotland. In this lake, a mixture of one-, two- and three-year-old trout from three different local strains was released. One of these strains was drawn from further downstream in the same water system.

Salmon have only been reintroduced in a few cases after liming, mainly in the form of roe and fry but also as smolts in certain cases. No remigration has been documented to date. Studies of unlimed watercourses using marked fish have shown that even if a large proportion (about two-thirds) of the fry released leave the water as smolt, only 0.5-1% of them return as adult fish, despite the fact that they are hatched from wild strains (Berg 1964, Altukhov and Salmenkova 1990). When hatchery-reared strains are used, the percentage falls to less than a tenth of this figure (Altukhov and Salmenkova 1990). As in the case of trout, the best results are obtained in waters which are devoid of fish. When a vigorous salmon strain becomes established after reintroduction or after fish-ways have been built, it will compete with existing populations of trout. This may mean that the trout is forced to take refuge in tributaries (Degerman *et al* 1990).

Grayling has been reintroduced in a few more waters than salmon, but there is a lack of documentation in this case too. When grayling is reintroduced, it is mostly in the form of fry or one-summer-old fish. One example of how grayling has been successfully introduced in a limed water is Lake Holmesjön (c. 40 ha) north-east of Gothenburg (Carlstrand, pers. comm.). The species was admittedly new to the water, but it still serves as an example of the procedure for reintroduction. The lake was limed in 1983 and grayling were implanted for the first time in 1984, after which 3,000 one-summer-old grayling were released each autumn for five years. Some years after the first release, spawning was observed downstream from Lake Holmesjön, together with a growing population of grayling in the lake.

One of the first species of fish to be reintroduced in limed lakes was Arctic char, and in several cases the results have been good (Lindström and Andersson 1981, Hasselrot *et al* 1984, Nyberg 1988). In the Fulufjäll area of central Sweden, Arctic char have been released in three limed lakes with very good results. In one of these, Lake Nedre Särnamannasjön, this species was reintroduced for the first time in 1973 with the release of one-summer-old fish from surrounding lakes. Further releases of one-year-old stock were undertaken in 1975 and 1976 (Lindström and Andersson 1981). The first release (1973) failed, probably because the water quality had not been fully restored after the initial liming treatment. Test fishing in the autumn of 1975 showed that the fish released during the summer of the same year had survived. The stock reproduced in the lake for the first time in 1976, and a few years later a vigorous and productive population was established (Fig. 3). Experimental fishing in Lake Nedre Särnamannasjön and the two other lakes has yielded test catches of Arctic char averaging 1.9 kg/net since reintroduction (Andersson, unpubl. data). In Lake Västra Skälsjön, southern Sweden, Arctic char have also been reintroduced with good results (Degerman *et al*, Chap. 8, Nyberg, Chap. 11). It may, however, take time to build up a viable population for fishing after reintroduction. Catches are often low during the first years after liming, even in lakes with residual populations (Nyberg *et al* 1986, Nyberg 1988).

3.2. Other species of fish

The non-salmonid fish which have been reintroduced in limed waters include roach, bream, pike-perch *(Stizostedion lucioperca)*, perch, pike and minnow. The main species is roach, but releases of bream are also relatively common. The other species have been reintroduced to a lesser extent.

Since roach is susceptible to acidification, it has died out in a large number of lakes and watercourses. Reproduction is already impaired at a pH of around 6.0. The most common way of reintroducing roach is to catch sexually mature fish in nearby lakes and watercourses and release them in the waters where they have died out. In isolated cases, experiments have been

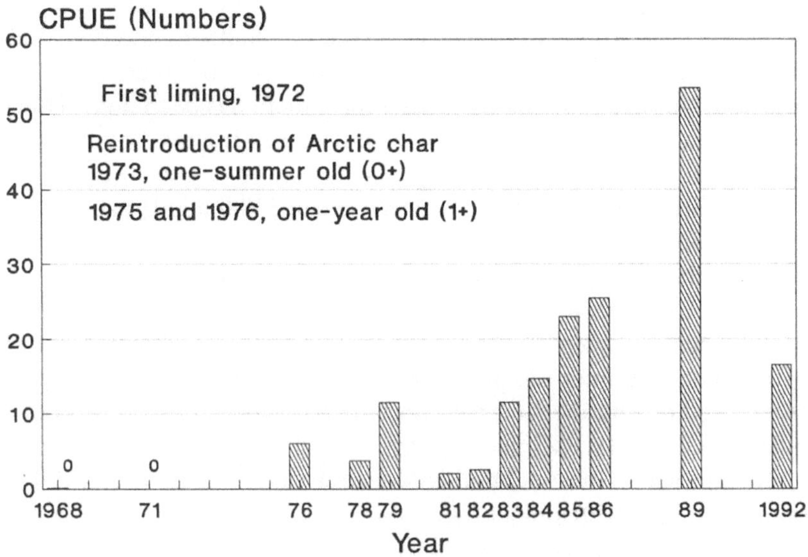

Figure 3. Catch per unit effort (CPUE) of Arctic char in Lake Nedre Särnamannasjön in the county of Kopparberg, central Sweden, between 1981 and 1992 after liming and reintroduction of one-summer-old (0+) and one-year-old Arctic char (1+).

conducted with releases of roe. Unauthorized releases of roach used as live bait have also occurred in lakes in conjunction with ice-fishing for pike. Results have varied considerably: in some cases, extensive restocking has failed to achieve any re-establishment, while small releases have met with quick success. Implants of roe have failed, while releases of adult fish have been more successful (Bergquist unpubl. data). The difference in the results obtained using adult fish is probably attributable to a combination of factors. One of them is the susceptibility of roach to stress and their high mortality rate in conjunction with capture, storage and transport. They should be held in captivity for as short a time as possible before release. In a Dutch study, Riemens (1984) reported a survival rate of approximately 70% when they were released directly after capture, but only 30% when they were kept in captivity for two to seven days before release. Competition and predation in the lake in which the fish are to be released are also important factors, as are vegetation and the nutrition status of the water - two factors which regulate the availability of protection and food for the roach fry. Finally, the genetic origin of the roach may also be of significance. Roach adapted to spawning in watercourses may have difficulties in adapting to spawning in lakes if there is a shortage of suitable watercourses.

Bream have been reintroduced in far fewer lakes than roach, but appear to be easier to re-establish, despite the fact that fewer fish have been released on each occasion. In two lakes studied in southern Sweden, Lake Gyslättasjön and Lake Nässjön, where limited numbers of bream have been released (less than 100 individuals), new recruitment has been noted (Bergquist unpubl. data). Like roach, bream have been reintroduced by releasing adult fish from nearby lakes. Lake Gyslättasjön is a good example of successful reintroduction of both roach and bream; 900 adult roach and 85 adult bream were released in 1988 and both species have since established themselves in the lake (Fig. 4; cf. Appelberg, Chap. 13).

Pike-perch have been reintroduced in a few limed waters, but no reproduction has yet been observed. In three lakes in northern Sweden, where 10,000 one-summer-old pike-perch have been released, very few fish were recaptured during the first year and the fish have failed to reproduce (Bergquist unpubl. data). Minnow, perch and pike have also been reintroduced in a number of waters, but no detailed studies exist to show how successful their re-establishment has been.

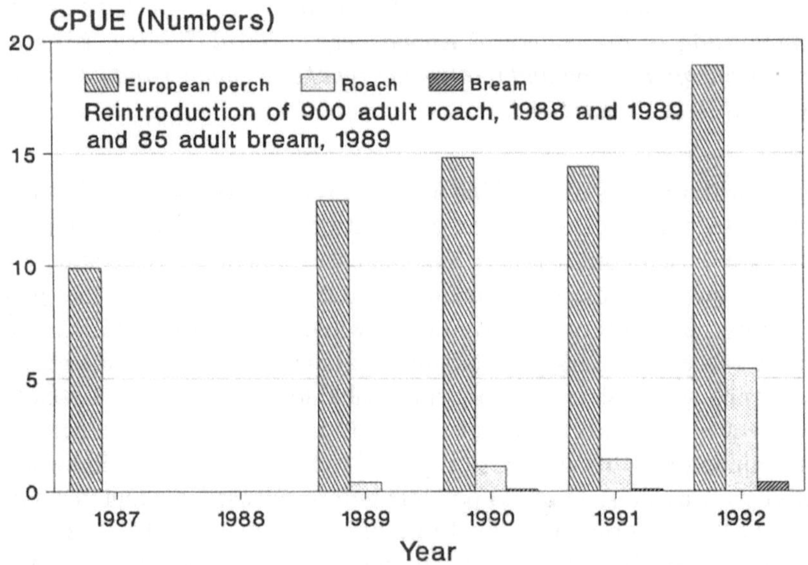

Figure 4. Catch per unit effort (CPUE) of perch, roach and bream in Lake Gyslättasjön in the county of Kronoberg, S Sweden, 1987-1992, after liming and reintroduction of roach and bream.

3.3. Crayfish and other invertebrates

Crayfish have been reintroduced in several lakes after liming. Both noble crayfish and signal crayfish were used up to 1991, but since 1991 state subsidies have only been available for releases of noble crayfish. Before the regulations for state subsidies were changed during the 1986-1991 period, a large proportion of the restocking consisted of signal crayfish, but only in waters which had been affected by crayfish plaque.

Reproducing populations of signal crayfish and noble crayfish have been established in several limed waters, but this is a very slow process (Appelberg and Bergquist, unpubl. data). Test fishing performed by the Institute of Freshwater Research indicates that density is still low several years after restocking. Even in limed waters where residual populations of crayfish still exist, population growth has been very slow (Appelberg and Odelström 1990). According to Appelberg and Odelström (1990), these results are attributable to the fact that the growth of the crayfish population is governed by factors other than water chemistry after liming. One such factor is probably the increase in numbers of predatory perch after liming.

Adult freshwater pearl mussels have been reintroduced in several limed watercourses in southern Sweden (Åkerman 1992, Henrikson unpubl. data). These mussels have survived, but no recruitment of small specimens has yet been documented. Other benthic faunas which have been reintroduced in a few waters after liming include the water louse *(Asellus aquaticus)*, the freshwater shrimp *(Gammarus pulex)*, the freshwater snail *Lymnea peregra* and the mayfly *Ephemera vulgata*. Hasselrot *et al* (1984) reported successful introduction of water louse in Lake Stora Holmevattnet in south-western Sweden. About 200 egg-bearing females were released in April 1980 and samples taken in 1981-1983 showed that the species had increased in number and spread to further areas *(op. cit)*. *G. pulex* and *E. vulgata* have been reintroduced in two streams connected with the Silen lakes in south-western Sweden. Initial sampling one year later showed that freshwater shrimp could still be found, but no mayflies were discovered (Medin, unpubl. data). The latter were, however, released in a very small number (about 150 at each site). Although adult individuals of the freshwater shrimp could be found, it was not possible at that time to confirm recruitment.

The fact that so little has been done to reintroduce the benthic fauna is very much due to the lack of knowledge concerning the measures required, and the taxonomic difficulties that exist. Several benthic species can easily be confused with others, and the risk of confusion together with the lack of background data has severely hampered active reintroduction of benthic species and other invertebrates, such as zooplankton.

4. Conclusions

As post-liming colonization and establishment problems are pronounced among benthic faunas and fish, there is a considerable need for supplementary measures in the form of biological restoration. However, it is unclear how many limed waters require such measures, since fulfillment of the biological objectives has only been studied in a small proportion of limed waters. Better monitoring of the biological effects of individual liming treatments is therefore essential if the need for supplementary measures is to be clarified.

Biological restoration of limed waters is best achieved by promoting natural colonization and re-establishment. Despite the fact that this requires more time, restoration is generally more comprehensive and lasting than if fish and benthic faunas are released directly. The difficulty of finding suitable specimens of locally extinct species for reintroduction also means that it is better to promote colonization and establishment than restocking the organisms concerned. It may, however, be necessary to reintroduce locally extinct species since they are unable to return on their own accord. The reintroduction of Arctic char in limed waters lakes is a typical example. The elimination of migration obstacles by demolishing dams and building fish passes has been given top priority, since such measures are essential if lakes and watercourses are to be recolonized by the species which have died out. In most limed waters where floatways have been cleared or waterways excavated, habitat improvement measures in the watercourses or their surroundings are essential to satisfactory re-establishment of locally extinct or depleted species.

Several restoration measures, such as the removal of migration obstacles or the release of trout and Arctic char, have shown good results in a number of restoration projects. There is a shortage of documentation in many cases, however, which makes general evaluation of biological compliance and cost-effectiveness difficult. A variety of measures have often been combined or implemented at an early stage after liming, moreover, which makes it difficult to evaluate the significance of each individual measure. By improving the monitoring of post-liming biological effects, particularly in waters where biological restoration measures have been implemented, better data can be obtained for future evaluations. In many waters, improved monitoring of water chemistry may also be necessary.

In order to make it possible to improve measures targeted at species of fish of lesser economic interest, lower fauna and vegetation, efforts should be made to achieve more wide-ranging solutions and restore entire water systems rather than individual populations.

5. References

Agerberg, A. 1990. Genetic variation in three species of freshwater crayfish, *Astacus astacus* L., *Astacus leptodactylus* Aesch., and *Pacifastacus leniusculus* (Dana), revealed by isozyme electrophoresis. - Hereditas 113: 101-108.

Åkerman, S.-E. 1992. Ett försök med återintroduktion av flodpärlmussla (*Margaritifera margaritifera*) i Silverån 1992. [Experimental reintroduction of freshwater pearl mussel (*Margaritifera margaritifera*) in the River Silverån 1992.] - Kalmar County Council,

Inform. (No. 384-10483-92).
Alenäs, I. 1986. Kalkningsprojektet Härskogen 1976-1986. Vattenkemisk och biologisk respons på kalkningsåtgärder i sju västsvenska sjöar. [The Härskogen liming project 1976-1986. Chemical and biological response to liming measures in seven lakes in western Sweden.] - Swedish Environm. Res. Inst., B 846.

Altukhov, Y.P. and Salmenkova, E.A. 1990. Introductions of distinct stocks of chum salmon, *Onchorhynchus keta* (Walbaum), into natural populations of the species. - J. Fish Biol. 37 (Suppl. A): 25-33.

Amundsen, P.-A. 1989. Effects of intensive fishing on food consumption and growth of stunted freshwater fish populations. - PhD thesis, Dep. Aquat. Biol., Univ. Tromsö.

Andersson, G. 1988. Restoration of lake Trummen, Sweden. Effects of sediment removal and fish manipulations. - In: Balvay, G. (Ed.) Eutrophication and lake restoration. Water quality and biological impacts. Inst. Limnol. Thonon-les-Bains, pp. 205-214.

Andersson, P., Grahn, P., Hörnström, E., Nyberg, P. and Dahlquist, K. 1989. Kalkning och gödsling i Rammsjön och Ämten, Örebro län 1980-1987. Ett försök att minska kvicksilverhalten i fisk. [Liming and fertilization of Lake Rammsjön and Lake Ämten, county of Örebro, 1980-1987. An attempt to decrease mercury levels in fish.] - Swedish Environm. Prot. Agency 3584.

Appelberg, M. and Degerman, E. 1991. Development and stability of fish assemblages after lime treatment. - Can. J. Fish. Aquat. Sci. 48: 545-554.

Appelberg, M. and Odelström, T. 1990. Kräftor i sura och kalkade vatten. [English summary: Effects of acidification and liming on freshwater crayfish *Astacus astacus*.] - Inform. Inst. Freshw. Res., Drottningholm (4): 1-25.

Appelberg, M., Degerman, E., Johlander, A. and Karlsson, L. 1989. Liming increases the catches of Atlantic salmon on the west coast of Sweden. - Nordic J. Freshw. Res. 65: 44-53.

Appelberg, M., Ekström, C. and Hörnström, E. 1990. Stora Härsjön - ett exempel på integrerad uppföljning av kalkningens effekter. [English summary: Lake Stora Härsjön; a case study on integrated monitoring of the effects of liming.] - Inform. Inst. Freshw. Res., Drottningholm (1).

Appelberg, M., Degerman, E. and Norrgren, L. 1992. Effects of acidification and liming on fish in Sweden. A review. - Finnish Fish. Res. 13: 77-91.

Berg, M. 1964. Utsettninger av laksyngel i vatn och tjern. [English summary: Introduction of salmon fry and parr to lakes and tarns.] - Direct. Nat. Managem. (Norway), Fisk og Fiskestell 4.

Berg, O.K. and Myhre, K.O. 1990. Handlingsplan. Fisketrapper og sikring av bestander. [Plan for action. Fishways and conservation of fish populations.] - Direct. Nat. Managem. (Norway), DN-notat 1990-4.

Berg, S. and Jørgensen, J. 1991. Stocking experiments with 0+ and 1+ trout parr, *Salmo trutta* L., of wild and hatchery origin; 1. Post-stocking mortality and smolt yield. - J. Fish Biol. 39: 151-169.

Bergquist, B.C. 1991. Extinction and natural colonization of fish in acidified and limed lakes. - Nordic J. Freshw. Res. 66: 50-62.

Bergquist, B., Engblom, E. and Lingdell, P.-E. 1992. Förekomst och kolonisation av bottenfauna i kalkade vatten. [English summary: Occurrence and colonization of benthos in limed waters.] - Inform. Inst. Freshw. Res., Drottningholm (4): 79-108.

Cooper, E.L. 1959. Trout stocking as an aid to fish management. - Pennsylvania State Univ., USA, Bull. 663: 1-21.

Craig, J.F. 1987. The biology of perch and related fish. - Croom Helm Ltd, London.

Cresswell, R.C., Harris, G.S. and Williams, R. 1984. Factors influencing the movements, recapture and survival of hatchery-reared trout released into flowing waters and their management implications. - In: Proceed. Symp., Stock enhancement in the management of freshwater fish, Budapest 1982. European Inland Fisheries Advisory Commission (EIFAC), Techn. paper 42/Suppl./1, pp. 129-142.

Crispin, V., House, R. and Roberts, D. 1993. Changes in instream habitat, large woody debris and salmon habitat after restructuring of a coastal Oregon stream. - North Amer. J. Fish. Managem. 13: 96-102.

Cueto, M. 1981. Vandløbsøkologi og Vandløbrestaurering. [The ecology and restoration of streams.] - The County Council of Nordjylland, Denmark.

Degerman, E. 1989. Redovisning av de enskilda projekten med sjöprovfiske inom Fiskeristyrelsens försöksverksamhet med kalkning av ytvatten. [Report on the test fishing in individual projects in the experimental liming programme for surface waters administrated by the National Board of Fisheries.] - Inst. Freshw. Res., Drottningholm, unpubl. rep.

Degerman, E. och Appelberg, M. 1992. The response of stream dwelling fish to liming. - Environm. Poll. 78: 149-155.

Degerman, E. and Nyberg, P. 1989. Effekter av sjökalkning på fiskbestånd. [English summary: Long-term effects of liming on fish populations in Sweden.] - Inform. Inst. Freshw. Res., Drottningholm (5).

Degerman, E., Sjölander, E., Johlander, A., Sjöstrand, P., Höglind, K., Thorsson, L. and Carlstrand, H. 1990. Kalkning för att motverka försurningspåverkan på fisk i rinnande vatten. [English summary: Liming of acidified running waters to improve conditions for fish.] - Inform. Inst. Freshw. Res., Drottningholm (4): 27-214.

Eriksson, F., Hörnström, E., Mossberg, P. and Nyberg, P. 1983. Ecological effects of lime treatment of acidified lakes and rivers. - Hydrobiologia 101: 145-164.

Ferguson, A. 1989. Genetic differences among brown trout, *Salmo trutta*, stocks and their importance for the conservation and management of the species. - Freshw. Biol. 21: 35-46.

Filipsson, O. 1989. Fiskets inverkan på fiskens storlek i fjällsjöar. [English summary: The effects of fisheries on the size of fish in mountain lakes] - Inform. Inst. Freshw. Res., Drottningholm (3).

Frissell, C.A. and Nawa, R.K. 1992. Incidence and causes of physical failure of artificial habitat structures in streams of western Oregon and Washington. - North Amer. J. Fish. Managem. 12: 182-197.

Gönzi, A.P., Henricsson, J. and Sjöberg, G. 1986. Fiskevård i älvmagasin. [Fishery management in reservoirs of regulated rivers.] - Final report from the project; Fishery management in river reservoirs (FÅK), part I, Inst. Freshw. Res., Drottningholm.

Håkanson, L., Andersson, P., Andersson, T., Bengtsson, Å., Grahn, P., Johansson, J.-Å., Jönsson, C.-P., Kvarnäs, H., Lindgren, G. and Nilsson, Å. 1990. Åtgärder mot höga kvicksilverhalter i insjöfisk. Slutrapport för kvicksilverdelen av projektet Kalkning-Kvicksilver-Cesium. [English summary: Measures to reduce mercury in lake fish. Final report from the project Liming-Mercury-Caesium.] - Swedish Environm. Prot. Agency 3818.

Hamrin, S.F., Bergstrand, E., Cronberg, G. and Romare, P. 1991. Sjörestaurering genom cyprinidreduktion. Ringsjöns utveckling sedan 1966 samt effekter av fiskreduktionen 1988-1990. [Restoration of lakes through decimation of cyprinides. The development in Lake Ringsjön since 1966, and effects of the fish reduction in 1988-1990.] - Dep. Ecol. Limnol., Univ. Lund, Rep.

Hannerz, L. and Degerman, E. 1984. Lax. En utredning beträffande förutsättningarna för det svenska laxfisket. [An official report concerning the prerequisites for the Swedish salmon fisheries.] - Ds Jo 1984:5, Liber, Norstedts, Stockholm.

Hasselrot, B., Andersson, B.I. and Hultberg, H. 1984. Ecosystem shifts and reintroduction of Arctic char (*Salvelinus salvelinus* (L.)) after liming of a strongly acidified lake in southwestern Sweden. - Rep. Inst. Freshw. Res., Drottningholm 61: 78-92.

Hindar, K., Ryman, N. and Utter, F. 1991. Genetic effects of cultured fish on natural fish populations. - Can. J. Fish. Aquat. Sci. 48: 945-957.

Hulbert, P.J. 1985. Post-stocking performance of hatchery-reared yearling brown trout. - New York Fish Game J. 32: 1-8.

Hume, J.M.B. and Parkinson, E.A. 1988. Effects of size at and time of release on the survival and growth of steelhead trout fry stocked in streams. - North Amer. J. Fish. Managem. 8: 50-57.

Hunt, R.L. 1969. Effects of habitat alteration on production, standing crops and yield of Brook trout in Lawrence Creek, Wisconsin. - In: Northcote, T.G. (Ed.) H.R. MacMillan Lectures in Fisheries, Symposium on salmon and trout in streams. Univ. British Columbia, Vancouver, pp. 281-312.

Karlström, Ö. 1977. Effects of human activities on the population densities of salmon and trout parr in Swedish rivers. - Dep. Zool., Univ. Uppsala, Rep.

Krueger, C.C., Gharrett, A.J., Dehring, T.R. and Allendorf, F.W. 1981. Genetic aspects of fisheries rehabilitation programmes. - Can. J. Fish. Aquat. Sci. 38: 1877-1881.

L'Abee-Lund, J.H. 1986. Fiskeforsterkningstiltak i norske vassdrag: Ørrets økologi og erfaringer fra kultiveringsarbeid. [English summary: Fishery management in Norwegian rivers: the ecology of trout and experiences from stocking and culturing of trout - a review.] - R. Norwegian Counc. Sci. Indust. Res. (NTNF), Rep. A6.

Langeland, A. 1986. Heavy exploitation of a dense resident population of Arctic char in a mountain lake in central Norway. - North Amer. J. Fish. Managem. 6: 519-525.

Lindgren, G. 1992. Projekt Kalkning-Kvicksilver-Cesium. Slutrapport för Gävleborgs län 1990. [Project Liming-Mercury-Caesium. Final report from the county of Gävleborg 1990.] - Gävleborg County Council, Rep. 1992:1.

Lindström, T. and Andersson, G. 1981. Population ecology of salmonid populations on the verge of extinction in acid environments. - Rep. Inst. Freshw. Res. Drottningholm 59: 81-96.

Linløkken, A. 1989. Foreløpig rapport om resultater av biotopforbedringsarbeidene i Tverrena og Letjenna i Glommavassdraget i Hedemark. [Report on the results of habitat improvement in Tverrena and Letjenna tributaries of the river Glomma in Hedemark.] - River Glomma project, Norwegian Inst. Water Res., Oslo, Rep.

Lowry, G.R. 1971. Effect of habitat alteration on brown trout in McKenzie Creek, Wisconsin. - Dep. Nat. Res., Madison, Wisconsin, Rep. 70.

Näslund, I. 1987. Effekter av biotopvårdsåtgärder på öringpopulationen i Låktabäcken. [English summary: Effects of habitat improvement on the brown trout (*Salmo trutta* L.) population of a north Swedish stream.] - Inform. Inst. Freshw. Res., Drottningholm (3).

Näslund, I. 1989. Effects of habitat improvement on the brown trout, *Salmo trutta* L., population of a northern Swedish stream. - Aquacult. Fish. Managem. 20: 187-198.

Näslund, I. 1990. Överlevnad, spridning och tillväxt hos naturdammsodlad, ensomrig öring i Låktabäcken, Lappland. [English summary: Survival, dispersal and growth in 0+ pond reared brown trout (*Salmo trutta* L.) released in a northern Swedish stream.] - Inform. Inst. Freshw. Res., Drottningholm (2).

Näslund, I. 1992. Öring i rinnande vatten. En litteraturöversikt av habitatkrav, täthetsbegränsande faktorer och utsättningar. [English summary: Brown trout (*Salmo trutta* L.) in running waters. Habitat requirements, density regulation and stocking - a review.] - Inform. Inst. Freshw. Res., Drottningholm (3).

Nyberg, P. 1984. Impact of *Chaoborus* predation on planktonic crustacean communities in some acidified and limed forest lakes in Sweden. - Rep. Inst. Freshw. Res., Drottningholm 61: 154-166.

Nyberg, P. 1988. Reclamation of acidified Arctic char (*Salvelinus alpinus* L.) lakes by means of liming. - Verh. Internat. Verein. Limnol. 23: 1732-1742.

Nyberg, P., Degerman, E., Ekström, C. and Hörnström, E. 1986. Försurningskänsliga rödingsjöar söder om Dalälven. [English summary: Acid-sensitive Arctic char (*Salve-*

linus salvelinus) lakes in southern and central Sweden.] - Inform. Inst. Freshw. Res., Drottningholm (6).

O'Grady, M.F. 1984. Observations on the contribution of planted brown trout (*Salmo trutta* L.) to spawning stocks in four Irish lakes. - Fish. Managem. 15: 117-122.

Olofsson, H., Blomqvist, P., Olsson, H. and Broberg, O. 1988. Restoration of the pelagic food web in acidified and limed lakes by gentle fertilization. - Limnologica 19: 27-35.

Olsson, H., Blomqvist, P. and Olofsson, H. 1992. Phytoplankton and zooplankton community structure after nutrient additions to the oligotrophic Lake Hecklan, Sweden. - Hydrobiologia 243/244: 147-156.

Riemens, R.G. 1984. Survival of roach caught with a seine net. - In: Proceed. Symp., Stock enhancement in the management of freshwater fish, Budapest 1982. European Inland Fisheries Advisory Commission (EIFAC), Techn. paper 42/Suppl./1: 33-39.

Ryman, N. 1991. Conservation genetics considerations in fishery management. - J. Fish Biol. 39 (Suppl. A): 211-224.

Saunders, J.W. and Smith, M.W. 1962. Physical alteration of stream habitat to improve brook trout production. - Trans. Amer. Fish. Soc. 91: 185-188.

Shearer, W.M. 1965. Recaptures from releases of hatchery-reared and wild smolts in the River Esk. - Internat. Counc. Explor. Sea (ICES), C.M. No. 59.

Sjöstrand, P. 1990. Kompletterande kalkningsuppföljning i Fylleån 1988. Redovisning av fiskeriundersökningar. [Complementary studies on the liming of River Fylleån 1988. Presentation of the fishery studies.] - Nat. Board Fish., Jönköping Reg. Fish. Off., Rep.

Sjöstrand, P. 1991. Intensivfiskeförsök i två sydsvenska insjöar. [Experiments on intensive fishing in two lakes in southern Sweden.] - Nat. Board Fish., Jönköping Reg. Fish. Off., Rep.

Strange, C.D. and Kennedy, G.J.A. 1979. Yield to anglers of spring and autumn stocked, hatchery reared and wild, brown trout (*Salmo trutta* L.). - Fish. Managem. 10: 45-52.

Svärdson, G. 1970. Effekter av hårt fiske. [Effects of intensive fishing] - Inform. Inst. Freshw. Res., Drottningholm (6).

Taggart, J.B. and Ferguson, A. 1986. Electrophoretic evaluation of a supplemental stocking programme for brown trout, *Salmo trutta* L.. - Aquacult. Fish. Managem. 17: 155-162.

Toivonen, J. 1975. Grunderna för beräkning av utplanteringsbehovet för vandringsfisk i Kemi älv. [Basics for calculation of stocking needs for migrating fish in River Kemi.] - Inst. Wildlife Fish. Res., Helsinki, Inform. No. 2: 1-22.

Toivonen, J. 1991. The fish populations and fisheries in small forest lakes in Finland. - Finnish Fish. Res. 12: 19-24.

Turnpenny, A.W.H. 1991. Loch Fleet: fishery restoration after liming. - In: Olem, H., Schreiber, R.K., Brocksen, R.W. and Porcella, D.B. (Eds.) International lake and watershed liming practices. The Terrene Inst., Washington, D.C., pp. 211-222.

Verta, M. 1990. Changes in fish mercury concentrations in an intensively fished lake. - Can. J. Fish. Aquat. Sci. 47: 1888-1897.

Waples, R.S. 1991. Genetic interactions between hatchery and wild salmonids; lessons from the Pacific Northwest. - Can. J. Fish. Aquat. Sci. 48 (Suppl. 1): 124-132.

Ward, B.R. and Slaney, P.A. 1979. Evaluation of in-stream enhancement structures for the production of juvenile steelhead trout and coho salmon in the Keogh river; progress 1977 and 1978. - British Columbia, Ministry of Environment. Fisheries Techn. Circ. 45.

Widarsson, L.-E. 1992. Kalkningsprojekt Fylleån. Lägesrapport för perioden september 1982 - maj 1988. [The River Fylleån liming project. Report of the situation during September 1982 - May 1988.] - Municipality of Halmstad, Rep.

Chapter 18

Socio-economic consequences of aquatic liming

Bo Bengtsson and Anders Bogelius

National Swedish Board on Fisheries, Box 2565, 40317 Gothenburg, Sweden

Summary

This chapter attempts to describe the various socio-economic consequences of liming activities. The main areas discussed are concerned with fish and fishing, although human health and the environment are also considered. Various examples of the socio-economic benefits of recreational fishing have been drawn from Swedish and non-Swedish studies which focus on the value of salmon fishing.

A section of this chapter is devoted to cost-benefit analyses of nine different liming projects. These projects were selected arbitrarily, so the results cannot be said to represent the Swedish liming effort as a whole. The cost to society is assumed to correspond to the cost of liming and fishery management, while the benefit is to be found in the value of fishing for recreational and household needs (current utilization value). The values of sustained biodiversity and other forms of utilization (such as bird-watching) which also result from liming are therefore not included. Despite this, the study shows that five of the projects could be considered profitable, one broke even and three were probably unprofitable.

The conclusion is that Swedish liming activities are an essential and profitable investment for society, with positive spin-off in terms of:
- recreational fishing, commercial fishing and fish farming;
- levels of mercury and caesium in fish;
- species and biotope conservation (biodiversity);
- increased insight into the aquatic environment and its organisms;
- environmental awareness among the general public;
- tourism.

Liming of Acidified Surface Waters
L. Henrikson and Y.W. Brodin (Eds.)
© Springer-Verlag Berlin Heidelberg 1995

1. Introduction

Fishery conservation associations and fishery management areas, various other organizations with environmental issues on their agendas, local authorities and even governments began to adopt a new attitude to the importance of the aquatic environment in the early 1970s, when the true extent of acidification damage began to emerge and demands for forceful and effective action gained political weight.

Action in the form of large-scale liming commenced in the early 1970s, and the Swedish state has been subsidizing liming treatments in acid and acidified lakes and watercourses since 1977. In the early days, one criterion for the 75-100% state subsidy was that the owner of the water and the fishing rights could prove that the fish were being harmed by acid deposition - and that she/he showed an interest in taking action.

Liming does not simply restore the ecosystem. It has a more far-reaching significance for Swedish society, in a wide variety of ways: scientific, economic, social and political alike. For owners of land and water rights and for those who fish for commercial or recreational reasons or for household use, the effects are very clear.

For the research community, the greatest interest lies in studying, describing and explaining post-liming processes in sediment and in forest and agricultural soils. In the early days, however, the negative impact of acidification on fish and fishing was the prime criterion for action, as it still is. For owners of land and water rights and for anglers, it is the expectation of restored, maintained or even improved fishing which determines whether or not liming is to be carried out. The more recent trend in projects aimed at species conservation, which has emerged as a result of increasing insight and knowledge of the other effects of acidification, is encouraging.

Studies focusing on the economic aspects of acidification are considerably less common than those which examine biological and chemical aspects. The OECD (1981) performed an overall cost-benefit analysis of reduced sulphur emissions in 1985 (Fig. 1). This study covers twelve Western European countries and indicates an overall benefit corresponding to 70% of the cost when 0.5% of the gross domestic product (GDP) is devoted to emission control and the best available technology is used. These figures apply for the most cautious assessment of the benefits. The upper limit and the margin of uncertainty, which primarily relates to health effects, show that the benefit is approximately six times greater (650 and 575% respectively) than the cost.

Butlin (1991) gives another example of the impact of air pollution on society. He has presented studies from the USA in which the cost of corrosion damage amounted to 4% of GDP. Measures corresponding to 1% of GDP would mean that 20-30% of this cost could be avoided. Air poll-

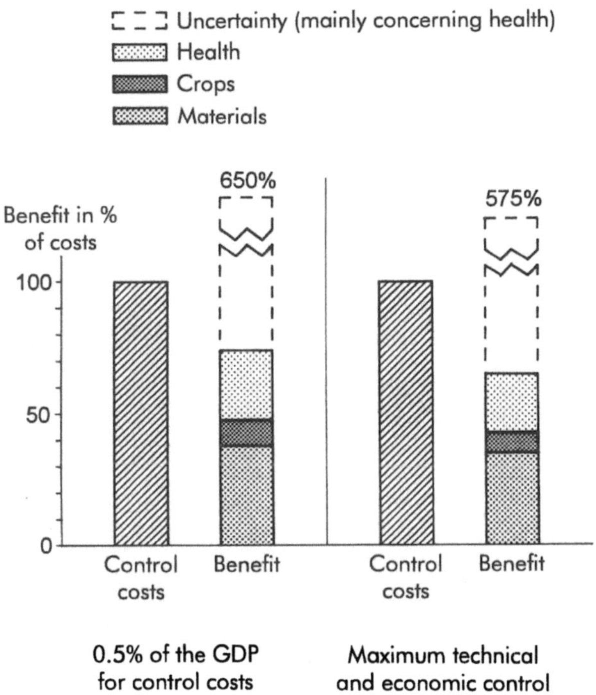

Figure 1. Cost-benefit analysis for the reduction in sulphur emissions in 12 Western European countries when 0.5% of GDP is devoted to emission control and when the best available technology is used (Wallenberg 1981).

ution is responsible for further costs amounting to 0.15% of GDP, approximately 4% of the total cost. According to Butlin (1991), British studies have shown that a 30% reduction in sulphur dioxide emissions over a period of 30 years would yield a benefit in excess of SEK 110 billion for modern and susceptible buildings in the UK.

Comprehensive socio-economic studies of air pollution and acidification are rare. The cost of the necessary measures is relatively well known, but little is known of the income side. The overall cost to Sweden of environmental protection activities amounted to SEK 1,800 million for the 1990/91 budget year (Statistics Sweden 1990b), with funds being made available to research, training and information, and direct action. International efforts via the UN and the Swedish International Development Authority also receive backing via the environmental protection budget.

Swedish industry is investing major sums of money in controlling air and water emissions. Environmental investments and ongoing costs amounted to SEK 2.5 billion in 1988 (Statistics Sweden 1990a). Most of this sum was used for water and air purification (Fig. 2). Measures have been implemented

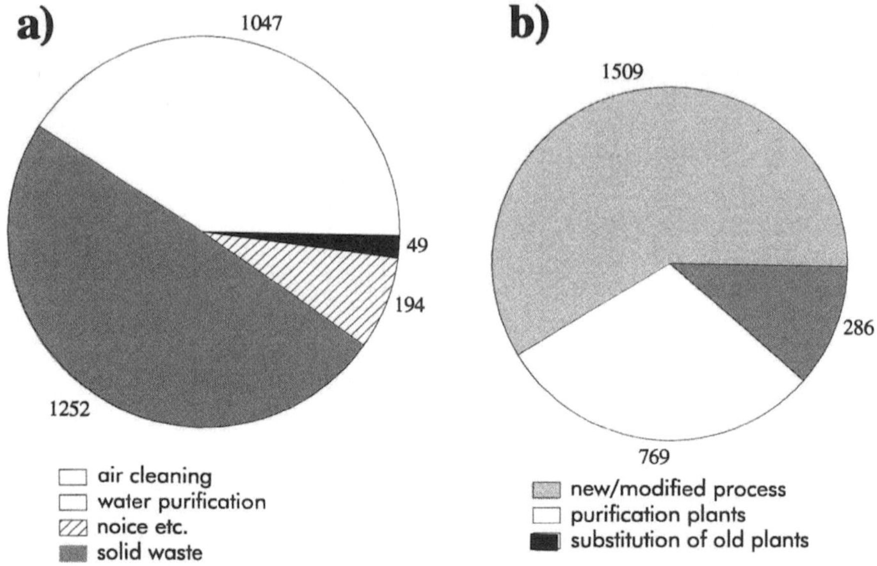

Figure 2. *Investments and costs allocated to environmental protection by Swedish industry in 1985; a) environmental issues, b) plants and processes.*

to limit emissions from traffic in the form of catalytic converters in cars, the introduction of unleaded petrol, and speed limits on certain roads in high-emission areas. Cost-intensive environmental measures have also been undertaken in forestry and agriculture, partly on the basis of public funds.

The nominal cost of liming to the nation in the form of state subsidies between 1977 and 1992 amounted to a total of SEK 1,000 million (Henrikson and Brodin, Chap. 1). Most of the grants cover 75 or 85% of the project cost, although many projects are now fully (100%) covered by state subsidies, particularly those in extensive (and thus expensive) areas of northern Sweden. No calculations have been made on the cost to the Swedish state and other interest groups within the framework of normal day-to-day operations (planning, projecting and administration). If the full cost of each project is taken into account, the total is estimated to be approximately SEK 1,150 million. This sum also includes grants for fish conservation projects. The cost of freshwater liming is thus low in relation to the cost of other measures against acidification.

The activity which is perhaps most affected by freshwater liming and which can be seen as providing "income" to society is fishing. This chapter therefore deals mainly with the socio-economic consequences of liming in terms of fishing and fishery conservation and associated activities. Liming activities preserve and restore the aquatic environment. Plant and animal species which are essential to the production of the fish species required for recreational fishing can thus survive or become re-established, assuming that there are no obstacles to re-establishment. There is no clear boundary between liming for the sake of fishing and for the preservation of species, of course. This is exemplified by salmon management measures. Sweden's genuine salmon strains are included in the Swedish Environmental Protection Agency list of endangered and rare species (Ahlén and Tjernberg 1992), but they are also of outstanding value in terms of commercial and recreational fishing. As a species, salmon is an eminent example of the value which may be lost as a result of acid deposition and which can be saved by liming. Other threats include toxic emissions to the air and water, exploitation of waters for hydro-electric power production and over-fishing or other mismanagement of resources.

NASCO (North Atlantic Salmon Conservation Organisation) (1991) has listed a number of parameters which illustrate the different "economic values" inherent in Atlantic salmon:

- it is a highly nutritional food;
- it provides income for the commercial fishing industry;
- it enhances property values for owners of land and water rights;
- it is valuable to tourism;
- it is generates jobs in a number of industries associated with fishing, e.g. the hotel and transportation industries and the fishing tackle manufacturing industry and retail trade.

In addition to these fairly quantifiable values, there are values which cannot be evaluated in monetary terms. These include relaxation, recreation and the pleasure experienced by the angler; the knowledge that the species exists; its need for clean water; its symbolic value in terms of vigour and strength; and the almost mythical qualities of its life cycle, living as it does in both freshwater and salt water with thousands of miles of migration. All these aspects contribute to the overall value of the existence of the species. Like trout, moreover, salmon is an intermediate host for the freshwater pearl mussel *(Margaritifera margaritifera)*, another endangered species, and is therefore crucial to its survival.

Apart from their immediate environmental effects, Swedish liming activities have probably had an impact in a number of other areas. They have probably played a quite significant role in promoting international agreements on emission controls and co-ordination of research and action programmes.

It is quite clear that, although detailed reports are available on the technical, economic and administrative aspects and the biological and chemical-physical results of liming, there is a lack of studies dealing with the overall socio-economic consequences. In order to remedy this serious shortcoming, projects and studies focusing on the cost-benefit aspect should be initiated and supported.

2. Methods

Air pollution causes acidification of soil and water. Although it involves no extra cost to the individual producers of the goods and services which cause the damage, it may have a far-reaching economic impact on other areas of society. The socio-economic value of an activity can only be estimated after all the effects have been taken into account, and the same applies to the benefits of measures such as emission control or liming of waters.

The purpose of this chapter is to highlight the socio-economic value of liming efforts, using external studies and a number of examples from the field. These studies focus on recreational fishing in areas which, without liming, would probably have become worthless. First, a general description is provided of the methods normally employed in estimating socio-economic values.

The foreign currencies in Section 3 have been converted according to the 1992 exchange rates prior to the depreciation of the Swedish krona and the nominal rates stated in the references.

How can the value of a natural resource be measured? What is the worth of a species? Is it at all possible to assess the value of the existence of a species of plant or animal in economic terms? What is the worth of the continued existence of a species as a part of the Swedish fauna and flora? Do we have any right whatsoever to eradicate a species?

It may be desirable to put a price on a natural resource in certain cases. One way of doing this is to estimate the willingness to pay (WTP) or willingness to accept (WTA) - a sort of "shadow" price for the resource. The former is a measure of how much an individual is willing to pay to avoid a negative impact, the latter of how much compensation the individual requires in order to accept a change.

The total value of a natural resource may be divided into the existence value and the utilization value (future and direct). The existence value is a value which the non-user attributes to the resource - the perceived benefit of the very knowledge that the natural resource exists. The utilization value can be divided into direct and future utilization value (option value). WTP and WTA depend completely upon the individual. As far as species which are generally known, familiar and appreciated (e.g. salmon, crayfish or dipper, *Cinclus cinclus*) are concerned, they can be reasonably considered to have an intrinsic value - their existence value. If they also have a commercial value, as in the case of salmon and crayfish, they may also be attributed an indirect value: they generate jobs, they promote sales of goods and services, etc. This also applies in a broader context to fishing waters and biotopes.

There are a number of methods which may be employed to calculate the value of a resource - a recreational fishing resource, for instance. Market-related methods do not provide a total value, simply a direct utilization value. WTP therefore tends to be used as a measure of value instead (Navrud 1988, 1990). WTP may be calculated using indirect methods such as the shadow price method, the unit day method and the travel/transport cost method (TKM) and direct methods such as the contingent valuation method (CVM) (Lundström 1992). Lundström considers that CVM is preferable in the evaluation of recreational fishing resources, his main reasons being that: 1) the total value can be estimated, 2) unlike TKM, CVM takes account of the fact that fishing is not the only purpose of a fishing trip, 3) the value of a local water (low travel cost) and seclusion can be estimated, which is difficult with TKM.

TKM is based on the notion that a person is unlikely to visit a fishing water, for instance, unless the visit is at least worth the cost of the trip. By studying how the number of people fishing varies with the distance to the fishing water, i.e. the travel cost, the consumer surplus can be calculated. The consumer surplus represents the difference between the maximum WTP and the actual travel cost. This difference is a measure of the net benefit of the fishing trip, which reflects the value of the fishing water. Society incurs a corresponding loss if the water becomes unusable for any reason (such as acidification).

Using CVM, the theoretically more correct measures of prosperity - CV (compensating variation) and EV (equivalent variation) - are obtained instead of the consumer surplus. The benefit is estimated by asking the angler about current expenses for fishing and the maximum he or she would be willing to pay (or what would be acceptable in compensation) in order to remain at the same benefit level if fishing conditions were to change.

When calculating the value of liming to society, the income side consists of the consumer surplus or the CV/EV and the cost side of the costs of liming and fishery conservation.

3. Examples of the value of recreational salmon fishing, the impact of acid rain and the benefit of liming

3.1. Canada

Toumi (1986) calculated that 54,000 active anglers spent SEK 234 million on salmon fishing during 1985, that SEK 270 million were devoted to long-term investments in fishing and that the value of fishing tackle amounted to SEK 732 million. It was estimated that the figures for recreational fishing corresponded to 2,100 full-time jobs/year and for commercial fishing to 163 full-time jobs each year.

Talhelm *et al* (1987) used a modified travelling-cost model and estimated the loss (fixed price according to the value of money in 1987) attributable to lake acidification in eastern Ontario over a period of fifty years at about SEK 40 million (CA$ 6,600,000) as a result of a reduction in fishing during the period, and at SEK 2,000 million (CA$ 328,000,000) if fishing in the region had been wiped out during the period.

3.2. Scotland

A study from 1982 (Anonymous 1982) showed that the cost of one day of recreational fishing corresponded to SEK 880-1170. On the basis of the findings of studies in three areas (the Kyle of Sutherland, the Tay and the Spey), the total expense for recreational salmon fishing in Scotland was estimated at SEK 220-460 million. Another study from 1989 (Anonymous 1989) puts the figure at SEK 504 million. It has been estimated that the gross expenditure on recreational fishing provides 3,360 full-time jobs each year.

3.3. England

The estimated net value of recreational fishing in certain English rivers in 1984 is shown in Table 1.

Table 1*. The value of recreational fishing in some rivers in England (Radford 1984).*

River	Value, SEK
Wye	280,716,000
Mawddoch	49,100,000
Tamar Lune	158,860,000

3.4. Iceland

In 1984, it was estimated that a rod-caught salmon corresponded to a value of about SEK 1,900. The total turnover for recreational salmon fishing in 1990 was stated to be SEK 150-200 million (North Atlantic Salmon Conservation Organization 1991).

3.5. Norway

Navrud (1991) has estimated the value of recreational fishing to some Norwegian waters in much the same way as was done in Scotland. The results are shown in Table 2.

Table 2. The value of recreational fishing in some Norwegian rivers (Navrud 1991).

River	Value, SEK
Gaula	17,225,000
Vikedalselva	698,755 - 1,011,990
Storelva	715,780 - 936,020
Audna	654,780

3.6. Sweden

As has been mentioned, there is an unfortunate lack of corresponding Swedish studies. Those which exist and which are considered to be worthwhile are discussed in the following sections, however. On the whole, they indicate just how extensive recreational fishing is in Sweden. A study by Statistics Sweden and the National Board of Fisheries (Fritidsfiske 90) (Statistics Sweden 1991) shows that fishing for recreational and household purposes represented a turnover of approximately SEK 2,000 million in 1989, not including the value of boats, caravans and property (leisure dwellings). Catches amounted to 34,000-53,000 tonnes, corresponding to one-sixth of the total commercial catch. Moreover, recreational fishing creates jobs, it is an important component in the tourist industry, it helps to keep the countryside alive and it is a leisure activity of particular social importance in terms of improving public health and quality of life, particularly among the elderly and disabled (Norling 1985).

The studies which have been published focus on the value of natural resources such as salmon and the value of associated activities such as recreational and commercial fishing. The next step is to look into the effects of liming by quantifying the impact of acidification (damage in SEK) and assuming that liming prevents such damage, and to present a number of examples which describe the results of liming in terms such as cost and income in SEK, catches in kg, number of visitors, number of overnight stays, etc., - a sort of reverse burden of proof. Most of the examples are drawn from fishing and fishing waters. These studies also discuss the value of conserving species, together with the value of liming in terms of ensuring the quality of drinking water and limiting the negative social and environmental effects of mercury and caesium in lakes and watercourses.

4. Cost-benefit studies of liming

Here, two Norwegian studies of liming activities are presented. One of these (Navrud 1988) deals with the pre-liming recreational value of fishing in River Vikedalselva, and the second one (Navrud 1990) deals with the cost/benefit to society of Norway's biggest liming project up to that time: River Audna in the Audnedal and Lindesnes municipalities in the south-west of the country. There are also some recent Norwegian cost-benefit studies on liming (Navrud 1993a,b,c). Those are, however, not considered in this chapter. The only Swedish study (Hjalte *et al* 1982) focuses on principles and was performed within the framework of liming activities which used Lake Vombsjön, an unacidified and thus unlimed lake, as the test object in order to illustrate the usefulness of the model.

4.1. The pre-liming recreational value of salmon and trout fishing in a Norwegian river

River Vikedalsälven has a drainage basin of 119 km^2 and a 10 km stretch populated by salmon and trout. The Norwegian Parliament decided in 1973 to protect the river from further exploitation. Since 1965, the water chemistry and biology have been monitored with state backing. A certain amount of liming began in 1987 to prevent excessive losses of smolt.

The objective of the study was to calculate the value of recreational salmon and trout fishing in the river in 1987, prior to regular liming of the water system.

The costs incurred by the project can be calculated relatively simply. They consist of the cost of installation, lime and labour, etc. As in the studies mentioned above, however, this study confirms the difficulty of estimating the recreational value of fishing, the value of the possibility of fishing and the value of conserving species. Navrud (1988) used both indirect (TKM) and direct (CVM) methods in order to calculate the recreational value of a day's fishing and obtained values of SEK 150 and 100-150 per day respectively. Using these input values, the total recreational value of the fish population in the river in 1987 was put at SEK 500,000-730,000.

4.2. The socio-economic value of liming in Audna

The biggest liming project in Norway prior to 1990 was Audna, a salmon river which enters the sea to the west of Kristiansand. Fifty kilometres of the river were previously inhabited by salmon and trout, but the salmon disappeared in 1985 and only a decimated population of sea trout remained.

Since liming began in the catchment area, catches have increased to above the 1970 level (Navrud 1990). Year-round treatments using dosers were supplemented with restocking and other fish conservation measures. The full effect of these measures is not expected to be felt until the mid-1990s, however. The benefit of these measures is to be found in:

- increased recreational value of fishing in the river and the sea;

- increased value of commercial fishing and fishing for household purposes in the river and the sea;
- a non-quantifiable increase in the "usefulness" of the river in terms of increased opportunities for fishing, etc.;
- increase in the value to the national economy in terms of income through tourism, spin-off for the regional economy, inflow of foreign currency;
- survival of species other than salmon and trout.

These benefits agree closely with those outlined by NASCO (see above). The parameters not mentioned here include health and quality of life and the general well-being of the local population. Only the increased recreational value has been calculated, while other values have been described on the basis of earlier studies and catch statistics, etc. Here, too, the CVM and TKM methods were used to determine the recreational value per fishing day. Using the mean value of SEK 160 per fishing day (CVM) the total value in 1988 was estimated to be SEK 750,000. This outcome does not simply indicate that the liming treatment was socio-economically profitable in itself (the cost of liming amounts to approximately SEK 2.4 million/year), but that the project as a whole showed a wide margin of profitability in terms of other, unspecified, social benefits.

5. Liming, commercial fishing, fish farming and aquaculture

5.1. Commercial fishing

Commercial fishing in Sweden today involves approximately 3,500 fishermen who land about 20,000 tonnes of fish each year with a first-hand value of approximately SEK 700 million. The trade value of Swedish fishing (consumption + export - import) in 1991 amounted to approximately SEK 7,200 million (Statistics Sweden 1992a). Most of this fishing is carried out at sea for species such as herring *(Clupea harengus)* and cod *(Gadus morrhua)*, but coastal and freshwater fishing is also conducted for species such as salmon, trout, Arctic char *(Salvelinus alpinus)*, whitefish *(Coregonus* sp.), pike-perch *(Lucioperca lucioperca)*, northern pike *(Esox lucius)* and European perch *(Perca fluviatilis)*. All these latter species are adversely affected by acid deposition.

Commercial fishing in inland lakes involves about 350 fishermen, mainly (79%) in the four largest lakes. In terms of volume, catches are dominated by vendace *(Coregonus albula)*, pike-perch, whitefish and northern pike (about 1,300 tonnes) and in terms of value by the first three of these species plus eel *(Anguilla anguilla)* (SEK 25 million).

The biggest lakes are not currently affected by acidification to the extent that liming is required. Liming is, however, being undertaken in the acidified catchment areas of Lakes Vättern, Lygnern and Bolmen, amongst others. Of the total surface area of Swedish lakes (3,850,000 ha), 68% is made up of lakes of more than 400 ha. According to Nyman (1978), this is the smallest size of lake which can support a commercial fisherman all year-round. Many lakes of this size have been limed and now possess the chemical and biological potential to support several of the species which are of importance to commercial fishing. This means that the conditions for commercial fishing in certain freshwaters are better or at least as good as they would have been without liming. The value lies mainly in the fact that job opportunities can be retained and that fishing can function as a part-time occupation to supplement agriculture, forestry and tourism. Commercial fishing in running waters is of little significance and is only carried out in the big rivers of northern Sweden, where salmon and trout are the main species caught. These waters have not yet been limed, either because it has not been considered necessary or for other reasons.

Coastal fishing, mainly for salmon and trout, has benefited from liming activities in the waters and lakes which produce these species. Such benefits have been seen on the south and west coasts. On the east coast, most of the fish is farmed, so changes in commercial catches cannot be related to liming efforts. In most cases, however, the effect on recreational fishing has been positive.

5.2. Fish farming and aquaculture

The scope of aquaculture in Sweden is limited. According to available statistics, current annual production (1990) amounts to about 6,400 tonnes of edible fish, 1,700 tonnes of mussels and 8 tonnes of crayfish (Statistics Sweden 1992b). Rainbow trout *(Salmo gairdneri)* and salmon are the main species. All salmon and mussel production and half of the production of rainbow trout take place in coastal farms.

Liming treatments have been required on several fish and crayfish farms in the most acidified areas. Crayfish, whether in natural waters or on farms, require good water quality. For optimum production, the pH should never fall below 6.0 at any time of the year (Appelberg and Odelström 1990). There is reason to believe that acidified water causes production problems (Dickson 1983). Fish farming operations have been cited in a number of applications for state liming subsidies. Together, these waters have received several million Swedish kronor for liming, which has played an active part in maintaining acceptable water quality in the fish farms and, naturally, in reducing the operating costs for the farms concerned.

6. The effects of liming on recreational fishing, leisure and tourism

6.1. Recreational fishing and liming in Sweden

Approximately two million anglers aged between 18 and 74 engage in recreational fishing for leisure and household requirements in Sweden each year, representing an annual turnover of approximately SEK 2,000 million (not including boats, leisure dwellings and other "heavier" equipment). Catches amount to 34,100-52,700 tonnes (Statistics Sweden 1991). Studies (e.g. Norling 1985) have shown that recreational fishing offers social, medical, ecological and, not least, economic benefits. The link made throughout this chapter between fish, fishing and related activities, on the one hand, and acidification and liming, on the other, is based on the assumption that, without liming, fishing will either be impaired or non-existent.

Swedish fishing for recreational and household purposes is very much dependent upon short- and long-term liming efforts. In the long term it is dependent upon effective emission restrictions. It can, therefore, be accepted that the socio-economic value of liming activities can be described with a reasonable degree of accuracy as the value of recreational fishing and other outdoor recreational activities which are restored or maintained as a result.

Liming activities began in 1977 in the form of trials conducted under the auspices of the National Board of Fisheries. In 1982, these trials became part of the normal operations for which the Swedish Environmental Protection Agency and the county councils bear the main responsibility. Liming has had an extremely stimulating effect on interest in fish conservation measures throughout the country, in practical and organizational terms alike (Andreasson *et al* 1990). This is largely due to the fact that the owners of fishing waters could, within the scope of state liming subsidies, also obtain state support for other fishery management measures such as biotope conservation, fish-ways and restocking. With subsidies covering 50-70% of the cost, these measures correspond to an overall cost of approximately SEK 25 million for the 1986-1991 period.

The state has also contributed a further SEK 50 million to corresponding activities in waters which do not necessarily require liming. Other sources in the state budget such as water charges and charges relating to environmental protection and franchise applications, etc. provide funds for specific fish conservation projects. For the 1990/1991 budget year, these amounted to approximately SEK 35 million. No figures have been given as to how much of these funds have been used for projects in limed waters. As a rough estimate, the nation must have devoted approximately SEK 5 million/year to fish conservation measures in limed acidified waters, in addition to liming, and thus to recreational fishing and the conservation of species. Is this a profitable investment? In all probability, despite the lack of studies, the answer is yes. The socio-economic benefit can be related to information published by the Swedish Tourist Board (1992) on leisure travel in Sweden. Five percent of such travel in Sweden is primarily undertaken for activities such as hunting and fishing and is estimated to have a value of SEK 30,000 million/year.

6.1.1. Forest Service fishing projects

Recently, the Forest Service has implemented two projects aimed at stimulating and extending the scope of recreational fishing in Sweden: the Crown fishing waters (Kronofisken) and the Forest Service fishing permit (Domänfiskekortet). Current liming operations have been crucial to both these projects. The very best fishing waters covered by the fishing permit, which is currently valid for more than 2,000 waters throughout the country, would not even have survived without liming. The Fulufjället district and the Hävlingen and Storån rivers in mid-Sweden are two examples.

Fulufjället, with its unique char lakes, has received full state liming subsidies since 1977, amounting to a total cost of SEK 1.4 million. Activities in the Hävlingen and Storån rivers began in 1987 and have cost SEK 3.1 million to date (Tosteby 1992, pers. comm.).

The two projects generate a total of 16,000 fishing days and 5,000 "bed-nights" a year. The Forest Service has calculated the total expense for an angler at SEK 196 per fishing day. Turnover for the two projects thus amounted to approximately SEK 3.1 million. The Swedish Tourist Board (1992) states that logistical costs for tourist activities amount to approximately 14% of the total cost. The value of bed-nights can thus be put at SEK 0.1 million and the total value of the project at SEK 3.2 million.

The Crown fishing waters consist of a smaller number of specially selected waters designed to meet the needs of different categories. Overnight accommodation is available in different price categories, for instance, and fishing opportunities exist for children, along with put-and-take fishing and fly-fishing in natural waters. Three of the Crown waters are situated in areas of the country where liming is essential to fishing. These waters represent a turnover of SEK 14.6 million/year. They generate 45,000 fishing days worth SEK 9.8 million. The cost of liming amounts to SEK 1.7 million/year. The Forest Service estimates that 50% of the fishing days would be lost without liming.

The Forest Service is also making state fishing waters available to the general public by way of permits for specific waters and areas. These waters represent an annual turnover of SEK 2.5 million and provide 51,000 fishing days. In this case, too, there would probably be a 50% loss without liming.

6.1.2. The Älvkarleby study

The Älvkarleby study is one of the few Swedish studies of the socio-economic effects of recreational fishing. Recreational salmon fishing was studied from the point of view of its total turnover, the effects on employment and tax revenue (Table 3). Revenue for the local authority was also estimated.

The main stream of the Dala River has not been limed, but several of the tributaries and lakes in the drainage basin have. Comprehensive fish management measures, including extensive restocking with salmon and trout, have also been undertaken in this fishery management area, which is one of the most well-managed and highly frequented in the country.

Table 3. *The value of recreational fishing in the Dala River at Älvkarleby, mid-Sweden, to the municipality of Älvkarleby (Uppsala County Administrative Board 1984).*

Value factor	Salmon fishing	Additional value
Total sales (SEK)	4,660,155	1,924,180
Employment (people)	10	4
Tax revenue (SEK)	217,440	53,580

6.1.3. The Blekinge County Council study

A more recent attempt to clarify the economic and social value of recreational fishing is to be found in a report by the Blekinge County Council (1991). Recreational fishing is the main tourist attraction in the county. Fishing conditions are good, there is plenty of variety and a long fishing season. More than 50,000 fishing permits were sold in 1989, worth SEK 5.1 million. Considering that coastal and sea fishing with hand-held tackle is subject to no restrictions, this is a good figure. The running waters account for 53% of the permits sold and 54% of the value. River Mörrumsån alone represents more than 97% of the total sales. Coastal fishing represented a turnover of SEK 45 million in 1989. The overall turnover for recreational fishing in the county in 1989 has been estimated at 75 million.

The report also endeavoured to clarify the social value of such extensive recreational fishing, but with no great success. Questionnaires were used to find out whether good opportunities for recreational fishing mean a better perception of leisure among the disabled (approximately 1% of the population). As in earlier studies (e.g. Norling 1985) it was found that recreational fishing is an equality sport in terms of age and sex, but that special facilities were required for the disabled, such as jetties, ramps, proximity to fishing waters and, not least, adequate information concerning the fishing opportunities which exist. Unfortunately, such measures failed to materialize, the stated reason being lack of funds.

6.1.4. The River Mörrumsån study

This study, i.e. the economic value of recreational fishing (Erlandsson-Hammargren *et al* 1991), follows much the same lines as the Blekinge study mentioned above. However, it focuses solely on the river's Crown salmon waters. This study best illustrates the value of liming to society, based as it is on the assumption that salmon and trout fishing in River Mörrumsån would be seriously impaired without liming.

The value of recreational fishing is estimated at SEK 18 million per annum, which corresponds to 28 full-time jobs/year. The value in terms of tourism of the 400,000-500,000 visitors to River Mörrumsån alone is immense. If every visitor spends SEK 50, it represents a revenue of SEK 22 million, bringing the total value up to SEK 40 million. This is twice the total (first-hand) value of Swedish commercial catches of salmon in the Baltic. It is worth mentioning, moreover, that commercial salmon and trout fishing also benefits from liming.

6.1.5. The West Coast study

Appelberg *et al* (1989) have described one good example of the importance of liming to fishing. The reproduction grounds for salmon in watercourses in western Sweden cover a total area of 1.3 km^2. Together, these areas have a potential production of 124,300-200,700 smolts/year. almost 60% of the area was probably acidified before the commencement of liming treatments. In total, the areas are considered to account for 39-46% of the production of wild smolts in the rivers on the west coast. Without the liming which has been carried out, as many as 48,800-92,290 smolts would probably have been lost each year. Salmon catches on the west coast have shown quite clearly that liming has produced results (Fig. 3). Up to the mid-1980s, catches of naturally produced salmon increased while catches of stock based on reared smolts remained unchanged. The decline towards the end of the 1980s is explained by the high mortality rate resulting from algal bloom *(Crysocromulina polyepis)* in the Skagerrak in 1988.

7. Examples of the value of liming to recreational fishing

7.1. Introduction

In the following, examples are given to illustrate the value of liming to recreational fishing. These examples are based on a questionnaire which the National Board of Fisheries circulated among seventy-seven respondents in county councils, local authorities and fishery conservation organizations in the winter of 1991/92. The purpose was to draw up an inventory of existing studies of liming, fishing and economic aspects. Six reports were re-

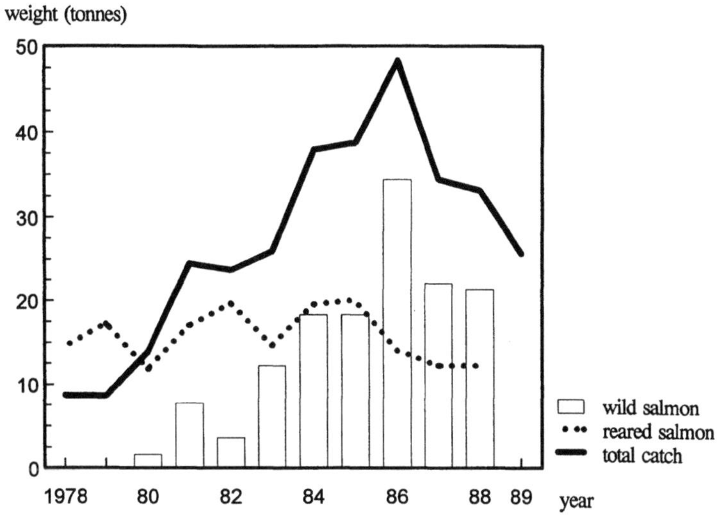

Figure 3. Swedish catch of Atlantic salmon in the county of Halland, south-western Sweden, 1978-1979 (Appelberg et al 1989).

ceived which provided appropriate material for the examples below. These examples are therefore neither randomly selected nor adequate in number to represent all limed waters in Sweden. The map (Fig. 4) shows their geographical locations.

The material provides no precise cost/benefit information and the figures should primarily be used as a basis for assessment of whether the liming costs are greater, equal to or lower than the benefit in socio-economic terms.

The net benefit of liming efforts is assumed to correspond to the recreational value minus the cost of liming, fishery conservation and other measures undertaken to promote fishing. The income from fishing permits has been assumed to be fed back into fishery conservation or measures to promote fishing and has therefore been counted as a cost. Public funds for fishery conservation have not been included since they are either minor or have been allocated in compensation (e.g. for water charges).

The recreational value is assumed to be equal to the consumer surplus calculated according to the WTP for recreational fishing. This provides an underestimate of the consumer surplus since the existence and option values are not included. In principle, individual studies of each water would be required in order to obtain adequate precision in the estimate of the socio-economic value. This has not been possible for the examples shown below. The consumer surplus from other studies (Navrud 1988, 1990, Rolfsen 1991, Silvander 1991) have been used instead after certain adjustment (deduction of cost of fishing permit).

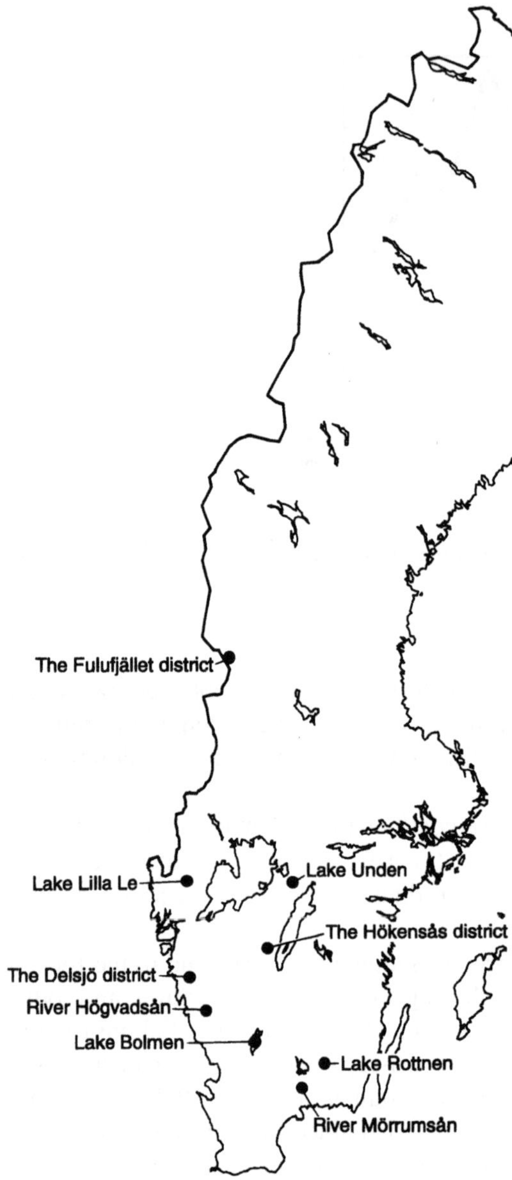

Figure 4. *Geographical locations of waters used as examples to illustrate the value of liming to recreational fishing (cf. text).*

Figure 5 shows the relationship between the consumer surplus and the fishing permit cost according to three Norwegian and one Swedish study. With the help of these relationships (which are admittedly tenuous) and information concerning the cost of fishing permit per day, a consumer surplus has been estimated for each example. If the fishing permit costs SEK

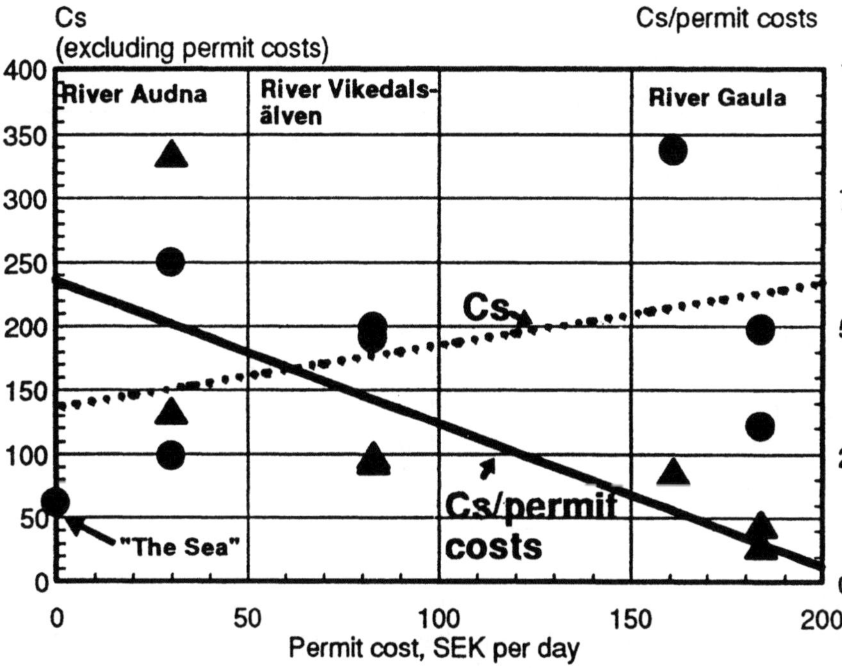

Figure 5. Consumer surplus (Cs) compared to fishing permit cost (triangles). The relationship is used to evaluate Cs by means of data on fishing frequency and income from fishing permits (circles), (cf. text).

30/day (Audna), the relationship between consumer surplus and permit cost is 5 and the consumer surplus is thus SEK 150/day (5 x 30). The total consumer surplus can be calculated in a corresponding manner. If fishing permit revenue amounts to SEK 15,000/year, for instance (with a cost of SEK 30/day), the consumer surplus will amount to SEK 75,000/year (15,000 x 5).

Two values are used to describe the liming cost, with a theoretical cost in addition to the actual cost. The theoretical cost has been calculated using the following formula:

$$\text{Cost (SEK)} = \text{Catchment area (km}^2\text{) x dose (t/km}^2\text{/year) x price (SEK/t)}$$

The theoretical cost refers to fully fledged liming activities in which the objectives are the same for all projects. The calculation of liming doses is based on the doses per area stated by the National Board of Fisheries (1982). The costs of liming treatment is based on information received from the county councils (Table 4).

Those responsible for fishery management and, in certain cases, county councils have provide information on fishing permit sales and visiting frequency, mainly for 1990.

Table 4. Examples of costs of liming treatment (liming agents + transport + spreading) for different Swedish counties and different methods. In order to correspond to 1990 prices, the costs have been reduced by 10%.

County council, reports from 1992	Cost (1990), SEK/t Boat	Aviation	Doser
Blekinge	684	720	-
Dalarna	450	810	450
Göteborg & Bohus	405	585	360
Halland	450	630	270
Jönköping	450	585	-
Kronoberg	405	765	315
Skaraborg	450	810	450
Älvsborg	383	630	360
Örebro	450	810	-

When calculating the value of fishing for recreational or household purposes, the point of departure is that all fishing will cease if liming is discontinued. In some cases, this is open to discussion in the short term, but if the high level of acid deposition continues, all the waters described here will most probably suffer. They are all situated in areas which, according to Bernes (1991), will eventually become severely acidified even if there were to be a 25% reduction in acid deposition.

7.2. Results and discussion

A compilation of the examples referred to in the following section is presented in Table 5.

7.2.1. River Högvadsån

River Högvadsån, county of Halland, is the main tributary of River Ätran, which it joins approximately 30 km from the sea. River Högvadsån drains an area of 476 km^2 and the mean flow at the mouth of the river is approximately 8.5 m^3/s. With an estimated smolt production of 25,000 per year, River Högvadsån is one of the most important reproduction grounds for salmon in the Ätran river system, accounting for approximately two-thirds of the total. In the mid-1970s, the production of young salmon had practically ceased as a result of acid deposition. Commercial fishing for salmon, like recreational fishing, was at its lowest level since 1954 when the control programme (which is still in progress) commenced. Recreational fishing in the river mainly takes place in the municipality of Falkenberg. During the 1980s, between 200 and 700 salmon were caught here each year, representing 30-50% of the total recreational fishing catches in the river. 1987 was

Table 5. Examples of the value of liming to recreational and household fishing in Sweden (cf. examples in Sect. 7). Cs = consumer surplus.

Example	Costs for permits SEK/day min/max	Cs/permit cost, SEK max/min	Income from permits, SEK/year	Benefit Cs, SEK/year min/max
Högvadsån	50/ 90	4.48/3.36	333,000	1,109,064/ 1,477,080
Mörrumsån	130/130	2.25/2.25	2,700,000	6,063,120/ 6,063,120
Lilla Le	10/ 10	5.59/5.59	12,000	67,094/ 67,094
Unden	20/ 20	5.31/5.31	18,000	95,623/ 95,623
Fulufjället	10/ 10	5.59/5.59	36,100	201,842/ 201,842
Bolmen	10/ 30	5.59/5.03	120,000	604,032/ 670,944
Hökensås	50/ 60	4.48/4.20	2,300,000	9,653,560/10,294,800
Delsjö area	70/ 80	3.92/3.64	114,000	414,914/ 446,698
Rottnen	10/ 30	5.59/5.03	40,000	201,344/ 223,648

	Cost Liming max/min	Fishery conservation	Net balance Value to society, SEK min/max
Högvadsån	1,700,000/ 516,000	330,000	-920,936/ 631,080
Mörrumsån	1,300,000/ 900,000	2,700,000	2,063,120/ 2,463,120
Lilla Le	3,500/ 10,000	12,000	51,594/ 45,094
Unden	370,000/ 275,000	18,000	-292,377/ -197,377
Fulufjället	140,000/ 81,000	41,600	20,242/ 79,242
Bolmen	2,700,000/2,000,000	120,000	-2,215,968/-1,449,056
Hökensås	20,000/ 12,000	2,300,000	7,333,560/ 7,982,800
Delsjö area	12,800/ 3,300	114,000	288,114/ 329,398
Rottnen	800,000/ 500,000	40,000	-638,656/ -316,352

a record year, with 2,993 salmon (a total of 9.6 tonnes) being caught in River Ätran - over 700 of them in Falkenberg itself (Lindgren 1991).

In addition to salmon, recreational fishing also includes pike, perch and roach *(Rutilus rutilus)*, etc. in some of the larger lakes in the limed area.

River Högvadsån has been limed since 1978, partly using two larger dosers and partly through direct lake liming. Since 1989, some wetland liming has also been performed. The theoretical liming cost according to formula 1 above amounts to SEK 1.7 million/year and the actual cost amounts to SEK 516,000/year (Degerman 1991).

Judging from the chemical and biological results achieved to date, these liming efforts have been successful. At the time of writing (1993), smolt production has returned to a natural level and recreational fishing in both the river and on the coast is enjoying normal catches.

Natural smolt production in River Högvadsån, which would have been lost without liming, is estimated at 45,000/year, with a value of approximately SEK 1 million/year (Degerman 1991). Bernes (1991) has estimated the annual benefit to society of liming in River Högvadsån to amount to SEK 600,000. The actual liming cost is approximately 500,000/year and the cost of replacing the natural smolt production in the river with farmed smolt is SEK 1,100,000. The actual liming cost is about 65% lower than that calculated using the formula. The difference is attributable to the fact that the model is based on the assumption that all lakes and watercourses in the drainage area are limed and that the treatments are performed by boat, which is not the case.

Revenue from fishing permits in River Ätran (Falkenberg), in River Högvadsån and in the larger lakes amounts to approximately SEK 330,000/year (Norell 1992, pers. comm.). Assuming that the cost of fishing permits is SEK 50-90/day, i.e. it is comparable with the cost in Audna (Navrud 1990), the consumer surplus will amount to SEK 1,109,064-1,477,080/year. The net value of liming therefore lies between -921,000 and +631,000.

7.2.2. River Mörrumsån

River Mörrumsån originates in the central regions of the county of Kronoberg and enters the Baltic in the west of the county of Blekinge. The drainage basin is 3,382 m^2 in size, with 236 m^2 of this lying downstream of Lake Åsnen. The average flow at the estuary is 27 m^3/s.

River Mörrumsån carries salmon and trout up to the power station in Fridafors, about 30 km from the estuary. Salmon production was estimated at 150,000 smolts/year in the mid-1970s (Häggström and Wendt 1975). The reproduction area covers about 65 ha, 21 ha of which are upstream of the migration barrier. Approximately 50,000-150,000 young salmon and 10,000-30,000 young sea trout are released each year (Statistics Sweden 1992b). Most of these releases are intended to compensate for expansion of hydroelectric power, while a small proportion (about 40,000 smolts) are paid for by the EC to compensate for salmon fishing by member states in the Baltic.

River Mörrumsån is best known for recreational salmon fishing, in both Mörrum and Svängsta. The Crown salmon water in Mörrum under the auspices of the Forest Service is one of the most popular in Sweden and has also attracted considerable international attention. During the late 1980s, recreational fishing accounted for catches of 400-1,300 salmon/year, together with 700-2,400 sea trout. Salmon fishing is also carried out upstream of the Crown water, but this area is not included in the study.

The River Mörrumsån water system includes Lakes Åsnen (150 km^2), Salen (2.2 km^2) and Helgasjön (50 km^2). These lakes support relatively extensive fishing for recreational and household purposes, which means that there is an intrinsic value in liming them.

According to the calculation model used, the cost of liming the entire

River Mörrumsån water system amounts of approximately SEK 8.5 million/year. The corresponding cost of liming the area downstream of Lake Åsnen, i.e. the area which is of the greatest interest in terms of salmon fishing, amounts to approximately SEK 0.6 million/year if the treatments are mainly performed by boat and SEK 0.9 million in the case of airborne liming, which is perhaps more realistic.

A total of 68 waters have been treated throughout the drainage basin, using a total of 21,400 tonnes of lime at a cost of SEK 11 million between 1977 and 1991. It is only recently that major liming activities have been implemented in the area downstream of Lake Åsnen. During 1988 and 1989, the wetland area was limed with a total of 6,500 tonnes at a cost of SEK 6.4 million (including the follow-up programme) (Blekinge County Council 1990). The effect of this treatment is expected to persist for about five years. The current annual cost is therefore about SEK 1.3 million.

Treatments targeted at the main river have been largely experimental and investigative, focusing on the effects of liming on land vegetation, metal mobility and the effects of aluminium on wild and farmed populations of salmon and trout, for instance.

Income from fishing permits in River Mörrumsån amount to approximately SEK 2.7 million/year for the Crown salmon water alone (Blekinge County Council 1991). Erlandsson-Hammargren *et al* (1991) have estimated the gross recreational value of River Mörrumsån at SEK 18 million/year (20,700 fishing days). With an average expense for fishing permits of SEK 130/day the total consumer surplus can be estimated at SEK 6 million/year. The net benefit is therefore in the region of SEK 2-2.5 million/year.

7.2.3. Lake Lilla Le

Lake Lilla Le is 53 m in depth and has a surface area of 83 ha. It is situated in Ed in the north-west of the county of Älvsborg. The water turnover time is about 20 years. The lake is drained through a diffuse outlet in the glacial ridge formation which provides a dam and separates it from the nearby (about 30 m lower) Lake Stora Le.

In terms of fishing, the main interest of the lake lies in its good population of char. Specimens weighing up to 8 kg have been caught (Rolandsson 1992, pers. comm.). European perch, roach, burbot *(Lota lota)*, northern pike, vendace and smelt *(Osmerus eperlanus)* are also found in this lake (Nyberg *et al* 1986).

Fishing in the lake is managed by the Lake Lilla Le fishery conservation association. There are about 1,500 visitors/year and fishing permit sales are worth about SEK 12,000/year (Lake Lilla Le fishery conservation association 1992, pers. comm.). The daily expense for fishing permits is about SEK 10, which gives an estimated consumer surplus of c. SEK 67,000/year.

Thanks to the long turnover time for the water, the lake does not require such frequent liming. According to the calculation model, only about 7 t/year are required, corresponding to a cost of about SEK 3,500/year. In practice, the lake has been limed on three occasions to date:

- 1978: 121 t, costing SEK 14,520 (36,700 at 1990 monetary value)
- 1985: 100 t, costing SEK 35,000 (47,200 at 1990 monetary value)
- 1989: 100 t, costing SEK 37,000 (40,800 at 1990 monetary value)

The County Council is planning to lime the lake again in 1993 and 1997, i.e. every fourth year, using 100 tonnes on each occasion. This means that the actual liming cost each year will be about SEK 10,000.

The net benefit of liming activities in Lake Lilla Le, based on the above values, is in the region of SEK 45,000-50,000/year.

7.2.4. Lake Unden

Lake Unden is 104 m deep and has a surface area of 95 km^2. It is a clear-water lake in the north-west of the county of Skaraborg. Parts of the catchment area lie within the Tiveden national park. The main species of fish are European perch, char and whitefish, although ruff *(Acerina cernua)*, roach, minnow *(Phoxinus phoxinus)*, gudgeon *(Gobio gobio)*, smelt, burbot and stickleback *(Gasterosteus aculeatus)* are also found (Nyberg *et al* 1986).

Lake Unden is one of the largest limed lakes in the world. Liming activities began in 1977 and are still under way, mostly in the form of treatment of lakes upstream and through dosers in the main stream. Char spawning grounds in the lake have also been limed. Upstream treatments have a certain intrinsic value, but this is not considered to have any major bearing on the calculations. According to the model, Lake Unden requires 820 t/year at a cost of SEK 370,000/year. In actual fact, the treatment amounts to 900 tonnes at a cost of SEK 275,000/year (Arwidsson 1992, pers. comm.).

The Lake Unden fishery conservation association manages fishing in the lake. Sales of fishing permits amount to approximately SEK 18,000/year (Arwidsson 1992, pers. comm.) and there are over 1,000 visitors/year (Skaraborg County Fishing Board 1987). The average expense for fishing permits is SEK 20/visit, giving a consumer surplus of approximately SEK 95,000. In addition, there is the value of fishing by approximately 70 holders of fishing rights (Arwidsson 1992, pers. comm.). The nature in the Lake Unden area is also of particular value since some of it is a national park. These values must be taken into account in a complete cost-benefit analysis, of course.

Taken as a whole, recreational fishing in Lake Unden does not bear the cost of liming in itself. The net cost amounts to SEK 200,000-300,000/year to society if the cost of liming is seen only in relation to the value of paying recreational fishing.

7.2.5. *Fulufjället*

Fulufjället is a sandstone plateau which is about 1,000 m above sea level and 400 km^2 in area in the north-west of the county of Dalarna on the Norwegian border. A large part of the area is a nature reserve. There are a large number of lakes and small tarns, and most of the lakes have relatively short turnover times (0.1-0.5 years). The main species of fish in the area are char, trout and burbot and, previously, also grayling *(Thymallus thymallus)*. Grayling was last caught in sample fishing in 1964 in Lake Stora Harrsjön and in 1976 in Lake Lilla Harrsjön (Andersson *et al* 1980). Char died out in Lake Nedre Särnamannasjön between 1967 and 1971 (Bergquist 1991).

Liming activities on Fulufjället started as far back as 1977 - on land, directly in the lakes and via dosers on a tributary to one of the lakes.

The char and trout stocks have recovered well in the limed lakes. The char required restocking in 1975 and 1976, while the trout recovered spontaneously (Bergquist 1991). The grayling population has not regained its former size, however.

The calculated lime requirement in the area is 100 t/year and the cost amounts to SEK 81,000/year. The actual quantity of lime is the same, but the cost amounts to approximately SEK 140,000/year (Dalarna County Council 1992).

Fishing is managed by the Forest Service and is included in their nationwide fishing permit. Since information is not available on the individual areas, it is not possible to determine exactly what proportion of sales of this permit can be attributed to fishing on Fulufjället. According to calculations of the Forest Service, the volume corresponds to approximately 4,000 fishing days/year (Tosteby 1992, pers. comm.). In 1991, 48,000 annual permits were sold, bringing in a total of SEK 5.2 million. According to Fritidsfiske '90, the average angler fishes twelve times each year. If this can be assumed to apply to the holder of an annual permit, the daily expense can be estimated at just under SEK 10/day and the total income to SEK 36,100/year, which gives a consumer surplus of approximately SEK 200,000/year.

Calculated in this way, liming activities on Fulufjället provide an annual socio-economic surplus of SEK 20,000-80,000.

7.2.6. *Lake Bolmen*

Lake Bolmen is the largest lake in the county of Kronoberg, with an area of 181 km^2 and a maximum depth of 37 m. The drainage basin is 637 km^2, of which 86% drains the two largest tributaries, River Storån and River Unneån. The theoretical water turnover time in the lake is about 2.8 years.

The main species of fish are pike-perch, whitefish, vendace, northern pike, and European perch (Kronoberg County Fishing Board 1986). Before the water level of the lake was adjusted, there was also a population of large trout which spawned at the outlet.

Fishing for recreational, household and commercial purposes takes place in the lake, together with aquaculture. According to the Kronoberg County Fishing Board (1986, 1987, 1988), recreational fishing corresponds to about 6,000 occasions/year while commercial fishing employed seven people on a full-time basis (1987) and seven on a part-time basis, while aquaculture employed one person with a permit to farm 10 tonnes of rainbow trout.

Liming activities in the catchment area of Lake Bolmen began as far back as 1977, i.e. before the water chemistry had become harmful to the fauna and flora. No liming has been performed in Lake Bolmen itself. All the treatments have been targeted at lakes and watercourses upstream. The cost of liming the entire drainage basin (1,637 km^2) is estimated at SEK 4.1 million/year. Since there is an intrinsic value in liming the River Unneån and River Storån water systems, these should not be included. The actual liming cost in the entire Lake Bolmen drainage basin amounts to approximately SEK 3 million/year (Kronoberg County Council 1991a, Jönköping County Council 1991). Since no treatments are performed in Lake Bolmen, it is difficult to calculate how much of the total cost is attributable to the lake itself. If the above area model is applied, however, a figure of approximately 66% is obtained, i.e. SEK 2 million/year.

In 1990, a fishery management area was formed, encompassing the entire lake. The previous structure of fishing rights meant that it had been difficult to obtain comprehensive information on revenue from sales of fishing permits. The cost of permits per fishing day is estimated, on the basis of other examples, to lie somewhere between SEK 10 and 30 per day, which means that the total revenue can be estimated at SEK 120,000 (60,000-180,000)/year and the consumer surplus at SEK 600,000/year.

There is also a value in the commercial fishing which is carried out in this lake. The eight commercial fishermen caught fish which, according to the Kronoberg County Fishing Board (1991) totalled 57 tonnes in 1990, corresponding to a wholesale value of approximately SEK 1.2 million. If the average price of freshwater fish to the consumer amounts to SEK 36.45/kg (Statistics Sweden 1992a) is assumed to apply, the consumption value amounts to approximately SEK 2.1 million/year.

The consumption value of fish (10 tonnes) from the aquaculture facilities can be estimated at SEK 0.5 million/year, if one assumes that it corresponds to the average price for salmonid fish during 1990, which was SEK 50/kg (Statistics Sweden 1992a).

The net cost of liming, on the basis of the above assumptions regarding the value of recreational fishing, amounts to SEK 1.4-2.2 million/year, which means that the value of recreational fishing alone does not correspond to the cost of liming and fish conservation measures.

7.2.7. Hökensås

The recreational fishing waters at Hökensås in Skaraborg county comprise about twenty nutrient-deficient lakes and tarns comprising about 200 ha (Wendt 1992, pers. comm.). The facilities are situated on the 100-km-long ridge of primary rock which runs along the western shore of Lake Vättern. Hökensås was the first recreational fishing area in Sweden to contain put-and-take lakes treated with rotenone, containing mature farmed fish. Activities began in 1956 and the number of anglers visiting the area has increased steadily. In 1990 there were about 40,000. The lakes are in an area which is susceptible to acidification and liming operations began as far back as 1959 in four lakes where the pH was between 4.2 and 4.6 (Karlén 1973). Between 1975 and 1982, the area was limed using a total of 200 tonnes (Skaraborg County Council 1984).

Liming activities in Hökensås have been quite important in terms of employment. Wendt (1992, pers. comm.) states that 10-15 people (depending on the time of year) are employed in direct conjunction with recreational fishing.

According to the County Council liming plan (Skaraborg County Council 1991), the limed drainage basin covers about 430 ha, the amount of lime amounts to approximately 25 t/year and the cost to approximately SEK 20,000/year. Treatments mainly take the form of lake liming, although some wetland liming is also performed. The calculated cost of more extensive liming is lower than the actual cost, about SEK 12,000/year.

Recreational fishing in Hökensås attracts about 40,000 visitors a year. Together, they pay approximately SEK 3.5 million in fishing charges (Wendt 1992, pers. comm.). The daily expense amounts to an average of SEK 814.50 and the consumer surplus is approximately SEK 12 million. The total net benefit of liming activities therefore amounts to approximately SEK 8.5 million.

7.2.8. The Delsjö area

The forests and hills around the Delsjö lakes on the eastern outskirts of Gothenburg are among the city's most important recreational areas. In addition to the two Delsjö lakes (139 and 61.5 ha respectively), there are a number of smaller tarns in the area, which have intermittently been used for put-and-take fishing since the 1960s. This example deals with three tarns: Västra and Östra Långevatten and Stora Björketjärn. These are 5.0, 4.0 and 1.5 ha respectively, and have drainage basins of 20, 30 and 30 ha. The main species is rainbow trout, but trout and brook char are also caught. Sales of fishing permits averaged 1,600/year during the 1983-1988 period.

The calculated cost of lime and liming amounts to only 5.6 tonnes and SEK 3,300/year. Between 1978 and 1986, the lakes were treated with a total of 76 tonnes of lime costing SEK 115,000 (1990) (National Board of Fisheries 1978, 1981). The annual cost amounts to approximately SEK 12,800 for approximately 8.4 t/year. Other fishery conservation costs have risen steadily and amounted to SEK 123,461 in 1988. Calculated according to the 1990 consumer price index, this corresponds to SEK 145,058.

Income from fishing permits in the three lakes in 1988 amounted to SEK 114,205 (Carlstrand 1992, pers. comm.) which corresponds to SEK 134,176 in 1990 figures. The daily expense for the average angler amounted to SEK 67. The total consumer surplus can be estimated at SEK 460,000 and the total net benefit at between SEK 310,000 and 320,000/year.

7.2.9. Lake Rottnen

Lake Rottnen is situated approximately 25 km south-east of Växjö in the southern part of Kronoberg County. The lake has an area of 3.3 km^2 and a depth of 18 m, with a drainage basin of 242 km^2. In terms of fishing, the lake is best known for its crayfish, with catches of between 40,000 and 100,000 crayfish/year (Kronoberg County Council Fishing Board 1986). It also supports relatively extensive recreational fishing: according to the Fishing Board (1986) the lake sold 2,030 permits in 1983. The main species fished are European perch, northern pike, whitefish and vendace, but roach, ruff, bream *(Abramis brama)* and white bream *(Blicca bjoercna)* are also caught in the lake (Lessmark 1990).

The first liming treatment was performed as long ago as 1978, mostly on forest, agricultural and meadow lands. The tributaries (100 tonnes) and the lake itself (492 tonnes) were subsequently limed during the autumn and winter of 1983 and 1984. The first substantial lake liming treatment, which also had an impact on the water chemistry, was performed during the spring of 1985 (3,850 tonnes). The lake is currently treated with dosers in a northern tributary (Lessmark 1989) and by way of lake and wetland liming in watercourses which flow in from the west (Kronoberg County Council 1991a). Between 1978 and 1985, a total of 11,402 tonnes of lime were distributed, i.e. an average of 1,400 t/year. The County Council (1991b) considers the current requirement to be about 1,200 tonnes and SEK 0.8 million/year.The calculated liming cost amounts to approximately SEK 0.5 million/year and the amount of lime to approximately 1,200/year.

The value of crayfish fishing to society has not been calculated since no comparable estimates of the recreational value are available. The following information about consumption and re-acquisition value may be of interest, however.

The value of crayfish fishing can be estimated at SEK 0.36-0.90 million/year for a yield of 40,000-100,000 crayfish/year (Kronoberg County Council Fishing Board 1986), a mean weight of 30 g/individual (Andersson *et al* 1983) and a price to the consumer of approximately SEK 300/kg.

Lake Rottnen was struck by crayfish plague in the mid-1980s and the entire crayfish population died out. With state backing, signal crayfish *(Pacifastacus leniusculus)* were introduced to the lake, with about 2 tonnes being released at a total cost of SEK 0.9 million. The economic re-acquisition value of the crayfish population can thus be said to be approximately SEK 0.9 million.

Recreational fishing in Lake Rottnen, considering the nature of the lake, the composition of fish species and permit fees, may be compared to Lake Bolmen in terms of permit expenses per fishing day (SEK 10-30/day). This means that the consumer surplus totals SEK 204,000-227,000. The cost of liming and fish conservation exceed this by SEK 313,000-636,000, which means a net cost to society.

8. Liming and mercury

During the 1900s, levels of mercury in soil, water and lake sediment have increased (Swedish Environmental Protection Agency 1991). In this context, the main waters of interest (about 400 lakes) are those formerly blacklisted by the National Food Administration, with mercury levels in excess of 1 mg/kg fish.

Intensive research, funded by the same grants as for liming activities, has indicated that more than 11,000 of the country's approximately 90,000 lakes are above the limit for mercury levels in fish. One objective of this research has been to find methods to reduce these levels through environmental action (Håkansson *et al* 1990, Meili, Chap. 16). Strategies tested include reducing the uptake of mercury in the nutrient chain and reducing the supply of mercury to the lake from the catchment area.

Unfortunately, the information available suggests that liming does not always reduce mercury levels in fish (Eriksson *et al* 1992, Meili, Chap. 16). Treatment of two west-coast lakes succeeded to the extent that the endangered roach population could be retained and restored, but it can be said to have failed in that the mercury levels have reached harmful levels in the fish which constitute the prey of piscivorous birds such as red-throated diver *(Gavia stellata)*.

9. Liming and caesium

In April 1986, Sweden was hit by radioactive fallout from the nuclear accident in Chernobyl in the former Soviet Union. Areas of northern Sweden suffered most, with fishing, hunting, reindeer farming, and fungus and berry picking being particularly affected. Figure 6 shows measured levels of caesium in the areas affected. These levels differed in the coastal areas, in running waters and in the lakes.

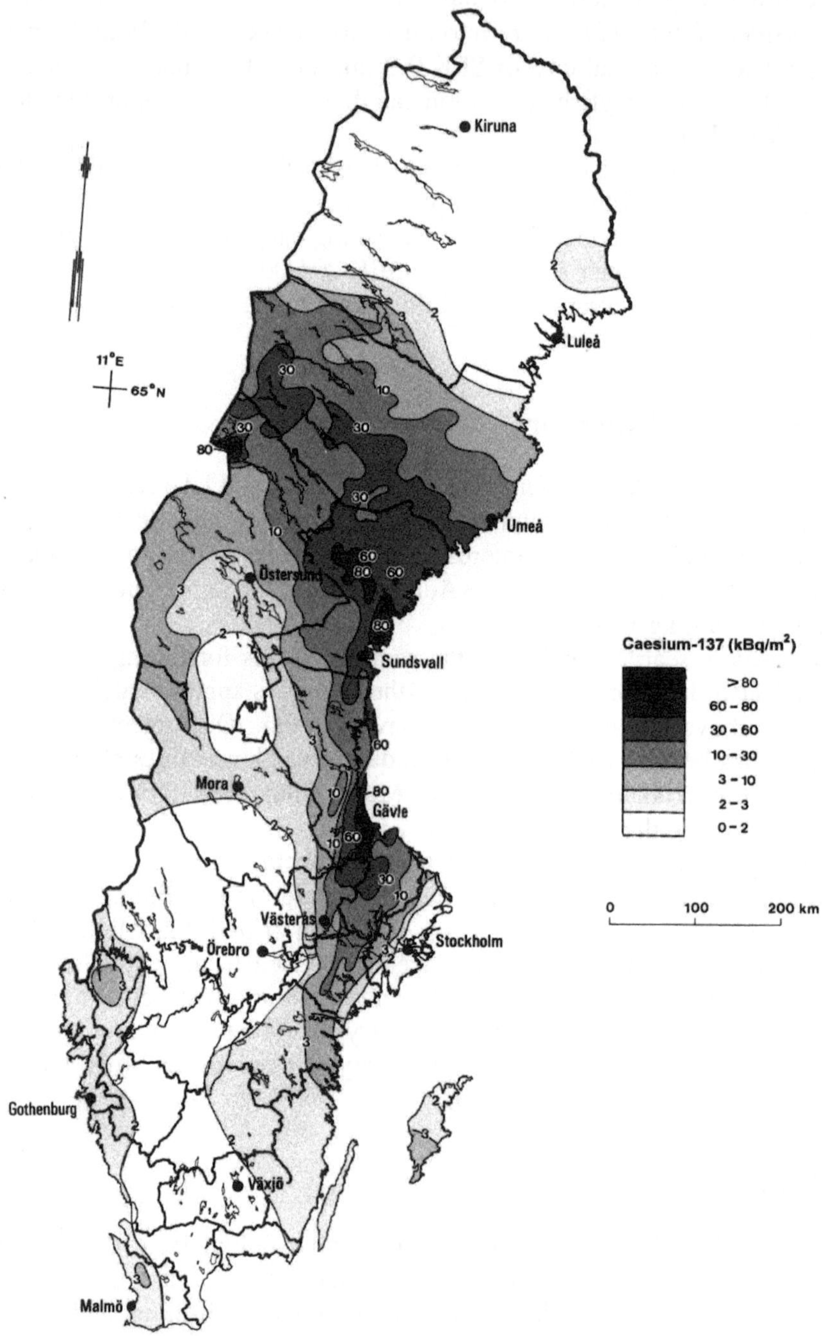

Figure 6. *Radioactive fallout of caesium in Sweden from the nuclear accident in Chernobyl, Ukraine, 1986. Assessment based on aerial measurements in 1986. (modified after Bengtsson et al 1991).*

The effects were mainly psychological in the coastal areas, as was the case in those parts of the country which escaped fallout. The reluctance of consumers to purchase fish caught in these coastal areas - or to purchase fish at all - became immediately apparent and, unfortunately, still persists after six years. Thanks to the rapid turnover of water, the running waters have recovered rapidly and levels of caesium in fish are acceptable today.

In the affected areas, lakes with low production and salt content were hardest hit. Sadly, the effects have persisted in these areas and the prognosis is not particularly encouraging (Bengtsson *et al* 1991).

Commercial fishing, fishing for household consumption and recreational fishing all suffered economically as a result of the reaction of the general public in Sweden and the restrictions and dietary recommendations (National Food Administration 1986) which the authorities issued after assessing the extent of the damage.

Therefore, it is not surprising that hopes focused on activities financed within the framework of an already long-standing project: Kalkning-kvicksilver-caesium (liming-mercury-caesium). The measures which were tested under this project consisted of lake liming supplemented by potash or selenium, wetland liming, full-scale liming, decimation fishing and fertilization.

On the whole, the effect of these measures was negligible (Andersson *et al* 1991). The high hopes of compensating for the disastrous effects of the caesium fallout in this way have so far been disappointing. The socio-economic values involved - the damage caused by the caesium fallout - are incalculable. Comprehensive evaluation is quite impossible.

10. Liming and environmental commitment

The value of liming to society in terms of the conservation of species and biotopes cannot be sufficiently emphasized. The fact that liming activities started in the first place was very much attributable to the general anxiety caused by reports and studies concerning the link between low pH in lakes and watercourses and the reduction or local extinction of fish populations (such as salmon and trout) in the early 1970s. The UN environmental conference in Stockholm in 1972 provided the impetus to international work on mapping out the effects of transboundary air pollution on plants, animals, human health and buildings. In the mid-1960s, acidification damage to lake faunas and floras began to receive increasing attention, particularly as a result of its impact on fishing. About ten years later, reports appeared about acidification damage to forests and forested land (Eriksson 1991). This gave the liming activities, which had been under way since 1977, a new dimension. It was no longer a matter of a threat to a few isolated fish populations - it was realized that entire ecosystems were under pressure and the slightest impairment of forest growth would involve losses to society amounting to billions of Swedish kronor.

Liming activities have done much to contribute to an increase in social awareness and commitment to conservation. They have done much to inform a wider public of the range of floras and faunas, their environmental needs and measures which can be undertaken to sustain or reintroduce them. This wider public has since expressed its complete political support for full-scale efforts in line with those which the 1977-1982 trials showed to be effective in fighting the impact of acidification in lakes and water-courses.

The main benefit of liming in terms of conserving species is probably the general heightening of environmental awareness throughout society together with the implementation of liming activities.

Fritidsfiske '90, a joint study published by Statistics Sweden and the National Board of Fisheries, has also shown that the environment is attributed a greater value than either the facilities or the catch potential, both among anglers and non-anglers, when it comes to evaluating the quality of a fishing water. In answer to the question of what is most important, 36% said the environment, 13% the catch and 1% the facilities. Among anglers, 54% said that fishing had deteriorated somewhat, or a lot, during recent years as a result of environmental impact.

Those who bear responsibility for liming activities at the county level (Rolandsson, county of Älvsborg, pers. comm.) consider, moreover, that the commitment of the general public to the preservation of species such as crayfish, salmon and trout has also led to stronger resistance to any exploitation of their biotopes. Such resistance also encompasses biotopes which are not threatened by acidification. This means that in recent years, it has become more expensive to exploit the Swedish environment. There are examples of development plans quite simply being shelved.

11. Acknowledgements

A number of people have been most helpful in providing material for this chapter. Sten Andreasson at the Fisheries Research Office in Härnösand provided information for the section on caesium. Rolf Persson at the fisheries department of the National Board of Fisheries provided us with statistics on commercial fishing and aquaculture. Lars Ask of the same department provided information on the objectives and scope of fishery conservation measures. Ivan Helgesson of the municipality of Lerum and Lars-Olof Ramnelid of the municipality of Uddevalla provided us with interesting data on municipal liming efforts. Vincent Rolandsson at the Älvsborg County Council gave us his views on changes in the environmental commitment of the general public and on the value of liming in terms of the conservation of species. Örjan Fritz of the Halland County

Council and Jörgen Hellberg, Ulla Bertills, Y.-W. Brodin, Kjell Johansson, George Moberg and others at the Swedish Environmental Protection Agency have provided data, reports and valuable comments.

I should like to extend my warmest thanks to all involved, including our patient editor Lennart Henriksson.

12. References

Anonymous 1982. A study of the economic value of sporting salmon fishing in three areas of Scotland. - Tourism and Recreational Research Unit, Univ. Edinburgh.

Anonymous 1989. Economic impact of salmon fishing and netting in Scotland. - Summary Rep. by Mackay Consultants to the Scottish Tourist Board and the Highland and Islands Development Board.

Ahlén, I. and Tjernberg, M. 1992. Artfakta, Sveriges hotade och sällsynta ryggradsdjur 1992. [Species information, endangered and rare mammals in Sweden, 1992.] - Data Collection of Endangered Species, Uppsala.

Andersson, B.O., Andersson, H.-G., Dahlqvist, K., Enderlein, O., Filipsson, O., Fürst, M., Hägglund, I., Nilsson, N.-A., Nyberg, P., Nyman, L., Olofsson, S., Ros, T., Sellerberg, G., Sjölund, T., Sjöström, T., Stube, M. and Ungsgård, Y. 1983. Utveckling och vård av kräftbestånd. [Development and management of crayfish populations.] - Inform. Inst. Freshw. Res., Drottningholm (8).

Andersson, G., Dickson, W., Filipsson, O., Lindström, T. and Öhman, R. 1980. Förändringar i södra fjällområdets fiskfauna - ett samspel mellan försurning och andra faktorer. [English summary: Changes in the fish populations of the southern parts of the Swedish mountain chain - interaction between acidification and other factors.] - Inform. Inst. Freshw. Res., Drottningholm (10).

Andersson, T., Håkansson, L., Kvarnäs, H. and Nilsson, Å. 1991. Åtgärder mot höga halter av radioaktivt cesium i insjöfisk. [English summary: Remedial measures against high levels of radioactive caesium in Swedish lake fish]. - Statens Strålskyddsinstitut, Rep. 91-07.

Andreasson, S., Ask, L. and Bengtsson, B. 1990. The organization of fishery management units in Sweden. - In: van Densen, W.L.T., Steinmetz, B. and Hughes, R.H. (Eds.) Management of freshwater fisheries. Proc. Symp. European Inland Fisheries Advisory Commission, Göteborg, Sweden, 31 May - 3 June 1988, Pudoc, Wageningen.

Appelberg, M. and Odelström, T. 1990. Kräftor i sura och kalkade vatten. [English summary: Effects of acidification and liming on the freshwater crayfish *Astacus astacus*.] - Inform. Inst. Freshw. Res., Drottningholm (4): 1-25.

Appelberg, M., Degerman, E., Johlander, A. and Karlsson, L. 1989. Liming increases the catches of Atlantic salmon on the west coast of Sweden. - Nordic. J. Freshw. Res. 65: 44-53.

Bengtsson, B., Norling, I. and Pierrou, U. 1991. National survey data on sportfishing, economics and environmental changes in Sweden. - Am. Fish. Soc. Symp. 12: 74-83.

Bergquist, B. 1991. Fiskevårdsåtgärder i kalkade vatten. [Fisheries management measures in limed waters.] - Nat. Board Fish., Rep. 1, 1990, pp. 80-89.

Bernes, C. 1991. Monitor 12. Acidification and liming of Swedish freshwaters. - Swedish Environm. Prot. Agency Informs.

Blekinge County Council 1990. Ansökan om statsbidrag för åtgärder mot försurning. [Application for state subsidies for measures against acidification.] - Application to Swedish Environm. Prot. Agency.

Blekinge County Council 1991. Sportfiske i Blekinge. [Sport fishing in Blekinge County.]

Butlin, R.N. 1991. Effects of air pollutants on buildings and materials. - Proceed. R. Soc. Edinburgh, 97B: 255-272.

Dalarna County Council 1992. Ansökan om statsbidrag till kalkning i Kopparbergs län. [Application for state subsidies for liming in the county of Kopparberg.] - Application to Swedish Environm. Prot. Agency.

Degerman, E. 1991. Kalkningseffekter på fisk och kräfta i vattendrag. [Effects of liming on fish and crayfish in watercourses.] - Nat. Board Fish., Rep. 1, 1990, pp. 68-73.

Dickson, W. 1983. Liming toxicity of aluminium to fish. - Vatten 39: 400-404.

Eriksson, M.O.G. 1991. Försurningspåverkan på fågel- och däggdjursbestånd. [English summary: Effects of acidification on bird and mammal population.] - Swedish Environm. Prot Agency 3969.

Eriksson, M.O.G., Johansson, I. and Ahlgren, C.-G. 1992. Levels of mercury in eggs of red-throated diver *Gavia stellata* and black-throated diver *G. arctica* in southwest Sweden. - Ornis Svecia 2: 29-36.

Erlandsson-Hammargren, E., Carlsson, B., Hedenskog, M. and Johnson, S. 1991. Sportfiskets ekonomiska värde. [The economic value of sport fishing.] - Folkhögskolan i Bräkne-Hoby, Ekologilinjen, Rep. 28.

Häggström, Å. and Wendt, C. 1975. Lax och öring i Mörrumsån. - Utredning till arbetsgruppen för Åsnen. [Salmon and trout in River Mörrumsån - study for the Åsnen Working Group. - Nat. Board Fish, unpubl. rep.

Håkanson, L., Andersson, P., Andersson, T., Bengtsson, Å., Grahn, P., Johansson, J.-Å., Jönsson, C.-P., Kvarnäs, H., Lindgren, G. and Nilsson, Å. 1990. Åtgärder mot höga kvicksilverhalter i insjöfisk. Slutrapport för kvicksilverdelen av projektet: Kalkning-Kvicksilver-Cesium. [Measures to reduce mercury in lake fish. Final report from the project Liming-Mercury-Caesium.] - Swedish Environm. Prot. Agency 3818.

Hjalte, K., Lidgren, K., Thelander, A.-L. and Wells, C. 1982. Ekonomiska konsekvenser av vattenkvalitetsförändringar i sjöar. En beräkningsmodell. [Economic consequences of changes in water quality in lakes. A calculation model.] - T.E.M., Univ. Lund, unpubl. rep.

Jönköping County Council 1991. Kalkningsplan för 1991/92. [Plan for liming, 1991/92.] - Jönköping County Council, Rep. 1991:1.

Karlén, A. 1973. PM angående kalkning av sjöar. [PM concering liming of lakes.] - Stockholm, 1973-02-16.

Kronoberg County Council 1991a. Kalkningsplan. [Plan for liming.] - Application to the Swedish Environm. Prot. Agency.

Kronoberg County Council 1991b. Åtgärder mot försurning under budgetåret 1991/92. [Measures against acidification during the fiscal year 1992/93.] - Application to the Swedish Environm. Prot. Agency.

Kronoberg County Fishing Board 1986. Fiskevårdsplan för Kronobergs län. [Plan for fishery management in the county of Kronoberg.] - Nat. Board Fish., unpubl. rep.

Kronoberg County Fishing Board 1987. Bolmen. [Lake Bolmen.] - Nat. Board Fish., unpubl. rep.

Kronoberg County Fishing Board 1988. Fångststatistik för sötvattensfiske år 1987 i Kronobergs län. [Catch statistics of freshwater fisheries during 1987.] - Nat. Board Fish., unpubl. rep.

Kronoberg County Fishing Board 1991. Fångsstatistik insjöfisket 1990 Kronobergs län. [Catch statistics of freshwater fisheries during 1990, county of Kronoberg.] - Nat. Board Fish., unpubl. rep.

Lagerman, B. 1983. Kalkningsprojektet Unden. [The Lake Unden liming project.] - Nat. Board Fish., unpubl. rep.

Lessmark, O. 1990. Kalkningar i Kronobergs län 1977-89. En utvärdering av sjökalkningars kemiska och biologiska effekt. [Liming in the county of Kronoberg 1977-1989. An evaluation of chemical and biological effects of lake liming.] - Miljöfakta i Kronobergs län 1990:1. The Kronoberg County Council.

Lindgren, B. 1991. Där laxen vandrar. [Where the salmon migrates.] - Settern Publ. Comp., Örkelljunga.

Lundström, R. 1992. Att mäta betalningsvilja - en metod att skatta sportfiskevattens rekreationsvärde. [How to measure the willingness to pay - a method of estimating the recreational value of fishing waters.] - Dep. Tourism, Univ. Östersund, unpubl. rep.

National Board of Fisheries 1978. Västra Långevatten m. fl. [Lake Västra Långevatten and other lakes.] - Application for subsidies, Nat. Board Fish., dnr. 304-586-78.

National Board of Fisheries 1981. Västra Långevatten m. fl. [Lake Västra Långevatten and other lakes.] - Application for subsidies, Nat. Board Fish., dnr 304-4374-81.

National Board of Fisheries 1982. Råd och riktlinjer för kalkning av sjöar och vattendrag. [Guidelines for liming of lakes and streams.] - Inform. Nat. Board Fish., 1, 1982.

National Food Administation 1986. Tjernobyl och våra livsmedel. [Chernobyl and our food.] - Vår Föda 38 No. 9-10.

Navrud, S. 1988. Rekreasjonsverdien av lakse- og sjøaurefisket i Vikedalselva i 1987 - før regelmessig kalking. [The recreational value of salmon and trout fishing in River Vikedalselva in 1987 before regular liming.] - Direct. Nat. Managem. (Norway), Rep.

Navrud, S. 1990. Nytte-kostnadsanalyse av vassdragskalking. En studie i Audna. [Cost-benefit analysis of liming of streams. A study in River Audna.] - Direct. Nat. Managem. (Norway), DN-notat 1990-5.

Navrud, S. 1991. The economic value of recreational fishing for Atlantic salmon - the Norwegian experience. - Paper, International Workshop on Licensed Recreational Fishing in Watercourses of the Kola Peninsula, Murmansk, USSR, April, 1991.

Navrud, S. 1993a. Sammfunnsøkonomisk lønnsomhet av å kalke mindre fiskevann. Nytte og kostnadsanalyse av kalking i innlandsfiskevassdrag. En studie av Lauvann og seks vann i Gjerstadskogene. [The economic value of liming of small fishing-lakes. Cost-benefit analyses of liming of fishing-lakes. A study of lake Lauvann and six lakes in the area of Gjerstadskogene.]- Direct. Nat. Managem. (Norway), Rep. 1993-3.

Navrud, S. 1993b. Sammfunnsøkonomisk lønnsomhet av å kalke Audna. [The economic value of liming of River Audna.] - Direct. Nat. Managem. (Norway), Rep. 1993-4.

Navrud, S. 1993c. Sammfunnsøkonomisk lønnsomhet av å kalke Vegår. [The economic value of liming of River Vegår.] - Direct. Nat. Managem. (Norway), Rep. 1993-5.

Norling, I. 1985. Sportfiske för alla. [Sport fishing for everybody.] - The Swedish Anglers Association, Stockholm.

North Atlantic Salmon Conservation Organization (NASCO) 1991. Economic value of Atlantic salmon. - NASCO Paper CNL (91) 29.

Nyberg, P., Degerman, E., Ekström, C. and Hörnström, E. 1986. Försurningskänsliga rödingsjöar i Syd- och Mellansverige. [English summary: Acid-sensitive Arctic char (*Salvelinus alpinus*) lakes in southern and central Sweden.] - Inform. Inst. Freshw. Res., Drottningholm (6).

Nyman, L. 1978. Avkastningspotentialen av konsumtionsfisk i svenska sjöar mot bakgrund av olika nyttjandeformer. [The potential yield of fish for consumption in Swedish lakes according to different forms of use.] - Inform. Inst. Freshw. Res., Drottningholm (11).

OECD 1981. The costs and benefits of sulphur oxide control, a methodological study. - OECD, Paris.

Radford, A.F. 1984. The economics and value of recreational salmon fisheries in England and Wales, an analysis of the Rivers Wye, Mawddach, Tamar and Lune. - Marine Resources Research Unit, Portsmouth Polytechnic.

Rolfsen, J. 1991. Rekreasjonsverdien av lakse- og sjøørretfiske på Tofta's soner i Gaula sesongen 1990. [The recreational value of salmon and lake trout fisheries in Totfa area in River Gaula during the season 1990.] - Inst. for økonomi og samfunnsfag & Noragric, Norges landbrukshøyskole.

Silvander, U. 1991. Värdering av kväveläckagets negativa effekter i Sverige. [English summary: Valuation of negative effects caused by nitrogen loss in Sweden.] - Dep. Economics, Swedish Univ. Agricult. Sci., Rep. 37.

Skaraborg County Council 1984. Plan för kalkning av sjöar och vattendrag inom Skaraborgs län 1984/85-1988/89. [Plan for liming of lakes and watercourses in the county of Skaraborg 1984/85-1988/89.] - Application to the Swedish Environm. Prot. Agency.

Skaraborg County Council 1991. Plan för åtgärder mot försurningen av sjöar och vattendrag i Skaraborgs län för budgetåren 1991/92 - 1996/97 med begäran om medel för budgetåret 1991/92. [Plan for measures against acidification of lakes and watercourses in Skaraborg County 1991/92-1996/97 and application for subsidies for the fiscal year 1991/1992.] - Application to the Swedish Environm. Prot. Agency.

Skaraborg County Fishing Board 1987. Unden. [Lake Unden.] - Nat. Board Fish., unpubl. rep.

Statistics Sweden 1990a. Industrins investeringar och kostnader för yttre miljöskydd 1985. [Environmental protection costs in Swedish industry 1988.] - Statistics Sweden, Örebro, Statistiska meddelanden Na 23 SM 9001.

Statistics Sweden 1990b. Statistical abstract of Sweden 1991. - Statistics Sweden, Örebro.

Statistics Sweden 1991. Fritidsfiske '90 [English summary: Recreational fishing '90.] - Statistics Sweden, Örebro.

Statistics Sweden 1992a. Fiske 1991 - En översikt. [Swedish sea-fisheries in 1991. Definitive data.] - Statistics Sweden, Örebro, Statistiska meddelanden, J 55 SM 9202.

Statistics Sweden 1992b. Vattenbruk 1991. [English summary: Aquaculture in Sweden 1991.] - Statistics Sweden, Örebro, Statistiska meddelanden, J 60 SM 9201.

Swedish Environmental Protection Agency 1991. Kvicksilver i Sverige, Problem och åtgärder. [Mercury in Sweden, problems and measures.] - Swedish Environm. Prot. Agency Informs.

Swedish Anglers Association 1978. Plan för kalkning av Västra Långevattnet, Östra Långevattnet och Stora Björketjärn. [Plan for liming of the lakes Västra Långevattnet, Östra Långevattnet and Stora Björketjärn.] - Application to the Nat. Board Fish., dnr 304-586-78.

Swedish Meteorological and Hydrological Institute 1948. De Svenska vattendragens arealförhållanden, 12. [Area conditions for Swedish watercourses.] - Meddelande, Serie C, Nr 2: (lagan).

Swedish Tourist Agency 1992. Sverigeturismen i siffror 1991. [Swedish tourism in figures.] - ISSN 1101-2870.

Talhelm, D., Hanna, J.H. and Victor, P. 1987. Product travel cost approach: estimating acid rain damage to sportfishing in Ontario. - Trans. Am. Fish. Soc. 116: 420-431.

Thörnelöf, E. 1986. Utdrag ur kalkningsregistret, 1986-02-28. [Excerpt from the liming register.] - Swedish Environm. Prot. Agency, PM.

Toumi, A.L.W. 1986. Canada's Atlantic salmon recreational fisheries and their future: an economic overview. - Atlantic Salmon Federation, St. Andrews, Canada.

Uppsala County Council 1984. Ekonomiska och sysselsättningsmässiga effekter av Älvkarleby laxfiske. [Effects on economy and employment of salmon fishery at Älvkarleby.]

Wallenberg, B. 1981. Presentation of the OECD study. The cost and benefits of sulphur oxide control. - Proc. Conf. on Acidification, Jönköping, 1981. Swedish Environm, Prot. Agency, unpubl. rep.

MIX
Papier aus verantwortungsvollen Quellen
Paper from responsible sources
FSC® C105338

If you have any concerns about our products,
you can contact us on
ProductSafety@springernature.com

In case Publisher is established outside the EU,
the EU authorized representative is:
Springer Nature Customer Service Center GmbH
Europaplatz 3, 69115 Heidelberg, Germany

Printed by Libri Plureos GmbH
in Hamburg, Germany